Technologies for Detection of DNA Damage and Mutations

Technologies for Detection of DNA Damage and Mutations

Edited by

Gerd P. Pfeifer
Beckman Research Institute of the City of Hope
Duarte, California

Plenum Press • New York and London

Library of Congress Cataloging-in-Publication Data

Technologies for detection of DNA damage and mutations / edited by
Gerd P. Pfeifer.
p. cm.
Includes bibliographical references and index.
ISBN 0-306-45237-5
1. Mutation (Biology)--Laboratory manuals. 2. Molecular genetics--Laboratory manuals. I. Pfeifer, Gerd P.
QH442.T375 1996
575.2'92--dc20 96-22324
CIP

ISBN 0-306-45237-5

A Division of Plenum Publishing Corporation
233 Spring Street, New York, N. Y. 10013

10 9 8 7 6 5 4 3 2 1

Printed in the United States of America

Contributors

Fernando Aguilar • Nestec SA, 1000 Lausanne 26, Switzerland

Steven A. Akman • Department of Medical Oncology and Therapeutics Research, City of Hope National Medical Center, Duarte, California 91010. *Present address*: Department of Cancer Biology, Wake Forest Cancer Center, Winston–Salem, North Carolina 27157.

Ivan A. Bespalov • Department of Microbiology and Molecular Genetics, The Markey Center for Molecular Genetics, The University of Vermont, Burlington, Vermont 05405

Vilhelm A. Bohr • Laboratory of Molecular Genetics, National Institute on Aging, National Institutes of Health, Baltimore, Maryland 21224

Anne-Lise Børresen • Department of Genetics, Institute for Cancer Research, The Norwegian Radium Hospital, N-0310 Oslo, Norway

Peter Cerutti • Nestec SA, 1000 Lausanne 26, Switzerland

Yiming Chen • Sealy Center for Molecular Science, University of Texas Medical Branch, Galveston, Texas 77555

Michael Dean • Laboratory of Viral Carcinogenesis, National Cancer Institute, Frederick Cancer Research and Development Center, Frederick, Maryland 21702

Johan G. de Boer • Center for Environmental Health, Department of Biology, University of Victoria, Victoria, British Columbia V8W 2Y2, Canada

Johan T. Den Dunnen • MGC-Department of Human Genetics, Leiden University, 3333 AL Leiden, The Netherlands

George R. Douglas • Mutagenesis Section, Environmental Health Directorate, Health Canada, Ottawa, Ontario K1A 0L2, Canada

Régen Drouin • Division of Biology, Beckman Research Institute of the City of Hope, Duarte, California 91010. *Present address*: Research Unit in Human and Molecular Genetics, Department of Pathology, Hospital Saint-François d'Assise, Laval University, Quebec G1L 3L5, Canada

Jörg Engelbergs • Institute of Cell Biology (Cancer Research) West German Cancer Center Essen, University of Essen Medical School, D-45122 Essen, Germany

Heather L. Erfle • Center for Environmental Health, Department of Biology, University of Victoria, Victoria, British Columbia V8W 2Y2, Canada

Michael F. Fenech • CSIRO, Division of Human Nutrition, Adelaide SA, 5000, Australia

Riccardo Fodde • MGC-Department of Human Genetics, Leiden University, Leiden, The Netherlands

Gerald Forrest • Division of Biology, Beckman Research Institute of the City of Hope, Duarte, California 91010

Shuwei Gao • Division of Biology, Beckman Research Institute of the City of Hope, Duarte, California 91010

Damjan Glavač • Institute of Pathology, Medical Faculty, Ljubljana, Slovenia

Barry W. Glickman • Centre for Environmental Health, Department of Biology, University of Victoria, Victoria, British Columbia V8W 2Y2, Canada

Keith A. Grimaldi • CRC Drug–DNA Interactions Research Group, Department of Oncology, University College London Medical School, London W1P 8BT, England

R. C. Gupta • Preventive Medicine and Environmental Health, University of Kentucky Medical Center, Lexington, Kentucky 40536

Philip C. Hanawalt • Department of Biological Sciences, Stanford University, Stanford, California 94305

John A. Hartley • CRC Drug–DNA Interactions Research Group, Department of Oncology, University of College London Medical School, London W1P 8BT, England

Frans B. L. Hogervorst • MGC-Department of Human Genetics, Leiden University, 3333 AL Leiden, The Netherlands

James Holcroft • Center for Environmental Health, Department of Biology, University of Victoria, Victoria, British Columbia, V8W 2Y2, Canada

Gerald P. Holmquist • Division of Biology, Beckman Research Institute of the City of Hope, Duarte, California 91010

Chris Huang • Molecular Genetics Section, Gerontology Division, Department of Medicine, Beth Israel Hospital and Harvard Medical School, Boston, Massachusetts 02215

George D. D. Jones • Centre for Mechanisms of Human Toxicity, University of Leicester, Leicester, LE1 9HN, England

Gopaul Kotturi • Centre for Environmental Health, Department of Biology, University of Victoria, Victoria, British Columbia V8W 2Y2, Canada

Yoke Wah Kow • Department of Microbiology and Molecular Genetics, The Markey Center for Molecular Genetics, The University of Vermont, Burlington, Vermont 05405

Wolfgang C. Kusser • Centre for Environmental Health, Department of Biology, University of Victoria, Victoria, British Columbia V8W 2Y2, Canada

Daizong Li • Molecular Genetics Section, Gerontology Division, Department of Medicine, Beth Israel Hospital and Harvard Medical School, Boston, Massachusetts 02215

Ching-I P. Lin • DNA Diagnostics Division, Bio-Rad Laboratories, Hercules, California 94547

Michel Liuzzi • Bio-Méga/Boehringer Ingelheim Research Inc., Laval, Quebec H7S 2G5, Canada

Monique Losekoot • MGC-Department of Human Genetics, Leiden University, Leiden, The Netherlands

Jimmie D. Lowery • DNA Diagnostics Division, Bio-Rad Laboratories, Hercules, California 94547

Simon R. McAdam • CRC Drug–DNA Interactions Research Group, Department of Oncology, University College London Medical School, London W1P 8BT, England

J. Justin McCormick • Carcinogenesis Laboratory, Departments of Microbiology and Biochemistry, The Cancer Center, Michigan State University, East Lansing, Michigan 48824

Veronica M. Maher • Carcinogenesis Laboratory, Departments of Microbiology and Biochemistry, The Cancer Center, Michigan State University, East Lansing, Michigan 48824

Jeffrey R. Mann • Department of Biology, Beckman Research Institute of the City of Hope, Duarte, California 91010

Robert J. Melamede • Department of Microbiology and Molecular Genetics, The Markey Center for Molecular Genetics, The University of Vermont, Burlington, Vermont 05405

David L. Mitchell • Department of Carcinogenesis, The University of Texas M. D. Anderson Cancer Center, Smithville, Texas 78957

Timothy R. O'Connor • Groupe 'Réparation des lésions radio- et chimioinduites,' URA147 CNRS, Institut Gustave-Roussy PRII, 94805 Villejuif Cedex, France. *Present address*: Division of Biology, City of Hope National Medical Center, Beckman Research Institute, Duarte, California 91010.

Gerd P. Pfeifer • Department of Biology, Beckman Research Institute of the City of Hope, Duarte, California 91010

Manfred F. Rajewsky • Institute of Cell Biology (Cancer Research), West German Cancer Center Essen, University of Essen Medical School, D-45122 Essen, Germany

Antonio A. Reyes • DNA Diagnostics Division, Bio-Rad Laboratories, Hercules, California 94547

Henry Rodriguez • Department of Medical Oncology and Therapeutics Research, City of Hope National Medical Center, Duarte, California 91010

Pauline A. M. Roest • MGC-Department of Human Genetics, Leiden University, 3333 AL Leiden, The Netherlands

Michael M. Seidman • OncorPharm, Gaithersburg, Maryland 20877

Frank Seiler • Institute of Cell Biology (Cancer Research), West German Cancer Center Essen, University of Essen Medical School, D-45122 Essen, Germany. *Present address*: Institute of Hygiene and Occupational Medicine, University of Essen Medical School, D-45122 Essen, Germany

Takao Sekiya • Oncogene Division, National Cancer Center Research Institute, Tokyo 104, Japan

Judith Singer-Sam • Department of Biology, Beckman Research Institute of the City of Hope, Duarte, California 91010

Narendra P. Singh • Bioelectromagnetic Research Laboratory, Center for Bioengineering, University of Washington, Seattle, Washington 98195

Charles A. Smith • Department of Biological Sciences, Stanford University, Stanford, California 94305

Piroska E. Szabó • Department of Biology, Beckman Research Institute of the City of Hope, Duarte, California 91010

Moon-shong Tang • Department of Carcinogenesis, University of Texas M. D. Anderson Cancer Center, Smithville, Texas 78957

Jürgen Thomale • Institute of Cell Biology (Cancer Research), West German Cancer Center Essen, University of Essen Medical School, D-45122 Essen, Germany

Silvia Tornaletti • Department of Biology, Beckman Research Institute of the City of Hope, Duarte, California 91010. *Present address*: Department of Biological Sciences, Stanford University, Stanford, California 94305

Luis Ugozzoli • Molecular Diagnostic Division, Bio-Rad Laboratories, Hercules, California 94547

Rob B. Van Der Luijt • MGC-Department of Human Genetics, Leiden University, 3333 AL Leiden, The Netherlands

Bennett Van Houten • Sealy Center for Molecular Science, University of Texas Medical Branch, Galveston, Texas 77555

Nathalie van Orsouw • Molecular Genetics Section, Gerontology Division, Department of Medicine, Beth Israel Hospital and Harvard Medical School, Boston, Massachusetts 02215

Jan Vijg • Molecular Genetics Section, Gerontology Division, Department of Medicine, Beth Israel Hospital and Harvard Medical School, Boston, Massachusetts 02215

R. Bruce Wallace • DNA Diagnostics Division, Bio-Rad Laboratories, Hercules, California 94547

Susan S. Wallace • Department of Microbiology and Molecular Genetics, The Markey Center for Molecular Genetics, The University of Vermont, Burlington, Vermont 05405

David Walsh • Center for Environmental Health, Department of Biology, University of Victoria British Columbia, Victoria, V8W 2Y2, Canada

Michael Weinfeld • Department of Radiobiology, Cross Cancer Institute, Edmonton, Alberta T6G 1Z2, Canada

F. Michael Yakes • Sealy Center for Molecular Science, University of Texas Medical Branch, Galveston, Texas 77555

Preface

Man-made carcinogens, natural genotoxic agents in the environment, as well as ionizing and ultraviolet radiation can damage DNA and are a constant threat to genome integrity. Throughout the evolution of life, complex DNA repair systems have developed in all living organisms to cope with this damage. Unrepaired DNA lesions can promote genetic alterations (mutations) that may be linked to an altered phenotype, and, if growth-controlling genes are involved, these mutations can lead to cell transformation and the development of malignant tumors. Proto-oncogenes and tumor suppressor genes may be critical targets for DNA damaging agents. In a number of animal model systems, correlations between exposure to a carcinogen, tumor development, and genetic changes in tumor DNA have been established. To understand mutagenesis processes in more detail at the molecular level, we need to know the type and frequency of DNA adducts within cells, their distribution along genes and specific DNA sequences, as well as the rates at which they are repaired. We also need to know what types of mutations are produced and which gene positions are most prone to mutagenesis.

This book provides a collection of techniques that are useful in mutagenesis research. The book is divided into three parts. In Part I, methods for DNA damage and repair analysis are provided. These methods include analysis of DNA damage frequencies in single cells or populations of cells, distribution of damage within specific genes and sequences, and various ways to quantitate specific DNA damage. A wide variety of different types of DNA damage is covered. In Part II, several methods for screening and sequence analysis of mutations are described. Part III includes chapters on complete mammalian mutagenesis systems using selectable genes in cell culture or living animals.

The field of DNA repair and mutagenesis research has progressed at a rapid pace over the last five years and, mostly through the advent of the polymerase chain reaction, many new technologies have been developed during this period. Some of these, as well as several more established techniques, have been incorporated into this book, although because of space limitations not all currently used techniques could be included. Most of the chapters are detailed experimental protocols that should be useful to the reader, even those without much prior experience. The book should appeal to undergraduate and postgraduate students as well as researchers who want to experiment with a technique with which they are unfamiliar. In addition, a few chapters review existing methodologies with the intent to provide the reader with guidance as to which particular method to choose.

I would like to thank all of the authors for their contributions and their enthusiasm in putting this book together.

Gerd P. Pfeifer

Duarte, California

Contents

Part I. Technologies for Detection of DNA Damage

Chapter 1. Microgel Electrophoresis of DNA from Individual Cells: Principles and Methodology

Narendra P. Singh

Chapter 2. The Cytokinesis-Block Micronucleus Technique

Michael F. Fenech

Chapter 3. Agarose Gel Electrophoresis for DNA Damage Analysis

Régen Drouin, Shuwei Gao, and Gerald P. Holmquist

Chapter 4. ^{32}P-Postlabeling for Detection of DNA Adducts

R. C. Gupta

Chapter 5. A Postlabeling Assay for Oxidative Damage

Michael Weinfeld, Michel Liuzzi, and George D. D. Jones

Chapter 6. Radioimmunoassay of DNA Damaged by Ultraviolet Light

David L. Mitchell

Chapter 7. Monoclonal Antibody-Based Quantification and Repair Analysis of Specific Alkylation Products in DNA

Jürgen Thomale, Jörg Engelbergs, Frank Seiler, and Manfred F. Rajewsky

Chapter 8. Detection of Oxidative DNA Base Damages: Immunochemical and Electrochemical Approaches

Robert J. Melamede, Yoke Wah Kow, Ivan A. Bespalov, and Susan S. Wallace

Chapter 9. Strategies for Measuring Damage and Repair in Gene-Sized Specific DNA Sequences

Charles A. Smith and Philip C. Hanawalt

Chapter 10. Methods to Measure the Repair of Genes

Vilhelm A. Bohr

Chapter 11. Mapping and Quantification of Bulky Chemical-Induced DNA Damage Using UvrABC Nucleases

Moon-shong Tang

Chapter 12. The Use of DNA Glycosylases to Detect DNA Damage

Timothy R. O'Connor

Chapter 15. Ligation-Mediated PCR for Analysis of UV Damage

Silvia Tornaletti and Gerd P. Pfeifer

Chapter 16. Ligation-Mediated PCR for Analysis of Oxidative DNA Damage

Régen Drouin, Henry Rodriguez, Gerald P. Holmquist, and Steven A. Akman

Chapter 19. Mutation Analysis by Denaturing Gradient Gel Electrophoresis (DGGE)

Riccardo Fodde and Monique Losekoot

Chapter 20. Constant Denaturant Gel Electrophoresis (CDGE) in Mutation Screening

Anne-Lise Børresen

Chapter 21. Single-Strand Conformation Polymorphism Analysis

Takao Sekiya

Chapter 24. The Protein Truncation Test (PTT) for Rapid Detection of Translation-Terminating Mutations

Johan T. Den Dunnen, Pauline A. M. Roest, Rob B. Van Der Luijt, and Frans B. L. Hogervorst

Chapter 29. Bacteriophage Lambda and Plasmid *lacZ* Transgenic Mice for Studying Mutations *in Vivo*

Jan Vijg and George R. Douglas

Chapter 30. The Use of *lacI* Transgenic Mice in Genetic Toxicology

Johan G. de Boer, Heather L. Erfle, David Walsh, James Holcroft, and Barry W. Glickman

Part I

Technologies for Detection of DNA Damage

Chapter 1

Microgel Electrophoresis of DNA from Individual Cells

Principles and Methodology

Narendra P. Singh

1. HISTORY

Rydberg and Johanson (1978) quantitated DNA damage using a novel approach that involved embedding cells in agarose, lysing them and denaturing their DNA with NaOH, then neutralizing the samples and staining the nucleoid thus formed with acridine orange. The ratio of red to green fluorescence from samples was used as a measure of DNA damage. Cells with less DNA damage showed higher ratios of green to red fluorescence and cells with more DNA damage showed lower ratios in photometric measurement. The technique is based on two assumptions: (1) a double-stranded DNA with more breaks denatures faster to single-stranded DNA in mild alkaline conditions and (2) acridine orange molecules in relatively high concentrations make a polymer along the length of single-stranded DNA. Polymer of the dye with single-stranded DNA yields a red fluorescence. Acridine molecules in low concentration intercalate in native double-stranded DNA and emit green fluorescence. In another important development, Ostling and Johanson (1984) exposed murine lymphoma cells to gamma rays and showed that lysed cells when electrophoresed in agarose on microscope slides in neutral conditions displayed a dose-dependent increase in DNA migration. These authors also used acridine orange, and claimed the sensitivity of their technique to detect DNA damage was as low as 50 rads.

In 1986 I started making microgels on plain microscope slides and electrophoresing under neutral conditions. However, this produced a decrease in length of DNA migration with increasing dose of X rays. This might have occurred because I employed entirely different conditions for lysis and electrophoresis than used by Ostling and Johanson (1984). I realized that DNA from lysed cells in agarose was in the form of a nucleoid and breaks in DNA led to relaxation of DNA supercoiling, which in turn retarded the migration of DNA in gel matrix. To prevent this problem, I started using alkaline solution that removes the DNA supercoiling and denatures the double-stranded DNA to single strands. With these modifications, X-radiation of

Narendra P. Singh • Bioelectromagnetic Research Laboratory, Center for Bioengineering, University of Washington, Seattle, Washington 98195.

Technologies for Detection of DNA Damage and Mutations, edited by Gerd P. Pfeifer, Plenum Press, New York, 1996.

human lymphocytes generated a linear dose–response curve with respect to length of DNA migration.

A major problem with the alkaline microgel electrophoresis technique was in keeping microgels adhered to regular microscope slides. Frosted slides minimized this problem. X rays and hydrogen peroxide produced a dose-dependent increase in DNA damage in human lymphocytes when 20 min each of alkaline unwinding and electrophoresis time were used. DNA migration was visualized by ethidium bromide (Singh *et al.*, 1988). This technique, however, is not sensitive enough to detect reliably minor differences in the levels of DNA single-strand breaks in lymphocytes from young and older human subjects (Singh *et al.*, 1990, 1991a). The electrophoresis time used was not sufficient to resolve small differences in DNA damage levels. However, longer electrophoresis time would cause excessive DNA damage due to generation of free radicals in alkaline electrophoretic solution. This prompted the idea of adding antioxidants to the electrophoresis solution. The procedure allowed electrophoresing samples for 1 hr or longer without generating additional strand breaks in control samples. Addition of radical scavengers to the electrophoretic solution and use of an intense fluorescent dye, YOYO-1, further enhanced the sensitivity severalfold (Singh *et al.*, 1994). The technique is easy to use and sensitive compared to other commonly used methods (Table I) to quantitate DNA damage. Several laboratories (Olive *et al.*, 1990; Gedik *et al.*, 1992; Vijaylaxami *et al.*, 1992; Barber and Pentel, 1993; Shafer *et al.*, 1994; Collins *et al.*, 1995; Hartmann and Speit, 1995; Hellman *et al.*, 1995) have used the microgel electrophoresis in one or another form. A review by Fairbairn *et al.* (1995) describes modifications of technique that various laboratories have adopted to meet their needs. Addition of antioxidants to the electrophoretic buffer, DNA precipitation, and use of a sensitive fluorescent dye (e.g., YOYO-1) have enhanced the sensitivity of the technique to screen for low levels of DNA damage in a variety of cells. We have applied these modifications to study DNA single-strand breaks in brain cells of rats exposed to low levels of microwaves (Lai and Singh, 1995) and to alternating electric current used for electroconvulsive therapy (Khan *et al.*, 1995). We also have applied alkaline and neutral microgel electrophoresis technique to human lymphocytes exposed to acetaldehyde (Singh and Khan, 1995) to study DNA single-strand and double-strand breaks, respectively. To attain enough sensitivity under neutral microgel electrophoresis conditions (Singh and Khan, 1995), relatively pure DNA (free from RNA and most proteins) from cells in microgels is a prerequisite. This was achieved by treating the microgels on slides with the enzymes ribonuclease A and proteinase K.

Table I

Sensitivity of Various Assays for Detection of DNA Single-Strand Breaks

Assays	Single-strand breaks in daltons of DNA	Lower limit of detection of DNA X-ray dose (in rads)[a]
Alkaline sucrose sedimentation (Lett *et al.*, 1970)	1 break/2–5 × 10^8	500
Nucleoid sedimentation (Lipetz *et al.*, 1982)	1 break/2 × 10^9	33
Alkaline elution (Kohn *et al.*, 1976)	1 break/2–3 × 10^9	30
Alkaline gel electrophoresis (Freeman *et al.*, 1986)	1 break/3 × 10^9	30
Alkaline unwinding (Rydberg, 1980)	1 break/6–9 × 10^9	10
Alkaline microgel electrophoresis (Singh *et al.*, 1994)	1 break/2 × 10^{10}	3.2

[a]Minimum amount of X-radiation needed to produce detectable increase in DNA damage.

2. BASIC MECHANISM

A theoretical model of DNA molecule movements through agarose during gel electrophoresis has been hypothesized (Deutch, 1988) based on the concept of "tube and chain." However, no such model is available for microgel electrophoresis of DNA from individual cells. This section briefly describes the basic mechanism with experimental evidence (Singh, submitted). Isolated human lymphocytes in 10 μl of phosphate-buffered saline (PBS) were mixed with 75 μl of high-resolution 3:1 agarose (Amresco, Solon, OH). Thirty microliters of this mixture was placed on a frosted microscope slide and covered with a 24 × 50-mm #1 cover glass. After a minute of cooling, the cover glass was removed by a gentle sliding motion in the plane of the slide. One hundred microliters of agarose was placed on top of this layer and covered with a cover glass to add one more layer of agarose. The cover glass was removed once more and the slide put in cold (4°C) lysing solution consisting of 2.5 M sodium chloride, 100 mM disodium EDTA, 1% sodium salt of *N*-laurylsarcosine, 10 mM Tris, pH 10. Then 1% Triton X-100 was added to the lysing solution just before an experiment by thorough stirring. This solution was cooled to below 4°C. After an hour of lysis in the cold, slides were treated with 10 μg/ml ribonuclease A (Boehringer-Mannheim Corp., Indianapolis, IN) in lysing solution (without Triton X, pH 7) for 2 hr at 37°C and then with 1 mg/ml proteinase K (Boehringer-Mannheim Corp.) for 2 hr at 37°C. Lysis in high salt and detergent and treatment with enzymes removed most of the proteins and RNA from the nuclear chromatin. Relatively free individual chromosomal DNA molecules embedded in microgels on slides were then electrophoresed in a solution having 100 mM Tris base, 300 mM sodium acetate, 500 mM sodium chloride, and 2.5% DMSO, pH 9.5, at 0.6 V/cm (300 mA) for 2 hr. Slides were then immersed for 10 min in 300 mM NaOH to degrade the remaining RNA and then neutralized for 30 min in 0.4 M Tris, pH 7.4. The neutralization step was repeated twice with fresh Tris and electrophoresed molecules were immobilized in agarose by dehydrating the slides in absolute ethanol and air drying. Slides were stained with 1 μM YOYO-1 (Molecular Probes, Eugene, OR) in distilled water.

Electrophoresed DNA molecules were observed using a fluorescence microscope equipped with a fluorescein isothiocyanate (FITC) filter combination for excitation and emission wavelengths. Figure 1.1 shows electrophoresed nuclear DNA molecules moving toward the positive pole. Moving ends (with variable migration toward the positive pole) give the impression that each strand belongs to a single human chromosomal DNA molecule (Sasaki and Norman, 1966). Looking at these molecules of various lengths, we can speculate that the lagging ends of each DNA molecule are adhered in the nucleus to a common site, possibly onto the inner nuclear membrane and the periphery of the nucleolus (Comings, 1980), and the loosely attached end moves first during electrophoresis. Such attachment to a common site was predicted by Rabl's model of chromosomal arrangements interphase nuclei. Intensely fluorescent dots, 2–5 μm apart along the length of DNA molecules (Fig. 1.1), are probably proteins of 54 to 68 kDa and are tightly bound to DNA (Sasaki and Norman, 1966; Krauth and Werner, 1979). These tightly bound proteins are clear along the leading segment of DNA, perhaps because of a more effective protein removal and unfolding of DNA. During neutral microgel electrophoresis DNA is double-stranded, but after electrophoresis it is treated with alkali to degrade RNA. This short alkali treatment may also partially unwind the DNA and remove some proteins while still maintaining tightly bound proteins (Fig. 1.1).

To prove that the dots along the length of DNA molecules are protein aggregates, lysed cells were first treated with 300 mM NaOH for an hour and then with proteinase K for two hr. This yielded pieces of DNA of unique sizes. It is not possible to measure exactly the actual size of these pieces because they are in dried agarose and have gone through an ethanol precipitation step. These pieces are 4 to 10 μm. It is assumed that these are the same pieces of DNA as have

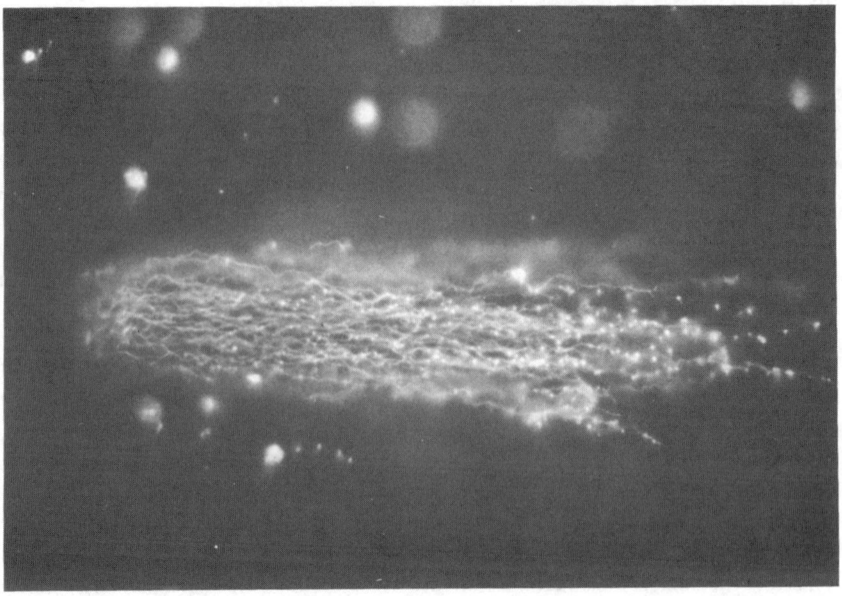

Figure 1.1. Photomicrograph of DNA molecules from control human lymphocyte lysed, treated with ribonuclease A and proteinase K, and electrophoresed in neutral conditions at 0.6 V/cm (~300 mA) for 2 hr. Note intense fluorescence from small particles, 2 to 5 μm apart along the length of individual double-stranded DNA molecules. (Magnification 430×, fluorescent dye, YOYO-1.)

been observed by previous investigators (Sasaki and Norman, 1966; Krauth and Werner, 1979). They found the size of these pieces to be 4 μm or 27 kb. Size of these pieces may be close to that of a replicon (Taylor and Hozier, 1976). These pieces can be easily visualized by subjecting them to neutral or alkaline microgel electrophoresis.

These experiments demonstrate that exposure of double-stranded DNA from lysed cells in microgels, first to 300 mM NaOH (40 min or longer) and then to proteinase K, can generate these small pieces of DNA. Addition of OH and H radical scavengers to alkaline (NaOH) minimizes the generation of these small pieces of DNA, thus reducing the basal levels of DNA strand breaks. Careful observation shows that DNA in the moving segment is mostly denatured after electrophoresis in alkaline conditions for 2 hr in the presence of OH and H scavengers (Singh *et al.*, 1994). Individual DNA molecules cannot be visualized clearly in their entire length without the dehydration and precipitation steps. Also, it is not known if the tightly bound proteins are still with DNA after electrophoresis in alkaline conditions. The single-stranded DNA after alkaline electrophoresis in microgels is easily broken into pieces of 4 to 10 μm by the dehydrating and precipitating effect of ethanol. Once broken, these pieces seem to be curled on themselves, forming bright fluorescent particles (Fig. 1.2). It appears that the intense fluorescence is the result of binding of dye to curled and precipitated DNA. Coprecipitation of some proteins with DNA probably also contributes to this intense fluorescence.

Thus, addition of free radical scavengers allows a longer electrophoresis time and in turn a higher resolution for detection of a difference in DNA damage between control and damaged cells. This resolution is completely lost if slides after lysis are treated with a strong agent such as ammonium sulfate for removal of proteins (2.5 M for 24 hr at room temperature) and then

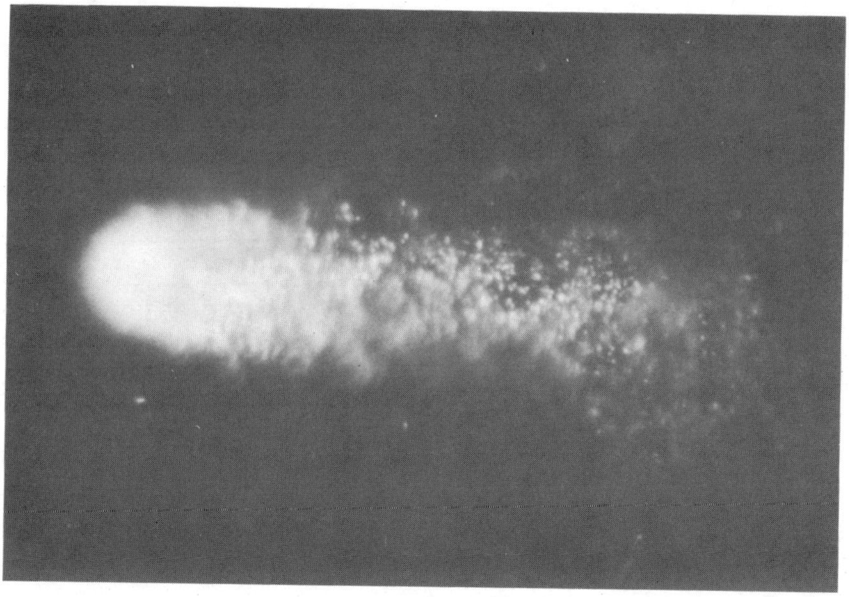

Figure 1.2. Photomicrograph of DNA molecules from a human lymphocyte lysed, treated with proteinase K, and electrophoresed in alkaline conditions at 0.6 V/cm (~300 mA) for 2 hr. Note that DNA molecules are broken in pieces of relatively uniform sizes. As these molecules are single-stranded, dehydration and precipitation (by ethanol) break them into small units. They seem to curl up and it is not possible to measure their length. (Magnification 430×, fluorescent dye, YOYO-1.)

electrophoresed in alkaline conditions, even in the presence of free radical scavengers. Thus, it is not possible to observe any dose effect of X rays on the length of DNA migration from cells that, after lysis and proteinase K treatment, are exposed to 2.5 M ammonium sulfate.

This study does not determine the reason for the relatively lower initial movement of the lagging end of DNA. These ends seem to be attached to a common site as predicted in Rabl's model of chromosomal arrangements in interphase nucleus (Rabl, 1885). The present protocol is unable to free them entirely from their attachments. These tightly bound ends may also be hard to reach by various chemical and enzymatic digestive procedures and thus do not move as fast initially as do the leading ends. Another reason for decreased movement of lagging DNA ends may be their association with positively charged proteins that exert a drag on the DNA in the opposite direction during electrophoresis.

This study also demonstrates the movement of individual chromosomal DNA molecules during gel electrophoresis and supports the idea that each chromosome has a single double-stranded DNA molecule (Sasaki and Norman, 1966; Kavenoff and Zimm, 1973) to which tightly bound proteins are attached at regular intervals (Werner *et al.*, 1980).

3. METHODOLOGY

This section describes the method for making reagents and other procedures needed for neutral (for detecting DNA double-strand breaks) and alkaline (for detecting DNA single-strand breaks) microgel electrophoresis.

3.1. Neutral Microgel Electrophoresis

3.1.1. Assay Steps

Suspend cells in 5 μl PBS
Mix in 75 μl agarose
Microgels are made on microscope slides using 30 μl of cell and agarose mixture
Cell lysis in microgel
Digestion with ribonuclease A
Digestion with proteinase K
Electrophoresis in neutral electrophoretic buffer
Alkali treatment of electrophoresed microgels to degrade RNA
Neutralization
Dehydration and DNA precipitation in microgels
Staining

3.1.2. Reagents

3.1.2a. PBS. 10×
NaCl 80.1 g/liter
Na_2HPO_4 11.5 g/liter
KCl 2.0 g/liter
KH_2PO_4 2.0 g/liter
Adjust pH to 7.4 by 1 N HCl or 1 N NaOH

Concentrated PBS should be stored at room temperature to avoid salt precipitation in the cold.

3.1.2b. Agarose 3:1. High-resolution agarose from Amresco: 0.5% in PBS is boiled and, for better adherence and less background on slides, agarose should be used within half an hour of melting and not be reused after solidification. Now, high-resolution agaroses are also available but enzymes are unable to reach the DNA of lysed cells within 2 hr in microgels made with them. This limits the efficacy of proteinase K, a self-digesting enzyme.

3.1.2c. Frosted Slides. Usually fully frosted slides (Erie Scientific Co., Portsmouth, NH) are clean and ready to be used. If cells, such as fibroblasts, need to be attached (Singh *et al.*, 1991b), these slides should be washed twice in tap water and then soaked in distilled water for at least 2 hr. After one more wash with distilled water, slides are dried and put in petri dishes for sterilization in a microwave oven or left overnight in a laminar flow hood under UV light.

3.1.2d. Lysing Solution.
2.5 M NaCl
100 mM EDTA disodium salt
1% laurylsarcosine sodium salt
10 mM Tris HCl pH 10

Just before use, mix 1% Triton X-100 or nonidet P-40 in lysing solution by stirring for at least 30 min at room temperature. Nonionic detergents are essential to maintain the shape of the microgels and also restrict the DNA from spreading in microgels during longer lysis.

3.1.2e. Ribonuclease A Treatment. Ribonuclease A (Boehringer-Mannheim Corp.) is used at a final concentration of 10 μg/ml. Make a solution of 10 μg/ml in lysing solution at pH 7 (without Triton X-100) and dispense in 1-ml aliquot in 1.5-ml microfuge tubes. Freeze at −20°C. Thaw just before use. Ribonuclease and proteinase K are somewhat expensive. Only 300 to 400 μl of ribonuclease and proteinase K in lysing solution need be used in the simple method that follows. Place a cover glass on the bench top and pipet 12.5 μl of agarose on all of its four corners, making small dots. Agarose solidifies within 1 min. After lysis, hold the slides vertical for a few seconds to quickly drain excess lysing solution (to avoid dilution of enzymes). The cover glass with agarose dots on four corners is then placed on the slide so that the dots sit on the microgel. This creates a small space 0.5 mm high on top of the microgel. Pipet in about 400 μl of ribonuclease or proteinase K solutions into this space. The problem of uneven enzyme action may arise if care is not taken to make a uniform space on top of the microgel. The above method is economical but laborious. Use of Coplin jars or shallow containers which can accommodate horizontal plastic trays with slides (Singh, 1996a) is recommended for enzyme treatment of slides.

3.1.2f. Proteinase K Treatment. DNase-free proteinase K (Boehringer-Mannheim Corp.) is used at a final concentration of 1 mg/ml. A solution of 1 mg/ml is made in lysing solution at pH 7.4 (without Triton X-100). A similar method as mentioned in section 3.1.2e for ribonuclease treatment, can be used for proteinase K treatment. For alkaline microgel electrophoresis where only proteinase K treatment is needed, proteinase K can be coated directly to slides by mounting each slide with 100 μl of a 5 mg/ml solution of this enzyme in 0.5% agarose. The coverglasses are removed and slides are dried at room temperature. These slides can be used immediately or even for up to a week. When proteinase K coated slides are used, lysis time should be minimized to 15 min to avoid dilution of proteinase K by the lysing solution. Effectiveness of proteinase K can easily be tested by using human sperm microgel slides and treating them for 2 hr or more with proteinase K at 37°C. Slides can be made in large quantities and kept for years in cold lysing solution with Triton X. Electrophoresing these slides under alkaline conditions after proteinase K treatment yields a DNA migration pattern indicative of single-strand breaks and/or alkali-labile sites (Singh *et al.*, 1989). If these slides have not been treated with proteinase K, DNA will not move during electrophoresis.

DMSO in lysing solutions reduces the efficacy of proteinase K to almost one-fourth.

3.1.2g. Neutral Electrophoretic Buffer. This solution is made fresh from stock of 3 M sodium acetate and 1 M Tris, pH 8.5 (adjusted with glacial acetic acid). Final concentrations are 300 mM sodium acetate and 100 mM Tris.

3.1.2h. Alkaline Solution. Ten-minute treatment with 300 mM NaOH in a Coplin jar is used for degrading remaining RNA in microgels electrophoresed under neutral conditions. This step not only degrades RNA but also makes most of the DNA single-stranded. Somehow, double-stranded DNA even after precipitation has a tendency to escape from the agarose.

3.1.2i. Neutralization Solution. A solution of 400 mM Tris pH 7.4 (adjusted by HCl or glacial acetic acid) is used to neutralize the alkali and, more importantly, to remove salts and detergents from microgels. Coplin jars or shallow containers (Singh, 1996a) are filled with 400 mM Tris, pH 7.4. After neutralization for 10 min this step is repeated twice to remove most of the detergents and salts. If salts and detergents are left in agarose, dehydration and drying steps will precipitate them permanently with agarose and this will result in a high background on slides.

3.1.2j. Stains. YOYO-1 (Molecular Probes) is a benzoxazolium-4-quinolinium dimer (Glazer and Rye, 1992). When bound to DNA it absorbs at 491 nm and emits at 509 nm. The filter combination used for FITC also works well for YOYO. Sybr green (Molecular Probes), a more sensitive dye that requires a similar filter combination, may also be used. However, when bound to DNA, this dye fades faster than does YOYO-1 and creates a problem in photographing and reexamining the slides. Although these dyes are expensive, only a small amount is needed for microgel staining. Other inexpensive dyes such as acridine orange, thiazole orange, and ethidium bromide can also be used. The concentrations of all of these dyes are critical: a minimum concentration should be used for a low background. Approximate concentrations are 100 to 1 μM for YOYO-1 and Sybr green, 0.25 μg/ml for acridine orange and thiazole orange, and 1 μg/ml for ethidium bromide.

3.1.3. General Protocol for Making Microgels on Frosted Slides

For detection of very low levels of DNA strand breaks, and particularly single-strand breaks, *only indirect incandescent light must be used during preparation of cell suspension and subsequent steps up to completion of electrophoresis*. Ten thousand to one million cells (lymphocytes, fibroblasts, tumor cell line, cells from various organs, etc.) are suspended in 5 μl of PBS and mixed with 75 μl of agarose. Microgels are made on microscope slides by placing 30 μl of the cell and agarose mixture on frosted slides and laying a cover glass on top to spread the agarose between the slide and cover glass. It is better to cut off 4 to 5 mm of the pipet tip before pipetting the agarose to avoid blockage due to quick solidification of agarose at the narrow ends. Slides are cooled immediately by placing them on ice for 60 sec. The cover glass is removed gently and 100 μl of agarose is quickly layered on top of the first layer and the cover glass is placed on top of the agarose to spread the second layer. After 60 sec of cooling, the cover glass is removed and slides are placed in cold lysing solution. Loss of microgels during transportation of slides, lysis, electrophoresis, etc. can be minimized by initially making a microgel on frosted slides with 100 μl of 0.5% agarose and drying it at room temperature. Apart from providing a firm attachment to subsequent layers of agarose, this dried layer can also be used for proteinase K treatment (usually overnight) by incorproating 5 mg/ml of agarose of this enzyme.

3.1.4. Dehydration and DNA Precipitation

Two slides are immersed per Coplin jar filled with absolute ethanol. DNA is precipitated and microgels are dehydrated at the same time in ethanol for 10 min at room temperature. This step is repeated twice. The DNA precipitation step is essential to the sensitivity of the technique. Slides are then left at room temperature to dry. The precipitated DNA, when bound with dye, gives intense fluorescence similar to double-stranded DNA.

3.1.5. Reexamination of the Slides

After examination, slides must not be exposed to absolute ethanol because the severe dehydrating action of ethanol will damage the DNA molecules anchored to agarose at various points along their length. Slides are simply left as such to dry at room temperature while cover glasses are still on. This will preserve the architecture of migrated DNA, particularly on slides that are electrophoresed under neutral conditions. Simply remove the cover glass if it has not fallen off by itself, and mount the slides in 25 μl of distilled water with a new cover glass (used cover glass is a source of high background). To reduce background or overstaining, 50–70% ethanol should be used.

3.2. Alkaline Microgel Electrophoresis

3.2.1. Assay Steps

Suspend 10^4 to 10^6 cells in 5 μl PBS
Mix in 75 μl agarose
Microgels are made by spreading 30 μl of this mixture on microscope slides
Digestion with proteinase K
Electrophoresis in alkaline electrophoretic buffer
Neutralization
Dehydration and DNA precipitation in microgels
Staining

In alkaline microgel electrophoresis, preparation of agarose microgels, lysis, and proteinase K treatment are similar to the steps as described in the neutral microgel electrophoresis section.

3.2.2. Alkaline Electrophoretic Buffer

Final concentration of NaOH is 300 mM. Make a 10 N solution and dispense in 30-ml aliquot in 50-ml centrifuge tubes. Store with caps tight because NaOH reacts with atmospheric CO_2 and converts to Na_2CO_3.

Make a 500 mM stock solution of EDTA disodium salt and adjust pH to 10 with 10 N NaOH. To make 1 liter of alkaline electrophoretic solution, add 1 gram of 8-hydroxyquinoline (0.1%), 20 ml of 500 mM EDTA (10 mM), 20 ml of DMSO (2%) and 30 ml of NaOH (300 mM) to distilled water. Stir gently for approximately 10 minutes to dissolve 8-hydroxyquinoline.

3.3. Electrophoretic Conditions for Both Neutral and Alkaline Microgel Electrophoresis

For studies mentioned below, we used an electrophoretic unit, HE-100 Super Sub™ from Hoefer Scientific Instruments (San Francisco, CA). Details of its modification are described elsewhere (Singh *et al.*, 1994). Any electrophoretic unit should work as long as both ends of the wire electrodes are connected to a power supply. This provides a more uniform electric field in the unit. Recirculation at low rate avoids accumulation of salt near the negative pole and thus creates an even electric field in the electrophoretic unit. Both of these measures reduce slide-to-slide variations in DNA migration pattern. The volume of electrophoretic solution depends on the electrophoretic unit size. A height between 6 and 7 mm above the surface of the slides is recommended because slight unevenness on the surface of the microgel will not make much difference in the DNA migration pattern. The lowest possible voltage that can move high-molecular-weight DNA will give the best resolution. A voltage gradient of 0.4 V/cm is recommended. Amperage will depend on the height of fluid, salt concentrations, and voltage applied. An ordinary power supply is not suitable for this application because the amperage used here is relatively high. A pure dc power supply without any ac component can be purchased from B&K Precision, Maxtec International Corporation, Chicago, IL.

3.4. Data Collection

Measurement of DNA migration length includes the diameter of the nucleus and the length of migrated DNA. A critical factor is to locate the leading edge of DNA. Criteria are determined by levels of the background on the slide. If background is fairly low, then the last three pixels of DNA in a row close together can be considered as a leading edge of migrating DNA. Length measurements are easy to gather and only require an eyepiece micrometer. Certain precautions must be taken when analyzing the slides: (1) Avoid areas near the edges of slides. (2) Select only slides of low background for data collection because it is difficult to determine the leading edge of migrating DNA on slides with high background. (3) Do not score cells in the line of a trapped air bubble as air does not allow the proper flow of electric current in that area and this would affect DNA migration.

4. APPLICATIONS

This section deals with several applications where microgel electrophoresis assay was used. Experiments with X rays and acetaldehyde are described in detail.

4.1. X Rays

To demonstrate the sensitivity of the alkaline and neutral microgel electrophoresis techniques, freshly isolated lymphocytes were exposed to 3.2, 6.4, 12.8, and 25.6 rads of X rays. DNA single- and double-strand breaks were measured using alkaline and neutral microgel electrophoresis, respectively.

4.1.1. Materials and Methods

Human lymphocytes were separated from whole blood of laboratory volunteers using a modification of the Ficoll–Hypaque centrifugation method of Boyum (1968). In the micromethod (Fig. 1.3, and Singh *et al.*, 1994), 20–100 μl of whole blood from a finger prick was mixed with ice-cold 0.5 ml RPMI 1640, free of phenol red (Life Technologies, Long Island, NY) in a 1.5-ml heparinized microfuge tube (Kew Scientific Inc., Columbus, OH). Then 100 μl of cold lymphocyte separation medium (LSM) was layered at the bottom of the tube using a pipetman. The sample was centrifuged at 4500 rpm for 3 min in a microfuge (Sorvall, Model Microspin 245) at room temperature. The lymphocytes, in the upper part of the Ficoll layer, were pipetted out. At this stage lymphocytes should be free of polymorphs and red blood cells which could contribute to DNA damage during cell lysis. Cells were washed twice in 0.5 ml RPMI 1640, using centrifugation for 3 min at 4500 rpm in the microfuge. The final pellet consisting of approximately $0.4–2.0 \times 10^5$ lymphocytes was suspended in PBS.

For X-ray dose–response experiments, 5 μl of the cell sample was mixed with 75 μl of 0.5%

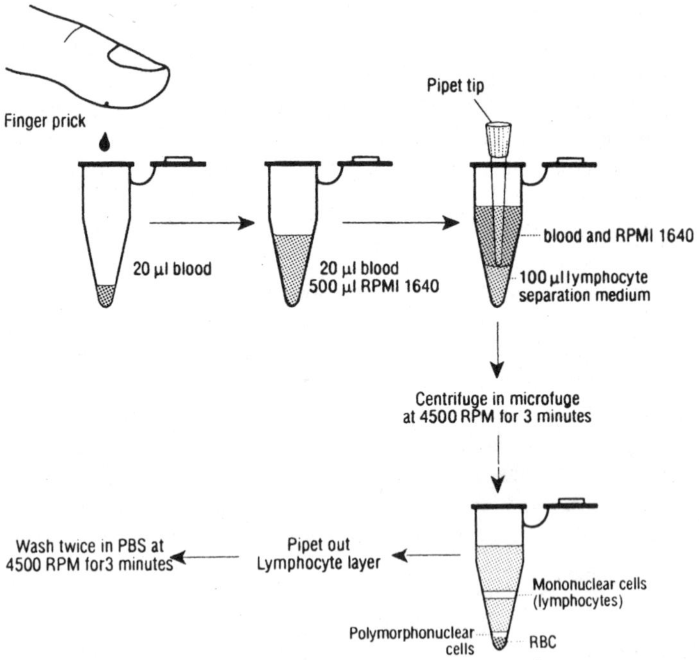

Figure 1.3. A diagrammatic outline of micromethod to isolate lymphocytes from human blood.

agarose (3:1 high-resolution agarose, Amresco) in PBS, maintained at 42°C. Then 30 μl of this mixture was pipetted on a fully frosted slide having low background fluorescence (Erie Scientific Co.) and immediately covered with a 24 × 50-mm square #1 cover glass (Corning Glass Works, Corning, NY). The slides were put in a cold steel tray kept on ice. The cover glasses were removed after 1 min and 100 μl of agarose was layered on top as before. These slides were put on top of an ice-cold plastic tray and exposed to various doses of X rays using a Kelley–Koett X-ray machine (Covington, CT) at a dose rate of 100 rads/min. Irradiation of samples was done below 4°C (on ice) to prevent DNA repair. Following irradiation, cover glasses were removed and slides were immersed in an ice-cold lysing solution (2.5 M NaCl, 1% sodium *N*-lauryl sarcosinate, 100 mM disodium EDTA, 10 mM Tris base, at pH 10) mixed thoroughly with Triton X-100.

For repair studies, approximately 10^5 lymphocytes were suspended in 10 μl of PBS, irradiated with 25.6 rads of X-rays, and suspended in 500 μl of RPMI 1640 medium supplemented with 10% fetal bovine serum (Hyclone, New Haven, CT). The cells, in 1.5-ml microfuge tubes, were incubated in a humidified incubator at 37°C having 5% CO_2. At different time points, cell viability was assessed using a combination of acridine orange and Hoechst 33258 (Singh and Stephens, 1986) and was found to be greater than 98%. At various repair time points cells were microfuged for 2 min at 4500 rpm and the pelleted cells were resuspended in approximately 10 μl of medium. In the next step 75 μl of 0.5% agarose, maintained at 42°C, was added and mixed with the cells. This cell suspension was pipetted onto fully frosted slides. The remainder of the slide preparation and lysing was performed as described above.

4.1.2. DNA Single-Strand Breaks

After an hour of lysis at 0°C, the slides were treated with 1 mg/ml DNase free proteinase K (Boehringer-Mannheim Corp.) in lysing solution for 2 hr at 37°C. Slides were put on the horizontal slab of an electrophoretic assembly. One liter of electrophoretic buffer (300 mM NaOH, 0.1% 8-hydroxyquinoline, 2% DMSO, 100 mM Tris, and 10 mM tetrasodium EDTA, pH 10) was gently poured into the assembly. After 20 min to allow for DNA unwinding, electrophoresis and recirculation were started simultaneously [electrophoresis at a voltage gradient of 0.4 V/cm for 60 min and recirculation at approximately 100 ml/min]. Following electrophoresis, the extra electrophoretic buffer was removed from the top of the slides, the slides were removed from the electrophoretic apparatus and placed in a Coplin jar (two slides per jar) containing Tris, pH 7.4, to neutralize NaOH in microgels. After 30 min the slides were transferred to another jar of Tris for the next 15 min. After one more change of Tris for 15 min, slides were immersed in absolute ethanol for 30 min to precipitate the DNA and dehydrate the gels. Slides were left vertical at room temperature to dry and stained with 25 μl of 1 nM YOYO-1 in distilled water.

4.1.3. DNA Double-Strand Breaks

After lysis, slides were treated with 10 μg/ml ribonuclease A in lysing solution (without Triton X-100) at 37°C for 2 hr followed by overnight treatment with 1 mg/ml proteinase K at 37°C. In the next step, slides were kept for 20 min in the electrophoretic unit containing 1 liter of electrophoretic buffer, 100 mM Tris, and 300 mM sodium acetate pH 8.5 (adjusted by glacial acetic acid). Slides were then electrophoresed at 0.4 V/cm for 1 hr while circulating the buffer at approximately 100 ml/min. After electrophoresis, slides are immersed in 300 mM NaOH for 10 min. Remaining steps were the same as described for DNA single-strand breaks.

4.1.4. Results

Figure 1.4 is a photomicrograph showing migration pattern of cells after alkaline (left panel: top is control and bottom is cells treated with 25 rads of X rays) and neutral microgel electrophoresis (right panel: top is control and bottom is cell treated with 25 rads of X rays). Figure 1.5 summarizes results of X-ray-induced DNA single-strand breaks and their repair. Each histogram shows the frequency distribution of length of DNA migration from 50 representative cells per X-ray dose and per repair time point after cells were exposed to 25 rads of X rays. In the dose–response study (Fig. 1.5, X–Y plot), a significant increase in length of DNA migration pattern was observed even when lymphocytes were exposed to the lowest dose (3.2 rads) of X rays. In the repair kinetics study (shown in X–Y plot), a significant decrease in DNA migration length was observed at the 15-min time point, indicating that repair of X-ray-induced single-strand breaks is fairly fast. Most of the damage seems to be repaired within 1 hr. Figure 1.6 summarizes results of X-ray-induced DNA double-strand breaks and their repair. Each histogram shows the frequency distribution of length of DNA migration from 50 representative cells per X-ray dose and per repair time point after cells were exposed to 25 rads of X rays. In the dose–response study (Fig. 1.6, X–Y plot), a significant increase in length of DNA migration pattern was observed only when lymphocytes were exposed to the highest dose (25.6 rads) of X rays. In the repair kinetics study (Fig. 1.6, shown in X–Y plot), a significant decrease in DNA migration length was observed at the 15-min time point, indicating that repair of X-ray-induced DNA double-strand breaks is also fairly fast. Most of the damage was repaired within 1 hr.

Thus, alkaline microgel electrophoresis can detect single-strand breaks in human lympho-

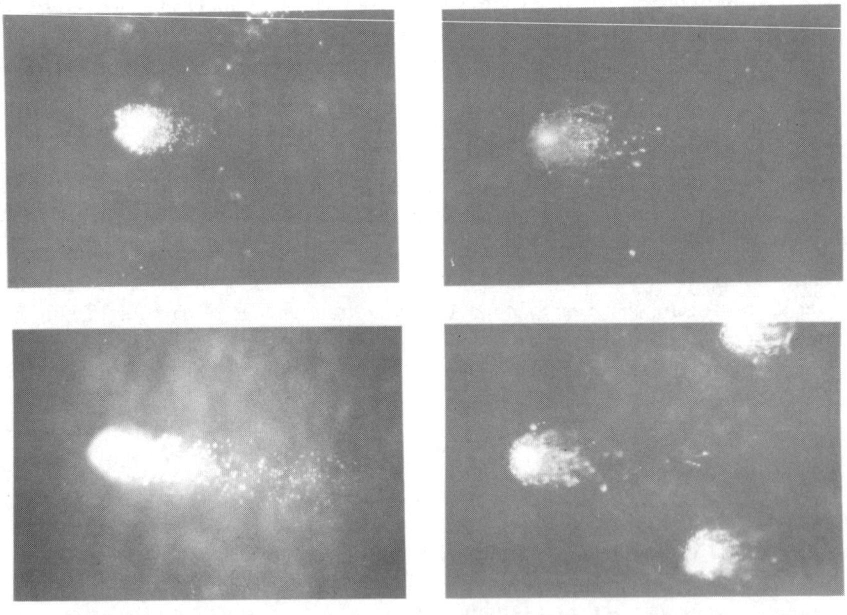

Figure 1.4. Photomicrographs of DNA migration patterns from human lymphocytes. DNA single-strand breaks (left panel): upper left is DNA migration from a control cell and lower left is DNA migration from cell exposed to 25 rads of X rays. DNA double-strand breaks (right panel): upper right is DNA migration from a control cell and lower right is DNA migration from cells exposed to 25 rads of X rays. (Magnification 430×, fluorescent dye, YOYO-1.)

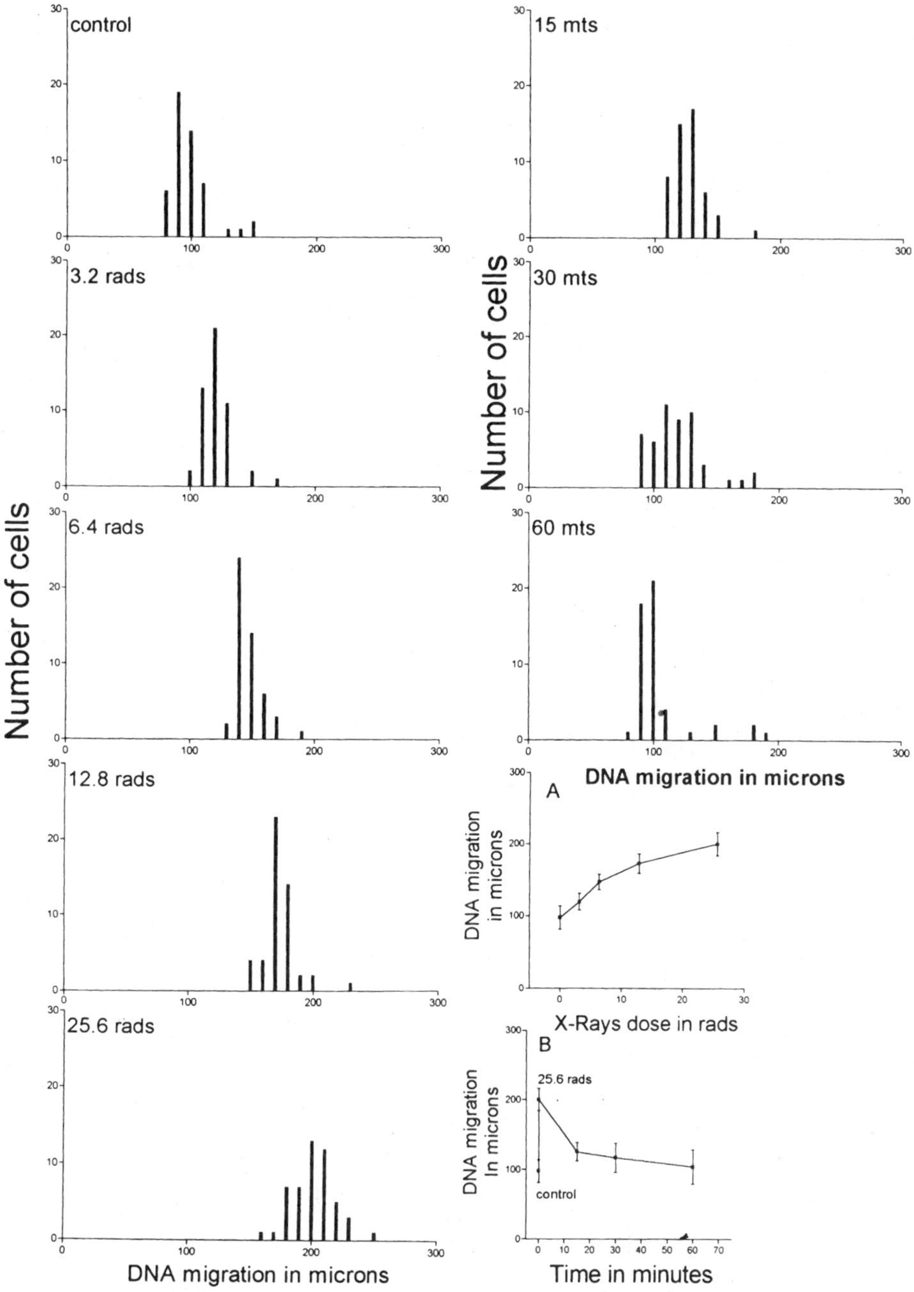

Figure 1.5. DNA single-strand breaks. Frequency distribution histograms of length of DNA migration from 50 representative cells (control, 3.2, 6.4, 12.8, and 25.6 rads) and a corresponding X–Y line plot (panel A) of means and standard deviations (error bars) of length of DNA migration from 50 representative cells exposed to various doses of X rays and assayed immediately after exposure. This figure also shows frequency distribution histograms of length of DNA migration from 50 representative cells and a corresponding X–Y line plot (panel B) of means and standard deviations (error bars) of length of DNA migration from 50 representative cells allowed to repair DNA (15, 30, and 60 min.) after exposure to 25 rads of X rays.

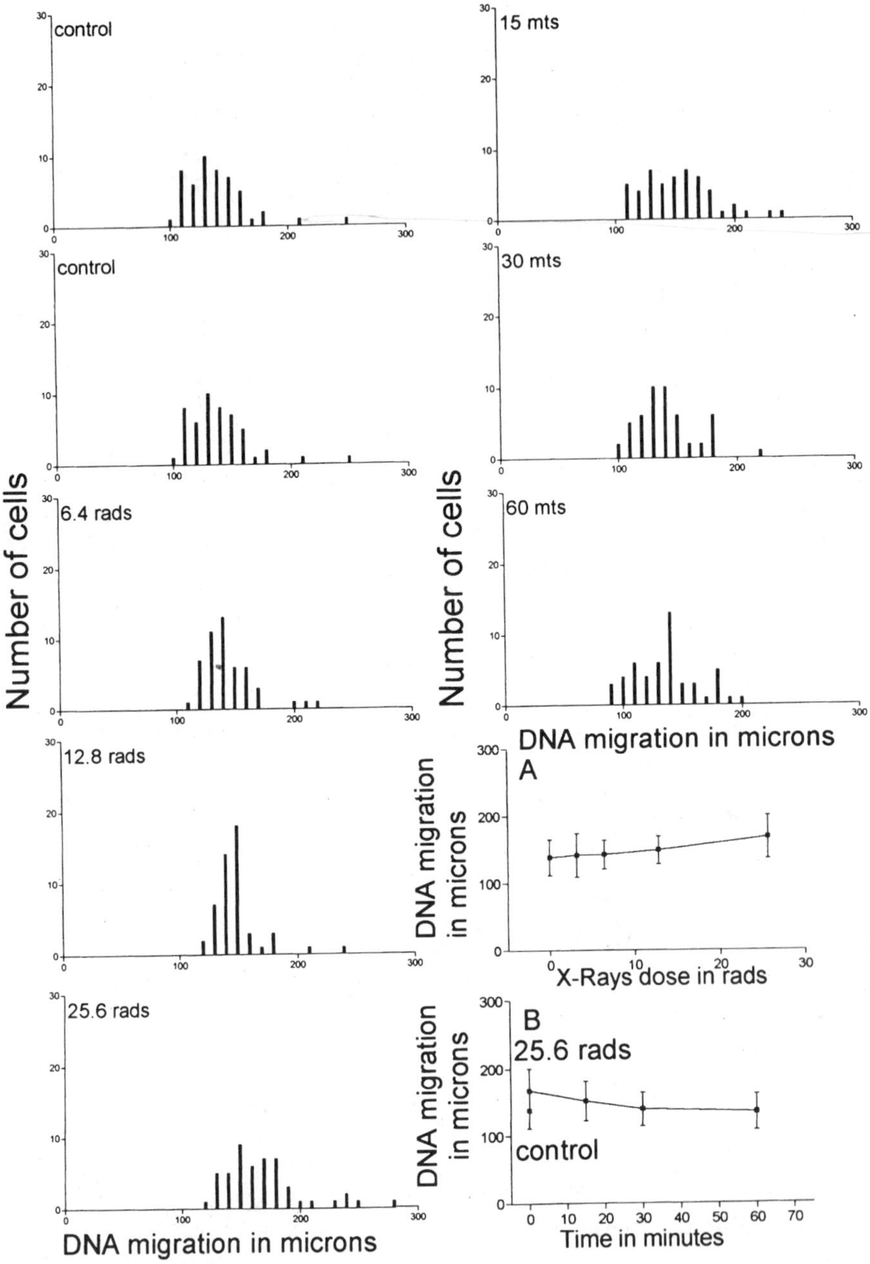

Figure 1.6. DNA double-strand breaks. Frequency distribution histograms of length of DNA migration from 50 representative cells (control, 3.2, 6.4, 12.8, and 25.6 rads) and a corresponding X–Y line plot (panel A) of means and standard deviations (error bars) of length of DNA migration from 50 representative cells exposed to various doses of X rays and assayed immediately after exposure. This figure also shows frequency distribution histogram of length of DNA migration from 50 representative cells and a corresponding X–Y line plot (panel B) of means and standard deviations (error bars) of length of DNA migration from 50 representative cells allowed to repair DNA (15, 30, and 60 min.) after exposure to 25 rads of X rays.

cytes caused by doses of X rays as low as 3.2 rads. Sensitivity of neutral microgel electrophoresis is only to 25 rads. Although it was possible to detect DNA double-strand breaks produced by 12.8 rads, their level at this dose did not reach significance (i.e., $p < 0.05$).

4.2. Ethanol and Acetaldehyde

We evaluated the DNA-damaging effects of ethanol and its major metabolite, acetaldehyde.

4.2.1. DNA Strand Breaks

Freshly isolated human lymphocytes from two volunteers were treated in a humidified incubator at 37°C with 5% CO_2. Incubation time was 1 hr with 0, 1.56, 6.25, 25, and 100 mM ethanol or acetaldehyde in complete medium. No antibiotics were used. After an hour of treatment, samples were pelleted. Microgel preparation and cell lysis were done as described earlier. In the dose–response part of the study, we evaluated DNA single- and double-strand breaks immediately after exposure. For the repair kinetics part of the study, following exposure to 100 mM of acetaldehyde, cells were washed four times with 0.5 ml of complete medium and resuspended in 0.5 ml of complete medium for 30, 60, and 120 min at 37°C. To quantitate cell viability and number, we used bisbenzimide, Hoechst 33258 (Behring Diagnostics, La Jolla, CA) as a vital dye in combination with acridine orange (Singh and Stephens, 1986). Evaluation of DNA single- and double-strand breaks was done as described for X-rays, except that 0.6 V/cm was used for electrophoresis instead of 0.4 V/cm.

4.2.2. Results

Ethanol did not significantly induce DNA strand breaks (Fig. 1.7). However, acetaldehyde produced significant increases of DNA single- (Fig. 1.8A) and double-strand (Fig. 1.8B) breaks. One-way ANOVA showed a statistically significant increase in DNA single-strand breaks after incubation with acetaldehyde ($F = 920.350$, $df = 1{,}8$, $p < 0.001$). Post-hoc t tests showed significant differences with each dose (1.56, 6.25, 25, and 100 mM) of acetaldehyde ($p < 0.001$). Index of linearity as measured by linear regression showed a significant dose–response relationship ($r^2 = 0.94$, $p < 0.001$). One-way ANOVA showed a statistically significant increase in DNA double-strand breaks with acetaldehyde concentrations ($F = 563.837$, $df = 1{,}8$, $p < 0.001$). Post-hoc t tests showed significant differences ($p < 0.001$) only between the highest concentration

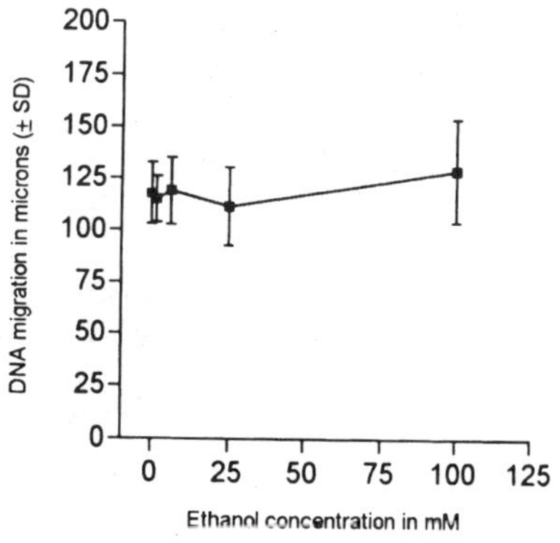

Figure 1.7. Level of DNA single-strand breaks in human lymphocytes treated with ethanol and assayed by alkaline microgel electrophoresis technique. Dose–response curve shows length of DNA migration (microns) in human lymphocytes as a function of ethanol concentration. Each point represents the mean of 50 cells. The range bars indicate standard deviations.

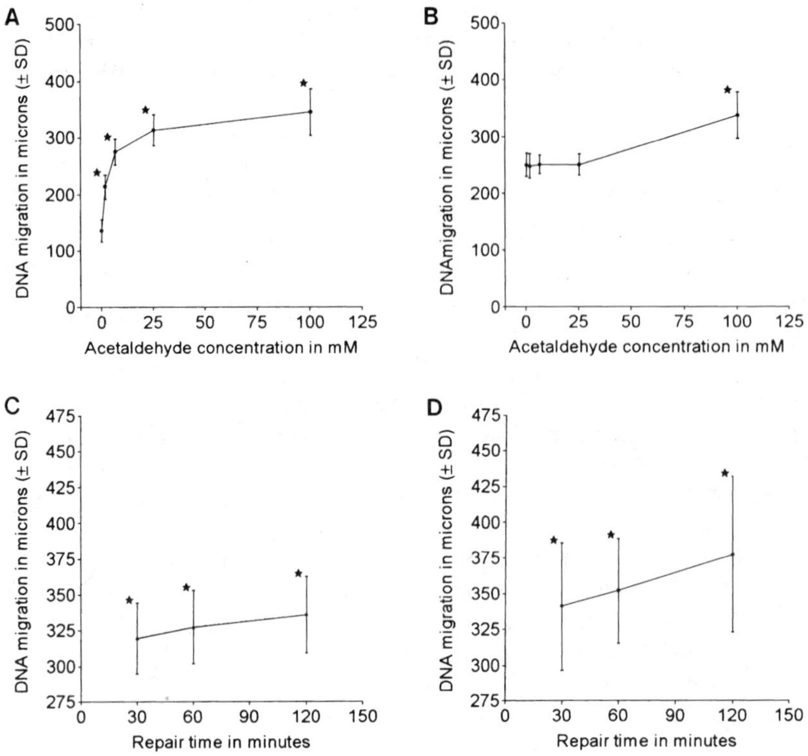

Figure 1.8. (A, B) Level of DNA strand breaks in human lymphocytes treated with ethanol and assayed by the microgel electrophoresis technique. Dose–response curve shows length of DNA migration (microns) in human lymphocytes as a function of acetaldehyde concentration. Each point represents the mean length of DNA migration for 50 representative cells from each subject. The error bars indicate standard deviations. (A) Single-strand breaks: asterisk indicates significantly different from control at $p < 0.001$. (B) Double-strand breaks: asterisk indicates significantly different from control at $p < 0.001$. (C, D) Level of DNA strand breaks in human lymphocytes treated with acetaldehyde and assayed by the alkaline microgel electrophoresis technique. Repair kinetics shows length of the DNA migration (microns) in human lymphocytes exposed to 100 mM acetaldehyde as a function of postexposure incubation time. Each point represents the mean of 50 cells from each subject. The error bars indicate standard deviations. (C) Single-strand breaks: asterisk indicates significantly different from control at $p < 0.001$. (D) Double-strand breaks: asterisk indicates significantly different from control at $p < 0.01$.

(100 mM) of acetaldehyde and the control level. There were no significant differences in the frequency of DNA double-strand breaks between any of the lower doses of acetaldehyde and the control. Furthermore, linear regression showed a lack of dose-relationship between concentration of acetaldehyde and DNA double-strand breaks ($r^2 = 0.53$, p = ns).

Figure 1.8C and 1.8D show repair kinetics of DNA single- and double-strand breaks, respectively, induced by incubation of lymphocytes with 100 mM acetaldehyde. DNA damage as measured by both single- and double-strand breaks was not repaired in the majority of lymphocytes analyzed at 30, 60, and 120 min postexposure compared to controls. Interestingly, damage in most cells increased as a function of time. One-way ANOVA showed a statistically significant increase in DNA single-strand breaks following postexposure repair time compared to control

($F = 387.679$, $df = 1{,}6$, $p < 0.001$). Similar results were found for DNA double-strand breaks ($F = 42.674$, $df = 1{,}6$, $p < 0.01$). Controls for DNA single- and double-strand breaks were unchanged during 2 hr of incubation. We observed heterogeneous patterns of DNA migration in individual lymphocytes in response to various doses of acetaldehyde and repair time. Distributions of DNA migration in 50 representative cells as a function of dose and repair time are shown in Figs. 1.9 and 1.10 for DNA single- and double-strand breaks, respectively. Figure 1.11 shows the

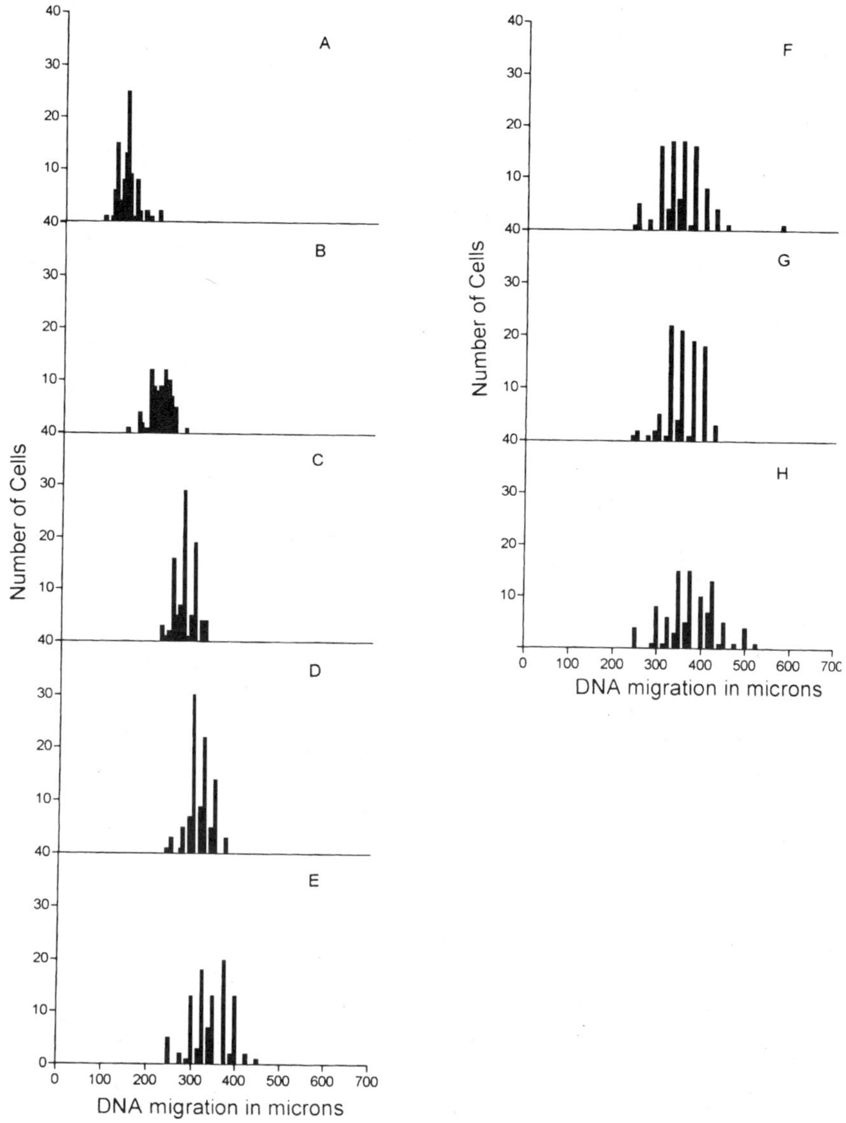

Figure 1.9. DNA single-strand break study: Distribution of lymphocyte DNA migration as a function of the dose of acetaldehyde (0, 1.56, 6.25, 25, and 100 mM, shown in panels A, B, C, D, and E, respectively) and as a function of repair time (30, 60, and 120 min, shown in panels F, G, and H, respectively).

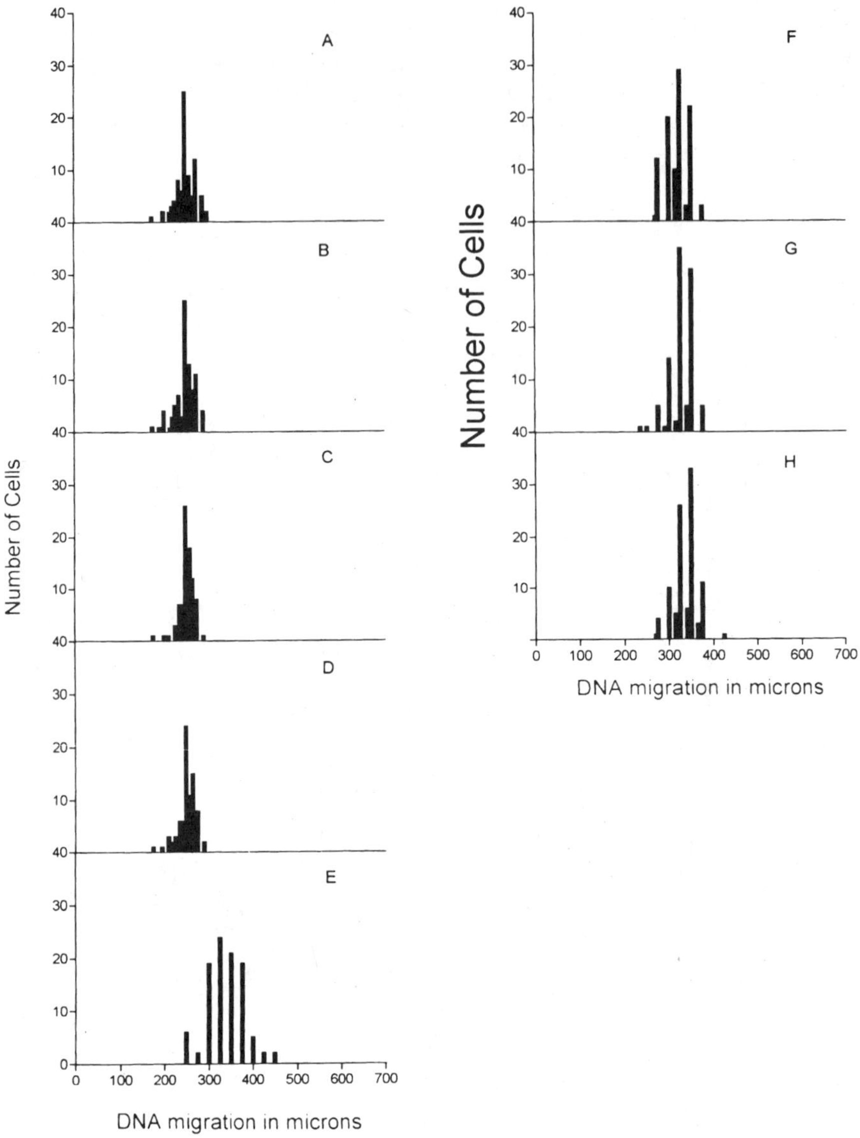

Figure 1.10. DNA double-strand break study: Distribution of lymphocyte DNA migration as a function of dose of acetaldehyde (0, 1.56, 6.25, 25, and 100 mM, shown in panels A, B, C, D, and E, respectively) and as a function of repair time (30, 60, and 120 min, shown in panels F, G, and H, respectively).

number of viable cells after incubation with 100 mM acetaldehyde for 1 hr. More than 80% of lymphocytes were not viable following exposure to 100 mM acetaldehyde for 2 hr. Figure 1.12 shows the DNA migration pattern from a late apoptotic cell demonstrating the presence of a majority of DNA in a band. No attempt was made to score such a cell as assessment of cell count was a better indicator of cellular toxicity than apoptotic cells in microgels. This was partly related

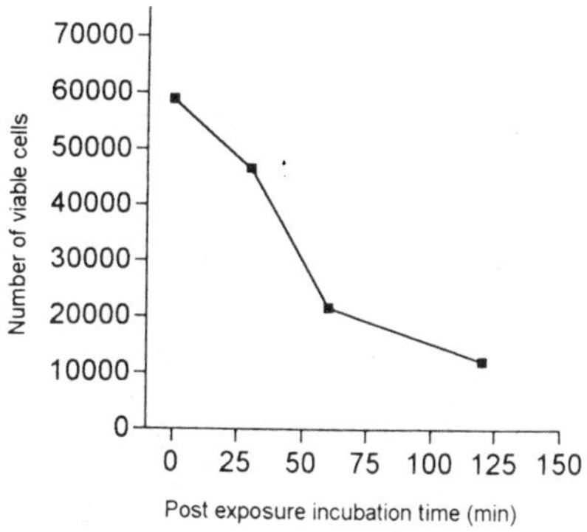

Figure 1.11. Loss of lymphocytes *in vitro* as a function of post exposure incubation time. Cells were exposed to 100 mM acetaldehyde for 1 hr at 37°C and counted at various time points after exposure.

to the use of higher voltage and longer electrophoretic time in this study. This resulted in excessive DNA migration from apoptotic cells and dilution of DNA in microgels, making it difficult to quantitate apoptotic cells accurately. Electrophoresis at lower voltage (0.4 V/cm) makes it possible to detect most of the apoptotic cells.

4.3. Applications of Microgel Electrophoresis to Other Cell Types

Cells from human, animal, and plant tissues can be dissociated using various methods. These dissociated cells can be mixed with agarose to make microgels. Some examples are described below.

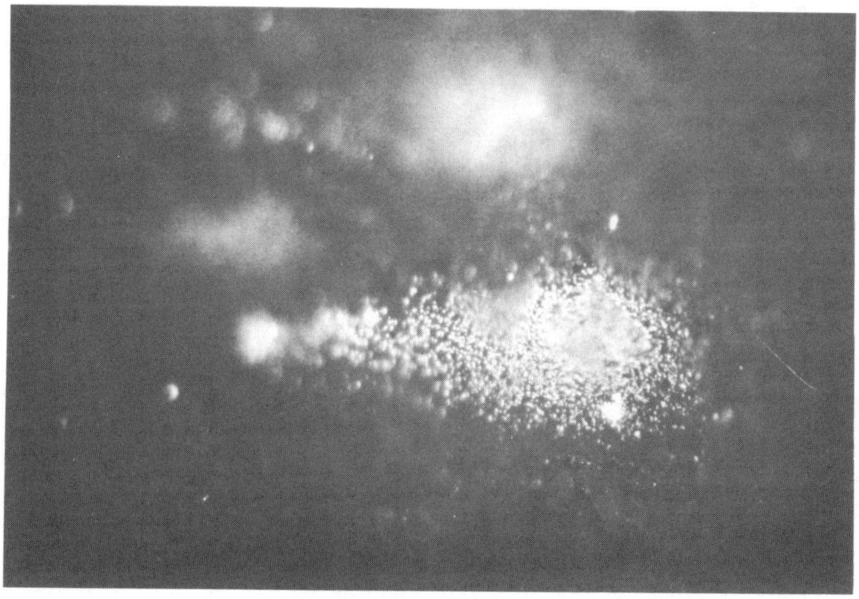

Figure 1.12. DNA migration from apoptotic lymphocyte. Cells were treated with 100 mM acetaldehyde and allowed to repair for 2 hr. (Magnification 430×, fluorescent dye YOYO-1.)

4.3.1. Brain Cells and Cells from Other Organs of Rats and Mice

Brain cells are easy to dissociate in single-cell suspension. However, they are very sensitive to environmental factors, with respect to DNA strand breaks, particularly single-strand breaks caused by stress to animals, light, air, and endonuclease which are very active above 4°C. *In vitro* plating of brain cells in a variety of culture media, sera, substrates, and gas conditions results in accumulation of a very large number of DNA single-strand breaks within 1 hr. However, *in vivo* basal levels of DNA single-strand breaks if assayed under ideal conditions (from 1-day-old to 3-month-old rats) are almost equal to those of lymphocytes. A brief protocol (Lai and Singh, 1995) to dissociate cells from solid animal tissues is described here. Immediately after dissection, tissue is immersed in ice-cold PBS with 200 mM *N-t*-butyl-α-phenylnitrone, a spin trap molecule. Tissue is washed four times with the same buffer to remove most of the red blood cells. A pair of sharp scissors is used to mince (approximately 200 times) the tissue in a 50-ml polypropylene centrifuge tube containing 5 ml of ice-cold PBS to obtain pieces of approximately 1 mm^3. Four more washings of these pieces (which settle very quickly at the bottom of the tube without centrifugation and PBS can be easily pipetted out from the top) with the cold buffer remove most of the remaining red blood cells. Finally, in 5 ml of PBS, these tissue pieces are dispersed into single-cell suspension using a 5-ml pipetman. This cell suspension consists of different types of cells from that tissue. Depending on the size of the tissue and thus the number of cells obtained, an appropriate minimum volume (to avoid too much dilution of agarose) of this cell suspension is mixed with agarose. Microgels are made as described in Sections 3.1.3 and 4.1.1.

4.3.2. Epithelial Cells

Cells from human buccal mucosa are scrapped with the edge of a wet frosted slide (to avoid quick dehydration) and gently spread over the rest of the slide. One hundred microliters of agarose is layered to make a microgel with the help of a cover glass as described in Sections 3.1.3. and 4.1.1. Similarly, urinary tract epithelial cells can be obtained by centrifugation of urine and used for DNA strand-break analysis.

4.3.3. Cultured Fibroblasts

Details of methodology for monolayer cultures are described elsewhere (Singh *et al.*, 1991b). Most of the cells adhere to clean frosted slides (Section 3.1.2c) within 1 hr of plating and show relatively normal levels of DNA single-strand breaks at approximately 8 hr. A low number of cells should be plated initially to facilitate the later analysis of DNA migration. Up to five slides (frosted surface up) can be used for plating in a 200-mm petri dish. Microgels can be made using 100 μl of agarose and a cover glass.

5. CONCLUSION

Sensitivity of the microgel electrophoresis technique has allowed detection of very low levels of DNA strand breaks induced by X rays and acetaldehyde in human lymphocytes. Only measurements of length of migrated DNA are used in these studies; other measurements of migrated DNA will probably increase the sensitivity of the technique. This technique also has been applied to screen basal and induced DNA strand breaks by a variety of agents in different cell types. With appropriate modifications, almost all cell types from living tissues can be subjected to the microgel electrophoresis technique to screen for low levels of DNA single- and double-strand breaks in individual cells. The technique is now well standardized for clinical applications and environmental mutagen monitoring. The methodology described here and in section 2 for stretching of chromosomal DNA molecules has a potential for use in mapping and quantitation of high resolution chromosomal aberrations like translocations.

ACKNOWLEDGMENTS. My sincere thanks to Dr. Henry Lai of the Bioelectromagnetic Research Laboratory, University of Washington, and Dr. William Reichert of Northwest Fisheries Science Center, NMFS/NOAA, Seattle, WA, for valuable comments. Also thanks to Sandy Marvinney for editing, and Leslie MacEven and Kevin Broderick for typing this manuscript.

REFERENCES

Barber, C. V., and Pentel, A. G. (1993). The role of oxygenation in embryotoxic mechanisms of three bioreducible agents. *Teratology* **47**:209–223.

Boyum, A. (1968). Isolation of mononuclear cells and granulocytes from human blood. *Scand. Clin. Lab. Invest.* **21**:77–89.

Collins, A. R., Ma, A. G., and Duthie S.J. (1995). The kinetics of repair of oxidative DNA damage (strand breaks and oxidized pyrimidines) in human cells. *Mutat. Res.* **336**:69–77.

Comings, D. E. (1980). Arrangement of chromatin in the nucleus. *Hum. Genet.* **53**:131–143.

Deutch, J. M. (1988). Theoretical studies of DNA during gel electrophoresis. *Science* **240**:922–924.

Fairbairn, D. W., Olive, P. L., and O'Neill, K. L. (1995). The comet assay: A comprehensive review. *Mutat. Res.* **339**:37–59.

Freeman, S. E., Blackett, A. D., Monteleone, D. C., Setlow, R. B., Sutherland, B. M., and Sutherland, J. C. (1986). Quantitation of radiation, chemical, or enzyme-induced single-strand breaks in nonradioactive DNA by alkaline gel electrophoresis: Application to pyrimidine dimers. *Anal. Biochem.* **158**:119–129.

Gedik, G. M., Ewen, S. W. B, and Collins, A. R. (1992). Single-cell gel electrophoresis applied to the analysis of UV-C damage and its repair in human cells. *J. Radiat. Biol.* **62**:313–320.

Glazer, A. N., and Rye, H. S. (1992). Stable dye–DNA interaction complexes as reagents for high-sensitivity fluorescence detection. *Nature* **359**:859–861.

Hartmann, A., and Speit, G. (1995). Genotoxic effects of chemicals in the single cell gel (SCG) test with human blood cells in relation to the induction of sister-chromatid exchanges (SCE). *Mutat. Res.* **346**:49–56.

Hellman, B., Vaghef, H., and Boström, B. (1995). The concepts of tail moment and tail inertia in the single cell gel electrophoresis assay. *Mutat. Res* **336**:123–131.

Kavenoff, R., and Zimm, B. H. (1973). Chromosome size molecules from Drosophila. *Chromosoma* **41**:1–27.

Khan, A., Lai, H., Nishimura,Y., Mirolo, M. H., and Singh, N. P. (1995). Effects of ECS on DNA-single strand breaks in rat brain cells. *Convuls. Ther.* **11**:114–121.

Kohn, K. W., Erickson, L. C., Ewig, R. A., and Friedman, C. A. (1976). Fractionation of DNA from mammalian cells by alkaline elution. *Biochemistry* **15**:4629–4637.

Krauth,W., and Werner, D. (1979). Analysis of the most tightly bound proteins in eukaryotic DNA. *Biochim. Biophys. Acta* **564**:390–401.

Lai, H., and Singh, N. P. (1995). Acute low intensity microwave exposure increases DNA single strand breaks in rat brain cells. *Bioelectromagnetics* **16**:207–210.

Lett, J. T., Klucis, E. S., and Sun, C. (1970). On the size of the DNA in the mammalian chromosome. *Biophys. J.* **10**:277–292.

Lipetz, P. D., Brash, D. E., Joseph, L. B., Jewett, H. D., Lisle, D. R., Lantry, L. E., Hart, R. W., and Stephens, R. E. (1982). Determination of DNA superhelicity and extremely low levels of DNA strand breaks in low numbers of nonradiolabeled cells by DNA-4′, 6-diamidino-2-phenylindole fluorescence in nucleoid gradients. *Anal. Biochem.* **121**:339–348.

Olive, P. L., Banath, J. P., and Durand, R. E. (1990). Detection of etoposide resistance by measuring DNA damage in individual Chinese hamster cells. *J. Natl. Cancer Inst.* **82**:779–783.

Ostling, O., and Johanson, K. J. (1984). Microelectrophoretic study of radiation-induced DNA damage in individual mammalian cells. *Biochem. Biophys. Res. Commun.* **123**:291–298.

Rabl, C. (1885). Über Zelltheilung, *Morphologisches Jahrbuch* **10**:214–330.

Rydberg, B. (1980). Detection of induced DNA strand breaks with improved sensitivity in human cells. *Radiat. Res.* **81**:492–495.

Rydberg, B., and Johanson, K. J. (1978). Estimation of DNA strand breaks in single mammalian cells, in: *DNA Repair Mechanisms* (P. C. Hanwalt and E. C. Friedberg, eds.), Academic Press, New York, pp. 465–468.

Sasaki, M.S., and Norman, A. (1966). DNA fiber from human lymphocyte nuclei. *Exp. Cell Res.* **66**:642–645.

Shafer, D. A., Xie, Y., and Falek, A. (1994). Detection of opiate-enhanced increases in DNA damage, HPRT mutants, and the mutation frequency in human HUT-78 cells. *Environ. Mol. Mutagen.* **23**:37–44.

Singh, N. P. (1996a). Sodium ascorbate induces DNA single-strand breaks in human cells in vitro. Submitted for publication.

Singh, N. P. (1996b). Investigation into basics of DNA microgel electrophoresis. Submitted for publication.

Singh, N. P., and Khan, A. (1995). Acetaldehyde: Genotoxicity and cytotoxicity in human lymphocytes. *Mutat. Res.* **337**:9–17.

Singh, N. P., and Stephens, R. E. (1986). A novel technique for viable cell determinations. *Stain Technol.* **61**: 315–318.

Singh, N. P., McCoy, M. T., Tice, R. R., and Schneider, E. L. (1988). A simple technique for quantitation of low levels of DNA damage in individual cells. *Exp. Cell Res.* **175**:184–191.

Singh, N. P., Danner, D. B., Tice, R. R., McCoy, M. C., Collins, G. D., and Schneider, E. L. (1989). Abundant alkali labile sites in DNA of human and mouse sperm. *Exp. Cell Res.* **184**:461–470.

Singh, N. P., Danner, D. B., Tice, R. R., Brant, L., and Schneider, E. L. (1990). DNA damage and repair with age in individual human lymphocytes. *Mutat. Res.* **237**:123–130.

Singh, N. P., Danner, D. B., Tice, R. R., Pearson, J. D., Brant, L., Morrel, C. H., and Schneider, E. L. (1991a). Basal DNA damage in individual human lymphocytes with age. *Mutat. Res.* **256**:1–6.

Singh, N. P., Tice, R. R., Stephens, R. E., and Schneider, E. L. (1991b). A microgel electrophoresis technique for the direct quantitation of DNA damage and repair in individual fibroblasts cultured on microscopic slides. *Mutat. Res.* **252**:289–296.

Singh, N. P., Stephens, R. E., and Schneider, E. L. (1994). Modifications of alkaline microgel electrophoresis for sensitive detection of DNA damage. *Int. J. Radiat. Biol.* **66**:23–28.

Taylor, J. H., and Hozier, C. (1976). Evidence for a four micron replicon unit in CHO cells. *Chromosoma* **57**: 341–350.

Vijayalaxmi, Tice, R. R., and Strauss, G. H. S. (1992). Assessment of radiation-induced DNA damage in human blood lymphocytes using the single-cell gel electrophoresis technique. *Mutat. Res.* **271**:243–252.

Werner, D., Krauth, W., and Hershey, H. V. (1980). Internucleotide protein linkers in Ehrlich ascites cell DNA. *Biochim. Biophys. Acta* **608**:243–258.

Chapter 2

The Cytokinesis-Block Micronucleus Technique

Michael F. Fenech

1. INTRODUCTION

The observation that chromosome damage can be caused by exposure to ionizing radiation or carcinogenic chemicals was among the first reliable evidence that physical and chemical agents can cause major alterations to the genetic material of eukaryotic cells (Evans, 1977). Although our understanding of chromosome structure is incomplete, evidence suggests that chromosome abnormalities are a direct consequence and manifestation of damage at the DNA level; for example, chromosome breaks may result from unrepaired double-strand breaks in DNA and chromosome rearrangements may result from misrepair of strand breaks in DNA (Savage, 1993). It is also recognized that chromosome loss and malsegregation of chromosomes (nondisjunction) are an important event in cancer and aging and that they are probably caused by defects in the spindle, centromere or as a consequence of undercondensation of chromosome structure before metaphase (Evans, 1990; Dellarco *et al.*, 1985; Guttenbach and Schmid, 1994).

In the classical cytogenetic techniques, chromosomes are studied directly by observing and counting aberrations in metaphases (Natarajan and Obe, 1982). Although this approach provides the most detailed analysis, the complexity and laboriousness of enumerating aberrations in metaphase and the confounding effect of artifactual loss of chromosomes from metaphase preparations have stimulated the development of a simpler system of measuring chromosome damage.

It was proposed independently by Schmid (1975) and Heddle (1973) that an alternative and simpler approach to assess chromosome damage *in vivo* was to measure micronuclei (MNi), also known as Howell–Jolly bodies to hematologists, in dividing cell populations such as the bone marrow. The micronucleus assay in bone marrow and peripheral blood erythrocytes is now one of the best established *in vivo* cytogenetic assays in the field of genetic toxicology; however, it is not a technique that is applicable to other cell populations *in vivo* or *in vitro* and methods have since been developed for measuring MNi in a variety of nucleated cells.

MNi are expressed in dividing cells that contain chromosome breaks lacking centromeres (acentric fragments) and/or whole chromosomes that are unable to travel to the spindle poles

---

Michael F. Fenech • CSIRO Division of Human Nutrition, Adelaide SA, 5000, Australia.

Technologies for Detection of DNA Damage and Mutations, edited by Gerd P. Pfeifer, Plenum Press, New York, 1996.

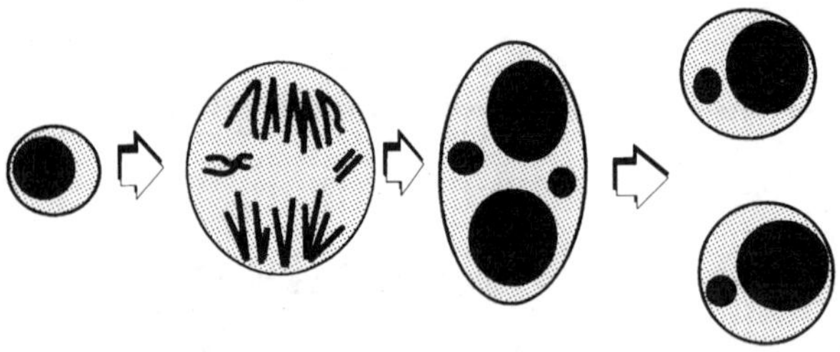

Figure 2.1. The origin of micronuclei from lagging whole chromosomes and acentric chromosome fragments at anaphase.

during mitosis. At telophase, a nuclear envelope forms around the lagging chromosomes and fragments, which then uncoil and gradually assume the morphology of an interphase nucleus with the exception that they are smaller than the main nuclei in the cell, hence the term "micronucleus" (Fig. 2.1). MNi, therefore, provide a convenient and reliable index of both chromosome breakage and chromosome loss.

It is evident from the above that MNi can only be expressed in dividing eukaryotic cells. In other words, the assay cannot be used efficiently or quantitatively in nondividing cell populations or in dividing cell populations in which the kinetics of cell division is not well understood or controlled. Consequently, there was a need to develop a method that could distinguish between cells that are not dividing and cells that are undergoing mitosis within a cell population. Furthermore, because of the uncertainty of the fate of MNi following more than one nuclear division, it is important to identify cells that have completed one nuclear division only. These requirements are also necessary because cells divide at different rates *in vivo* and *in vitro* depending on the various physiological, genetic, and micronutrient conditions.

Several methods have been proposed based on stathmokinetic, flow cytometric, and DNA labeling approaches, but the method that has found most favor because of its simplicity and lack of uncertainty regarding its effect on baseline genetic damage is the cytokinesis-block micronucleus (CBMN) assay (Fenech and Morley, 1985a,b, 1986).

In the CBMN assay, cells that have completed one nuclear division are blocked from performing cytokinesis using cytochalasin-B (CYT-B) and are consequently readily identified by their binucleated appearance (Fig. 2.2). CYT-B is an inhibitor of actin polymerization required for the formation of the microfilament ring that constricts the cytoplasm between the daughter nuclei during cytokinesis (Carter, 1967). The use of CYT-B enables the accumulation of virtually all dividing cells at the binucleate stage in dividing cell populations regardless of their degree of synchrony and the proportion of dividing cells. MNi are then scored in binucleated cells only, which enables reliable comparisons of chromosome damage between cell populations that may differ in their cell division kinetics. The method was initially developed for use with cultured human lymphocytes (Fenech and Morley, 1985a,b) but has now been adapted to various cell types (Masunaga *et al.*, 1991; Odagiri *et al.*, 1994). Furthermore, new developments have occurred that allow (1) MNi originating from whole chromosomes to be distinguished from MNi originating from chromosome fragments (Parry *et al.*, 1995) and (2) the conversion of excision-repaired sites to MNi within one cell division (Fenech and Neville, 1991). The standard CBMN assay and a method for identifying whole chromosomes within MNi are described in detail in the next section.

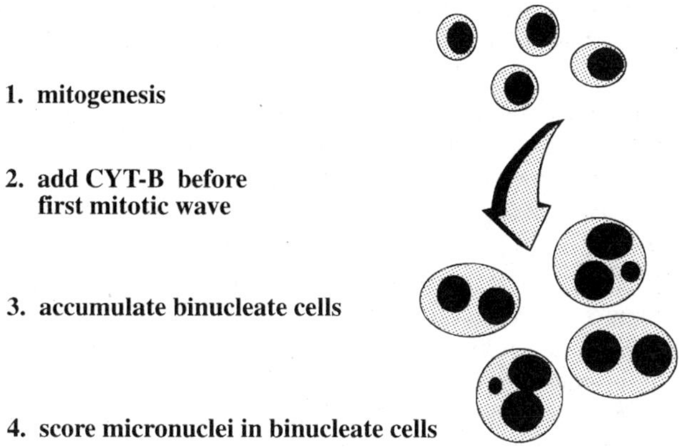

1. mitogenesis
2. add CYT-B before first mitotic wave
3. accumulate binucleate cells
4. score micronuclei in binucleate cells

Figure 2.2. The basic elements of the cytokinesis-block micronucleus assay. CYT-B, cytochalasin-B.

2. METHODS

The method described in this section is mainly applicable to cultured human lymphocytes. Notes on modification of the assay for application to other cell types are included.

2.1. Standard CBMN Assay for Isolated Human Lymphocytes

In this technique MNi are scored only in those cells that have completed one nuclear division following phytohemagglutinin (PHA) stimulation. These cells are recognized by their binucleated appearance after they are blocked from performing cytokinesis by CYT-B which should be added before the first mitotic wave. All equipment should have biosafety features to protect the operator and solutions used in this procedure should be filter sterilized.

2.1.1. Lymphocyte Isolation, Cell Culture, and Cell Harvesting

1. Fresh blood is collected by venepuncture in tubes with lithium heparin as anticoagulant.
2. The blood is then diluted 1:1 with isotonic (0.85%) sterile saline and gently inverted to mix.
3. The diluted blood is overlaid gently on Ficoll–Paque (Pharmacia) density gradients using a ratio of approximately 1:3 (e.g., 2 ml Ficoll–Paque to 6 ml diluted blood), being very careful not to disturb the interface.
4. The gradient is then spun in a centrifuge at 400*g* for 25 min at 22°C after carefully balancing the tubes.
5. The lymphocyte layer at the interface of Ficoll–Paque and diluted plasma is collected with a sterile plugged Pasteur pipet and added to 3× volume of Hanks' balanced salt solution (HBSS) at 22°C and the resulting cell suspension is centrifuged at 280*g* for 5 min.
6. The supernatant is discarded, the cells resuspended in 2× volume of HBSS and centrifuged at 180*g* for 5 min.
7. The supernatant is discarded and the cells resuspended in 1 ml McCoy's culture medium modified according to Pike and Robinson (1970).

8. Cell concentration is then measured using a Coulter counter or hemocytometer and the concentration adjusted by the percentage of viable cells measured using trypan blue exclusion assay.
9. The cells are resuspended in modified McCoy's medium containing 8% heat-inactivated fetal calf serum at 1×10^6 cells/ml and cultured in 1 ml volume in round-bottom tissue culture tubes (10-mm width).
10. Lymphocytes are then stimulated to divide by adding PHA (MUREX Diagnostics) to each culture tube at 10 μl/ml (from a stock solution in H_2O of 22 μg/ml) and incubated at 37°C with loose lids in a humidified atmosphere containing 5% CO_2.
11. Forty-four hours after PHA stimulation, 4.5 μg CYT-B is added to each milliliter of culture (use gloves and fume hood): a 100-μl aliquot of CYT-B stock solution in DMSO (600 μg/ml) is thawed, 900 μl culture medium added and mixed. Seventy-five microliters of the mixture is added to each milliliter of culture to give a final concentration of 4.5 μg CYT-B/ml. Culture tubes are then reincubated with loose lids.
12. Twenty-eight hours after adding CYT-B, cells are harvested by cytocentrifugation (Shandon Elliot). One hundred microliters of the culture medium is removed without disturbing the cells and then cells are gently resuspended in their tubes. One hundred to one hundred and twenty microliters of cell suspension is transferred to cytocentrifuge cups and centrifuged to produce two spots per slide (set the cytocentrifuge as follows: time, 5 min; speed, 600 rpm). Slides are removed from the cytocentrifuge and allowed to air-dry for 10–12 min only and then fixed for 10 min in absolute methanol.
13. The cells can be stained using a variety of techniques that can clearly identify nuclear and cytoplasmic boundaries. In our experience the use of "Diff Quik" (Lab-Aids, Australia), a commercial ready-to-use product, provides rapid and optimal results.
14. After staining, the slides are air-dried and coverslips placed over the cells using Depex mounting medium. This procedure is carried out in the fume hood and the slides are left to set in the fume hood and then stored indefinitely until required.

2.1.2. Examination of Slides and Assessment of MN Frequency

Slides are examined at 1000× magnification using a light microscope. Slides should be coded before analysis so that the scorer is not aware of the identity of the slide. For each slide the following information can be derived:

1. The number of MNi in at least 1000 binucleate (BN) cells should be scored and the frequency of MNi per 1000 BN cells calculated. The number of cells scored should be determined depending on the level of change in the micronucleus index that the experiment is intended to detect and the expected standard deviation of the estimate. The criteria for scoring MNi in BN cells is detailed below.
2. The distribution of MNi among the BN cells; the number of MNi in BN cells normally ranges from 0 to 3 in lymphocytes of healthy individuals but can be greater than 3 on occasion depending on genotoxin exposure and age.
3. The frequency of micronucleated BN cells in at least 1000 BN cells.
4. The proportion of mono-, bi-, tri-, and tetranucleated cells per 500 cells scored. From this information the nuclear division index (explained below) can be derived.
5. The number of dead or dying cells caused by apoptosis or necrosis per 500 cells may also be scored on the same slide.

The basic elements of a typical score sheet are listed in Table I.

Table I
Information That Should Be Included on a Score Sheet for the Cytokinesis-Block Micronucleus Assay

1.	Code number of each slide
2.	Number of BN[a] cells scored
3.	Proportion of BN cells with 0, 1, 2, 3, or more MNi in at least 1000 BN cells
4.	Total number of MNi in BN cells
5.	Frequency of MNi in 1000 BN cells
6.	Frequency of micronucleated BN cells in 1000 BN cells
7.	Proportion of mino-, bi-, tri-, and tetranucleated cells in at least 500 cells
8.	Frequency of BN cells in a total of 500 cells
9.	Nuclear division index
10.	Proportion of cells that are undergoing apoptosis/necrosis

[a]BN, binucleate; MNi, micronuclei.

2.1.3. Criteria for Scoring MNi in Cytokinesis-Blocked Cells

2.1.3a. Criteria for Cytokinesis-Blocked Cells. The cytokinesis-blocked cells that may be scored for MNi frequency should have the following characteristics:

1. The cells should be binucleated.
2. The two nuclei in a binucleated cell should be of approximately equal size.
3. The two nuclei within a BN cell may be attached by a fine nucleoplasmic bridge which is no wider than one-third of the nuclear diameter (these nucleoplasmic connections are probably the result of dicentric bridges and are observed in cells treated with ionizing radiation and other free-radical-generating systems).
4. The two main nuclei in a BN cell may touch or overlap each other especially in preparations in which the cytoplasm has been preserved.

Examples of the types of cytokinesis-blocked cells that may or may not be scored are illustrated diagrammatically in Fig. 2.3.

2.1.3b. Criteria for Scoring Micronuclei. MNi are morphologically identical to but smaller than nuclei. They also have the following characteristics:

1. The diameter of MNi in human lymphocytes usually varies between 1/16 and 1/3 of the diameter of the main nuclei.
2. MNi are nonrefractile and can therefore be readily distinguished from artifact such as staining particles.
3. MNi are not linked to the main nuclei but they may overlap the main nuclei.
4. MNi usually have the same staining intensity as the main nuclei but occasionally staining may be more intense.

Examples of cellular structures that should not be classified as MNi are illustrated in Fig. 2.4.

2.1.4. Nuclear Division Index

This index is calculated according to the method of Eastmond and Tucker (1989). Analyze 500 cells to score the frequency of cells with one, two, three, or four nuclei and calculate the NDI using this formula: NDI = $(M1 + 2\times M2 + 3\times M3 + 4\times M4)/N$, where M1–M4 represent the number of cells with one to four nuclei and N is the total number of cells scored. This index and

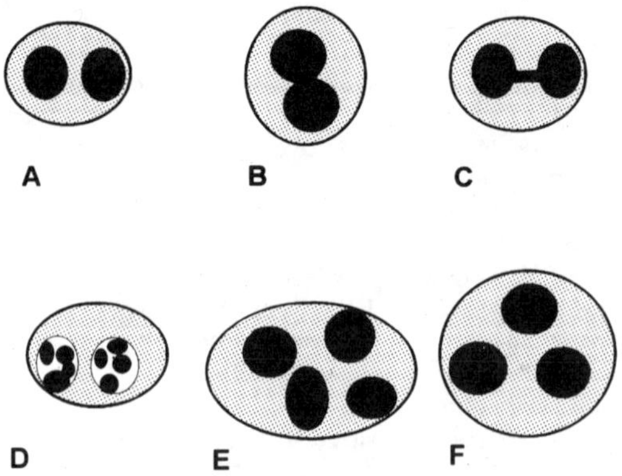

Figure 2.3. Criteria for choosing binucleate cells in the cytokinesis-block micronucleus assay. A, ideal binucleate cell; B, binucleate cell with overlapping nuclei; C, binucleate cell with nucleoplasmic bridge between nuclei; D, binucleate cell undergoing apoptosis; E, tetranucleated cell; F, trinucleated cell. Cell types A–C can be selected for the micronucleus count but cell types D–F should not be considered in the micronucleus assay. Apoptotic cells may appear to contain micronuclei due to nuclear disintegration, but they should not be scored. Tri- and tetranucleated cells have MNi frequencies that are much higher than seen in binucleated cells possibly due to aberrant spindle formation.

the proportion of binucleated cells are useful parameters for comparing the mitogenic response of lymphocytes and cytostatic effects of agents examined in the assay.

2.1.5. Adaptation of the Standard CBMN Assay to Whole Blood Cultures and Murine Lymphocytes

2.1.5a. Whole Blood Cultures for Human Lymphocytes. The CBMN assay in human lymphocytes can also be examined using whole blood cultures (Surrales *et al.*, 1992). Typically 0.5 ml of whole blood is added to 4.5 ml of culture medium (e.g., RPMI 1640) supplemented with fetal calf serum containing L-glutamine, antibiotics, and PHA. CYT-B is added at 44 hr post-PHA stimulation; the optimal concentration of CYT-B for whole blood cultures appears to be 6 μg/ml. The binucleated lymphocytes are harvested 28 hr later, hypotonically treated to lyse red blood

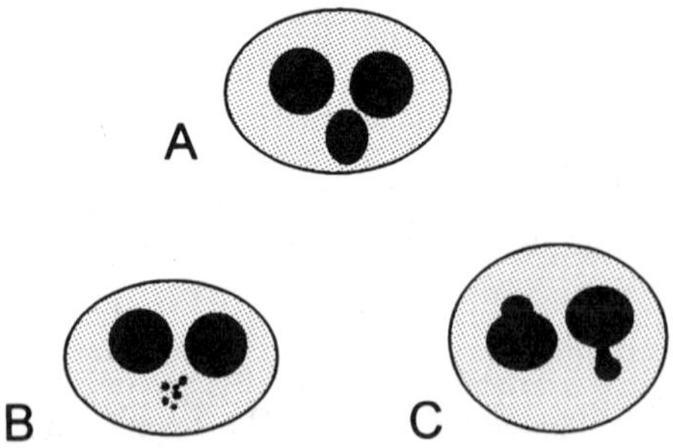

Figure 2.4. Occasionally, binucleated cells may contain structures that resemble micronuclei. These situations include: (A) trinucleated cells in which one of the nuclei is relatively small but has a diameter greater than one-third the diameter of the other nuclei; (B) dense stippling in a specific region of the cytoplasm; and (C) nuclear blebs that have an obvious nucleoplasmic connection with the main nucleus.

cells and fixed with methanol–acetic acid prior to transfer to slides and staining. As an alternative, it is also possible to isolate the binucleated lymphocytes directly from the culture using Ficoll gradients and then transfer cells to slides by cytocentrifugation prior to fixation and staining (unpublished observation) which precludes the requirement for hypotonic treatment and enables optimal preservation of the cytoplasm.

2.1.5b. Murine Lymphocyte Cultures. Lymphocytes are isolated from either the spleen or peripheral blood and cultured according to the procedures described by Fenech *et al.* (1991). Because murine lymphocytes have shorter cell cycles, it is essential to add CYT-B no later than 18 hr after stimulation by mitogen and to harvest the cells 20 hr later.

2.1.5c. Adaptation of the Standard CBMN Assay to Other Cell Types. The CBMN assay can be readily adapted to other cell types to assess DNA damage induced *in vitro*, *in vivo*, or *ex vivo*. The most important points to remember are (1) to ensure that MNi are scored in the first nuclear division following the genotoxic insult and (2) to perform preliminary experiments to determine the concentration of CYT-B at which the maximum number of dividing cells will be blocked at the binucleate stage. It is also also important to remember that CYT-B may take up to 6 hr before it starts to exert its cytokinesis-blocking action (unpublished observation). When using established or primary cell lines from dividing cell populations, it is usual to add CYT-B immediately to the initial culture medium to capture all cells undergoing their first nuclear division; this usually requires an incubation period of about 24 to 48 hr, depending on the cell cycle time, before harvesting the cells. Attached cells can be trypsinized and then prepared by cytocentrifugation as described for human lymphocytes. Specific methods have been described for use with nucleated bone marrow cells (Odagiri *et al.*, 1994), lung fibroblasts (Heddle *et al.*, 1990), skin keratinocytes (He and Baker, 1989), and a variety of cell lines such as MCL-5 cells (White *et al.*, 1992) and primary tumor cell cultures (Masunaga *et al.*, 1991). It is generally more practical to assess *in vivo* induction of micronuclei by blocking cytokinesis in dividing cells after the cells have been isolated from the animal and placed in culture medium in the presence of CYT-B; this approach has proven to be successful with a variety of cell types including fibroblasts, keratinocytes, and nucleated bone marrow cells.

2.2. Techniques That Distinguish between Micronuclei Originating from Whole Chromosomes and Acentric Chromosome Fragments

To take full advantage of the ability of the CBMN assay, it is essential to distinguish between MNi originating from whole chromosomes or from acentric fragments. This is best achieved by using probes that are specific for the centromeric DNA or antibodies that bind to the kinetochore proteins that are assembled at the centromeric regions of active chromosomes. The use of micronucleus size as a discriminant is not recommended for human cells or other cell types in which the size of chromosomes is heterogeneous because a small micronucleus may contain either a fragment of a large chromosome or a whole small chromosome. The simplest and least expensive technique to use is the antikinetochore antibody method (Vig and Swearngin, 1986), but this approach does not distinguish between unique chromosomes and may not detect chromosome loss occurring due to absence of kinetochores on inactive centromeres (Earnshaw and Migeon, 1985). The use of *in situ* hybridization (ISH) to identify centromeric regions is more expensive and laborious but it can provide greater specificity; for example, centromeric probes for unique chromosomes can be used which also enables the detection of nondisjunctional events (i.e., unequal distribution of homologous chromosomes in daughter nuclei) in binucleated cells (Farooqi *et al.*, 1993). In this chapter, only the kinetochore antibody method will be described. For details on the use of centromere detection by ISH, refer to the papers by Farooqi *et al.* (1993) and Hando *et al.* (1994). The types of results that can be expected with the various techniques are illustrated in Fig. 2.5.

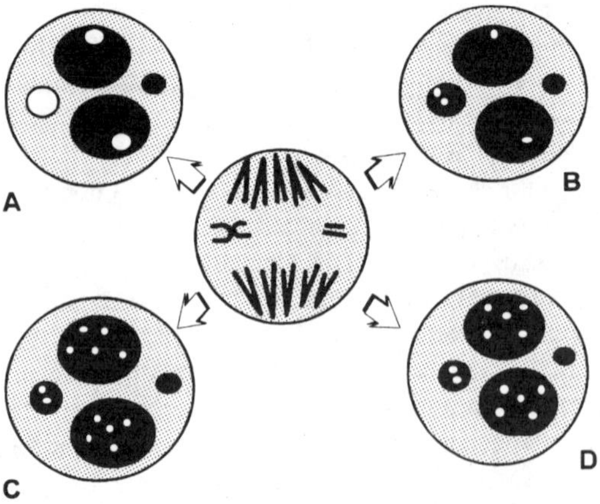

Figure 2.5. The results of four different techniques for distinguishing micronuclei containing whole chromosomes from micronuclei containing acentric fragments. (A) Whole chromosome painting by ISH using a probe cocktail specific to the whole chromosome included in the large micronucleus; (B) detection by ISH using centromeric DNA probes specific to the whole chromosome included in the large micronucleus; (C) detection by ISH using nonspecific centromeric DNA probes; (D) Detection of whole chromosome in micronucleus using antikinetochore antibodies. The example shown is for a hypothetical cell with three pairs of chromosomes.

2.2.1. Kinetochore Detection in MNi in the CBMN Assay

2.2.1a. Slide Preparation. In this technique BN cells are accumulated as described in the standard CBMN assay, transferred to a slide using a cytocentrifuge, air-dried for 5 min and fixed in methanol for 10 min and air-dried again. At this stage slides may either be processed immediately or stored for a maximum of 3 months in a sealed desiccated box in a nitrogen atmosphere above liquid nitrogen. For detection of kinetochores the stored slides are removed from the nitrogen atmosphere and allowed to equilibrate at room temperature within the sealed box.

2.2.1b. Kinetochore Detection. The antikinetochore sera may be obtained commercially or from an immunology clinic that has serum samples from scleroderma patients of the CREST subtype (Moroi *et al.*, 1981). Use of the latter sera would require Human Ethics approval and consent from the donor patient. The sera should be tested on slides of metaphase spreads of cultured cells using a rabbit FITC-conjugated secondary anti-human IgG antibody and examined by fluorescence microscopy. Only sera that appear to react exclusively with kinetochores on metaphase chromosomes should be selected for the assay.

The use of FITC-conjugated secondary antibody to visualize kinetochores is a direct technique but requires a fluorescence microscope and nonpermanent slide preparations; the fluorescence technique has been described in detail elsewhere (Vig and Swearngin, 1986). An alternative precedure is to use an immunoperoxidase staining method that allows permanent slide preparations to be obtained (Fenech and Morley, 1989) which is more practical for routine screening and is described in the next paragraph.

In the immunoperoxidase technique, fixed slides are incubated overnight at 20°C in a humidity chamber with the primary antikinetochore antibody diluted 1/40 in Tris-saline buffer, pH 7.6 (6.055 g Tris base/liter saline). Negative control slides are exposed to the diluted serum of a normal healthy individual. The following day the slides are washed by dipping for 30 sec in the same Tris-saline buffer used to dilute the antibody. Slides are then drained without drying, and incubated for 3 hr with peroxidase-labeled rabbit anti-human IgG. Again, slides are drained without drying in preparation for the peroxidase histochemical reaction.

The histochemical method that gives best contrast is the nickel chloride/imidazole modification of the standard diaminobenzidine (DAB) reaction which produces a black precipitate (Straus, 1982; Scopsi and Larsson, 1986). The reaction mixture consists of the following: 1 ml of DAB (1 mg/ml in Tris base buffer stock, 60.55 g/liter, pH 7.6), 3 ml of Tris-base buffer stock pH 7.6 (60.55 g/liter), 25 μl of $NiCl_2$ solution (8% solution in Tris-base buffer stock prepared immediately before use), 40 μl of 0.1 M imidazole, and 10 μl of 30% hydrogen peroxide solution.

The reaction mixture is prepared just before use and applied immediately to slides through a 0.22-μm filter to minimize nonspecific precipitation on the slides. Slides should be stained in batches including a slide with the negative control serum. The reaction is allowed to proceed for 1 min at 20°C and then stopped by draining the slides and rinsing in water. The slides are then air-dried, counterstained with the nuclear stain Neutral Red (0.1% in distilled water) for 30 sec, washed in water, air-dried, and mounted to give permanent preparations.

2.2.1c. Scoring Procedure. Scoring of kinetochore status of MNi is restricted to those binucleated cells in which a minimium of 20 kinetochores within each nucleus is observed. A minimum of 100 MNi should be classified according to whether they contain kinetochores or not and the number of kinetochores within each MN should be noted. The final value for the proportion of MNi with kinetochores is determined by the following formula:

$$(P_s - P_c)/(1 - P_c)$$

where P_c is the proportion of MNi that has a positive peroxidase reaction in slides exposed to normal control serum and P_s is the proportion of MNi that has a positive peroxidase reaction in slides exposed to antikinetochore serum.

3. DISCUSSION

Although the CBMN assay is a relatively simple method to employ, interpretation of the data obtained has to be done with care especially when used for biomonitoring of individuals or populations exposed to genotoxins. The most important point is to understand or to be aware of the main confounders that could influence the index scored. In a series of ongoing studies on MNi in human peripheral blood lymphocytes, we have been able to determine that age and sex are the most important variables influencing the results: MN frequency increases steadily with age in both sexes but the MN frequency in females is approximately 1.5 times higher than that for males across all decades between 20 and 90 years (Fenech *et al.*, 1994) possibly due to the loss of the inactive X chromosome (Hando *et al.*, 1994). Other factors influencing the MN frequency are the number of cigarettes smoked per day (Fenech, 1993) and preliminary data suggest that in young adults (less than 30 years old) MN frequency correlates negatively with plasma folate and plasma vitamin B_{12} in females, and negatively with plasma vitamin B_{12} but positively with vitamin C status in young males (Fenech and Rinaldi, 1994). Consequently, it is important to account for all of these potential confounders when designing a cytogenetic/epidemiological study of exposure to genotoxic agents or dietary factors. Because it may not always be possible to use an individual as their own control, it is useful to establish a historical data base of baseline MN frequencies which includes at least 20 individuals of each sex per decade. It is technically feasible to develop such data bases on several hundred individuals within 1 year with few resources as has been demonstrated by our group (Fenech and Rinaldi, 1995) and others (Ban *et al.*, 1993).

Our experience with the CBMN assay suggested that the method was relatively insensitive, *in vitro*, to genotoxins that mainly produced DNA adducts (e.g., monofunctional methylating agents) or base modifications (UV radiation) rather than strand breakage or chromosome loss. It was evident from studies with methylnitrosourea (MNU), UV, and X rays that the extent of cell death relative to MN induction was very low for X rays but much higher for UV and MNU (Fenech and Neville, 1991). We reasoned that the base damage induced by UV and MNU was not being translated efficiently into MNi and that the method could be improved if excision-repair sites, which occur following DNA adduction or other base modification, were converted to double-stranded breaks (DSBs) by inhibiting the gap-filling step in excision repair using cytosine arabinoside. DSBs result in acentric chromosome fragments which are readily lost at anaphase and efficiently expressed as MNi (Littlefield *et al.*, 1989; Ramalho *et al.*, 1988). Results with this method showed that the CBMN assay's sensitivity to UV and MNU could be improved at least 40-fold indicating that it could be used to detect DNA-damaging events that mainly induce excision-repairable lesions (Fenech and Neville, 1991).

There are several avenues for research with the CBMN assay that may further refine its applicability to biomonitoring and prediction of radiocurability of tumors. There is also a need to understand in more detail the actual fate of micronucleated cells *in vivo*. Automated scoring of slides from the CBMN assay is a real possibility using image analysis and considerable progress has been achieved in this area (Tates *et al.*, 1990; Castelain *et al.*, 1993); the benefits of these systems will be evident once the CBMN technique becomes more widely adopted for routine biomonitoring. The positive correlation between colony forming ability and MNi in cytokinesis-blocked tumor cells (Shibamoto *et al.*, 1991) is a promising indication that the method could be used clinically to customize radiotherapy protocols to individual tumor radiosensitivity and assess radiosensitivity of normal tissues (Huber *et al.*, 1989). However, there also appears to be a need to further understand the significance of MNi in relation to cell death especially in tumors that are p53 negative as compared to tumors that are p53 positive (Fisher, 1994). The recent observation that p53 may be uniquely expressed in MNi (Unger *et al.*, 1994) suggests that the classification of p53-positive and p53-negative MNi may be of relevance to the eventual survival of the cell and should be considered in future research.

ACKNOWLEDGMENTS. The development of the CBMN assay was the result of research performed at the Medical School of the Flinders University of South Australia in Professor Alec Morley's laboratory and the CSIRO Division of Human Nutrition with the support of the Anti-Cancer Foundation of the Universities of South Australia. I also acknowledge the important role of Ms. J. Rinaldi, Ms. C. Aitken, and Ms. S. Neville who have contributed significantly to the more recent research effort.

REFERENCES

Ban, S., Cologne, J. B., Fujita, S., and Awa, A. A. (1993). Radiosensitivity of atomic bomb survivors as determined with a micronucleus assay. *Radiat. Res.* **134**:170–178.

Carter, S. B. (1967). Effects of cytochalasins on mammalian cells. *Nature* **213**:261–264.

Castelain, P., Van Hummelen, P., Deleneer, A., and Kirsch-Volders, M. (1993). Automated detection of cytochalasin-B blocked binucleated lymphocytes for scoring micronuclei. *Mutagenesis* **8(4)**:285–293.

Dellarco, V. L., Mavournin, K. H., and Tice, R. R. (1985). Aneuploidy and health risk assessment: Current status and future directions. *Environ. Mutagen.* **7**:405–424.

Earnshaw, W. C., and Migeon, B. R. (1985). Three related centromere proteins are absent from the inactive centromere of a stable dicentric chromosome. *Chromosoma* **92**:290–296.

Eastmond, D. A., and Tucker, J. D. (1989). Identification of aneuploidy-inducing agents using cytokinesis-blocked human lymphocytes and an antikinetochore antibody. *Environ. Mol. Mutagen.* **13(1)**:34–43.

Evans, H. J. (1977). Molecular mechanisms in the induction of chromosome aberrations, in: *Progress in Genetic Toxicology* (D. Scott, B. A. Bridges, and F. H. Sobels, eds.), Elsevier/North-Holland, Amsterdam, pp. 57–74.

Evans, H. J. (1990). Cytogenetics: Overview. *Prog. Clin. Biol. Res.* **340B**:301–323.

Farooqi, Z., Darroudi, F., and Natarajan, A. T. (1993). Use of fluorescence in situ hybridisation for the detection of aneugens in cytokinesis-blocked mouse splenocytes. *Mutagenesis* **8**:329–334.

Fenech, M. (1993). The cytokinesis-block micronucleus technique: A detailed description of the method and its application to genotoxicity studies in human populations. *Mutat. Res.* **285**:35–44.

Fenech, M., and Morley, A. A. (1985a). Solutions to the kinetic problem in the micronucleus assay. *Cytobios* **43:**233–246.

Fenech, M., and Morley, A. A. (1985b). Measurement of micronuclei in lymphocytes. *Mutat. Res.* **147**:29–36.

Fenech, M., and Morley, A. A. (1986). Cytokinesis-block micronucleus method in human lymphocytes: Effect of in vivo ageing and low-dose x-irradiation. *Mutat. Res.* **161**:193–198.

Fenech, M., and Morley, A. A. (1989). Kinetochore detection in micronuclei: An alternative method for measuring chromosome loss. *Mutagenesis* **4(2)**:98–104.

Fenech, M., and Neville, S. (1991). Conversion of excision-repairable DNA lesions to micronuclei within one cell cycle in human lymphocytes. *Environ. Mol. Mutagen.* **19(1)**:27–36.

Fenech, M., and Rinaldi, J. (1994). The relationship between micronuclei in human lymphocytes and plasma levels of vitamin C, vitamin E, vitamin B_{12} and folic acid. *Carcinogenesis* **15(7)**:1405–1411.

Fenech, M., and Rinaldi, J. (1995). A comparison of lymphocyte micronuclei and plasma micronutrients in vegetarians and non-vegetarians. *Carcinogenesis* **16(2)**:223–230.

Fenech, M. F., Dunaiski, V., Osborne, Y., and Morley, A. A. (1991). The cytokinesis-block micronucleus assay as a biological dosimeter in spleen and peripheral blood lymphocytes in the mouse following acute whole body irradiation. *Mutat. Res.* **263**:119–126.

Fenech, M., Neville, S., and Rinaldi, J. (1994). Sex is an important variable affecting spontaneous micronucleus frequency in cytokinesis-blocked lymphocytes. *Mutat. Res.* **313**:203–207.

Fisher, D. E. (1994). Apoptosis in cancer therapy: Crossing the threshold. *Cell* **78**:539–542.

Guttenbach, M., and Schmid, M. (1994). Exclusion of specific human chromosomes into micronuclei by 5-azacytidine treatment of lymphocyte cultures. *Exp. Cell Res.* **211**:127–132.

Hando, J. C., Nath, J., and Tucker, J. D. (1994). Sex chromosomes, micronuclei and aging in women. *Chromosoma* **103**:186–192.

He, S., and Baker, R. S. U. (1989). Initiating carcinogen, triethylenemelamine, induces micronuclei in skin target cells. *Environ. Mol. Mutagen.* **14(1)**:1–5.

Heddle, J. A. (1973). A rapid in vivo test for chromosome damage. *Mutat. Res.* **18**:187–192.

Heddle, J. A., Bouch, A., Khan, M. A., and Gingerich, J. D. (1990). Concurrent detection of gene mutations and chromosomal aberrations induced *in vivo* in somatic cells. *Mutagenesis* **5(2)**:179–184.

Huber, R., Braselmann, H., and Bauchinger, M. (1989). Screening for interindividual differences in radiosensitivity by means of the micronucleus assay in human lymphocytes. *Radiat. Environ. Biophys.* **28**:113–120.

Littlefield, G. L., Sayer, A. M., and Frome, E. L. (1989). Comparison of dose-response parameters for radiation-induced acentric fragments and micronuclei observed in cytokinesis-arrested lymphocytes. *Mutagenesis* **4**:265–270.

Masunaga, S., Ono, K., and Abe, M. (1991). A method for the selective measurement of the radiosensitivity of quiescent cells in solid tumours—Combination of immunofluorescence staining to BrdU and micronucleus assay. *Radiat. Res.* **125**:243–247.

Moroi, Y., Hartman, A. L., Nakane, P.K., and Tan, E. M. (1981). Distribution of kinetochore antigen in mammalian cell nuclei. *J. Cell Biol.* **90**:254–259.

Natarajan, A. T., and Obe, G. (1982). Mutagenicity testing with cultured mammalian cells: Cytogenetic assays, in: *Mutagenicity: New Horizons in Genetic Toxicology* (J. A. Heddle, ed.), Academic Press, New York, pp. 171–213.

Odagiri, Y., Takemoto, K., and Fenech, M. (1994). Micronucleus induction in cytokinesis-blocked mouse bone-marrow cells in vitro following in vivo exposure to X-irradiation and cyclophosphamide. *Environ. Mol. Mutagen.* **24**:61–67.

Parry, E. M., Henderson, L., and Mackay, J. M. (1995). Guidelines for testing of chemicals. Procedures for the detection of chemically induced aneuploidy: Recommendations of a UK Environmental Mutagen Society working group. *Mutagenesis* **10(1)**:1–14.

Pike, B. L., and Robinson, W. A. (1970). Human bone-marrow colony growth in agar gel. *J. Clin. Physiol.* **76**: 77–84.

Ramalho, A., Sunervic, I., and Natarajan, A. T. (1988). Use of the frequencies of micronuclei as quantitative indicators of X-ray-induced chromosomal aberrations in human peripheral blood lymphocytes: Comparison of two methods. *Mutat. Res.* **207**:141–146.

Savage, J. R. K. (1993). Update on target theory as applied to chromosomal aberrations. *Environ. Mol. Mutagen.* **22**:198–207.

Schmid, W. (1975). The micronucleus test. *Mutat. Res.* **31**:9–15.

Scopsi, I., and Larsson, L. I. (1986). Increased sensitivity in peroxidase immunochemistry. A comparative study of a number of peroxidase visualisation methods employing a model system. *Histochemistry* **84**:221–230.

Shibamoto, Y., Streffer, C., Fuhrmann, C., and Budach, V. (1991). Tumour radiosensitivity prediction by the cytokinesis-block micronucleus assay. *Radiat. Res.* **128**:293–300.

Straus, W. (1982). Imidazole increases the sensitivity of the cytochemical reaction for peroxidase with diaminobenzidine at neutral pH. *J. Histochem. Cytochem.* **30**:491–493.

Surralles, J., Carbonell, E., Marcos, R., Degrassi, F., Antoccia, A., and Tanzarella, C. (1992). A collaborative study on the improvement of the micronucleus test in cultured human lymphocytes. *Mutagenesis* **7(6)**: 407–410.

Tates, A. N., van Welie, M. T., and Ploem, J. S. (1990). The present state of the automated micronucleus test for lymphocytes. *Int. J. Radiat. Biol.* **58**:813–825.

Unger, C., Kress, S., Buchmann, A., and Schwarz, M. (1994). Gamma-irradiation-induced micronuclei from mouse hepatoma cells accumulate high levels of the tumour suppressor protein p53. *Cancer Res.* **54(14)**:3651–3655.

Vig, B. K., and Swearngin, S. E. (1986). Sequence of centromere separation: Kinetochore formation in induced laggards and micronuclei. *Mutagenesis* **1**:464–465.

White, N. H., de Matteis, F., Davies, A., Smith, L. L., Crofton-Sleigh, C., Venitt, S., Hewer, A., and Phillips, D. H. (1992). Genotoxic potential of tamoxifen and analogues in female Fischer F344/n rats, DBA/2 and C57BL/6 mice and in human MCL-5 cells. *Carcinogenesis* **13(12)**:2197–2203.

Chapter 3

Agarose Gel Electrophoresis for DNA Damage Analysis

Régen Drouin, Shuwei Gao, and Gerald P. Holmquist

1. INTRODUCTION

Physical and chemical mutagens induce frank breaks in DNA which reduce its single-strand molecular weight. Other nonbreak lesions in the DNA can often be converted into strand breaks by chemical and enzymatic means. Using agarose gel electrophoresis along with various cleavage schemes, the average density of breaks and various lesion classes along mammalian DNA can be determined.

For base damage lesions, both glycosylase activity and lyase activity are required to produce the strand break. For example, after treatment with dimethyl sulfate to induce as lesions the modified bases 7-methylguanine and 3-methyladenine, the AlkA protein, 3-methyladenine glycosylase, from *E. coli*, will hydrolyze the *N*-glycosidic bond of the alkylated purines. Subsequent cleavage of the resulting abasic sites with the lyase activity of Nth protein from *E. coli* will cleave the phosphate backbone at the abasic site (O'Connor and Laval, 1990; Doetsch and Cunningham, 1990). Some endonucleases combine a glycosylase and lyase activities (viz., Nth or Fpg protein from *E. coli* excises the oxidized pyrimidines or purines and then incises the phosphate backbone at the abasic site it has created) (Wallace, 1988; Boiteux, 1993).

Agarose gels (Sambrook *et al.*, 1989) have a wide range of separation; they can resolve a mixture of DNA fragments between 50 bp (3.0% agarose) and 500,000 bp (0.1% agarose). For single-strand gel electrophoresis, DNA must remain denatured during electrophoretic separation. Denaturing agents such as formaldehyde, formamide, methylmercuric hydroxide, sodium hydroxide, and urea can be added to the gel to prevent DNA renaturation (McMaster and Carmichael, 1977; Ogden and Adams, 1987; Sambrook *et al.*, 1989). Methylmercuric hydroxide and formaldehyde are toxic and either formamide or urea alters the properties of agarose (Maniatis *et al.*, 1982). Alkaline agarose gels are most commonly used for single-strand DNA electrophoresis. However, many lesion classes are alkali-labile. For example, abasic sites are cleaved by beta elimination in alkaline agarose conditions and 8-oxoguanine bases are labile

Régen Drouin, Shuwei Gao, and Gerald P. Holmquist • Division of Biology, Beckman Research Institute of the City of Hope, Duarte, California 91010. *Present address of R.D.*: Research Unit in Human and Molecular Genetics, Department of Pathology, Hospital Saint-François d'Assise, Laval University, Quebec G1L 3L5, Canada.

Technologies for Detection of DNA Damage and Mutations, edited by Gerd P. Pfeifer, Plenum Press, New York, 1996.

above pH 11 (Schneider *et al.*, 1990). Glyoxal-agarose gel electrophoresis of single-stranded DNA at neutral pH will be described. It size fractionates DNA as in an alkali-agarose gel electrophoresis but while retaining alkali-labile sites intact (Drouin *et al.*, 1996).

Glyoxalation of single-stranded nucleic acids allows molecules to remain denatured during electrophoresis, resulting in accurate separation by molecular weight (Carmichael and McMaster, 1980; Ogden and Adams, 1987; Sambrook *et al.*, 1989). Glyoxal or ethanedial, which stabilizes the denatured regions, is a reagent with a large relative selectivity for guanine (Broude and Budowsky, 1971). Glyoxal reacts with the free amino groups of adenosine, cytidine, and guanosine (Shapiro and Hachmann, 1966); the guanosine adduct is by far the most stable (Nakaya *et al.*, 1968; Shapiro *et al.*, 1969, 1970; Broude and Budowsky, 1971). Glyoxalation of DNA in the presence of a denaturing agent such as dimethyl sulfoxide (DMSO) introduces an additional ring onto guanine bases, thus sterically hindering its pairing with cytosine bases and preventing the renaturation of the DNA molecules (Shapiro and Hachmann, 1966; Hutton and Wetmur, 1973). Glyoxalation is reversible and allows efficient DNA and RNA hybridization (Carmichael and McMaster, 1980; Thomas, 1980). It has been used as a tool for DNA denaturation mapping (Johnson, 1975).

2. MATERIALS AND METHODS

2.1. Alkaline Gel Electrophoresis

The gel is solidified in a neutral solution and subsequently equilibrated in an alkaline buffer because NaOH inhibits gelation by hydrolysis of the polysaccharide of the agarose (Maniatis *et al.*, 1982).

- Prepare the agarose gel in 50 mM NaCl and 4 mM EDTA, let it set completely.
- Presoak the gel in the alkaline electrophoresis buffer (30 mM NaOH and 2 mM EDTA) for at least 1 hr. Usually, we prepare the gel and let it soak overnight at room temperature or at 4°C.
- Mix 3 μl of bidistilled water containing 1 to 3 μg of DNA with 3 μl of freshly prepared 100 mM NaOH/4 mM EDTA buffer and 4 μl of denaturing gel loading buffer (1 M NaOH, 50% glycerol, 0.05% bromocresol green). The total volume is 10 μl.
- Incubate at room temperature for 15–20 min prior to loading.
- Run the horizontal gel with an overlay of 1 to 2 mm of buffer. Voltages should not be over 0.25 V/cm. Alkaline gels draw more current than neutral gels at similar voltages and ohmic heating must be avoided. To avoid accumulation of air bubbles under the gel tray, the gel should be cast in a tray shorter than the platform of the gel apparatus on which the tray is placed.
- Carry out electrophoresis until the dye has migrated 40 to 60% of the length of the gel.
- Neutralize the gel in a tray containing 500 ml of 500 mM Tris-HCl pH 8 at room temperature for at least 30 min while shaking.
- Stain with 1 μg/ml of ethidium bromide in water for 30 min.
- Destain in water for 10 to 30 min or as needed.

2.2. Glyoxal Gel Electrophoresis

Methods were previously described for electrophoresis of single-stranded RNA (McMaster and Carmichael, 1977; Carmichael and McMaster, 1980; Ogden and Adams, 1987; Sambrook *et al.*, 1989). We have slightly modified these for the purpose of neutral agarose gel electrophoresis of glyoxal/dimethylsulfoxide (DMSO)-denatured genomic DNA (Drouin *et al.*, 1996). Glyoxal or ethanedial should be purchased as 6 M stock solutions or 40% aqueous solutions (available from Sigma, St. Louis, MO). Because glyoxal is readily air-oxidized, it should be deionized before use by treatment with AG-501-X8(D) ion exchange resin (1 g resin/ml glyoxal, Bio-Rad) until the pH of the glyoxal stock is > 5.0. Use a pH color indicator strip for estimating the pH. DMSO is also deionized with the type of ion exchange resin, 1 g/ml DMSO. Glyoxal and DMSO should be stored at −70°C in aliquots in tightly capped tubes and thawed only once.

- Prepare the agarose gel (agarose type I: low EEO, catalog # A6013, Sigma) in 10 mM sodium phosphate pH 7. Allow about 45 min prior to loading of the DNA samples for the gel to set.
- Dissolve DNA pellets (5 to 12 μg) in the following mix: 4.5 μl of distilled water, 2 μl of 100 mM sodium phosphate pH 7, 3.5 μl of 6 M glyoxal (1 M final concentration), 10 μl DMSO [50% (v/v) final concentration].
- Incubate at 50°C for 1 hr.
- Add to the DNA samples 4 μl of loading buffer (10 mM sodium phosphate pH 7, 50% glycerol, 0.25% bromophenol blue, 0.25% xylene cyanol FF) before loading.
- Run the gel submerged in 10 mM sodium phosphate pH 7, at 3–4 V/cm with constant recirculation of the buffer. This is required to maintain the pH at 7.0 and is critical because glyoxal readily dissociates from DNA at pH 8.0 or higher. Stop electrophoresis when the bromophenol blue has migrated 65 to 75% of the length of the gel or the xylene cyanol FF has migrated 30 to 40% of the length of the gel.
- Stain the gel with acridine orange, 1 to 30 μg/ml (10 mg/ml stock solution in water) in running buffer at room temperature for 45 to 60 min.
- Destain at room temperature overnight or as needed, then photograph the gel.

Figure 3.1 shows an example of an alkaline gel for estimating the global frequency of cyclobutane pyrimidine dimers (CPD) and pyrimidine (6–4) pyrimidone photoproducts (6–4 photoproducts) as a function of the UV dose. CPD have been enzymatically converted to single-strand breaks with T4 endonuclease V and 6–4 photoproducts have been chemically converted to single-strand breaks with hot piperidine. Since 6–4 photoproducts and other minor photoproducts are alkali-sensitive, when estimating the CPD frequency (Fig. 3.1, lanes 2–10), it should be kept in mind that a small proportion of the single-strand breaks is not due to CPD conversion. Because CPD are alkali-resistant, the other samples (Fig. 3.1, lanes 11–19) only show the frequency of 6–4 photoproducts and other hot piperidine-sensitive photoproducts. Piperidine treatment involves high temperature (>80°C), alkaline conditions, and creates a high background of depurination. Hot piperidine cleaves one normal base per 4 to 5 kb (Fig. 3.1, lane 11), whereas T4 endonuclease V cleaves only one normal base per 25 to 40 kb (Fig. 3.1, lane 2).

Figure 3.2 shows a glyoxal gel for distinguishing various classes of breaks and lesions induced by reactive oxygen species. Treated DNA shows a reduction in molecular weight due to frank breaks (control lane), breaks + abasic sites (T4 endonuclease V and endonuclease IV lanes), breaks + abasic sites + ring-saturated or ring-opened pyrimidine bases [Nth protein (endonuclease III)], and breaks + abasic sites + ring-saturated or ring-opened pyrimidine and purine bases (Nth + Fpg proteins).

3. DISCUSSION

To approximate the average break frequency from the DNA weight distribution (ethidium fluorescence intensity) along a gel, one need know only the molecular weight of the peak fraction.

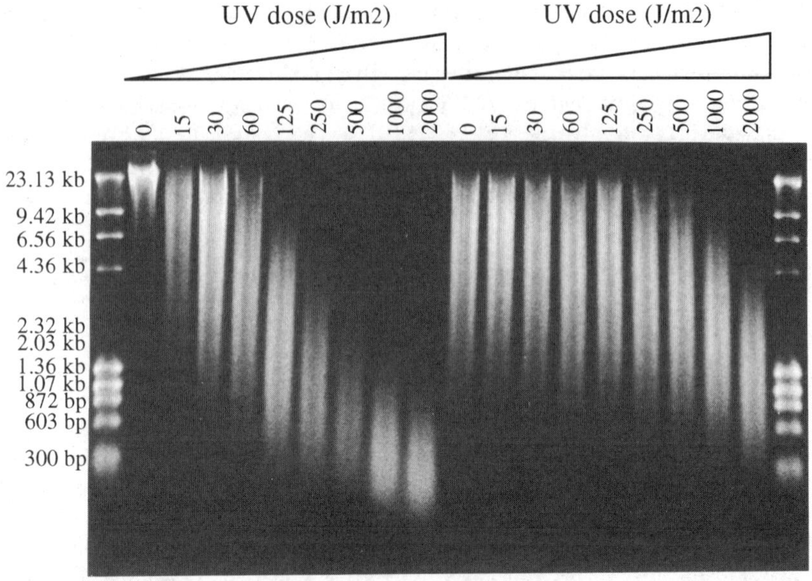

Figure 3.1. Alkaline gel display of photolesion frequency following enzymatic (lanes 2–10) or chemical (lanes 11–19) conversion of photoproducts. Purified human male fibroblast DNA was irradiated *in vitro* (after DNA extraction) with different UV doses (0 to 2000 J/m^2). After UV irradiation, cyclobutane dimers were converted into single-strand breaks with T4 endonuclease V (lanes 2–10) (Pfeifer *et al.*, 1992), and 6–4) photoproducts were converted into single-strand breaks with hot piperidine (lanes 11–19) (Pfeifer *et al.*, 1991). Three micrograms of DNA per lane was denatured in 30 mM sodium hydroxide solution, then electrophoresed through an alkaline 1.2% agarose gel. The gel was stained with ethidium bromide (1 μg/ml). The first and last lanes of this gel contain *Hin*dIII lambda bacteriophage + *Hae*III-digested ϕX174 DNA molecular weight standards.

When an infinitely long polymer is partially cleaved so that each bond is broken with the same probability P, then the weight distribution of the resulting molecules (Tanford, 1961) quickly approaches a certain distribution called "Kuhn's approximation to the Montroll and Simha equation" by Hamer and Thomas (1975) who applied it to cumulative mass distributions of *Drosophila* DNA digested with restriction endonucleases. This distribution (Fig. 3.3) is like a Poisson distribution in that one number P or its reciprocal $\overline{M}_n$ determines the entire normalized distribution. P is the probability of a bond being broken, $\overline{M}_n$ is the number-average molecular weight, and $\overline{M}_n$ = half of the weight-average molecular weight $\overline{M}_w$ (Tanford, 1961). When such a randomly cleaved distribution is fractionated along a gel, the mobility of the each fragment is proportional to the log of the molecular weight throughout the middle of the mobility range (Willis *et al.*, 1988). When labeled DNA from a restriction digest is gel fractionated, the gel sliced, and cpm determined for each slice, a Kuhn distribution can be fitted to the data in the log-linear region of the mobility function (Holmquist,1988). The value P was varied until a best fit was obtained and $\overline{M}_n = 1/P$. During these analyses, we recognized that the molecular weight corresponding to mobility of the mass distribution's peak was approximately equal to $\overline{M}_w$ (Holmquist, 1988) and Fig. 3.3 shows this is a very exact approximation for gels. The number-average molecular weight, $\overline{M}_n$, is the average break frequency in a population of gel-fractionated molecules and this is shown as lesions per 10 kb for each lane in Fig. 3.2. Here, for example, the Cu/ascorbate/H_2O_2 treatment induced 0.8 break per 10 kb above a 0.2 break/10 kb background and Cu/ascorbate/H_2O_2-treated endonuclease IV-cleaved DNA revealed 0.5 abasic site/10 kb in addition to the single-strand breaks.

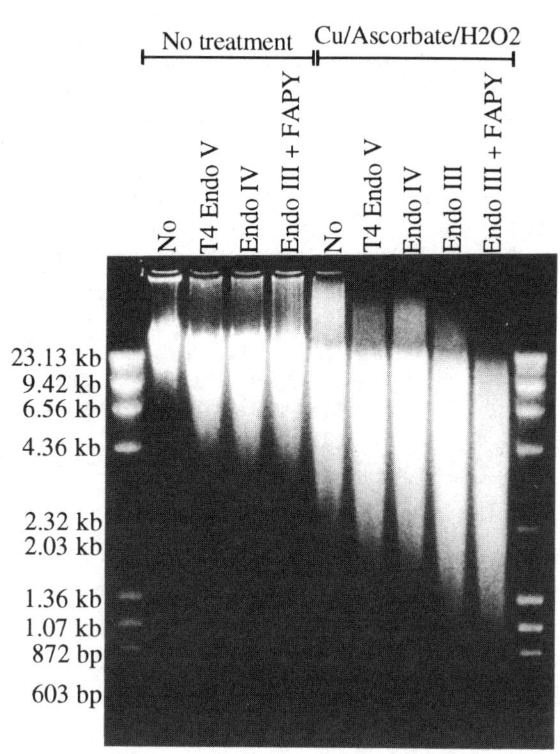

Figure 3.2. A glyoxal/DMSO gel display of the various lesion classes generated by Cu/ascorbate/ H_2O_2 treatment. Ten micrograms per lane of human male fibroblast DNA was treated either with buffer alone (lanes 2–5) or with 50 μM Cu(II), 100 μM ascorbate, 5 mM H_2O_2 for 30 min at 37°C (lanes 6–10). After treatment, DNA was digested with either buffer alone (lanes 2 and 6), 1050 ng of T4 endonuclease V (lanes 3 and 7), 5100 ng of endonuclease IV (lanes 4 and 8), 250 ng of endonuclease III (lane 9), or 250 ng of endonuclease III plus 400 ng of Fpg protein (lanes 5 and 10). The DNA was electrophoresed through a neutral 1.2% agarose gel. The gel was stained with acridine orange (3 μg/ml). Lanes 1 and 11 are DNA size standards.

Alkaline gels are easier to run than are glyoxal/DMSO gels and require five times less DNA for an accurate determination of the frequency of DNA strand breaks. The best results, in terms of resolution and accurate sizing, are obtained by running alkaline and glyoxal gels very slowly (< 5 V/cm). Alkaline gels must be run more slowly than glyoxal gels and a very thin layer of buffer must cover the gel. To avoid creating unacceptable H^+ gradients during electrophoresis, buffer circulation is mandatory for glyoxal gels and advisable for alkaline gels. Glyoxalation is reversed if the pH exceeds 8 (Broude and Budowsky, 1971). Glyoxal may decompose at temperatures higher than 50°C. Glyoxal treatments using greater than 1 M glyoxal concentrations do not affect the electrophoretic mobility of the DNA but treatments using lower concentrations may result in subsequent partial renaturation of the DNA. Unless deionized, glyoxal solution may bring about DNA degradation caused by contaminating oxidation products and chemically related substances, such as glycolic acid, glyoxylic acid, and formic acid (Carmichael and McMaster, 1980). For alkaline gels, we recommend type II low-EEO (electroendosmosis) agarose whereas for glyoxal gel purposes, we recommend type I low-EEO agarose.

Alkaline gels are best stained by ethidium bromide after thorough neutralization of the gel. Ethidium bromide is not recommended for glyoxal gels. Staining with acridine orange (AO) is a more sensitive method to visualize single-stranded glyoxalated DNA (McMaster and Car

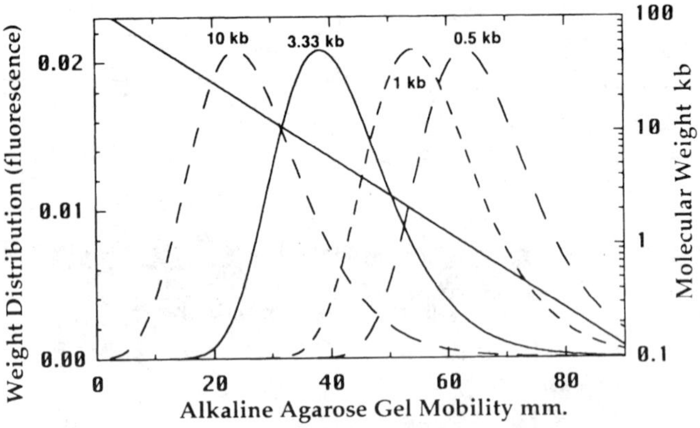

Figure 3.3. Theoretical mass distributions of randomly cleaved DNA using actual mobility of markers through a 0.6% alkaline gel. The log-linear best fit to the lambda-*Hind*III DNA molecular weight markers was $\log M_r = -\mu/30 + 2.1$ where μ is mobility in millimeters. The weight fraction of molecules exactly W_t bases long is $W_t = P^2t(1 - P)^{x-1}$ and the fraction of DNA molecules from 0 to t bases is $\int_0^t W_t dt = 1 - (1 + Pt)e^{-Pt}$ (Hamer and Thomas, 1975). The mass distributions were calculated for 0.5-mm mobility intervals using the formula

$$\text{mass fraction in mobility interval} = \int_0^{t+\Delta t/2} W_t dt - \int_0^{t-\Delta t/2} W_t dt$$

Distributions of number-average molecular weight 10 kb, 3.33 kb, 1 kb, and 0.5 kb corresponding to the probability any one phosphodiester bond is broken with $P = 0.0001$, 0.0003, 0.001, and 0.002 are shown. The peak of each distribution is $2\overline{M}_n$.

michael, 1977; Carmichael and McMaster, 1980). Glyoxal may react with ethidium bromide, changing its spectral properties (Johnson, 1975). As do many aniline dyes, AO exhibits the phenomenon of metachromasy which is a characteristic effect from π-bond interactions when the dye stacks along single-stranded DNA. The fluorochrome AO binds, as a monomeric molecule, with native, double-stranded DNA producing an orthochromatic yellow-green fluorescence (515–575 nm); it binds in a stacked form to single-stranded DNA or to RNA and emits a metachromatic red-orange fluorescence (600–650 nm) (Rigler, 1966; Kasten, 1967). Hence, its metachromasy not only visually confirms glyoxal denaturation but also makes it a powerful reagent for rapid determination of DNA strandedness after gel electrophoresis. Black-and-white photographs using Polaroid 105 positive/negative or 107C positive film can be taken with a red filter whereas a yellow filter is necessary for color photographs.

ACKNOWLEDGMENTS. We thank Steven Lloyd for T4 endonuclease V, Richard Cunningham for endonuclease IV, and Timothy O'Connor for Nth and Fpg proteins.

REFERENCES

Boiteux, S. (1993). Properties and biological functions of the NTH and FPG proteins of Escherichia coli: Two DNA glycosylases that repair oxidative damage in DNA. *J. Photochem. Photobiol.* **19**:87–96.

Broude, N. E., and Budowsky, E. I. (1971). The reaction of glyoxal with nucleic acid components. III. Kinetics of the reaction with monomers. *Biochim. Biophys. Acta* **254**:380–388.

Carmichael, G. G., and McMaster, G. K. (1980). The analysis of nucleic acids in gels using glyoxal and acridine orange. *Methods Enzymol.* **65**:380–391.

Doetsch, P. W., and Cunningham, R. P. (1990). The enzymology of apurinic/apyrimidinic endonucleases. *Mutat. Res.* **236**:173–201.

Drouin, R., Rodriguez, H., Gao, S., Gebreyes, Z., O'Connor, T. R., Holmquist, G. P., and Akman, S. A. (1996). Cupric ion/ascorbate/hydrogen peroxide-induced DNA damage: DNA-bound copper ion primarily induces base modifications. *Free Radical Biol. Med.* (in press).

Hamer, D. H., and Thomas, C. A., Jr. (1975). The cleavage of *Drosophila melanogaster* DNA by restriction endonucleases. *Chromosoma* **49**:243–255.

Holmquist, G. P. (1988). DNA sequences in G-bands and R-bands, in: *Chromosomes and Chromatin* (K. W. Adolph, ed.), CRC Press, Boca Raton, FL, pp. 75–121.

Hutton, J. R., and Wetmur, J. G. (1973). Effect of chemical modification on the rate of renaturation of deoxyribonucleic acid. Deaminated and glyoxalated deoxyribonucleic acid. *Biochemistry* **12**:558–563.

Johnson, D. (1975). A new method of DNA denaturation mapping. *Nucleic Acids Res.* **2**:2049–2054.

Kasten, F. H. (1967). Cytochemical studies with acridine orange and the influence of dye contaminants in the staining of nucleic acids. *Int. Rev. Cytol.* **21**:141–202.

McMaster, G. K. and Carmichael, G. G. (1977). Analysis of single- and double-stranded nucleic acids on polyacrylamide and agarose gels by using glyoxal and acridine orange. *Proc. Natl. Acad. Sci. USA* **74**:4835–4838.

Maniatis, T., Fritsch, E. F., and Sambrook, J. (1982). *Molecular Cloning: A Laboratory Manual*, Cold Spring Harbor Laboratory Press, Cold Spring Harbor, NY.

Nakaya, K., Takenaka, O., Horinishi, H., and Shibata, K. (1968). Reactions of glyoxal with nucleic acids, nucleotides and their component bases. *Biochim. Biophys. Acta* **161**:23–31.

O'Connor, T. R., and Laval, J. (1990). Isolation and structure of a cDNA expressing a mammalian 3-methyladenine-DNA glycosylase. *EMBO J.* **9**:3337–3342.

Ogden, R. C., and Adams, D. A. (1987). Electrophoresis in agarose and acrylamide gels. *Methods Enzymol.* **152**: 61–87.

Pfeifer, G. P., Drouin, R., Riggs, A. D., and Holmquist, G. P. (1991). In vivo mapping of a DNA adduct at nucleotide resolution: Detection of pyrimidine (6–4) pyrimidone photoproducts by ligation-mediated polymerase chain reaction. *Proc. Natl. Acad. Sci. USA* **88**:1374–1378.

Pfeifer, G. P., Drouin, R., Riggs, A. D., and Holmquist, G. P. (1992). Binding of transcription factors creates hot spots for UV photoproducts in vivo. *Mol. Cell. Biol.* **12**:1798–1804.

Rigler, R. (1966). Microfluorometric characterization of intracellular nucleic acids and nucleoproteins by acridine orange. *Acta Physiol. Scand.* **67(Suppl. 267)**:1–122.

Rodriguez, H., Drouin, R., Holmquist, G. P., O'Connor, T. R., Boiteux, S., Laval, J., Doroshow, J. H., and Akman, S. A. (1995). Mapping of copper/hydrogen peroxide-induced DNA damage at nucleotide resolution in human genomic DNA by ligation-mediated PCR. *J. Biol. Chem.* **270**:17633–17640.

Sambrook, J., Fritsch, E. F., and Maniatis, T. (1989). *Molecular Cloning: A Laboratory Manual*, 2nd ed., Cold Spring Harbor Laboratory Press, Cold Spring Harbor, NY.

Schneider, J. E., Price, S., Maidt, L., Gutteridge, J. M. C., and Floyd, R.A. (1990). Methylene blue plus light mediates 8-hydroxy 2′-deoxyguanosine formation in DNA preferentially over strand breakage. *Nucleic Acids Res.* **18**:631–635.

Shapiro, R., and Hachmann, J. (1966). The reaction of guanine derivatives with 1,2-dicarbonyl compounds. *Biochemistry* **5**:2799–2807.

Shapiro, R., Cohen, B. I., Shiuey, S.-J., and Maurer, H. (1969). On the reaction of guanine with glyoxal, pyruvaldehyde, and kethoxal, and the structure of the acylguanines. A new synthesis of N2-alkylguanines. *Biochemistry* **8**:238–245.

Shapiro, R., Cohen, B. I., and Clagett, D. C. (1970). Specific acylation of the guanine residues of ribonucleic acid. *J. Biol. Chem.* **245**:2633–2639.

Tanford, C. (1961). *Physical Chemistry of Macromolecules*, Wiley, New York.

Thomas, P. S. (1980). Hybridization of denatured RNA and small DNA fragments transferred to nitrocellulose. *Proc. Natl. Acad. Sci. USA* **77**:5201–5205.

Wallace, S. S. (1988). AP endonucleases and DNA glycosylases that recognize oxidative DNA damage. *Environ. Mol. Mutagen.* **12**:431–477.

Willis, C. K., Willis, D. G., and Holmquist, G. P. (1988). An equation for DNA electrophoretic mobility. *Appl. Theor. Electrophor.* **1**:11–18.

Chapter 4

^{32}P-Postlabeling for Detection of DNA Adducts

R. C. Gupta

1. INTRODUCTION

It is generally believed that the formation of DNA adducts by covalent interaction of electrophilic species of carcinogens (ultimate carcinogens) with macromolecules, particularly DNA, is an essential first step in the multistage process of carcinogenesis (Miller and Miller, 1981). Until the early 1980s, use of radiolabeled carcinogens was the main method to determine the binding of chemical carcinogens to DNA. Since then several methods have been utilized for nonradiolabeled carcinogens. The assays that are practiced currently are based on specific antibodies, fluorescence properties of adducts, gas chromatography/mass spectrometry, and ^{32}P-postlabeling. These assays require a few micrograms to several hundred micrograms of DNA, with a detection limit of 1 adduct per 10^6 to 10^{10} nucleotides, depending on the method (Beach and Gupta, 1992). The ^{32}P-postlabeling assay has emerged as the major tool for measuring DNA adducts because of its ultrasensitivity and applicability to theoretically any DNA-damaging agent, irrespective of its chemical nature, including unknowns. Over 100 individual agents or complex mixtures have been tested in rodents and aquatic systems *in vivo* and rodent and human cells *in vitro* (Beach and Gupta, 1992).

The ^{32}P-postlabeling is comprised of several steps (Fig. 4.1). Step I involves degradation of DNA to 3′-monophosphates of normal deoxynucleosides (Np) and adducts (Xp). Adducts resulting from aromatic and/or lipophilic carcinogens (Xp) are enriched by removal of Np by either selective extraction of Xp in 1-butanol (Gupta, 1985), by conversion of Np to N selectively by hydrolysis of the digest with nuclease P1 (Reddy and Randerath, 1986), or by C18 reverse-phase HPLC (Dunn and San, 1988) or TLC (Spencer and Gupta, 1992) (Step II). The enriched adducts are 5′-^{32}P-labeled by T4 polynucleotide kinase-catalyzed phosphorylation in the presence of [γ-^{32}P]-ATP (Step III). The resultant labeled adducts ([5′-^{32}P]pXp) are separated by multidirectional, anion exchange polyethyleneimine (PEI)-cellulose TLC (Step IV). The presence of extra spot(s) in the treated as compared to untreated DNA processed in parallel shows the presence of adduct(s), as detected by autoradiography and quantitated by measurement of the adduct radioactivity (Step V). Thus, lipophilic adducts can be detected at a level of 1 per 10^{10}

R. C. Gupta • Preventive Medicine and Environmental Health, University of Kentucky Medical Center, Lexington, Kentucky 40536.

Technologies for Detection of DNA Damage and Mutations, edited by Gerd P. Pfeifer, Plenum Press, New York, 1996.

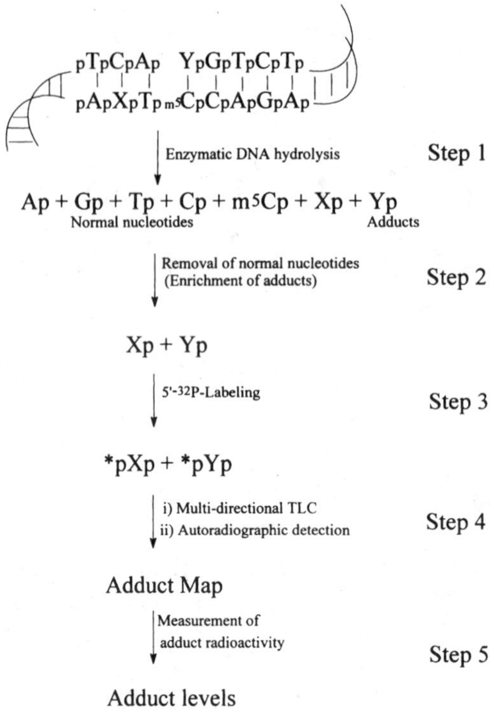

Figure 4.1. Principle of the ^{32}P-postlabeling technique.

nucleotides. "Small" adducts resulting from alkylating agents can also be detected but with a detection limit that is three to four orders of magnitude lower. The lower assay sensitivity for such adducts reflects the inapplicability of the above enrichment procedures and/or insufficient resolution of the adducts from normal nucleotides. Selective procedures based on HPLC or immunoaffinity have been applied for enrichment of small adducts prior to ^{32}P-labeling.

In this chapter, we describe precise conditions for each step of the assay which we consider optimal. After describing experimental conditions, a brief commentary is also presented providing explanations and necessary precautions. It is to be noted that the conditions described below for isolation of DNA and ^{32}P-postlabeling adduct analysis are somewhat modified as compared to those described elsewhere (Beach and Gupta, 1992; Gupta, 1993).

2. MATERIALS AND GENERAL METHODS

2.1. Enzymes

Enzymes are purchased from the sources indicated below: Micrococcal nuclease (MN) (~110 U/mg protein; grade VI), RNase A (type III-A), RNase T1 (500,000 U/mg) from Sigma Chemical Co. (St. Louis, MO); nuclease P1 (480 U/mg protein) from Calbiochemicals (San Diego, CA); proteinase K and spleen phosphodiesterase (SPD) (2 U/mg) from Boehringer-Mannheim Corp. (Indianapolis, IN); and cloned T4 polynucleotide kinase (10 U/μl) from US Biochemical Co. (Cleveland, OH). All enzymes are dissolved in glass-distilled or HPLC-grade water and used without further purification, except for SPD (spleen phosphodiesterase), MN, and

RNase T1 which are dialyzed against three changes of 1 liter distilled water at 4°C for 6–8 hr, and RNase A, which is dissolved in 100 mM sodium acetate, pH 5, incubated at 80°C for 10 min and cooled in ice water to inactivate contaminating DNases. Solutions of MN and SPD are mixed at a concentration of 1 μg/μl. This enzyme mixture and nuclease P1 (1 μg/μl), RNase A (10 μg/μl), RNase T1 (200 U/μl), and proteinase K (10 μg/μl) are stored in aliquots at −20°C (preferably −80°C) for greater than 1 year.

2.2. Buffers

Buffers are prepared at 10 times the concentrations required: 10× tissue homogenizing buffer = 500 mM Tris·HCl, pH 8.0/100 mM EDTA, Na_2 (10× TE). 10× MN-SPD buffer = 100 mM sodium succinate, pH 6.0/50 mM $CaCl_2$; and 10× kinase buffer = 300 mM Tris·HCl, pH 9.5/100 mM $MgCl_2$/100 mM dithiothreitol/10 mM spermidine.

2.3. [γ-^{32}P]-ATP

Carrier-free [γ-^{32}P]-ATP (3000–6000 Ci/mmole) can be purchased from New England Nuclear (Irvine, CA), ICN Pharmaceuticals (Costa Mesa, CA), or Amersham (Arlington Heights, IL) in aqueous form. We prepare this material in our laboratory by transferring $^{32}P_i$ from ^{32}P-labeled 1,3-diphosphoglycerate to ADP as a four-step enzymatic conversion of L-glycerol-3-phosphate to 3-phosphoglycerate (Johnson and Walseth, 1979), after modification (Gupta *et al.*, 1982; Gupta, 1985). This is conveniently done by mixing 1 vol of carrier-free, HCl-free $^{32}P_i$ (10–40 μl of 400–800 mCi/ml; 2–8 mCi) (New England Nuclear, Irvine, CA) and 1 vol of a pre-prepared enzyme–buffer mix (Gupta *et al.*, 1982; Gupta, 1985). The enzyme-buffer mix can be stored in 20- to 40-μl aliquots at −80°C for greater than 1 year and has provided 96–98% yields of [γ-^{32}P]-ATP. Its specific activity, however, varies, depending on the extent of traces of phosphate contamination in the reagents. We usually obtain specific activity of [γ-^{32}P]-ATP at ≥3000 Ci/mmole. This preparation has been used fresh or after storage at −20°C for about a week, without purification. The procedure is rather simple and involves little work, once the enzyme–buffer mix has been prepared; it is also highly economical (3–10 times less expensive than commercial [γ-^{32}P]-ATP).

2.4. Chromatography Supplies

Normal and adducted nucleotides have been separated largely on PEI (polyethyleneimine)-cellulose thin layers. While commercial PEI-cellulose has been used in many laboratories, we prefer using layers prepared in the laboratory for both economic and better resolution purposes. "Homemade" sheets have been prepared by spreading a slurry of 3.0% PEI-HCl solution (pH 6.0 by pH paper) over planished, rigid vinyl sheets (21 × 131 cm), as described (Randerath and Randerath, 1967; Gupta *et al.*, 1982). High-molecular-weight PEI is available from Sigma Chemical Co. (St. Louis, MO), Polysciences Inc. (Warrington, PA, Cat. #6090), or Virginia Chemicals (Portsmouth, VA; brand name Corcat P-600). The sheets are allowed to air-dry overnight, cut into desired sizes (13 × 20 or 20 × 20 cm), and stored at 4°C (preferably at −20°C). Sheets for adduct separation have been used without purification; if stored long-term (months), they are purified by ascending development in water to the top of the sheet and air-dried prior to use. Sheets used for normal nucleotide separation are preequilibrated by ascending development in 100 mM ammonium formate, pH 3.5, and air-dried. Sheets can also be washed by soaking in 500 ml H_2O or 100 mM ammonium formate, pH 3.5, per 2–3 sheets for ≥ 10 min.

Standard glass tanks with lids have been used for TLC. We also use specially designed acrylic TLC racks by mounting horizontal acrylic bars and an acrylic or stainless-steel tray. This rack is placed in an acrylic tank. This assembly allows development of up to 15 sheets (i.e., 30 samples) collectively.

2.5. Autoradiography

Cassettes equipped with one or two intensifying screens (DuPont Lightning Plus or Kyokko HS) and X-ray films (Kodak XAR-5 or DuPont Cronex 4) are from a local supplier; the DuPont films are more economical but about three times less sensitive than the Kodak films. Radioactive inks are prepared by adding the following amounts of $^{32}P_i$ per milliliter of black ink: 30–50 μCi (>30 min exposure); 4–6 μCi (2–8 hr exposure); and 2–4 μCi (overnight exposure). The amount of radioactivity is adjusted every 2 weeks to allow for the decay (which decays by ~50% every 2 weeks). Alternatively, stable isotopes can be used: 100–250 μCi ^{14}C red ink (ICN) per ml black ink for long exposures (>8 hr) or 50–250 μCi $NH_4{}^{99}TcO_4$ (New England Nuclear) per ml black ink for short exposures (a few minutes to several hours). Several institutions require segregation of radioactive waste based on the isotope. ^{32}P-labeled ink offers advantage in such a situation, besides being economical.

2.6. Handling of ^{32}P-Labeled Materials

All operations involving synthesis of [γ-^{32}P]-ATP, ^{32}P-labeling, chromatography of ^{32}P-labeled nucleotides, and excision of nucleotide spots from the chromatograms are performed behind ⅜- to ½-inch-thick portable, transparent acrylic plates or fixed glass plates. Use of slidable glass plates mounted on a workbench provides a convenient means for most of these operations. For all other equipment and operations, readers are directed to our previous articles (Gupta, 1985; Gupta and Randerath, 1988).

2.7. Isolation of DNA

Purity of DNA in terms of RNA and protein contamination is of vital importance. Many laboratories employing the ^{32}P-postlabeling assay appear to have followed the rapid solvent extraction procedure in which proteins are first removed by digestion with proteinase K, followed by rapid solvent extractions. The resultant nucleic acids (DNA and RNA) are then treated with RNases and solvent extracted to recover pure DNA (Gupta, 1984). We have now established a simpler/shorter version of this method in which isolated crude nuclei or cells are treated sequentially with RNases and proteinase K, followed by solvent extractions to remove RNA and proteins, respectively (Gupta, 1993). The method is as follows:

- Suspend 0.1–0.5 g frozen tissue in 3 ml 1×TE buffer for 5–10 min (20-ml glass scintillation vials are convenient for homogenization).
- Homogenize with a Polytron (one-half the speed on the dial; 5–10 sec) or any other homogenizer. Transfer homogenate to a 5-ml polypropylene tube. Keep all tubes on ice until all samples are homogenized.
- Sediment tissue homogenates using an unrefrigerated or refrigerated, tabletop centrifuge (<3000 rpm; 10 min). Collect nuclear pellet by gently pouring off supernatant or using a Pasteur pipet.
- Suspend pellet in 1.5 ml 1×TE buffer by either gently vortexing (mixing on a vortex mixer) or using a Pasteur pipet.
- Add 22 μl RNase A (10 μg/μl) and 8 μl RNase T1 (200 U/μl) and incubate at 37°C for 30–40 min (final concentrations ~150 μg/ml RNase A and 1 U/μl RNase T1).
- Add 75 μl 10% sodium dodecyl sulfate (SDS) and 22 μl proteinase K (10 μg/ml) and

continue incubation for additional 30 min (final concentrations 0.5% SDS and 150 μg/ml proteinase K).

- Extract reaction mixture sequentially with 1×TE buffer-saturated distilled phenol (1.5 ml), phenol:Sevag (chloroform:isoamyl alcohol, 24:1) (0.75 ml:0.75 ml), and, finally, with Sevag (1.5 ml). For extractions, place up to 15 tubes in a plastic or glass beaker and shake manually by inverting the beaker back and forth for 5 min, followed by separation of the aqueous and nonaqueous phases using a tabletop microfuge (9000 rpm) for 7, 5, and 3 min, respectively. Collect aqueous phase with a Pasteur pipet and transfer it to a fresh tube.
- Add 150 μl 5 M sodium chloride (≈0.1 vol) and 1.5 ml −20°C-chilled ethanol (≈1 vol), invert tubes several times, centrifuge using a tabletop microfuge (9000 rpm; 7 min), and collect DNA precipitate by pouring off supernatant (at a concentration of 0.1–0.3 mg/ml, DNA is precipitated almost instantly as a "lump"; at lower concentration, keep tubes at −20°C for several hours or overnight prior to centrifugation).
- To remove traces of salt, add 1–1.5 ml 70% ethanol, gently invert tubes, and pour off supernatant. Repeat this step one or two more times (centrifugation is used for 2–3 min if the pellet is loose or floats prior to removing supernatant).
- Remove residual ethanol by draining or with a cotton swab.
- Dissolve still-moist pellet in 200 μl–1000 μl 1/100 SSC-EDTA (1.5 mM NaCl and 0.15 mM sodium citrate–0.1 mM EDTA) at a concentration of 0.6–1 mg/ml.
- Pass DNA solution two or three times through a 1-ml syringe with a very fine needle (No. 25G 5/8, Beckton-Dickinson) to shear DNA prior to storage.

The shearing of DNA avoids its possible polymerization during freezing or thawing unless DNA solution is quickly frozen (in dry ice–acetone) and quickly thawed (in warm water).

- Measure DNA concentration spectrophotometrically: dilute 5–15 μl DNA to 300 μl with water and record absorbance at various wavelengths (230, 260, and 280 nm). DNA concentration in μg/μl = 15 × A_{260}/5, assuming 5 μl DNA was used for dilution. Store DNA solution at −20°C (preferably at −80°C). Store DNA in lyophilized form if it is to be analyzed after several weeks or months.

Procedure for isolating DNA from cells is essentially the same as for tissue nuclei, except that cells (up to 50×10^6) are suspended in 600 μl 1×TE in 1.5- to 2-ml Eppendorf-type tubes and sheared by passing two or three times through a fine needle as described above for shearing DNA instead of homogenizing prior to incubation with RNases.

This procedure is rapid (2–3 hr) resulting in DNA preparations essentially free of RNA (99%) and proteins, in high yields (≈2 and 3.5 mg/g rat liver and lung, respectively). The method has been applied for tissues of rodents, aquatic organisms, and humans and other organisms (e.g., bacteria, yeast, plants). Optimal A_{230}/A_{260} and A_{260}/A_{280} ratios of 0.40 ± 0.04 and 1.80 ± 0.05 suggest that the DNA preparation is essentially free of proteins and RNA, respectively. Certain tissues (fish liver, rodent tumors, some human tissues) have sometimes resulted in DNA preparations containing greater than 5% RNA contamination, as reflected by a higher A_{260}/A_{280} ratio (>1.90). In such cases, DNA is further treated with RNases A and T1; the inclusion of RNase T2 (10 U/ml) has particularly proved useful for more complete removal of RNA. Occasionally, some DNA preparations have been found to be turbid as a result of coprecipitation of other macromolecules (e.g., glycogens), thus resulting in a high A_{230}/A_{260} ratio. This contamination, however, does not appear to interfere in subsequent adduct analyses, except that DNA concentration estimated spectrophotometrically is higher than it actually is. The actual concentration may be determined by ^{32}P-labeling of normal nucleotides and TLC and compared with known amount of calf thymus DNA digest labeled in parallel, and used to correct DNA concentrations.

DNA adducts resulting from PAHs (polycyclic aromatic hydrocarbons) have been found stable during storage of both tissues and isolated DNAs at −80°C for several weeks to months; aromatic amine-derived adducts are somewhat labile and the lability increases with duration; therefore, it is suggested that such adducts are analyzed without excessive storage (J. M. Arif and R. C. Gupta, unpublished results).

3. PROTOCOL FOR A STANDARD ^{32}P-POSTLABELING ASSAY

3.1. Step I: Enzymatic DNA Digestion

- Mix 12 μg DNA and 3 μl 10× MN-SPD buffer in a 1.5-ml Eppendorf-type polypropylene tube and make the total volume to 27.6 μl. Add 2.4 μl MN-SPD (1 μg/μl each). Mix reaction mixture by briefly vortexing (2 sec) and centrifugation (3000 rpm; 2 sec). Incubate at 37°C for 5 hr, and process the digest immediately or after storage at −20°C for up to overnight (final concentrations, 0.4 μg/μl DNA and 0.08 μg/μl each of MN and SPD).
- Dilute an aliquot of the digest (5 μl; 2 μg) to 6000 μl with water (0.33 ng/μl, final concentration) for labeling of total nucleotides in Step IIIA and use the remaining undiluted digest for adduct enrichment in Step II.

Most adducted DNAs can be hydrolyzed completely under these conditions. The enzyme:DNA ratio can be regulated at the expense of incubation time, e.g., a 3-hr incubation may be sufficient for an enzyme:DNA ratio of 1:3, up to overnight incubation for an enzyme:DNA ratio of 1:10–40; however, this should be tested in a pilot experiment as certain adducts may be labile to such long incubations. Correspondingly more DNA is digested for replicate adduct analyses. Vehicle-treated DNA as a negative control and reference adducted DNA as a positive control are processed in parallel with the test DNA. Additional negative controls in which DNA or enzymes or both are omitted should be included in initial experiments to ascertain that spots designated as adducts are DNA-derived. Several laboratories appear to depend on the specific activity of ATP and spectrophotometrically estimated DNA concentration to calculate adduct levels. In our experience, however, spectrophotometrically estimated DNA concentration for certain tissue DNAs may be significantly lower than the actual concentration due to coprecipitation of glycogen or other contaminants which may underestimate adduct levels.

3.2. Step II: Adduct Enrichment

Butanol extraction or nuclease P1 treatment is used commonly for enrichment of adducts for structurally diverse aromatic and/or lipophilic carcinogens (Beach and Gupta, 1992). Since adducts of most or all arylamines are lost almost completely in the nuclease P1 procedure (Gupta and Earley, 1988; Gallagher *et al.*, 1989), it is imperative that both butanol and enzymatic procedures are employed, particularly for unknown adducts, in pilot experiments to ensure maximal adduct recoveries. The enzymatic enrichment procedure is preferred for its ease if both procedures furnish comparable adduct recoveries.

Butanol Extraction

This is a solvent–solvent microanalytical partitioning procedure in which aromatic/lipophilic adducts are preferentially extracted in the butanol phase from the acidic aqueous phase in the presence of the phase-transfer agent, tetrabutylammonium chloride (TBA).

- Mix DNA digest (0.1–40 μg) with 20 μl each of 100 mM ammonium formate, pH 3.5, and 10 mM TBA in a 1.5-ml Eppendorf-type tube and add distilled water to make the total volume to 200 μl (final concentrations 0.5 ng–0.2 μg/μl DNA, 1 mM TBA (tetrabutyl ammonium chloride) and 10 mM ammonium formate).
- Extract mixture with 200 μl water-saturated, double distilled 1-butanol by vortexing (1 min), followed by centrifugation in a microfuge (9000 rpm; 1–2 min).
- Collect butanol (upper) phase in a fresh 1.5 ml tube and repeat extraction of aqueous phase with butanol.
- Extract combined butanol phase with 360 μl butanol-saturated water by vortexing (20–30 sec) and centrifugation (9000 rpm; <1 min).
- Repeat this step once (<2 μg DNA), twice (<10 μg DNA), or thrice (<40 μg DNA), depending on the amount of DNA.
- Add 3 μl 200 mM Tris·HCl, pH 9.5, and evaporate solution to dryness under reduced pressure (e.g., Speed-vac concentrator, Savant Instruments Inc., Farmingdale, NY). Use residue in Step III.

1-Butanol should be distilled once, preferably twice, and saturated with water prior to use. The distillation is necessary to remove trace acidity present in the analytical-grade butanol which otherwise appears to interfere during subsequent polynucleotide kinase labeling. After distillation, 1-butanol is saturated with distilled water by adding 30–40% water, mixing vigorously in a separatory funnel or a bottle, followed by centrifugation to separate the two phases. This results in two phases: water-saturated butanol and butanol-saturated water. These solutions can be stored at room temperature for more than 1 year.

Nuclease P1 Enrichment

This enrichment strategy takes advantage of the preferential conversion of normal nucleotides (Np) to nucleosides by the 3′-activity of nuclease P1, while adducted nucleotides (Xp) may be resistant, depending on their structure. In subsequent ^{32}P-labeling (Step III), the resultant nucleosides are not substrates for 5′- phosphorylation. Though adducts of PAHs are largely or completely resistant to hydrolysis, arylamine adducts are mostly sensitive to the enzymatic hydrolysis. Thus, while PAH adducts are recovered substantially or completely, arylamine adducts are largely lost (Gupta and Earley, 1988; Gallagher *et al.*, 1989).

Nuclease P1 enrichment is performed as described by Reddy and Randerath (1986), except that (1) pH of the DNA digest (Step I) is adjusted neither before nor after the nuclease P1 treatment; and (2) a lower amount of nuclease P1 per microgram DNA digest is employed.

- Add 3 μl 1 mM $ZnCl_2$ and 3.3 μl nuclease P1 (1 μg/μl) to 25 μl (10 μg) of Step I DNA digest (final concentrations, 0.32 μg/μl DNA, 0.1 mM $ZnCl_2$, and 0.1 μg/μl nuclease P1).
- After incubation at 37°C for 40 min, evaporate digest to dryness under reduced pressure (Speed-vac concentrator). Use residue for ^{32}P-labeling in Step III. Use digest without evaporation if ≤ 5 μg (≤ 15 μl) digest is to be labeled.

Up to 5 μg DNA is sufficient to detect 1 adduct per $<10^9$ nucleotides. Higher amount of DNA (10–15 μg) is employed to increase the assay sensitivity to 1 adduct per $<10^{10}$ nucleotides. The ^{32}P-labeling recipe described in Step III can accommodate up to 15 μg nuclease P1-enriched DNA digest without any interference of the enzymes and co-factors present; up to 40 μg DNA digest can be accommodated if enriched by butanol extraction. It is recommended that nuclease P1 digest be used immediately without storage for Step III, as trace phosphatase/nucleotidase activity present in some nuclease P1 preparations may result in partial loss of adducts during storage of the digest.

3.3. Step III: ^{32}P-Labeling of Adducts

Enriched adducted nucleotides (Xp) are converted to 5′-^{32}P-labeled 3′,5′-bisphosphates ([5′-^{32}P]pXp) (Fig. 4.1) using carrier-free, high-specific-activity [γ-^{32}P]-ATP (>3000 Ci/mmole). Wear two or three layers of disposable gloves during the ^{32}P-labeling step; check gloves frequently using a Geiger counter and change the top glove if found contaminated.

- Reconstitute residue from Step II in 15 μl water by tapping and agitating the tube manually, vortexing and brief centrifugation.
- Place tubes, without caps, in a specially designed acrylic test-tube rack with a slidable lid (Gupta, 1985) or any appropriate acrylic test-tube rack (e.g., beta-blocks, Research Products International Corp., Mt. Prospect, IL).
- Add 5 μl "hot mix" [2 μl 10× kinase buffer, 0.4 μl T4 polynucleotide kinase (10 U/μl), 1 μl [γ-^{32}P]-ATP (~3000 Ci/mmole; 100 μCi) and 1.6 μl water] to 15 μl adduct solution. Stir solution with the same pipet tip that was used to add hot mix (final concentrations, 30 mM Tris·HCl, pH 9.5, 10 mM $MgCl_2$, 10 mM dithiothreitol, 1 mM spermidine, 0.2 U/μl polynucleotide kinase, and 1.6 μM [γ-^{32}P]-ATP).
- Following incubation at room temperature (22°C or higher, up to 37°C) for ≥40 min, use labeled solution for adduct chromatography in step IV, usually without storage.

Adducts are usually stable when the labeled digest is stored at −20°C, except for aromatic amine-derived adducts which appear to be labile during the storage presumably due to depurination of the adducts. It is, therefore, suggested that the labeled solution, particularly for unknown adducts, is chromatographed without storage as much as possible (once bound to the thin layer matrix, most or all adducts have been found stabilized).

The kinase labeling incubation at room temperature has been found to furnish as efficient a labeling as incubation at 37°C as described previously (Gupta *et al.*, 1982; Gupta, 1985). The incubation of radioactive digests at room temperature minimizes handling of radioactive tubes. It is necessary that [γ-^{32}P]-ATP be used at >1 μM to achieve maximal labeling of aromatic/lipophilic adducts. Even though a lower concentration of [γ-^{32}P]-ATP (20–50 μCi; 3000 Ci/mmole) may still be present in molar excess in the reaction mixture, this concentration may not always provide maximal adduct labeling. Certain adducts (e.g., adducts of styrene oxide, diaziquone) may require ATP concentration from 2 to 10 μM for maximal adduct labeling. In such cases, either additional [γ-^{32}P]-ATP (100–200 μCi) or appropriate amount of nonradioactive ATP or both must be added to raise the total ATP concentration. Optimal ATP concentration should be predetermined in pilot experiments for unknown adducts, though adducts of most polyaromatics appear to provide maximal labeling under the specified conditions. For highly adducted DNAs (1 adduct per 10^4–10^6 nucleotides), we use [γ-^{32}P]-ATP of lower specific activity (300–600 Ci/mmole) by adding 10–20 μCi [γ-^{32}P]-ATP (3000 ci/mmole) and 25–30 pmole nonradioactive ATP.

Some laboratories still use the apyrase treatment to hydrolyze unused [γ-^{32}P]-ATP to $^{32}P_i$ as described originally (Gupta *et al.*, 1982). We no longer consider this step necessary in most cases because of a more efficient D1 purification step.

3.4. Step IIIA: ^{32}P-Labeling of Total Nucleotides

To calculate adduct levels, total nucleotide radioactivity is determined by ^{32}P-labeling of dilute DNA digest in parallel with adducts.

- Add 3 μl hot mix prepared in Step III to 9 μl dilute DNA digest (3 ng) from Step I, incubate at room temperature for ≥40 min and dilute to 150 μl with 10 mM Tris·HCl, pH 9.5/5 mM EDTA (0.02 ng/μl), and chromatograph in Step IVA immediately or after storage at −20°C.

3.5. Step IV: Chromatography of Adducts

The labeled adducts are separated by multidirectional PEI-cellulose TLC (Fig. 4.2) in two stages: in the first stage relatively polar, residual normal nucleotides and other radioactive contaminants are removed by development of the chromatogram in a high-salt solution. Under these conditions, adducts being lipophilic are retained at or close to the origin of the chromatogram. Adducts are then displaced in the second stage by a two-directional development of the chromatogram in high-salt solutions in the presence of a denaturant such as high concentration of urea (7–8.5 M), mixture of organic solvents and dilute ammonium hydroxide, or dilute ammonium hydroxide alone. All developments are performed in ascending fashion.

Transfer Efficiency

Prior to adduct separation, a rapid TLC test is frequently performed to determine the extent of ^{32}P transferred from [γ-^{32}P]-ATP to nucleotides to ensure that adducts were labeled in the presence of molar excess of ATP.

- Apply 0.1–0.3 μl labeled solution to a PEI-cellulose layer (6 × 20 cm) (mark origins 0.8–1 cm apart on the longer side of the sheet with an extrasoft pencil either on the layer side or the plastic side at 1.5 cm above the bottom edge; thus, 15–20 samples can be applied per sheet).
- Develop sheet in 4.5 M ammonium formate, pH 3.5, to top.
- Dry chromatogram in a current of warm air.
- Detect radioactive compounds by autoradiography as in Step V.

This results in the appearance of three spots, the fastest migrating major spot is a mixture of $^{32}P_i$ and residual C, A, and T; the middle spot, usually present in traces, is G; and the slowest-

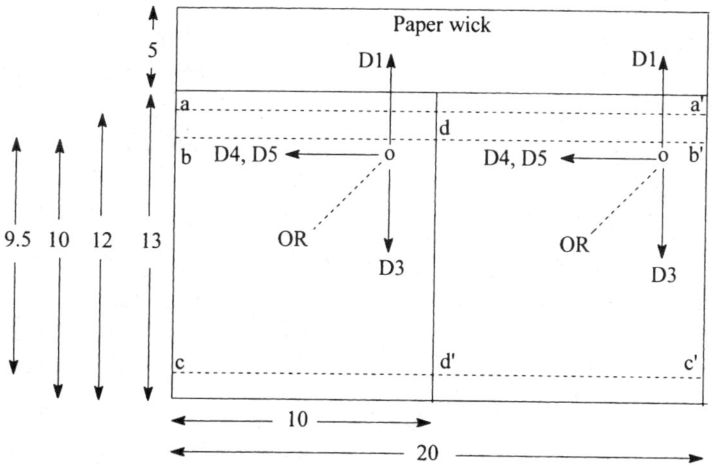

Figure 4.2. Schematic design of the multidirectional PEI-cellulose TLC for the separation of ^{32}P-labeled DNA adducts.

migrating major spot is ATP (see Fig. 4.4A). A visual comparison of the intensities of slowest and fastest spots reveals the approximate amount of unused ATP. In both the butanol and nuclease P1 enriched samples, 60–80% of the added ATP usually remains unused. A negative control (lacking DNA) processed in parallel results in predominantly or exclusively one spot, which allows identification of ATP.

Adduct Separation

Before applying labeled solution onto a PEI-cellulose thin layer (13 × 20 cm), a template is drawn with a pencil on the plastic side of the plate (Fig. 4.2). Up to two samples can be chromatographed per plate. Sheets are numbered on the top, right-hand corner on the layer side. A 5-cm-long Whatman 17 chrome paper wick [or two pieces of Whatman No. 3 paper or TLC saturation pads (Analtech, Inc.)] is attached to the top of the sheet by stapling (Fig. 4.2). Developments in directions 1–5 follow the original nomenclature (Gupta *et al.*, 1982; Gupta, 1985).

- Apply an aliquot (4–10 μl; 20–50 μCi) or the entire labeled solution (~18 μl; ≥90 μCi) slowly, but continuously, without intermittent drying to the origin of the layer. Use an acrylic shielded spotter during application of the labeled solution (Gupta, 1985). Collect all sheets in a glass or acrylic tank covered with a lid until all samples have been applied (keep tubes capped, if labeled solution is to be used after storage).
- Develop sheets in 1.0 M sodium phosphate, pH 6.0 (solvent 1, Table I), in direction 1 (D1) overnight (8–15 hr).
- Using a long tweezer and a long scissor, excise paper wick at the broken line, aa′ (Fig. 4.2). Discard the wick containing the bulk (>99%) of the radioactivity appropriately.
- To remove salt prior to development in the next direction, immerse up to four chromatograms in 1 liter deionized or distilled water in an acrylic, glass, or steel tray. After agitating trays occasionally for 5 min, pour off the water and repeat the washing step.

We usually place 10–15 chromatograms vertically in a custom-made acrylic rack which is then placed in an acrylic tank with continuous in- and outflow of water for 3–4 min.

- After draining water, dry chromatograms under a current of lukewarm air for 7–10 min.

Regular hair dryers are economic and convenient for this purpose. We usually place the rack containing wet chromatograms under a specially designed wooden stand mounted with up to four hair dryers.

- Develop chromatograms in direction 3 (D3) by first, predevelopment of sheets to ≤1 cm from the bottom edge in 1 M ammonium formate, pH 3.5, followed by development in 4 M lithium formate/7 M urea, pH 3.5 (Solvent 2, Table I) to top of the sheet. Alternatively, develop chromatograms in a nonurea solvent, i.e., 0.3–0.35 M ammonium hydroxide (Solvent 3, Table I), diluted freshly; development in direction 2 (D2) is omitted.
- Wash chromatograms in water to remove salt and urea and dry them as above; omit washing step if the development is performed in ammonium hydroxide.
- Excise chromatograms at dotted lines, bb′ and cc′, then at dd′, to give two chromatograms from one sheet.
- Develop chromatograms in direction 4 (D4) by predevelopment in 0.5 M Tris·HCl, pH 8, to ≤1 cm, followed by development to top of the sheet in 0.8 M lithium chloride/0.5 M Tris·HCl/7 M urea, pH 8 (Solvent 4, Table I). Alternatively, develop chromatograms in a

nonurea solvent, i.e., isopropanol:4 M ammonium hydroxide (1.2:1) (Solvent 5, Table I) to 2 cm onto a 5-cm-long Whatman No. 1 paper wick attached to the top of the sheet by stapling.

- Dry chromatograms after washing in water (for Solvent 4) or without washing (Solvent 5), as above.

Chromatograms are generally excised at 0.7–0.9 cm to the right-hand side of the origin prior to a final "cleanup" development to remove the residual background radioactivity (Gupta, 1993). This excision is necessary only when Solvent 5 is used in direction 4, since a large amount of the background radioactivity is retained at the D3 origin, which otherwise tends to diffuse in subsequent D5 development (see Fig. 5 of Gupta, 1993). If the adducts migrate close to the origin, as determined in pilot experiments, then either avoid the trimming or increase the ammonium hydroxide content to increase adduct mobilities.

- Develop chromatograms finally in direction 5 (D5) in 1 or 1.7 M sodium phosphate, pH 6.0, to 4–5 cm on a Whatman No. 1 paper wick.
- Wash chromatograms in water and dry as above. Chromatograms are now ready for detection of adduct spots in Step V.

Most PAH and aromatic amine adducts are retained at or close to the origin of the chromatogram during the D1 development under the indicated conditions. However, certain adducts, depending on their polarity, may migrate several centimeters away from the origin, and thus may be lost onto the paper wick. In such cases, either a higher concentration of sodium phosphate, pH 6.0 (1.7 M) is used in D1 development or the paper wick is attached at the opposite end of the sheet ("D1 in D3 template") such that D1 development can be performed in D3 direction, although such alterations may result in slightly increased background noise. For unknown compounds, we recommend that these altered conditions be tried first to ensure that

Table I
Solvents for Multidirectional PEI-Cellulose TLC of ^{32}P-Labeled Aromatic/Lipophilic DNA Adducts[a]

Solvent No.	Composition	TLC direction
1	1 M sodium phosphate, pH 6.0	D1
2	4 M lithium formate/7 M urea, pH 3.5	D3
3	0.3–0.35 M ammonium hydroxide[b]	D3
4	0.8 M lithium chloride/0.5 M Tris·HCl/7 M urea, pH 8.0	D4
5	Isopropanol:4 M ammonium hydroxide (1.2:1)[c]	D4
6	1 M sodium phosphate/7 M urea, pH 8.0	D4
7	0.2 M sodium bicarbonate/7 M urea	D4
8	0.5 M H_3BO_3/0.5 M Tris·HCl/10 mM EDTA/1.3 M sodium chloride/8 M urea, pH 8.0	D4
9	0.8 M sodium bicarbonate/0.15 M Tris·HCl/8 M urea, pH 9.5	D4
10	1.7 M sodium phosphate, pH 6.0[d]	D5

[a]Directions 1–5 follow the original nomenclature (Gupta *et al.*, 1982; Gupta, 1985); D2 development is usually omitted (See Fig. 4.2).
[b]Extremely lipophilic adducts such as aralkylated DNA adducts of benzo[*a*]pyrene require as much as 2 M ammonium hydroxide (Stansbury *et al.*, 1994).
[c]The ratio may be varied from 0.8:1 to 2:1, depending on the lipophilicity of adducts. Isopropanol may also be substituted for methanol or ethanol for less lipophilic adducts.
[d]Sodium phosphate at a concentration $>$ 1.7 M can be prepared by lowering the pH to 5.5–5.8.

adducts are not lost during the D1 development. For still less lipophilic adducts that may migrate sufficiently in the D1, the D3 development can be completely avoided, D4 being performed immediately after the D1 development.

Experience with commercial Machery-Nagel PEI-cellulose thin layers in several laboratories, including our own, suggest that these sheets provide similar separations as homemade sheets, provided ionic strengths of the D3 and D4 solvents are increased by 30–40% to compensate for higher anionic strength of the commercial PEI-cellulose layers.

The urea solvents (Solvents 2 and 4) described above are used commonly in many laboratories. Numerous alternate urea (Solvents 6–9) and nonurea (Solvents 3 and 5) solvents have been described, particularly for D4 (Table I). Further, a change in pH of the urea solvents, particularly for Solvents 2 and 6, from pH 3.5 to pH 3.7 and pH 8.0 to pH 6.4, respectively, results in different mobility of adducts. Solvent 7 is by far the most powerful of the urea solvents and is applicable for lipophilic adducts of diolepoxides of PAHs. However, extremely lipophilic adducts such as those resulting from aralkylation of benzo[*a*]pyrene require a much higher concentration and higher pH (i.e., 0.8 M sodium bicarbonate/7 M urea, pH 9.5, Solvent 9) than for Solvent 7; such adducts do not migrate away from the origin in commonly used Solvents 2 and 4 (Stansbury *et al.*, 1994).

The nonurea solvents (Solvents 3 and 5) are quite distinct from all of the high-salt, high-urea solvents, in that the presence of dilute ammonium hydroxide alone or ammonium hydroxide coupled with organic solvents such as isopropanol results in the migration of adducts away from the origin. Adducts spots are generally better resolved and are sharper than in the urea solvents. Further, the ammonium hydroxide-based solvents are easy to prepare, economical, and chromatograms do not require intermittent water washing. Most importantly, background noise is substantially reduced with nonurea solvents, thus increasing signal-to-noise ratio. The presence of ammonium hydroxide perhaps neutralizes the anionic charge on the PEI-cellulose so that the latter behaves more like cellulose partition TLC, thus resulting in adduct migration distinct from those obtained with urea solvents (Gupta, 1993; Spencer *et al.*, 1993, 1996). Because of the volatility of ammonium hydroxide, however, several precautions need to be taken: (1) store ammonium hydroxide in airtight bottles; (2) 0.3–0.35 M ammonium hydroxide should be diluted freshly from the commercially available concentrated solution; and (3) cover TLC tank with a plastic wrap (e.g., Saran wrap) during chromatography.

To expedite the development with isopropanol:4 M ammonium hydroxide (1.2:1) in D4, isopropanol may be substituted with a 1:1 mixture of isopropanol and acetonitrile. A much higher proportion of acetonitrile is found unsuitable as the plastic of the homemade sheets and acrylic equipment tend to get softened. The ratio of isopropanol:4 M ammonium hydroxide can be regulated from 0.8 to 2:1, depending on the lipophilicity of adducts, but for most aromatic amine and PAH adducts, the conditions specified above are optimal. For more polar adducts, isopropanol can also be substituted with methanol or ethanol; this solvent mixture cuts down the development time by two- to threefold as compared to isopropanol:4 M ammonium hydroxide mixture.

3.6. Step IVA: Chromatography of Normal Nucleotides

- Apply 5 μl dilute labeled digest (~2 μCi) onto a 100 mM ammonium formate, pH 3.5-preequilibrated PEI-cellulose thin layer (20 cm long) and develop the sheet, with or without drying origins, in 1.2 M ammonium formate, pH 3.5. Alternatively, apply digest on a water-washed thin layer and develop sheet in 0.27 M sodium phosphate, pH 6, or 0.27 M ammonium sulfate.
- Dry chromatograms without washing in water and proceed for autoradiograpy in Step V.

3.7. Step V: Detection and Quantitation of Adducts

Detection

Adduct and normal nucleotide spots are detected by autoradiography. The less sensitive X-ray films (DuPont Cronex) are used for the detection of normal nucleotides and transfer efficiency analyses and for those adducts that are present at relatively high levels (1 adduct per 10^5–10^7 nucleotides); the more sensitive films (Kodak XAR-5) are used for lower adduct levels (1 adduct per 10^7–10^{10} nucleotides). The selection of films and exposure times are determined by estimating the amount of radioactivity on chromatograms by probing with a Geiger counter.

- Place up to six chromatograms side by side on a plane paper (20×30 cm) by using a small piece of double-stick tape (or glue) on the plastic side of the chromatograms.
- Mark radioactive ink dots with a fine nib holder on corners of each chromatogram to align the autoradiogram with the chromatogram.
- Place chromatogram(s) in a cassette equipped with intensifying screen(s), then on top place an X-ray film, close, and keep the cassette at room temperature or -70 to -85°C, depending on the amount of radioactivity (autoradiographic exposure coupled with the intensifying screens at such low temperatures accelerates detection by about 50-fold). Process films in an automatic film processor (e.g., Kodak RP X-OMAT, Model M7) or manually.

A 5- to 10-min exposure is sufficient for detecting normal nucleotides and spots in the transfer efficiency analysis. Exposure time for adducts varies, depending on their levels and X-ray film types: with Kodak XAR-5 film = 1–2 hr (1 adduct per 10^5–10^6 nucleotides), 2–6 hr (1 adduct per 10^6–10^7 nucleotides), overnight (>15 hr) (1 adduct per 10^8–10^9 nucleotides), and 1–2 days (1 adduct per 10^{10} nucleotides). Exposure duration is increased by about twofold with the DuPont Chronex X-ray film.

Alternatively, electronic image processors can be used to to detect and quantitate radioactive spots. Two types of computer-based detection technologies are in use: one is based on preexposure of chromatograms to phosphorus-based screens, followed by laser scanning of the stored energy on the screens (e.g., Molecular Dynamics, Bio-Rad). The second is based on direct measurement of β particles emitted by ^{32}P-labeled compounds (from Betagen, Packard, etc.). We use Packard Instant-Imager for measuring the radioactivity in adduct and normal nucleotides on PEI-cellulose chromatograms.

Figure 4.3 shows autoradiographic detection of DNA adducts in human primary peripheral blood lymphocytes exposed in cell culture to indicated carcinogenic aromatic amines and PAHs. As evident from the panels, each carcinogen shows its characteristic adduct pattern. The aromatic amines show predominantly or exclusively one to two adducts, while the PAHs show multiple adducts, except for 1-nitropyrene which shows one adduct. Vehicle-treated lymphocyte DNA shows no detectable spot. Levels of individual adducts are determined by measurement of the radioactivity present in each spot as described below.

Figure 4.4B shows separation of normal nucleotides in 1.2 M ammonium formate, pH 3.5, in the following order: $^{32}P_i > C > A > T > G$; ATP stays at the origin. Sometimes, $^{32}P_i$ and C tend to migrate together or very close to each other. The order of separation in 0.27 M ammonium sulfate or 0.27 M sodium phosphate, pH 6.0, is different than found with the ammonium formate development: $^{32}P_i > T > C > A > G > ATP$.

Quantitation

Align autoradiogram with chromatogram to locate the adduct or normal nucleotide spots on the chromatogram and mark desired areas with a pencil on the plastic side of the chromato-

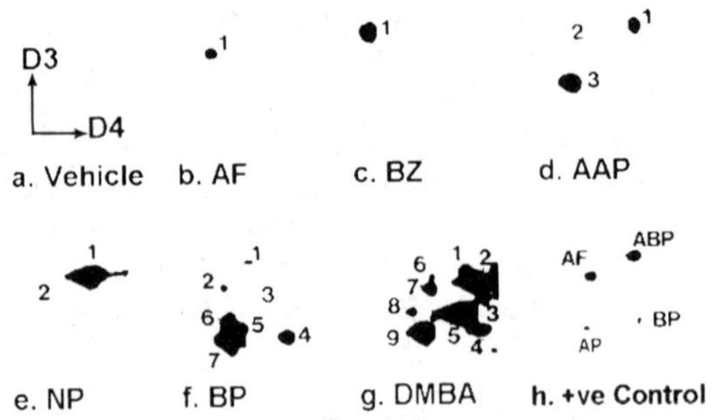

Figure 4.3. ^{32}P-adduct maps of DNA of freshly isolated peripheral blood lymphocytes exposed in cell culture to indicated carcinogens (b–g) or vehicle only (a). A mixture of DNA containing adducts of 2-aminofluorene (AF), 4-aminobiphenyl (ABP), 2-aminophenanthrene (AP), and benzo[*a*]pyrene (BP) at a level of 1 per 10^8 nucleotides served as a positive control. Adducts were analyzed following the butanol-enrichment version of the ^{32}P-postlabeling assay. BZ, benzidine; AAP, 2-acetylaminophenanthrene; NP, 1-nitropyrene; DMBA, 7,12-dimethyl-benzanthracene. TLC solvents were as follows: D1 = Solvent 1; D3 = Solvent 2; D4 = Solvent 5; D5 = Solvent 10; D2 was omitted (Table I). (Modified from Gupta *et al.*, 1988.)

gram. Spots are excised from the chromatogram with a scissor and placed in a plastic scintillation vial using a tweezer and counted in a scintillation counter for 1–10 min, depending on the amount of radioactivity; longer counting is desired for low amount of radioactivity. A blank area of the chromatogram is also excised to determine the background radioactivity. The radioactivity in an individual adduct spot is determined by subtracting the blank. To determine total normal nucleotide radioactivity, we usually measure the radioactivity in one of the four major spots (the G or A spot) and calculate the total normal nucleotide radioactivity by multiplying the cpm in G with 4.76 (the multiplying factor is based on the known level of G in mammalian DNA, i.e., 21%; the multiplying factor would be 3.45 for the A spot).

- Calculate adduct levels as relative adduct labeling (RAL) based on the adduct and normal nucleotide radioactivity as follows:

$$\text{RAL} = \frac{\text{cpm in adduct(s)}}{\text{cpm in normal nucleotides}} \times \frac{1}{\text{dilution factor}}$$

Adduct chromatography usually involves 5–10 μg DNA, and normal nucleotides ~0.1 ng DNA (therefore, dilution factor = 5000–10000). RAL values are expressed as number of adducts per 10^9 nucleotides (Beach and Gupta, 1992). In our experience adduct levels measured in experimental animals following treatment with PAHs and aromatic amines range from 1 adduct per 10^5–10^8 nucleotides, depending on the nature of carcinogen, dose, animal species, tissue type, and time after treatment. In humans exposed environmentally, occupationally, or by life-style (cigarette smokers), adducts have been detected at significantly lower levels (1 adduct per 10^7–10^{10} nucleotides).

4. DISCUSSION

No single adduct detection assay is applicable to all carcinogens under one standardized condition due to the diversity of compounds that are present in the environment. ^{32}P-postlabeling assay in fact has worked exceedingly well for most, if not all, aromatic or lipophilic carcinogens

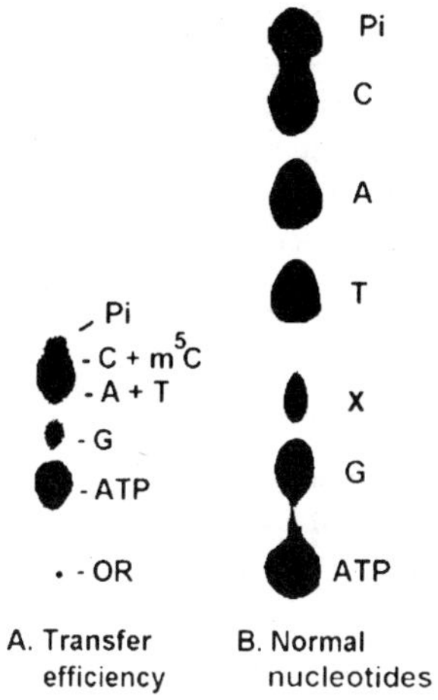

Figure 4.4. (A) Separation of residual normal nucleotides and ATP by PEI-cellulose TLC in 4.5 M ammonium formate, pH 3.5 (transfer efficiency analysis); a butanol-enriched labeled DNA digest was applied onto the layer. (B) Separation of 5′-^{32}P-labeled normal nucleotides by PEI-cellulose TLC. Development was in 1.2 M ammonium formate, pH 3.5.

such as aromatic amines and homocyclic-, heterocyclic-, and nitroPAHs under one set of standardized conditions. A list of aromatic carcinogens or complex mixtures that have been tested by ^{32}P-postlabeling in rodent and aquatic organisms *in vivo* and human cells *in vitro* has been published elsewhere (Gupta and Randerath, 1988; Beach and Gupta, 1992). This procedure has been applied to a larger number of carcinogens than perhaps all other procedures combined. It should be noted, however, that the assay may not apply to some carcinogens or certain adducts such as mycotoxins which produce adducted di- and trinucleotides (Reddy *et al.*, 1985) that are resistant to enzymatic digestion. Further, due to large differences in lipophilicity of adducts resulting from one carcinogen, solvents of different strengths may have to be used to elute all adducts, e.g., adducts of benzo[*a*]pyrene originating from diolepoxide versus aralkylation mechanisms (Stansbury *et al.*, 1994). For most lipophilic carcinogens, however, one solvent system is applicable for eluting all adducts collectively.

Despite the multidirectional aspect of the PEI-cellulose TLC, some adduct spots may still not represent single entities. This is particularly important when establishing chromatographic identities of unknown adducts with reference adducts. It is suggested that three to four or more solvents of different properties (e.g., urea versus nonurea solvents; Table I) be employed to establish chromatographic identities. Adduct spots can also be eluted from the layers (in 4 M pyridinium formate, pH 4, or 2 M ammonium hydroxide) and then analyzed by TLC or HPLC to further establish resemblance of unknown adducts with reference adducts.

An obvious limitation of the ^{32}P-postlabeling is its inability to identify adducts. Identification is largely based on cochromatography with reference adducts that can be prepared *in vitro* by reaction of electrophilic metabolites with DNA or individual deoxynucleoside 3′-monophosphates (e.g., Beach and Gupta, 1994); certain reference PAH adducts are available from the National Cancer Institute Carcinogen Repository. This strategy does not necessarily apply to DNA adducts detected in human tissues where etiological agent(s) are largely unknown. A comparison of butanol extraction and nuclease P1 enrichment procedures has gėnerally proved fruitful in determining whether adducts in human tissues were derived from arylamines or other PAHs, as adducts resulting from arylamines are largely lost in the nuclease P1 procedure (Chacko and Gupta, 1988; Talaska *et al.*, 1991). Use of a variety of suspected arylamine and PAH candidate DNA adducts has usually proved fruitful in identifying DNA adducts in human tissues. The classical approach of identifying adducts by spectroscopic techniques is not practical since adducts isolated from typical chromatograms are present in amounts (attomole to femtomole) that are several orders of magnitude lower than typically needed for spectroscopic analyses (which are tens to hundreds of nanograms).

In conclusion, the ^{32}P-postlabeling assay has emerged as the method of choice because (1) it allows detection of extremely low levels of adducts (1 per 10^{10} nucleotides); (2) it requires microgram quantities of DNA, thus readily applicable to human biopsy samples; (3) it does not require prior knowledge of test adducts, making it applicable to unknown carcinogens that may be encountered in daily life; (4) it can detect adducts resulting from different compounds collectively such as found in complex mixtures; and (5) it is able to detect adducts resulting from interaction of presumably, but unknown endogenous electrophiles. Despite its inability to identify adducts, unless coupled with reference adducts, this assay has been applied to a wide variety of biopsy and autopsy specimens of humans exposed environmentally, occupationally, clinically, or by life-style (e.g., cigarette smoking) to assess human exposure to carcinogens.

ACKNOWLEDGMENTS. Our work was supported by American Cancer Society grant CN-67 and USPHS grant CA 30606. Dr. Jamal Arif, Glenda Spencer-Beach, Wendy Smith, and Udaysankar Devanaboyina are acknowledged for their useful comments.

REFERENCES

Beach, A. C., and Gupta, R. C. (1992). ^{32}P-postlabeling assay and human biomonitoring. *Carcinogenesis* **13**:1053–1074.

Beach, A. C., and Gupta, R. C. (1994). DNA adducts induced *in vivo* by the ubiquitous environmental contaminant cyclopenta[*cd*]pyrene. *Carcinogenesis* **15**:1065–1072.

Chacko, M., and Gupta, R. C. (1988). Evaluation of DNA damage in the oral mucosa of tobacco users and nonusers. *Carcinogenesis* **9**:2309–2314.

Dunn, B. P., and San, R. H. C. (1988). HPLC enrichment of hydrophobic DNA–carcinogen adducts for enhanced sensitivity of ^{32}P-postlabeling analysis. *Carcinogenesis* **9**:1055–1060.

Gallagher, J. E., Jackson, M. A., George, M. H., Lewtas, J., and Robertson, I. G. C. (1989). Differences in detection of DNA adducts in the ^{32}P-postlabeling assay after either 1-butanol extraction or nuclease P1 treatment. *Cancer Lett.* **45**:7–12.

Gupta, R. C. (1984). Nonrandom binding of the carcinogen N-hydroxy-2-acetylaminofluorene to repetitive sequences of rat liver DNA *in vivo*. *Proc. Natl. Acad. Sci. USA* **81**:6943–6947.

Gupta, R. C. (1985). Enhanced sensitivity of ^{32}P-postlabeling analysis of aromatic carcinogen–DNA adducts. *Cancer Res.* **45**:5656–5662.

Gupta, R. C. (1993). ^{32}P-postlabeling of bulky aromatic adducts, IARC Publication No. 124, pp. 11–24.

Gupta, R. C., and Earley, K. (1988). ^{32}P-adduct assay: Recoveries of structurally diverse DNA adducts in the various enhancement procedures. *Carcinogenesis* **9**:1687–1693.

Gupta, R. C., and Randerath, K. (1988). Analysis of DNA adducts by ^{32}P-labeling and thin-layer chromatography, in: *DNA Repair*, Volume 3 (E. C. Friedberg and P. C. Hanawalt, eds.), Dekker, New York, pp. 401–420.

Gupta, R. C., Reddy, M. V., and Randerath, K. (1982). ^{32}P-postlabeling analysis of nonradioactive aromatic carcinogen DNA adducts. *Carcinogenesis* **3**:1081–1092.

Gupta, R. C., Earley, K., and Sharma, S. (1988). Use of human peripheral blood lymphocytes to measure DNA binding capacity of chemical carcinogens. *Proc. Natl. Acad. Sci. USA.* **85**:3513–3517.

Johnson, R. A., and Walseth, T. F. (1979). The enzymatic preparation of [α-^{32}P]ATP, [α-^{32}P]GTP, [^{32}P]cAMP, and [^{32}P]cGMP, and their use in assay of adenylate and guanylate cyclases, and cyclic nucleotide phosphodiesterases. *Adv. Cyclic Nucleotide Res.* **10:**135–167.

Miller, E. C., and Miller, J. A. (1981). Mechanisms of chemical carcinogenesis. *Cancer* **47**:1055–1064.

Randerath, K., and Randerath, E. (1967). Thin-layer separation methods for nucleic acid derivatives. *Methods Enzymol.* **12A**:323–347.

Reddy, M. V., and Randerath, K. (1986). Nuclease P1-mediated enhancement of ^{32}P-postlabeling test for structurally diverse DNA adducts. *Carcinogenesis* **7**:1543–1551.

Reddy, M. V., Irvin, T. R., and Randerath, K. (1985). Formation and persistence of sterigmatocystin-DNA adducts in rat liver determined via ^{32}P-postlabeling analysis. *Mutat. Res.* **152**:85.

Spencer, G. G., and Gupta, R. C. (1992). A C18 TLC-mediated enhancemnent of ^{32}P-postlabeling for detecting DNA adducts of varying degrees of lipophilicity. *Proc. Am. Assoc. Cancer Res.* 33, Abstr. 720.

Spencer, G. G., Beach, A. C., and Gupta, R. C. (1993). Enhanced thin-layer chromatographic separation of ^{32}P-postlabeled DNA adducts. *J. Chromatogr.* **612**:295–301.

Spencer, G. G., Beach, A. C., and Gupta, R. C. (1996). High-resolution anion-exchange and partition thin-layer chromatography for complex mixtures of ^{32}P-postlabeled DNA adducts. *J. Chromatogr.* in press.

Stansbury, K., Flesher, J. W., and Gupta, R. C. (1994). Mechanism of aralkyl–DNA adduct formation from benzo[*a*]pyrene *in vivo. Chem. Res. Toxicol.* **7**:254–259.

Talaska, G., Al-Juburi, A. Z. S. S., and Kadlubar, F. F. (1991). Smoking-related carcinogen–DNA adducts in biopsy samples of human urinary bladder: Identification of N-(deoxyguanosine-8-yl)-4-aminobiphenyl as a major adduct. *Proc. Natl. Acad. Sci. USA* **88**:5350–5354.

Chapter 5

A Postlabeling Assay for Oxidative Damage

Michael Weinfeld, Michel Liuzzi, and George D. D. Jones

1. INTRODUCTION

The ^{32}P-postlabeling assay was originally devised by Randerath *et al.* (1981) to measure carcinogen–DNA adducts. In the procedure, DNA is first digested by micrococcal nuclease and calf spleen phosphodiesterase to give nucleoside 3′-monophosphates (normal and modified) that are subsequently labeled by incubation with [γ-^{32}P]-ATP and T4 polynucleotide kinase. The radiolabeled compounds are then separated by two-dimensional TLC. The assay has two important advantages. First, there is no requirement for prelabeling the DNA, which makes the assay useful for the study of DNA lesions in tissues. Second, because of the availability of [γ-^{32}P]-ATP of high specific activity, the assay permits detection at the femtomole level. There are, however, two major drawbacks: (1) the polynucleotide kinase must be able to act on the modified nucleoside 3′-monophosphate and (2) the resulting labeled modified nucleoside diphosphate must be separable from the high background of normal nucleoside diphosphates. These problems are well illustrated in the reports of efforts to detect thymine glycols, a well-known oxidative base lesion, in irradiated DNA (Reddy *et al.*, 1991; Hegi *et al.*, 1989). Furthermore, lesions that involve base loss, such as abasic sites and deoxyribose fragments, cannot be detected by this approach.

In this laboratory we have developed an alternative approach that overcomes the problems noted above. The assay is based on the observation that certain DNA lesions prevent hydrolysis by snake venom phosphodiesterase (phosphodiesterase I) and DNase I of the adjacent 5′-internucleotide phosphodiester linkage (Weinfeld *et al.*, 1993; Weinfeld and Buchko, 1993; Liuzzi *et al.*, 1989; Stuart and Chambers, 1987). Thus, digestion of damaged DNA with these enzymes and an alkaline phosphatase generates lesion-containing dinucleoside monophosphates, or for some lesions trinucleoside diphosphates, which are good substrates for polynucleotide kinase because the nucleoside at the 5′-end is unmodified (Fig. 5.1). The undamaged bases are recovered as mononucleosides, which are not substrates for this kinase and so remain unlabeled.

This procedure has been used to quantify cyclobutane pyrimidine dimers (Weinfeld *et al.*,

Michael Weinfeld • Department of Radiobiology, Cross Cancer Institute, Edmonton, Alberta T6G 1Z2, Canada. **Michael Liuzzi** • Bio-Méga/Boehringer Ingelheim Research Inc., Laval, Quebec H7S 2G5, Canada. **George D. D. Jones** • Centre for Mechanisms of Human Toxicity, University of Leicester, Leicester LE1 9HN, England.

Technologies for Detection of DNA Damage and Mutations, edited by Gerd P. Pfeifer, Plenum Press, New York, 1996.

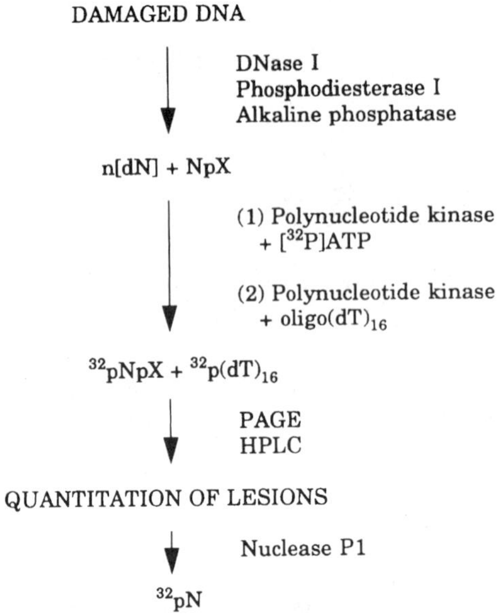

Figure 5.1. Scheme of the postlabeling assay. Damaged DNA is digested to mononucleosides (dN) and "dinucleoside" monophosphates (NpX) where X represents a damaged base or deoxyribose. The NpX species are radioactively phosphorylated (^{32}pNpX) and, after transferring most of the excess label to a large oligonucleotide, the ^{32}pNpX products can be analyzed by polyacrylamide gel electrophoresis (PAGE) and/or HPLC. The labeled 5′-nearest neighbor can be released by nuclease P1-mediated cleavage of the phosphodiester bond in the purified ^{32}pNpX molecules and identified by HPLC.

1989a), abasic sites (Weinfeld *et al.*, 1990) and radiation-induced thymine glycol and phosphoglycolate termini (Weinfeld and Soderlind, 1991). It has also been employed to examine the influence of nitrogen, oxygen, and nitroimidazole radiosensitizers on radiogenic DNA damage (Buchko and Weinfeld, 1993), to identify enzyme activity that removes phosphoglycolate groups from bleomycin-treated DNA (Winters *et al.*, 1992), to detect phosphoglycolate groups produced in cellular DNA by oxidative stress (Bertoncini and Meneghini, 1995), and to compare oxidative damage produced by Fenton chemistry and ionizing radiation (Fig. 5.2). Variations of this approach have been used to evaluate *cis*-platin adducts in calf thymus DNA (Försti and Hemminki, 1994) and psoralen-type DNA adducts in HeLa cells (Gillardeaux *et al.*, 1994), and UV-photo products (Bykov *et al.*, 1995).

2. METHODS

2.1. Materials

2.1.1. Reagents

Oligo(dT)$_{16}$: purchased from Pharmacia (catalog #27-7609-01) and is dissolved in distilled water to a concentration of 5 AU$_{260}$ U/ml.

Calf thymus DNA (type 1, sodium salt): purchased from Sigma Chemical Co. (catalog #D1501).

[γ-^{32}P]-ATP (Redivue™, 3000 Ci/mmole):obtained from Amersham Corporation (catalog # AA0066).

Acrylamide/methylene bisacrylamide: prepared as a 38:2% solution in distilled water and stored at 4°C in a dark glass bottle. Use electrophoresis-grade chemicals. Care must be taken in handling this solution because acrylamide is a neurotoxin and cancer suspect agent.

Ammonium persulfate: prepared as a 3% wt/vol solution in distilled water. The solution can be kept for up to 2 weeks at 4°C.

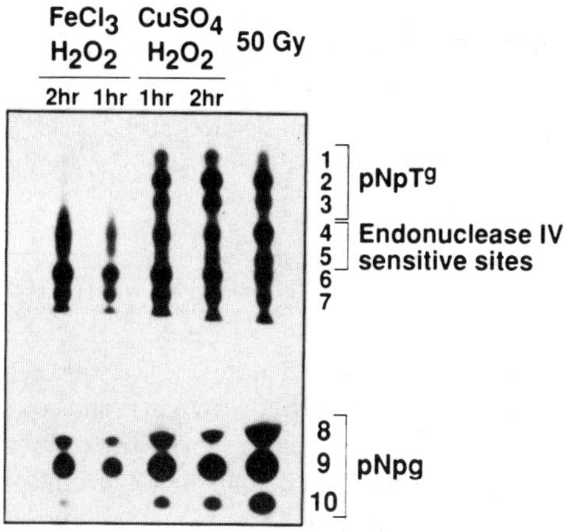

Figure 5.2. Autoradiogram of polyacrylamide gel showing postlabeled products following oxidative DNA damage generated by two different metal ion/hydrogen peroxide systems and γ-radiation (50 Gy). Some of the detectable products generated by ionizing radiation have been identified (Weinfeld and Soderlind, 1991). Bands 1–3 are the labeled thymine glycol-containing dinucleotides; $^{32}pGpT^{g}$ is in band 1, $^{32}pApT^{g}$ and $^{32}pTpT^{g}$ are in band 2, and $^{32}pCpT^{g}$ is in band 3. Bands 8–10 are the labeled phosphoglycolate species ($^{32}pNpg$), resulting from hydroxyl radical attack and oxidation at the C-4′ carbon atom in deoxyribose; band 8 contains the labeled deoxyguanosine 3′-phosphoglycolate ($^{32}pGpg$), band 9 contains the $^{32}pApg$ and $^{32}pTpg$, and band 10 contains $^{32}pCpg$. The lesions in bands 4 and 5 have yet to be identified, but they can be removed from damaged DNA by Escherichia coli endonuclease IV. Bands 6 and 7 are seen with the untreated control. Not shown in this figure are three bands containing inorganic mono-, di-, and triphosphate that migrate faster than the phosphoglycolate molecules, and oligo$(dT)_{16}$ that remains near the top of the gel.

2.1.2. Buffers

TE: 10 mM Tris-HCl (pH 7.4), 1 mM EDTA (pH 8.0)

DNA digestion buffer: TE plus 4 mM $MgCl_2$

10× T4 polynucleotide kinase reaction buffer: 0.5 M Tris-HCl (pH 7.6), 100 mM $MgCl_2$, 100 mM 2-mercaptoethanol. This buffer is supplied with the enzyme by United States Biochemical.

10× TBE: 900 mM Tris base, 900 mM boric acid, 25 mM EDTA. This buffer can be conveniently purchased in the form of ready-to-dissolve premixed powder from ICN Biomedicals.

Gel loading buffer: 90% formamide, 0.02% bromphenol blue, and 0.02% xylene cyanol in 1× TBE.

Nuclease P1 buffer: 10 mM sodium acetate (pH 5.3), 1 mM $ZnCl_2$

2.1.3. Enzymes

DNase I (type I from bovine pancreas) is purchased from Sigma (catalog # D4763) and dissolved in distilled water to a concentration of 60 U/μl, aliquoted, and stored at −20°C.

Snake venom phosphodiesterase (Type IV from *Crotalus atrox*, Sigma Chemical Co. catalog # P4506) is diluted with water to 0.011 U/μl, aliquoted, and stored at −20°C. (Care should be taken in handling this material since it is a relatively crude preparation of snake venom.)

Shrimp alkaline phosphatase (1 U/μl) is obtained from United States Biochemical (catalog # 70092).

T4 polynucleotide kinase (30 U/μl) can be purchased from United States Biochemical (catalog # 70031). It is supplied with a 10× buffer (see above).

Nuclease P1 (Pharmacia catalog # 27-0852-01) is resuspended at 1 mg/ml (~600 U/ml) in 10 mM sodium acetate (pH 5.3), aliquoted, and stored at −20°C.

Definitions of the units of these enzymes are those given by the suppliers.

2.2. Preparation of Polyacrylamide Gels

2.2.1. Apparatus

We employ a power supply, capable of delivering at least 1000 V, together with a Hoefer SE 620 electrophoresis unit (Hoefer Scientific Instruments, San Francisco, CA), using 18-cm (width) × 32-cm (length) glass plates (catalog # SE 6202), 1.5-mm spacers (catalog # SE 6219-2-1.5), and a 20-tooth comb (catalog # SE 511-20-1.5). This unit can run two gels simultaneously. If running only one gel, do not forget to attach the acrylic block to the unused side of the upper buffer chamber.

2.2.2. Preparation

1. Set up the glass plates with the spacers and clamps in the gel caster as described by the manufacturer. Ensure that the glass plates are clean and have no cracks.
2. Add together 8 ml of 10× TBE, 40 ml of the polyacrylamide:methylene bisacrylamide solution, and 31.5 g of urea (electrophoresis grade). Stir the mixture at room temperature until the urea is completely dissolved (30–45 min). Then add in 1.6 ml of ammonium persulfate solution and sufficient distilled water to make a total volume of 80 ml.
3. Filter the whole mixture through a 0.22-μm filter into a clean 100-ml bottle and hold on ice. (We use a Falcon 7105 bottle top filter from Becton Dickinson Labware.)
4. Add 25 μl of tetramethylethylenediamine (TEMED) to the gel solution, allow it to mix thoroughly, and pour the gel solution at a steady flow rate between the glass plates. (We use a 50-ml syringe barrel to direct the flow of the solution close to one of the spacers in order to reduce formation of bubbles.)
5. Stop pouring the solution when it is about 2 cm from the top of the glass plates. Insert the comb in between the glass plates. (Try to avoid forming bubbles under the teeth of the comb.) As the solution polymerizes, the volume shrinks, so continue adding gel solution to maintain the level close to the top of the plates. After the gel solidifies, it can be kept at room temperature for 2–3 days.
6. Before using the gel, remove the comb, scrape off any gel on top of the glass plates, and attach the upper buffer chamber to the gel (do not forget the gasket). Release the gel from the casting stand and place it in the main (lower buffer) chamber. (We normally keep 6–7 liters of 1× TBE in the lower buffer chamber for up to 3 months. This serves to cool the gel during running.) Pour 1× TBE into the upper buffer chamber and ensure that there is no leakage of buffer from this chamber. Clean out debris from the wells of the gel using a 5-ml syringe with a 19-gauge 1.5-inch needle.
7. Attach the lid to the chamber and prerun the gel for 45 min at 800 V. (A few microliters of gel loading buffer can be run in the outer wells to ensure that the gel is running properly.)
8. Load the samples into the wells using either a standard air displacement pipettor with 200-μl fine pipette tips or a Hamilton syringe. Avoid using the outer three wells on either side of the gel since samples applied in these wells will migrate to the edge of the gel and be lost.

2.3. Postlabeling Assay

1. Digest damaged and undamaged DNA samples (10 μg) to mononucleosides and damage-containing "dinucleoside" monophosphates by an overnight incubation at 37°C with 0.4 U of DNase I, 0.044 U of snake venom phosphodiesterase, and 0.4 U of shrimp alkaline phosphatase in 40 μl DNA digestion buffer.
2. Precipitate the enzymes by addition of 3 vol of ice-cold ethanol and storage at −20°C for 1 hr. Remove the precipitate by centrifugation (10,000*g*, 15 min, 4°C) in a microcentrifuge.
3. Evaporate the supernatants to dryness and dissolve the resulting residues in distilled water (100 μl). Heat the samples in screw-top microcentrifuge tubes at 100°C for 15 min to inactivate residual nuclease and phosphatase activity. The samples can then be stored at −20°C.
4. To phosphorylate the damage-containing "dinucleoside" monophosphates in 5 μl of digested DNA, add 1 μl of 10× T4 polynucleotide kinase buffer, 0.5 μl (5 μCi, 1.67 pmole) [γ-^{32}P]-ATP, 7.5 U of T4 polynucleotide kinase, and distilled water to a total volume of 10 μl, and incubate the samples for 1 hr at 37°C. (If several samples are to be phosphorylated simultaneously, we premix the 10× buffer, ATP, enzyme, and water, and then add 5 μl of this mixture to each sample. When handling the polynucleotide kinase, keep it on ice at all times since the enzyme loses activity fairly rapidly.) To consume the excess ATP, add 1 μl of oligo$(dT)_{16}$ (5 AU_{260} U/ml) and 3.75 U of T4 polynucleotide kinase to each sample and incubate at 37°C for a further 30 min.
5. Add 10 μl of gel loading buffer to each sample, and run half of the reaction mixture on a 20% polyacrylamide/7 M urea gel (prepared as described above). The other half of the reaction mixture can be stored at −20°C. For standard conditions in which all of the radioactivity is retained on the gel (including the fast-migrating inorganic phosphate bands), electrophoresis is carried out at 800 V until the bromphenol blue marker has migrated 11–12 cm (3–4 hr). Gels can of course be run longer to increase separation between bands, but inorganic phosphate will run off the gel.
6. After the electrophoresis is complete, remove one gel plate, and wrap the remaining plate and gel in Saran Wrap. Apply two or three fluorescent markers (prepared by marking 1 cm^2 pieces of colored tape with an UltEmit™ pen sold by NEN, DuPont) to areas of the wrapped gel where no radioactivity is expected, e.g., corners of the gel, and expose the gel to fluorescent light for 1 min. Visualize the radiolabeled products by autoradiography on Kodak X-Omat K film (30 × 24 cm). Autoradiography usually takes 45–90 min. Alternatively, the radiolabeled products can be detected and quantified by a phosphorimager.
7. Locate the radioactive products in the gel using the fluorescent markers to position a template over the gel and excise the radioactive bands from the gel. (The template can be generated by placing a transparent acetate sheet on the autoradiogram, marking the location of the fluorescent markers and cutting holes in the acetate sheet at the sites of the radioactive bands.) The radioactivity in each gel fragment can be counted without addition of scintillant.
8. If all of the radioactivity has been retained on the gel, the molar quantity of material in a particular band can be quantified by multiplying the fraction of the total radioactivity in the lane found in that band by the molar quantity of ATP used in the reaction. However, before this can be translated into the molar quantity of the lesion in the DNA, it is necessary to determine (a) the resistance of the phosphodiester bond 5′ to the lesion to digestion by snake venom phosphodiesterase (see Discussion), (b) the labeling efficiency of the dinucleoside monophosphate (see Discussion), and (c) the amount of DNA in the digested mixture (see step 9).

9. The amount of DNA recovered following digestion of the DNA and removal of the protein (step 3) can be accurately determined by HPLC by comparing the nucleoside content of the digested DNA to standards. With a diode array UV detector it is possible to quantify as little as 0.1–1 nmole of each nucleoside.

2.4. HPLC Analysis of Radioactive Bands and Nearest-Neighbor Analysis

The bands in the gel often contain more than one radiolabeled product. These can be isolated by eluting the radioactive products from the gel and resolving them by reversed phase HPLC. The labeled products have the general structure ^{32}pNpX, i.e., the lesion is attached 3′ to a normal nucleotide, and it is the latter that bears the labeled phosphate group. The normal nucleotide can of course be any one of the four nucleotides found in DNA. To establish the nature of the nucleotide neighboring the lesion, the molecules isolated by HPLC can be digested by nuclease P1 to release the labeled 5′-mononucleotides, which can then be identified by reversed-phase HPLC.

2.4.1. Eluting Radioactive Material from Polyacrylamide Pieces

1. Mash up the individual polyacrylamide fragments in their scintillation vials and then add 0.5–1 ml of water. (Do not add the water first because the gel becomes too slippery to mash.) Shake the gel vigorously overnight at 4°C.
2. Remove the particulate by filtration and/or centrifugation and rinse the debris with a further 0.5 ml of water. Combine the filtrate and count the recovered radioactivity. (Approximately 50% of the radioactivity should be recoverable.)

2.4.2. HPLC Analysis

2.4.2a. Instrumentation. A standard instrument is required, capable of pumping a solvent gradient, with on-line UV detector and radioactivity monitor. In our laboratory, this consists of a Spectra Physics P2000 pump, coupled to a Spectra Physics FOCUS UV/Vis detector and a Berthold LB504 radioactivity monitor. (The latter is set with a 30 sec time constant and a cpm range of 3000 or 10,000 cpm.) If an on-line radioactivity monitor is not available, fractions can be collected and counted. We have used reversed-phase C_{18} columns purchased from a variety of suppliers (Waters, Supelco, and Phenomenex) and found that they perform similarly. It is useful to have injection loops with different loading capacities, including larger volumes such as 1 or 2 ml.

2.4.2b. Elution Conditions. Prior to injection onto the column, dilute 5000–50,000 cpm of recovered material to 1 ml with water. (The samples eluted from the polyacrylamide gel contain a large quantity of urea and the radioactive samples should be diluted to reduce the disruptive influence of urea on hydrophobic interactions.) Using a Waters μBondapak C_{18} RCM 8 × 10 RadialPak cartridge the gradient conditions are as follows: 100% buffer A (50 mM NaH_2PO_4, pH 4.5) and 0% buffer B [100 mM NaH_2PO_4, pH 4.5/methanol (1:1 v/v)] for 1 min, followed by a linear gradient to 20% buffer A/80% buffer B over 30 min at a flow rate of 1 ml/min. Fractions can be collected every 0.25–0.5 min.

2.4.3. Nuclease P1 Digestion

Having collected the desired fractions from the HPLC, raise the pH to 7.0 by careful addition of 1 M NaOH, dry the solution down under vacuum, and redissolve the residue in 0.5 ml of distilled water. Desalt the sample by passage through a C_{18} Sep-Pak cartridge (Waters), i.e., apply the sample to an activated (2 ml of methanol) and washed (5 ml of distilled water) cartridge, elute the salt with 1.5 ml of distilled water, and elute the radioactive material with 2–3 ml of 50% aqueous methanol. (The best way to monitor the elution of radioactive material is with a Geiger counter.) Remove the solvent and redissolve the residue in 50 μl of nuclease P1 buffer. Add 2 μl of an aqueous solution containing a UV-detectable quantity of unlabeled dinucleotide, to serve as a control of successful digestion of the labeled material, and 1 μl (0.6 U) of nuclease P1. Allow the reaction to proceed at 37°C for 1 hr, and then analyze the digestion products by reversed-phase HPLC using the same system as that described above.

2.5. Marker Compounds

The preparation of marker compounds for products detectable by this postlabelling methodology has been described in several articles. Below is a list of compounds and the references detailing their synthesis:

- Dinucleoside monophosphate containing a thymine glycol (NpT^g)—Baleja *et al.* (1993) and Weinfeld and Soderlind (1991)
- Nucleoside 3′-phosphoglycolate (Npg)—Urata and Akagi (1993) and Henner *et al.* (1983)
- "Dinucleoside" monophosphate containing an abasic site (NpS) or methoxyamine-modified abasic site (NpM)—Weinfeld *et al.* (1989b, 1990)
- Trinucleotide containing a cyclobutyl thymine dimer—Liuzzi *et al.* (1989)

3. DISCUSSION

3.1. Notes on the Protocols (Troubleshooting)

1. Heat inactivation of proteins. We have found that simple ethanol precipitation does not always entirely remove enzyme activity, especially phosphodiesterase activity, and consequently the heat inactivation step is required. [Failure to inactivate residual phosphodiesterase activity results in the appearance of a ladder stemming from partially digested $oligo(dT)_{16}$.] If, however, the lesions to be detected are heat-labile, an alternative to heat inactivation is the use of micro-ultrafiltration units such as those supplied by Millipore or Micron Separations Inc. Removal of protein by these units can be based on passage of the digested DNA fragments through a small-pore (<10 kDa) ultrafiltration membrane or by selective binding of protein to a hydrophobic membrane such as PVDF (Harley and Vaziri, 1991).

2. Failure to label products efficiently. This can result if (a) there is insufficient ATP due to its decomposition or (b) the kinase has little or no activity in the reaction mixture. The latter can occur if the protein itself is inactive or if there is an inhibitor of kinase activity. We have observed that it is important to keep the total salt concentration, including Tris-HCl, in the kinase reaction low, preferably <80 mM.

3. Spurious bands in untreated DNA. (a) There can be genuine damage already present in the DNA. We have observed high levels of damage, including abasic sites, in commercial preparations of herring sperm DNA. (b) We have observed that the use of some commercial preparations of PNK increases the likelihood of spurious bands. (c) Old preparations of phenol used in the purification of cellular DNA or divalent metal ion contamination of DNA may give rise to oxidative damage.

3.2. Quantitative Aspects of the Assay

Quantitative parameters arising from the dependence of postlabeling assays on several enzymes have been critically reviewed by Hemminki *et al.* (1993). In our procedure the key enzymes are snake venom phosphodiesterase and polynucleotide kinase. With the former, the question arises as to whether the phosphodiester bond immediately 5′ to the lesion is totally refractory to the enzyme. We have used "dinucleotide" models with the conditions described above, and in the case of thymine glycol and phosphoglycolate species have observed excellent recovery, i.e., virtually no digestion of these molecules. However, as pointed out by Hemminki *et al.* (1993), results obtained with dinucleotides do not necessarily reflect results obtained with longer oligonucleotides and DNA. In the case of the labeling reaction, it is important to have at least a three- to fivefold excess of ATP in the reaction mixture. Lower levels of ATP can result in preferential labeling of certain substrates. For example, we have observed that the phosphoglycolate-containing species appear to be better substrates for polynucleotide kinase than the thymine glycol-containing species. Clearly without control experiments using standard oligonucleotides or DNA with a known number of lesions the technique must be considered semiquantitative. Although the technique has the potential to detect lesions down to 1 fmole/μg DNA or ~3 lesions per 10^7 nucleotides, we would argue that with the current technology a more realistic detection limit is closer to 10–20 fmole/μg DNA.

3.3. General Comments on the Assay (Advantages and Disadvantages)

Certain advantages of our approach to postlabeling are mentioned in the Introduction, including the absence of labeling of undamaged DNA and the improved labeling of lesion-containing molecules because of the normal nucleotide 5′ to the lesion. There are, however, several drawbacks. (1) In common with other postlabeling techniques a significant problem is that unless marker compounds are available, the assay provides no structural information to help identify the detectable lesions. (2) Since the damaged base or sugar moiety is isolated with the neighboring 5′-nucleotide, there are always four species generated for each lesion. (3) Only a limited number of oxidative lesions can be detected by this technique. 8-Oxoguanine and 5-hydroxymethyluracil are among the common lesions not detectable by our approach. Nonetheless, the assay is an important addition to existing postlabeling methodologies because of its capacity to detect lesions that cannot be measured by the other approaches or are detected with very low efficiency.

ACKNOWLEDGMENTS. We wish to acknowledge the people who have contributed to the development of this assay, particularly Dr. Malcolm Paterson, Krista-June Soderlind, Dr. Garry Buchko, Dr. Gu Ruiqi, and Jane Lee. Work described here was supported by the National Cancer Institute of Canada with funds from the Canadian Cancer Society and an Alberta Cancer Board postdoctoral fellowship to G.D.D.J.

REFERENCES

Baleja, J. D., Buchko, G. W., Weinfeld, M., and Sykes, B. D. (1993). Characterization of γ-radiation induced decomposition products of thymidine-containing dinucleoside monophosphates by NMR spectroscopy. *J. Biomol. Struct. Dynam.* **10**:747–762.

Bertoncini, C. R. A., and Meneghini, R. (1995). DNA strand breaks produced by oxidative stress in mammalian cells exhibit 3′-phosphoglycolate termini. *Nucleic Acids Res.* **23**:2995–3002.

Buchko, G. W., and Weinfeld, M. (1993). Influence of nitrogen, oxygen, and nitroimidazole radiosensitizers on DNA damage induced by ionizing radiation. *Biochemistry* **32**:2186–2193.

Bykov, V. J., Kumar, R., Försti, A., and Hemminki, K. (1995). Analysis of UV-induced DNA photoproducts by ^{32}P-postlabelling. *Carcinogenesis* **16**:113–118.

Försti, A., and Hemminki, K. (1994). A ^{32}P-postlabelling assay for DNA adducts induced by cis-diammine-dichloroplatinum(II). *Cancer Lett.* **83**:129–137.

Gillardeaux, O., Périn-Roussel, O., Nocentini, S., and Périn, F. (1994). Characterization and evaluation by ^{32}P-postlabelling of psoralen-type DNA adducts in HeLa cells. *Carcinogenesis* **15**:89–93.

Harley, C. B., and Vaziri, H. (1991). Deproteinization of nucleic acids by filtration through a hydrophobic membrane. *GATA* **8**:124–128.

Hegi, M. E., Sagelsdorff, P., and Lutz, W. (1989). Detection by ^{32}P-postlabeling of thymidine glycol in γ-irradiated DNA. *Carcinogenesis* **10**:43–47.

Hemminki, K., Försti, A., Löfgren, M., Segerbäck, D., Vaca, C., and Vodicka, P. (1993). Testing of quantitative parameters in the ^{32}P-postlabelling method, in: *Postlabelling Methods for the Detection of DNA Damage*, IARC Scientific Publications No. 124 (D. H. Phillips, M. Castegnaro, and H. Bartsch, eds.), IARC Publications, Lyon, pp. 51–63.

Henner, W. D., Rodriguez, L. O., Hecht, S. M., and Haseltine, W. A. (1983). γ ray induced deoxyribonucleic acid strand breaks. *J. Biol. Chem.* **258**:711–713.

Liuzzi, M., Weinfeld, M., and Paterson, M. C. (1989). Enzymatic analysis of isomeric trithymidylates containing UV light-induced cyclobutane pyrimidine dimers: Nuclease Pl-mediated hydrolysis of the intradimer phosphodiester linkage. *J. Biol. Chem.* **264**:6355–6363.

Randerath, K., Reddy, M. V., and Gupta, R. C. (1981). ^{32}P-postlabeling test for DNA damage. *Proc. Natl. Acad. Sci. USA* **78**:6126–6129.

Reddy, M. V., Bleicher, W. T., and Blackburn, G. R. (1991). ^{32}P-postlabeling detection of thymine glycols: Evaluation of adduct recoveries after enhancement with affinity chromatography, nuclease Pl, nuclease S1, and polynucleotide kinase. *Cancer Commun.* **3**:109–117.

Stuart, G. R., and Chambers, R. W. (1987). Synthesis and properties of oligodeoxyribonucleotides with an AP site at a preselected position. *Nucleic Acids Res.* **15**:7451–7462.

Urata, H., and Akagi, M. (1993). A convenient synthesis of oligonucleotides with a 3′-phosphoglycolate and 3′-phosphoglycoaldehyde terminus. *Tetrahedron Lett.* **34**:4015–4018.

Weinfeld, M., and Buchko, G. W. (1993). Postlabelling methods for the detection of apurinic sites and radiation-induced DNA damage, in: *Postlabelling Methods for the Detection of DNA Damage*, IARC Scientific Publications No. 124 (D. H. Phillips, M. Castegnaro, and H. Bartsch, eds.), IARC Publications, Lyon, pp. 95–103.

Weinfeld, M., and Soderlind, K.-J. (1991). ^{32}P-postlabeling detection of radiation-induced DNA damage: Identification and estimation of thymine glycols and phosphoglycolate termini. *Biochemistry* **30**:1091–1097.

Weinfeld, M., Liuzzi, M., and Paterson, M. C. (1989a). Enzymatic analysis of isomeric trithymidylates containing UV light-induced cyclobutane pyrimidine dimers: II. Phosphorylation by phage T4 polynucleotide kinase. *J. Biol. Chem.* **264**:6364–6370.

Weinfeld, M., Liuzzi, M., and Paterson, M. C. (1989b). Selective hydrolysis by exo- and endonucleases of phosphodiester bonds adjacent to an apurinic site. *Nucleic Acids Res.* **17**:3735–3745.

Weinfeld, M., Liuzzi, M., and Paterson, M. C. (1990). Response of phage T4 polynucleotide kinase toward dinucleotides containing apurinic sites: Design of a ^{32}P-postlabeling assay for apurinic sites in DNA. *Biochemistry* **29**:1737–1743.

Weinfeld, M., Soderlind, K.-J., and Buchko, G. W. (1993). Influence of nucleic acid base aromaticity on substrate reactivity with enzymes acting on single-stranded DNA. *Nucleic Acids Res.* **21**:621–626.

Winters, T. A., Weinfeld, M., and Jorgensen, T. J. (1992). Human HeLa cell enzymes that remove phosphoglycolate 3′-end groups from DNA. *Nucleic Acids Res.* **20**:2573–2580.

Chapter 6

Radioimmunoassay of DNA Damaged by Ultraviolet Light

David L. Mitchell

1. INTRODUCTION

Quantitative imunoassays such as radioimmunoassay (RIA) and enzyme-linked immunosorbant assay (ELISA) are sensitive and reliable procedures used to measure UV photoproducts in purified sample DNA (Eggset *et al.*, 1983; Strickland, 1985; Mitchell and Nairn, 1989; Matsunaga *et al.*, 1990; Wani and Arezina, 1991; Vink *et al.*, 1993). Competitive immunoassays have distinct advantages over other procedures in the analysis of DNA damage in human and environmental samples. For instance, unlike spectroscopic analyses, immunoassays are not limited to certain classes of adducts, and unlike HPLC and ^{32}P-postlabeling, immunoanalysis of DNA damage does not require hydrolysis of sample DNA which can decrease the signal-to-noise ratio and alter chemical structure. ELISAs and RIAs require minimal sample manipulation and have been performed on crude cell lysates. In addition, sample DNA does not require prelabeling with a radioactive tracer and antibody binding is quasi-independent of molecular weight (i.e., DNA degradation). Immunoassays are thus readily applied to a variety of biological materials and have typically been used to measure DNA damage in cell and organ cultures, normal tissues and tumor biopsies, and various other samples including buccal cells, bone marrow aspirates, and peripheral blood lymphocytes. RIA is a sensitive and facile technique for measuring genotoxic damage in DNA.

RIA is a competitive binding assay between an unlabeled and a radiolabeled antigen for binding to antibody raised against that antigen. For convenience, the radiolabeled antigen is referred to as the "probe" and the unlabeled competitor as "sample" or "standard." The amount of radiolabeled antigen bound to antibody is determined by separating the antigen–antibody complex from free antigen by, for example, secondary antibody or high-salt precipitation. The amount of radioactivity in the antigen–antibody complex in the presence of known amounts of competitor (i.e., standards) can then be used to quantify the amount of unknown sample present in the reaction. The sensitivity of the RIA is determined by the affinity of the antibody and specific activity of the radiolabeled antigen. Using high-affinity antibody and probe labeled to a

David L. Mitchell • Department of Carcinogenesis, The University of Texas M. D. Anderson Cancer Center, Smithville, Texas 78957.

Technologies for Detection of DNA Damage and Mutations, edited by Gerd P. Pfeifer, Plenum Press, New York, 1996.

high specific activity, the reaction can be limited to such an extent that extremely low levels of damage in sample DNA can be detected.

2. METHODS

2.1. Antiserum

Few commercial sources are available for the purchase of anti-UV DNA antisera and include Kamiya Biomedical Co., Thousand Oaks, CA (CPD monoclonal antibody) and Dermigen, Inc., Smithville, TX [CPD, (6–4) photoproduct, and custom rabbit antisera]. Following is a procedure for producing polyclonal antisera in rabbits. The production of high-affinity rabbit polyclonal antisera is dependent on the following factors: (1) the "foreignness" of the immunogen to the host (e.g., steric and distortive deviations from normal DNA structure), (2) the number of antigenic determinants presented to the host immune system (i.e., concentration, dose), (3) accessibility of the host immune system to the antigen (e.g., strandedness of the DNA), (4) stability of the immunogen in the host animal, and (5) host variability. For production of anti-UV DNA antisera we typically immunize four rabbits and select the animals that show the greatest immune response for exsanguination (Fig. 6.1).

UVC radiation (240–290 nm) produces several types of dimeric damage in DNA, including the cyclobutane dimer (CPD), (6–4) photoproduct, and Dewar pyrimidinone (Cadet and Vigny, 1990). In a mixture of antigenic determinants, the lesion with the greatest immunogenicity will elicit the greatest immune response. In UVC-irradiated DNA this determinant is the (6–4) photoproduct and the antibody subpopulation elicited against this photoproduct is 10- to 100-fold more active than that recognizing the less distortive (i.e., less "foreign") cyclobutane dimer. Because of this, UVC-DNA antisera can be diluted to such an extent that binding to the minor CPD antibody subpopulation is undetectable and the RIA is specific for the (6–4) photoproduct.

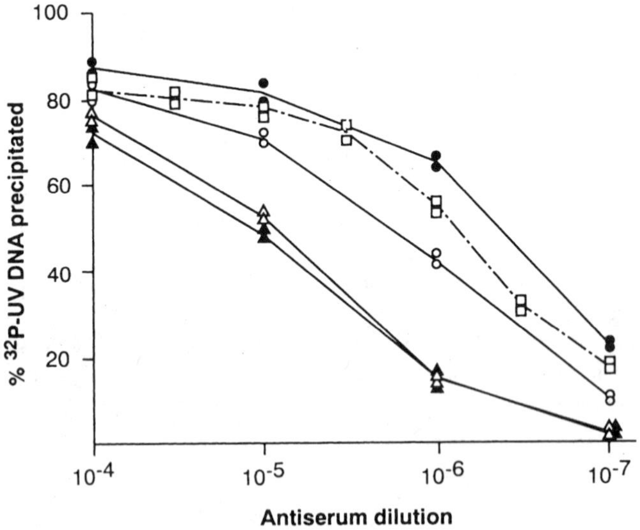

Figure 6.1. Host variability in the immune response. Binding curves were generated in four rabbits receiving two immunizations of UVC-irradiated DNA (500 μg each) at 14-day intervals.

In order to develop RIAs for the cyclobutane dimer and other types of UV damage with lower immunogenicities, other strategies are required. The following protocols are used for preparing DNA substrates for the production of rabbit antisera to use in immunoassays specific for dimeric UV photodamage:

A. For (6–4) photoproduct antisera, DNA is irradiated with 100 kJ/m^2 UVC light. Our UVC source consists of a bank of five Philips Sterilamp G8T5 bulbs emitting predominantly 254-nm light. At a distance of ~20 cm the fluence rate is ~14 J/m^2 per sec measured with an International Light IL1400 photometer coupled to an NS254-nm probe. At this fluence rate the average duration of exposure is ~2 hr.

B. For CPD antisera it is necessary to produce DNA containing this photoproduct exclusively. Two strategies have been utilized: (1) DNA is irradiated with UVB light (290–320 nm) in the presence of a triplet sensitizer (e.g., 2×10^{-2} M acetophenone; 10% acetone) to produce predominantly *cis,syn*-cyclobutane dimers (Lamola and Yamane, 1967). Our UVB source consists of four Westinghouse FS20 sunlamps filtered through cellulose acetate. This filter (Kodacel) has a wavelength cutoff of 290 nm and can be obtained from Kodak. Emission spectra for sunlamp irradiation through various filters can be found in Rosenstein (1984). The DNA is irradiated with ~75 kJ/m^2 UVB light measured with the IL1400 photometer coupled to an SCS 280 probe. At a distance of ~10 cm the fluence rate is ~5 J/m^2 per sec. Hence, exposure times of ~4 hr are required for adequate CPD induction. The DNA is extensively dialyzed postirradiation to remove the sensitizer. An alternate strategy is to produce a mixture of CPDs and (6–4) photoproducts by UVC irradiation, convert the (6–4) photoproducts to Dewar pyrimidinones by UVB photoisomerization (see below), and digest the Dewar pyrimidinones by 4-hr incubation in 0.4 N NaOH at 23°C, leaving the CPDs intact.

C. For Dewar pyrimidinone antisera, DNA is irradiated to produce (6–4) photoproducts which are then photoisomerized to the Dewar pyrimidinone by irradiation with Westinghouse FS20 sunlamps filtered through Mylar 500D (available from DuPont). The most effective wavelengths for Dewar formation correspond to the peak absorbance of the (6–4) photoproducts (i.e., 310–320 nm). Although Mylar 500D has a wavelength cutoff of 320 nm, enough stray UVB light is present to photoisomerize the Dewar pyrimidinones without inducing additional CPDs and (6–4) photoproducts. For action spectra of photoproduct induction and photoisomerization using immunoassays, see Rosenstein and Mitchell (1987), Mitchell and Rosenstein (1987), and Matsunaga *et al.* (1993). At 10 cm from the light source, 80 to 90% photoisomerization is complete within ~4 hr.

Native DNA is irradiated as above at 1 mg/ml in H_2O, heat-denatured, and coupled to methylated bovine serum albumin (Plescia *et al.*, 1964). Methylated BSA is added dropwise until the UV-irradiated DNA solution turns cloudy (~50 μl/ml). For the initial immunization, the UV-DNA is mixed with an equal volume of Freund's complete adjuvant and 1.0 ml is injected subcutaneously at ten sites. Rabbits are injected at 2-week intervals with 0.5 mg DNA and after two injections, 1 ml serum is drawn and evaluated using immunoprecipitation. The latter consists of antiserum diluted at approximately half-log increments (e.g., 1/100, 1/300, 1/1000, 1/3000, etc.) incubated with ^{32}P-labeled UV-DNA and precipitated with goat anti-rabbit IgG (GaRGG) in the presence of normal rabbit serum (NRS). The amount of radioactivity present in the immune pellet is a measure of antiserum titer. Depending on cost considerations, rabbits that display the weakest immune response after the second injection may be culled. When the antiserum titer reaches acceptable levels (i.e., does not increase further), rabbits are exsanguinated, typically yielding 50–80 ml antiserum. Because repeated freezing and thawing greatly reduces antibody activity, care is taken to minimize the number of times the antiserum will be frozen. Antiserum may be frozen in 1-ml aliquots at either −20 or −70°C. For subsequent use (in RIA) an aliquot is thawed slowly (overnight at 4°C), dispensed into 30-μl aliquots in 0.5-ml Eppendorf tubes, and refrozen.

Specificities of antisera are determined using either direct immunoprecipitation, competitive RIA, or mobility-shift immunoassay. In the first case, the ^{32}P-labeled probe is treated to produce or remove specific types of damage; in the second case, the unlabeled sample DNA is similarly treated. Specificity of antisera for (6–4) photoproducts or Dewar pyrimidinones is determined by irradiating UV-DNA with photoisomerizing light and monitoring antibody binding. Under these light conditions, reduced binding indicates specificity for (6–4) photoproducts; increased binding indicates specificity for Dewar photoproducts (Rosenstein and Mitchell, 1987; Mitchell and Rosenstein, 1987). CPD-specific antisera will not be affected by photoisomerizing light but will show reduced binding after treatment with *Escherichia coli* photoreactivating enzyme or T4 endonuclease V. Alternatively, triplet-sensitized UVB irradiation of DNA will produce predominantly CPDs with negligible production of (6–4) photoproducts and will show binding to CPD antisera but not (6–4) photoproduct antisera (Mitchell *et al.*, 1989). We have recently developed a technique for the visualization of antibody specificity using mobility-shift immunoassays (Ghosh *et al.*, in press). In this procedure, purifed IgG is incubated with radiolabeled oligonucleotides containing specific types of UV-induced damage and electrophoresed on agarose or polyacrylamide gels. Binding of antibody to different types of photodamage is seen as gel retardation of the ^{32}P-endlabeled oligonucleotide.

Should the antiserum of interest show nonspecific binding or activity toward more than one photoproduct (e.g., antisera raised against UVC-irradiated DNA), it can be preadsorbed to DNA-Sepharose. For some, particularly *in situ*, applications (e.g., immunohistochemistry), we routinely preadsorb CPD antisera to unirradiated DNA–Sepharose and (6-4) antisera to triplet-sensitized, UVB-irradiated DNA–Sepharose. Should further purification of IgG be necessary, facile procedures are currently available (e.g., ammonium sulfate precipitation, protein A column chromatography).

2.2. Radiolabeled Probe

The sensitivity and specificity of an RIA is determined not only by the affinity of the antibody but also by the character of the labeled DNA probe. High-specific-activity probes allow lower concentrations of antibody and, hence, increased sensitivity. DNA is nick translated with ^{32}P-dCTP and ^{32}P-TTP to give specific activities of ~5×10^8 to 10^9 cpm/μg. The greater the specific activity of the probe, the less antiserum can be used in the RIA. We try to use 10 pg labeled probe in the RIA containing 5000 to 10,000 cpm. The antigenicity of the probe (i.e., its capacity to bind antibody) is determined by the UV dose, AT:GC content, and strandedness of the DNA. In Fig. 6.2 DNAs with different AT:GC contents irradiated with the same fluence of UVC light (i.e., 30 kJ/m^2) were compared and indicate that the greater the proportion of thymine bases in the probe, the better the antibody binding. In Fig. 6.3 *Clostridium perfringens* DNA (Sigma) was irradiated with increasing fluences of UVC light and immunoprecipitated with antisera for cyclobutane dimers and (6–4) photoproducts. From this dose response we determined that ~30 kJ/m^2 gives optimal binding. Even though UV photoproducts are considerably more antigenic in single-stranded DNA, denaturation subsequent to UVC irradiation using heat, alkali, or formamide had no effect on antibody binding. From these results, we concluded that nick translation coupled with the high UVC irradiation resulted in a single-stranded DNA probe, at least in the vicinity of the damage. In Fig. 6.4 the size of the radiolabeled probe before and after immunoprecipitation is shown. *C. perfringens* DNA and ^{32}P-endlabeled molecular weight standards (Boehringer-Mannheim) were UV-irradiated, incubated with anti-(6–4) photoproduct serum, and precipitated with GaRGG/NRS. These data showed that the average length of the precipitated probe was ~150 bases and that all of the UV-DNA > 150 bases was efficiently precipitated.

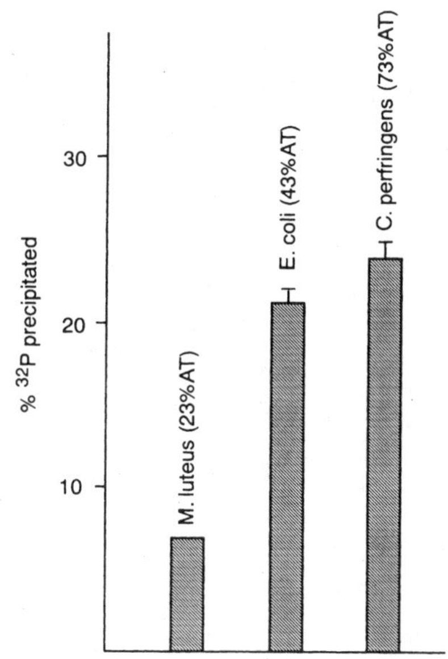

Figure 6.2. The effect of thymine content on the antigenicity of the radiolabeled probe. Binding of radiolabeled DNAs with increasing AT:GC content to antibody raised against UVC-irradiated DNA [(6–4) photoproduct)] was determined.

2.3. Sample and Standard Preparation

One of the major attributes of RIA is its ability to measure photoproducts in DNA that has not been extensively purified and most DNA isolation protocols are suitable for RIA sample preparation. For tissue culture, cells are plated 2–3 days prior to an experiment in medium containing 0.005–0.010 μCi/ml [^{14}C]-TdR (prelabeling facilitates equilibrating the DNA for the

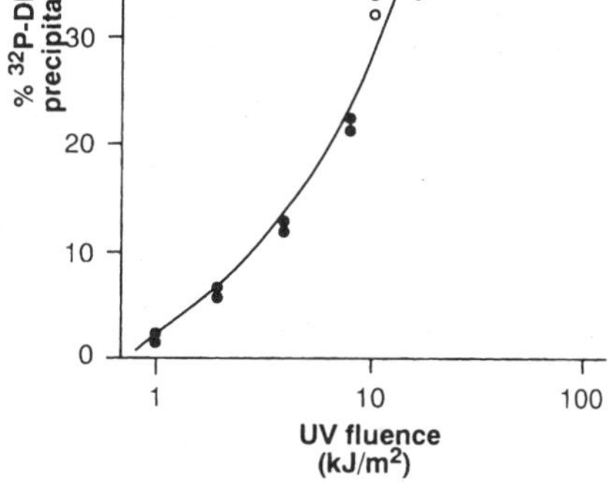

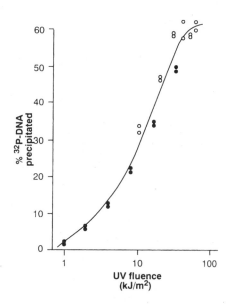

Figure 6.3. Antibody binding to the radiolabeled probe after increasing fluences of UVC light. Antibody binding was determined for antisera specific for cyclobutane dimers (closed circles) and (6–4) photoproducts (open circles).

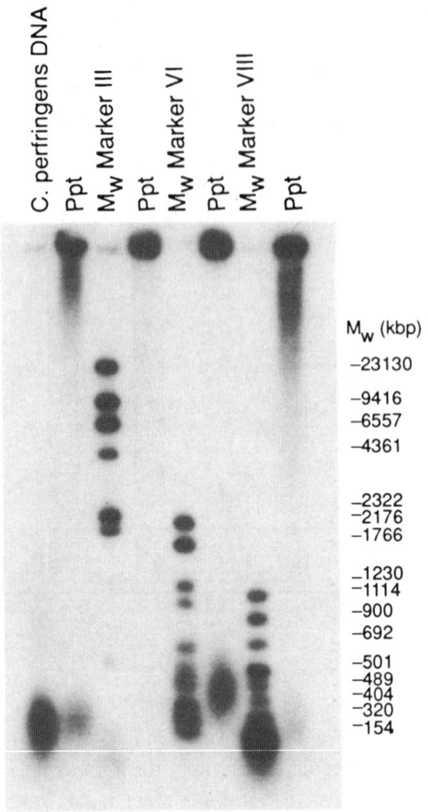

Figure 6.4. Molecular weight of radiolabeled probe and the effect of fragment size on immunoprecipitation. *C. perfringens* DNA and molecular weight markers (Boehringer-Mannheim) were endlabeled with ^{32}P-ATP, irradiated with 30 kJ/m^2 UVC light, and precipitated with anti-(6–4) photoproduct serum, GaRGG, and normal rabbit serum. Fragment sizes before and after precipitation were determined by electrophoresis in 0.8% agarose.

RIA). The medium is removed, cells are washed twice with phosphate-buffered saline, and are covered with 4 ml TES (10 mM Tris, pH 8, 1 mM EDTA, 150 mM NaCl) prior to UV irradiation. Cells are scraped from the plates and DNA is isolated by lysis with 0.02% SDS (final concentration) followed by overnight digestion in proteinase K (30 μg/ml) at 37°C. The protein is extracted with an equal volume of chloroform/isoamyl alcohol (24:1), then ethanol precipitated after the addition of 0.4 vol of 5 M ammonium acetate. (Note: Ammonium acetate is filter steriized and kept at 4°C). Digestion with 20 μg/ml RNase A (Sigma) for 15 min at 37°C is optional, depending on the experiment (see below). The DNA pellet is redissolved in sterile H_2O (or 0.1× TE buffer) at a concentration of 50–100 μg/ml. For tissues or whole organisms, DNA is isolated using the same procedure as used for cells or by high-salt extraction (Stratagene). Briefly, the sample is lysed in 50 mM Tris, pH 8, 20 mM EDTA, chilled on ice for 10 min, and gently mixed with ~1/3 vol saturated NaCl for 5 min. The precipitates are pelleted by centrifugation for 15 min at 2000*g* at 4°C and supernatants treated with 20 μg/ml RNase A (Sigma) for 15 min at 37°C. The DNA is precipitated overnight with 2 vol absolute ethanol at −20°C, centrifuged, and resuspended as above.

For RIA of prelabeled DNA in which relative rather than absolute amounts of photodamage are desired, it is not necessary to treat the samples with RNase and quantify the DNA content of all samples. By titrating the binding activity of one sample in a preliminary competitive RIA (within a set of samples that should have the same specific activity), the ^{14}C counts can be used to adjust the volumes of each sample such that they all contain the same amount of DNA. In other samples in which absolute numbers of photoproducts are desired, accurate determinations of DNA concentrations are critical. Absorbance is appropriate for native or denatured samples that have been digested with RNase (the $A_{260/280}$ should be $>$ 1.7–1.8 for reliable measurements). Spectrofluorometry using a Hoechst or DapI dye is suitable for double-stranded DNA that has not been extensively purified or treated with RNase. When in doubt, native or denatured DNA can be visualized and quantified by electrophoresis on agarose gels. Because damage is significantly more antigenic in single-stranded DNA, the standard and sample are heat-denatured prior to analysis. RIA is a competitive inhibition assay that uses a standard curve through which unknown sample inhibitions are extrapolated for antigen (photoproduct) determinations. The standard curve consists of incubation of fixed amounts of labeled probe and antibody with different concentrations (or UV fluences) of unlabeled antigen. For experiments that compare relative amounts of damage (e.g., repair protocols), the standard curve is simply a dilution of the sample to which the other samples are to be compared (e.g., DNA from cells harvested immediately after irradiation for repair experiments). For absolute determinations, a standard of known concentration is diluted to give a linear inhibition curve within which the unknown samples are expected to fall. The amount of damage in the standard should be comparable to that anticipated in the samples. In other words, the standard used for RIA analysis of biological samples should be treated (UV-irradiated) with doses comparable to that received by the sample (Santella *et al.*, 1988). For RIA of photodamage we routinely use a standard curve irradiated with 3, 10, 30, and 100 J/m^2 UVC light. In assembling an RIA protocol, the sensitivity can be estimated using standard DNA. Antibody binding curves like those used to evaluate antibody titer during immunization (see above) are determined and an antiserum dilution that gives 30–70% binding is selected for subsequent competitive RIAs. The greater the dilution of antiserum, the greater the sensitivity of the RIA (Fig. 6.5). An acceptable RIA [for (6–4) photoproducts or CPDs] should show ~50% inhibition for 1–3 μg DNA extracted from cells irradiated with 10–20 J/m^2 UVC light.

2.4. RIA Protocol

1. RIA buffer (1 ml) is dispensed into 12 × 75-mm test tubes. RIA buffer consists of TES supplemented with 0.5% gelatin (Sigma) and carrier DNA. Gelatin is added to buffer and heated to 40°C. (Note: Overheating the buffer at this stage can affect the antibody–antigen reaction and

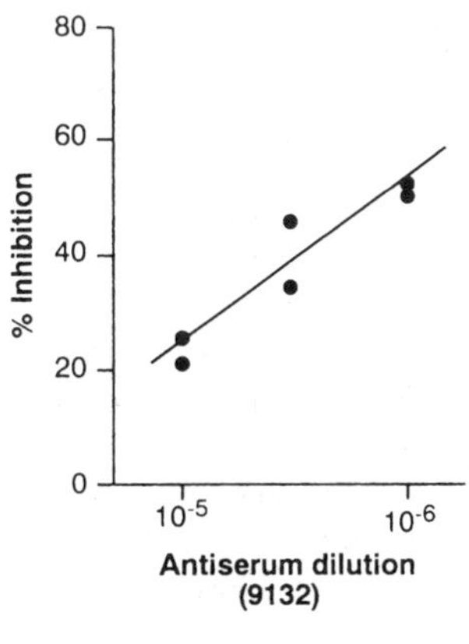

Figure 6.5. The effect of antiserum dilution on sensitivity. The percent inhibition in a (6–4) photoproduct RIA was determined with increasing dilutions of antiserum.

the reliability of the RIA.) We include carrier denatured DNA at 20 μg/ml to mitigate any nonspecific binding effects and, for many antisera, increase binding. The amount of carrier DNA suitable for each antiserum should be determined in both direct and indirect (competitive) binding assays.

2. Equivalent amounts of denatured standard or sample DNA are added. Since the concentrations of the DNA samples may vary, different volumes may be added for each sample. The differences in volume should not influence the RIA results unless these differences exceed 5% of the total volume (i.e., there is > 50 μl difference in total volume after the sample is dispensed). In these cases, buffer should be removed or added to compensate for differences in volume. For sample volumes > 200 μl the amount of each reagent, including buffer, is doubled.

3. Diluted antiserum (100 μl) is added and mixed well. The reaction can be incubated at this stage or, if desired, after addition of the radiolabeled probe. There is some evidence that preincubation of antibody and unlabeled sample DNA prior to the addition of the radiolabeled probe increases the sensitivity of the RIA. We incubate samples for 2–4 hr at 37°C followed by overnight incubation at 4°C.

4. Radiolabeled probe (100 μl) is added and mixed well. The probe may be added at the same time as the unlabeled sample DNA if desired, and incubated as above. (Note: Incubation times and temperatures can vary somewhat without significantly affecting the RIA. These parameters may be determined more by the time schedule of the investigator rather than the dictates of the reaction.)

5. The antibody–DNA complex is precipitated by the addition of GaRGG (Calbiochem) and NRS as carrier. We routinely add 50 μl of each reagent; GaRGG is added directly from a reagent vial to which 9–10 ml of RIA buffer was added; NRS is diluted 1:50. The amount of GaRGG and NRS that give the highest number of counts in the precipitate (not necessarily the largest precipitate) should be determined during RIA development.

6. The second antibody precipitation is incubated for 1 to 2 days at 4°C, then centrifuged at ~4000*g* for 30 to 45 min. After the supernatant is decanted, the immune pellet is dissolved in 100 μl NCS tissue solubilizer (supplemented with 10% H_2O) (Amersham) and incubated at 37°C for > 20 min on a rotating platform. (Note: It is important that the immune pellet be completely solubilized but not allowed to dry.) Two milliliters of Scintiverse or Scintisafe (Fisher) is added, mixed well, and decanted into a 20-ml scintillation vial. The test tube is washed twice with 4 ml Scintiverse and counted in a liquid scintillation counter. To eliminate chemoluminescence caused by the tissue solubilizer, acetic acid (1 ml/liter) is added to the scintillation cocktail.

7. Microsoft Excel is used for quantifying RIA data . A sample of the spreadsheet (with calculations) is shown in Fig. 6.6. In addition to the unlabeled samples and standard curve, the RIA contains duplicate (or triplicate) tubes to which no antibody has been added and to which antibody, but no competitor, has been added. These tubes give the background and total binding, respectively.

3. DISCUSSION

The ability to selectively detect and quantify xenobiotic damage in DNA is probably the most powerful attribute of antibody-based assays. Antibodies have been produced against a broad spectrum of genotoxic chemicals and pharmaceuticals, including aflatoxin B1 (Groopman *et al.*, 1992; Hsieh and Hsieh, 1993; Nakatsuru *et al.*, 1990; Santella *et al.*, 1993; Brylawski *et al.*, 1991), 4-aminobiphenyl (Roberts *et al.*, 1988), *N*-methyl-, *N*-ethyl-, and *N*-[*n*-butyl]-*N*-nitrosourea (Seiler *et al.*, 1993; Hochleitner *et al.*, 1991), dimethylnitrosamine (Ozaki *et al.*, 1993), *N*-nitrosodimethylamine (Fadlallah *et al.*, 1994), vinyl chloride (Eberle *et al.*, 1989), acrolein

Tube #	Dimer	sample		ul/(x)ug	ug/ml	cpm	cpm-bkg	%inhibition	T<>T/107D		
1		-Ab									
							=AVERAGE(G2, G3)				
2	T<>T	+Ab					=G4-$H$3				
							=G5-$H$3	=AVERAGE(H4, H5)			
			UV dose								
3	T<>T	standard	10	=(x/F7)*1000	y		=G7-$H$3	=(1-($H7/$I$5))*100	=LOG(D7*0.245)	=RSQ(I7:I12,J7:J12)	Corr.Coef.
			10				=G8-$H$3	=(1-($H8/$I$5))*100	=LOG(D8*0.245)		
4	T<>T	standard	30	=(x/F9)*1000	y		=G9-$H$3	=(1-($H9/$I$5))*100	=LOG(D9*0.245)	=SLOPE(I7:I12,J7:J12)	Slope
			30				=G10-$H$3	=(1-($H10/$I$5))*100	=LOG(D10*0.245)		
5	T<>T	standard	100	=(x/F11)*1000	y		=G11-$H$3	=(1-($H11/$I$5))*100	=LOG(D11*0.245)	=INTERCEPT(I7:I12,J7:J12)	Y-intercept
			100				=G12-$H$3	=(1-($H12/$I$5))*100	=LOG(D12*0.245)		
6	T<>T	sample		=(x/F13)*1000	z		=G13-$H$3	=(1-($H13/$I$5))*100	=10^(($I13-$K$11)/$K$9)		
							=G14-$H$3	=(1-($H14/$I$5))*100	=10^(($I14-$K$11)/$K$9)		
1		-Ab									
							=AVERAGE(G16,G17)				
2	(6-4)	+Ab					=G18-$H$3				
							=G19-$H$3	=AVERAGE(H18,H19)			
			UV dose								
3	(6-4)	standard	10	=(x/F21)*1000	y		=G21-$H$3	=(1-($H21/$I$5))*100	=LOG(D21*0.056)	=RSQ(I21:I26,J21:J26)	Corr.Coef.
			10				=G22-$H$3	=(1-($H22/$I$5))*100	=LOG(D22*0.056)		
4	(6-4)	standard	30	=(x/F23)*1000	y		=G23-$H$3	=(1-($H23/$I$5))*100	=LOG(D23*0.056)	=SLOPE(I21:I26,J21:J26)	Slope
			30				=G24-$H$3	=(1-($H24/$I$5))*100	=LOG(D24*0.056)		
5	(6-4)	standard	100	=(x/F25)*1000	y		=G25-$H$3	=(1-($H25/$I$5))*100	=LOG(D25*0.056)	=INTERCEPT(I21:I26,J21:J26)	Y-intercept
			100				=G26-$H$3	=(1-($H26/$I$5))*100	=LOG(D26*0.056)		
6	(6-4)	sample		=(x/F27)*1000	z		=G27-$H$3	=(1-($H27/$I$5))*100	=10^(($I27-$K$11)/$K$9)		
							=G28-$H$3	=(1-($H28/$I$5))*100	=10^(($I28-$K$11)/$K$9)		

Figure 6.6. Microsoft Excel spreadsheet showing calculations used for RIA analyses.

(Foiles *et al.*, 1989), crotonaldehyde (Foiles *et al.*, 1987), benzo[*a*]pyrene (Lee and Strickland, 1993; Paules *et al.*, 1985; Newman *et al.*, 1990; Paleologo *et al.*, 1992; Rojas *et al.*, 1994; Santella *et al.*, 1992; Roggeband *et al.*, 1994; van Schooten *et al.*, 1992), benz[*a*]anthracene (Lee and Strickland, 1993; Newman *et al.*, 1990), 3-methylcholanthrene (Lee and Strickland, 1993), chrysene (Lee and Strickland, 1993), mixed polyaromatic hydrocarbons (Schoket *et al.*, 1993; Weston *et al.*, 1990; Yang *et al.*, 1988), acetylaminofluroene (Culp *et al.*, 1993; Poirier *et al.*, 1979; Huitfeldt *et al.*, 1990, 1994; Olivero *et al.*, 1990; Leng *et al.*, 1978), *cis*-platinum (Chao *et al.*, 1991; Fichtinger-Schepman *et al.*, 1989; Motzer *et al.*, 1994; Poirier *et al.*, 1992; Blommaert *et al.*, 1993), photoactivated psoralens (Yang *et al.*, 1988; Gasparro and Santella, 1988), thymine glycol (Leadon and Hanawalt, 1983), and 8-hydroxyadenine (West *et al.*, 1982). It is conceivable that antibodies can be produced that bind most DNA adducts of choice.

Different types of DNA damage have different immunogenicities and may not be suitable for the production of antibodies in rabbits or mice. For example, we have been unsuccessful in raising rabbit antisera against DNA cross-linked with mitomycin C, DNA irradiated with UVA light or treated with alkylating agents (e.g., methylmethane sulfonate), and DNA homopolymers [e.g., poly(dA):poly(dT)] irradiated with UVC light. These failures may have resulted from the low antigenicity of the damage itself (e.g., UVA, mitomycin C), the stability of the immunogen in the host (e.g., digestion of oligonucleotides or polynucleotides), or host variability (i.e., not enough rabbits). To produce antisera to antigens with low immunogenicity it may be productive to construct haptens covalently coupled to protein carriers. One drawback with this approach is that the antibody may bind the hapten (in hydrolyzed DNA) quite well but be unable to recognize the damage in DNA. Antibodies that bind haptenic damage exclusively have been successfully used to develop sensitive analytical approaches that combine immunoaffinity chromatography and HPLC (see Chapter 7).

Specificity of a particular antiserum may also present some problems for immunoassay development. Antisera may bind (cross-react) with known or unknown antigenic determinants. Two cases from our own experience are considered: (1) Antisera were raised in rabbits against Dewar photoproducts [i.e., UVC-irradiated DNA that had been quantitatively photoisomerized with UVB light (see above)]. Mobility-shift immunoassays showed that this antiserum recognized the (6–4) photoproduct with greater affinity than the Dewar photoproduct (i.e., it cross-reacted). Attempts to eliminate the cross-reactivity by preadsorbing to DNA–Sepharose containing only (6–4) photoproducts removed all of the binding activity. These results suggest that the antiserum was elicited against a common epitope of the (6–4) photoproduct and Dewar pyrimidinone. (2) Antisera were raised in rabbits against DNA treated with the oxidizing agent, osmium tetroxide (OsO_4). These antisera showed significant binding to DNA treated with OsO_4, but negligible binding to UVA-irradiated DNA or DNA treated with ionizing radiation. These results suggest either that the antisera did not recognize the major damage produced by treatment with OsO_4 and ionizing radiation (i.e., thymine glycol) or that thymine glycol is not the major base damage produced by these treatments.

RIA and ELISA are powerful techniques for the sensitive and specific detection of genotoxic damage in DNA. Other techniques, such as quantitative immunocytochemistry, immunohistochemistry, and immunoelectron microscopy, have been designed to detect DNA damage *in situ*, thus visualizing the distribution of damage in tissues and cells. One of the more recent and potentially useful applications of antibody technology is the separation of damaged from undamaged DNA fragments by immunoprecipitation, immunoaffinity chromatography, or nitrocellulose binding. Enrichment of damaged DNA fragments using antibodies has been combined with PCR amplification and Southern analysis to determine the genomic distribution of DNA damage (Hochleitner, *et al.*, 1991) and with HPLC (or ligation-mediated PCR) to increase the resolution and, hence, sensitivity of detection (Groopman *et al.*, 1992). Lastly, immunoseparation techniques can be used to purify oligonucleotides of defined sequence containing site-specific damage. Such substrates can be used for studies designed to examine the toxicity and mutagenicity of specific adducts and structures at the sequence level using *in vitro* DNA replication and repair protocols. In addition, such purified substrates could aid in identifying and purifying DNA damage-specific binding proteins (Reardon *et al.*, 1993) and help elucidate sequence effects of specific lesions on DNA–protein interactions (i.e., nucleosome reassembly and transcription factor binding). It is evident that antibodies raised against DNA damage will continue to provide an invaluable and innovative resource for molecular biological approaches to problems of cancer biology and environmental toxicology.

REFERENCES

Blommaert, F. A., Michael, C., Terheggen, P. M., Muggia, F. M., Kortes, V., Schornagel, J. H., Hart, A. A., and den Engelse, L. (1993). Drug-induced DNA modification in buccal cells of cancer patients receiving carboplatin and cisplatin combination chemotherapy, as determined by an immunocytochemical method: Interindividual variation and correlation with disease response. *Cancer Res.* **53**:5669–5675.

Brylawski, B. P., Cordeiro-Stone, M., and Kaufman, D. G. (1991). The use of rabbit polyclonal antibodies for the isolation of carcinogen-adducted DNA by immunoprecipitation. *Mol. Carcinogen.* **4**:315–321.

Cadet, J., and Vigny, P. (1990). The photochemistry of nucleic acids, in: *Bioorganic Photochemistry*, Volume I (H. Morrison, ed.), Wiley, New York, pp. 1–272.

Chao, C. C., Shieh, T. C., and Huang, H. (1994). Use of a monoclonal antibody to detect DNA damage caused by the anticancer drug cis-diamminedichloroplatinum (II) in vivo and in vitro. *FEBS Lett* **354**:103–109.

Culp, S. J., Poirier, M. C., and Beland, F. A. (1993). Biphasic removal of DNA adducts in a repetitive DNA sequence after dietary administration of 2-acetylaminofluorene. *Environ. Health Perspect.* **99**:273–275.

Eberle, G., Barbain, A., Laib, R. J., Ciroussel, F., Thomale, J., Bartsch, H., and Rajewsky, M. J. (1989). 1,N6-etheno-2′-deoxyadenosine and 3,N4-etheno-2′-deoxycytidine detected by monoclonal antibodies in lung and liver DNA of rats exposed to vinyl chloride. *Carcinogenesis* **10**:209–212.

Eggset, G., Volden, G., and Krokan, H. (1983). U.v.-induced DNA damage and its repair in human skin *in vivo* studied by sensitive immunohistochemical methods. *Carcinogenesis* **4**:745–750.

Fadlallah, S., Lachapelle, M., Krzystyniak, K., Cooper, S., Denizeau, F., Guertin, F., and Fournier, M. (1994). O6-methylguanine–DNA adducts in rat lymphocytes after in vivo exposure to N-nitrosodimethylamine (NDMA). *Int. J. Immunopharmacol.* **16(7)**:583–591.

Fichtinger-Schepman, A. M., Baan, R. A., and Berends, F. (1989). Influence of the degree of DNA modification on the immunochemical determination of cisplatin–DNA adduct levels. *Carcinogenesis* **10**:2367–2369.

Foiles, P. G., Chung, F.-L., and Hecht, S. S. (1987). Development of a monoclonal antibody-based immunoassay for cyclic DNA adducts resulting from exposure to crotonaldehyde. *Cancer Res.* **47**:360–363.

Foiles, P. G., Akerkar, S. A., and Chung, F.-L. (1989). Application of an immunoassay for cyclic acrolein deoxyguanosine adducts to assess their formation in DNA of Salmonella typhimurium under conditions of mutation by acrolein. *Carcinogenesis* **10**:87–90.

Gasparro, F. P., and Santella, R. M. (1988). Immunoassay of DNA damage. *Photochem. Photobiol.* **48**:321–328.

Groopman, J. D., Zhu, J. Q., Donahue, P. R., Pikul, A., Zhang, L. S., Chen, J. S., and Wogan, G. N. (1992). Molecular dosimetry of urinary aflatoxin–DNA adducts in people living in Guangxi Autonomous Region, People's Republic of China. *Cancer Res.* **52**:45–52.

Hochleitner, K., Thomale, J., Nikitin, A. Y., and Rajewsky, M. F. (1991). Monoclonal antibody-based, selective isolation of DNA fragments containing an alkylated base to be quantified in defined gene sequences. *Nucleic Acids Res.* **19**:4467–4472.

Hsieh, L. L., and Hsieh, T. T. (1993). Detection of aflatoxin B1–DNA adducts in human placenta and cord blood. *Cancer Res.* **53**:1278–1280.

Huitfeldt, H. S., Brandtzaeg, P., and Poirier, M. C. (1990). Reduced DNA adduct formation in replicating liver cells during continuous feeding of a chemical carcinogen. *Proc. Natl. Acad. Sci. USA* **8**:5955–5958.

Huitfeldt, H. S., Beland, F. A., Fullerton, N. F., and Poirier, M. C. (1994). Immunohistochemical and microfluorometric determination of hepatic DNA adduct removal in rats fed 2-acetylaminofluorene. *Carcinogenesis* **15**:2599–2603.

Lamola, A. A., and Yamane, T. (1967). Sensitized photodimerization of thymine in DNA. *Proc. Natl. Acad. Sci. USA* **58**:443–446.

Leadon, S. A., and Hanawalt, P. C. (1983). Monoclonal antibody to DNA containing thymine glycol. *Mutat. Res.* **112**:191–200.

Lee, B. M., and Strickland, P. T. (1993). Antibodies to carcinogen–DNA adducts in mice chronically exposed to polycyclic aromatic hydrocarbons. *Immunol. Lett.* **36**:117–123.

Leng, M., Sage, E., Fuchs, R. P. P., and Daune, M. P. (1978). Antibodies to DNA modified by the carcinogen N-acetoxy-N-2-acetylaminofluorene. *FEBS Lett.* **92**:207–210.

Matsunaga T., Mori T., and Nikaido, O. (1990). Base sequence specificity of a monoclonal antibody binding to (6–4) photoproducts. *Mutat. Res.* **235**:187–194.

Matsunaga, T., Hatakeyama, Y., Ohta, M., Mori, T., and Nikaido, O. (1993). Establishment and characterization of a monoclonal antibody recognizing the Dewar isomers of (6–4) photoproducts. *Photochem. Photobiol.* **57**:934–940.

Mitchell, D. L., and Nairn, R. S. (1989). The biology of the (6–4) photoproduct. *Photochem. Photobiol.* **49**: 805–819.

Mitchell, D. L., and Rosenstein, B. S. (1987). The use of specific radioimmunoassays to determine action spectra for the photolysis of (6–4) photoproducts. *Photochem. Photobiol.* **45**:781–786.

Mitchell, D.L., Adair, G. M., and Nairn, R. S. (1989). Inhibition of gene expression in a triplet-sensitized, UVB-irradiated plasmid transfected into rodent cells. *Photochem. Photobiol.* **50**:639–646.

Motzer, R. J., Reed, E., Perera, F., Tang, D., Shamkhani, H., Poirier, M. C., Tsai, W. Y., Parker, R. J., and Bosl, G. L. (1994). Platinum–DNA adducts assayed in leukocytes of patients with germ cell tumors measured by atomic absorbance spectrometry and enzyme-linked immunosorbant assay. *Cancer* **73**:2843–2852.

Nakatsuru, Y., Qin, X. S., Masahito, P., and Ishikawa, T. (1990). Immunological detection of in vivo aflatoxin B1–DNA adduct formation in rats, rainbow trout and coho salmon. *Carcinogenesis* **11**:1523–1526.

Newman, M. J., Weston, A., Carver, D. C., Mann, D. L., and Harris, C. C. (1990). Serological characterization of polycyclic aromatic hydrocarbon diolepoxide–DNA adducts using monoclonal antibodies. *Carcinogenesis* **11**:1903–1907.

Olivero, O. A., Huitfeldt, H., and Poirier, M. C. (1990). Chromosome site-specific immunohistochemical detection of DNA adducts in N-acetoxy-2-acetylaminofluorene-exposed Chinese hamster ovary cells. *Mol. Carcinog.* **3**:37–43.

Ozaki, K., Kato, T., Asamoto, M., Wild. C. P., Montesano, R., Nagao, S., Iwase T., Matsumoto K., and Tsuda, H. (1993). Decreased dimethylnitrosamine-induced O6- and N7-methyldeoxyguanosine levels correlate with development and progression of lesions in rat hepatocarcinogenesis. *Jpn. J. Cancer Res.* **84**:1245–1251.

Paleologo, M., van Schooten, F. J., Pavanello, S., Kriek, E., Zordan, M., Clonfero, E., Bezze, C., and Levis, A. G. (1992). Detection of benzo[a]pyrene-diol-epoxide–DNA adducts in white blood cells of psoriatic patients treated with coal tar. *Mutat. Res.* **281**:11–16.

Paules, R. S., Poirier, M. C., Mass, M. J., Yuspa, S. H., and Kaufmann, D. G. (1985). Quantitation by electron microscopy of the binding of highly specific antibodies to benzo[a]pyrene–DNA adducts. *Carcinogenesis* **6**: 193–198.

Plescia, O. J., Braun, W., and Palczuk, N. C. (1964). Production of antibodies to denatured deoxyribonucleic acid (DNA). *Proc. Natl. Acad. Sci. USA* **52**:279–285.

Poirier, M. C., Dubin, M. A., and Yuspa, S. H. (1979). Formation and removal of specific acetylaminofluorene–DNA adducts in mouse and human cells measured by radioimmunoassay. *Cancer Res.* **39**:1377–1381.

Poirier, M. C., Reed, E., Litterst, C. L., Katz, D., and Gupta-Burt, S. (1992). Persistence of platinum-ammine–DNA adducts in gonads and kidneys of rats and multiple tissues from cancer patients. *Cancer Res.* **52**:149–53.

Reardon, J. T., Nichols, A. F., Keeney, S., Smith, C. A., Taylor, J.-S., Linn, S., and Sancar, A. (1993). Comparative analysis of binding of human damaged DNA-binding protein (XPE) and Escherichia coli recognition protein (UvrA) to the major ultraviolet photoproducts: T[c,s]T, T[t,s]T, T[6-4]T, and T[dewar]T. *J. Biol. Chem.* **268**:21301–21308.

Roberts, D. W., Benson, R. W., Groopman, J. D., Flammang, T. J., Nagle, W. A., Moss, A. J., and Kadlubar, F. F. (1988). Immunochemical quantitation of DNA adducts derived from the human bladder carcinogen 4-amminobiphenyl. *Cancer Res.* **48**:6336–6342.

Roggeband, R., Van den Berg, P. T., Van der Wulp, C. J., and Baan, R. A. (1994). Detection of DNA adducts in basal and non-basal cells of the hamster trachea exposed to benzo(a)pyrene in organ culture. *J. Histochem. Cytochem.* **42**:1427–1434.

Rojas, M., Alexandrov, K., van Schooten, F. J., Hillebrand, M., Kriek E., and Bartsch, H. (1994). Validation of a new fluorometric assay for benzo[a]pyrene diolepoxide–DNA adducts in human white blood cells: Comparisons with ^{32}P-postlabeling and ELISA. *Carcinogenesis* **15**:557–560.

Rosenstein, B. S. (1984) Photoreactivation of ICR 2A frog cells exposed to solar UV wavelengths. *Photochem. Photobiol.* **40**:207–213.

Rosenstein, B. S. and Mitchell, D. L. (1987). Action spectra for the induction of pyrimdine(6–4)pyrimidone photoproducts and cyclobutane pyrimidine dimers in normal human skin fibroblasts. *Photochem. Photobiol.* **45**:775–781.

Santella, R. M., Weston, A., Perera, F. P., Trivers, G. T., Harris, C. C., Young, T. L., Nguyen, D., Lee, B. M., and Poirier, M. C. (1988). Interlaboratory comparison of antisera and immunoassays for benzo[a]pyrene-diol-epoxide-I-modified DNA. *Carcinogenesis* **9**:1265–1269.

Santella, R. M., Grinberg-Funes, R. A., Young, T. L., Dickey, C., Singh, V. N., Wang, L. W., and Perera. F. P (1992). Cigarette smoking related polycyclic aromatic hydrocarbon–DNA adducts in peripheral mononuclear cells. *Carcinogenesis* **13**:2041–2045.

Santella, R. M., Zhang, Y. J., Chen, C. J., Hsieh, L. L., Lee, C. S., Haghighi, B., Yang, G. Y., Wang, L. W., and Feitelson, M. (1993). Immunohistochemical detection of aflatoxin B1–DNA adducts and hepatitis B virus antigens in hepatocellular carcinoma and nontumorous liver tissue. *Environ. Health Perspect.* **99**:199–202.

Schoket, B., Phillips, D. H., Poirier, M. C., and Vincze, I. (1993). DNA adducts in peripheral blood lymphocytes from aluminum production plant workers determined by ^{32}P-postlabeling and enzyme-linked immunosorbent assay. *Environ. Health Perspect.* **99**:307–309.

Seiler, F., Kirstein, U., Eberle, G., Hochleitner, K., and Rajewsky, M. F. (1993). Quantification of specific DNA O-alkylation products in individual cells by monoclonal antibodies and digital imaging of intensified nuclear fluorescence. *Carcinogenesis* **14**:1907–1913.

Strickland, P. T. (1985). Immunoassay of DNA modified by ultraviolet radiation: A review. *Environ. Mutagen.* **7**:599–607.

van Schooten, F. J., Hillebrand, M. J., Scherer, E., den Engelse, L., and Kriek, E. (1992). Immunocytochemical visualization of DNA adducts in mouse tissues and human white blood cells following treatment with benzo[a]pyrene or its diol epoxide. A quantitative approach. *Carcinogenesis* **12**:427–433.

Vink, A. A., Berg, R. J., de Gruijl, F. R., Lohman, P. H., Roza, L., and Baan, R. A. (1993). Detection of thymine dimers in suprabasal and basal cells of chronically UV-B exposed hairless mice. *J. Invest. Dermatol.* **100**: 795–799.

Wani, A. A., and Arezina, J. (1991). Immunoanalysis of ultraviolet radiation induced DNA damage and repair within specific gene segments of plasmid DNA. *Biochim. Biophys. Acta* **1090**:195–203.

West, G. J., West, I. W.-L., and Ward, J. F. (1982). Radioimmunoassay of 7,8-dihydro-8-oxoadenine (8-hydroxyadenine). *Int. J. Radiat. Biol.* **42**:481–490.

Weston, A., Newman, M. J., Mann, D. L., and Brooks, B. R. (1990). Molecular mechanics and antibody binding in the structural analysis of polycyclic aromatic hydrocarbon-diol-epoxide–DNA adducts. *Carcinogenesis* **11**:859–864.

Yang, X.Y., Delohery, T., and Santella, R. M. (1988). Flow cytometric analysis of 8-methoxypsoralen–DNA photoadducts in human keratinocytes. *Cancer Res.* **48**:7013–7017.

Chapter 7

Monoclonal Antibody-Based Quantification and Repair Analysis of Specific Alkylation Products in DNA

Jürgen Thomale, Jörg Engelbergs, Frank Seiler, and Manfred F. Rajewsky

1. INTRODUCTION

Exposure of cells to DNA-reactive agents (exogenous and endogenous carcinogens and mutagens; cancer chemotherapeutic compounds) results in a variety of potentially mutagenic and/or cytotoxic modifications of genomic DNA (Singer and Grunberger, 1983; Hemminki and Ludlum, 1984; Rajewsky, 1989; Loeb, 1989). The molecular nature of specific DNA lesions (e.g., carcinogen–DNA adducts, DNA modifications caused by UV light or oxygen radicals) is dictated by the structure and chemical reactivity of the causative agent and, therefore, represents a "genomic fingerprint" (Singer and Grunberger, 1983; Basu and Essigmann, 1988; Rajewsky, 1989). The analysis of agent-specific DNA modifications is of considerable importance for the molecular epidemiology of carcinogen exposure as well as for the pre- and intratherapeutic dosimetry of exposure to anticancer agents (see, e.g., Umbenhauer *et al.*, 1985; Huh *et al.*, 1989; Groopman *et al.*, 1991; Müller *et al.*, 1994).

Specific DNA lesions are not formed at random throughout the genome of target cells; rather, the probability of their formation depends on DNA sequence and conformation, and on the transcriptional status of target genes (Nehls and Rajewsky, 1985a,b; Sendowski and Rajewsky, 1991). Moreover, the conversion of premutational DNA lesions into mutations through DNA replication is counteracted by a complex network of DNA repair pathways with differing, partly overlapping lesion specificity and often preferential action on transcriptionally active genes (Rajewsky, 1989; Bohr *et al.*, 1989; Friedberg *et al.*, 1995). Proteins involved in DNA repair are

Jürgen Thomale, Jörg Engelbergs, Frank Seiler, and Manfred F. Rajewsky • Institute of Cell Biology (Cancer Research), West German Cancer Center Essen, University of Essen Medical School, D-45122 Essen, Germany. *Present address of F.S.*: Institute of Hygiene and Occupational Medicine, University of Essen Medical School, D-45122 Essen, Germany.

Technologies for Detection of DNA Damage and Mutations, edited by Gerd P. Pfeifer, Plenum Press, New York, 1996.

differentially expressed in distinct cell types and—in an unpredictable manner—in malignant cells and their precursors (Thomale *et al.*, 1994b). Therefore, while defined DNA lesions may serve as qualitative indicators of exposure to specific agents, their use as quantitative indicators (dosimetry) requires prior determination of the kinetics of lesion-specific repair in a given type of target cell and in defined gene sequences. The latter information *per se* is of particular importance to human cancer therapy, where the resistance of individual tumors or leukemias toward particular cytotoxic agents may be linked to the capacity of cells for repair of critical DNA lesions in specific genes.

The identification and quantification of very low concentrations of defined DNA lesions, and the measurement of their repair kinetics in multicell samples, individual cells, or specific gene sequences, require ultrasensitive methodology. This laboratory has chosen an immunoanalytical approach based on the development of lesion-specific, high-affinity monoclonal antibodies (MAbs). The first of these MAbs were produced in 1979 (Rajewsky *et al.*, 1980). A large collection of MAbs specific for different DNA lesions has since been established and is being systematically expanded (Müller and Rajewsky, 1981; Adamkiewicz *et al.*, 1982, 1984, 1985, 1986; Eberle, 1989; Eberle *et al.*, 1989, 1990; Glüsenkamp *et al.*, 1993). The procedures used for the synthesis of haptens (e.g., alkyl-ribonucleosides) and their coupling to carrier proteins (e.g., keyhole limpet hemocyanin) for immunization, for the generation of mouse × mouse or rat × rat hybridomas, and for the characterization of MAbs, have been published (Müller and Rajewsky, 1981; Adamkiewicz *et al.*, 1982, 1984; Eberle *et al.*, 1989, 1990; Glüsenkamp *et al.*, 1993).

MAbs presently contained in our collection (antibody affinity constants of up to 3×10^{10} liters/mole) are directed specifically against DNA *O*- or *N*-alkylation products as well as etheno adducts, including O^6-alkyl-2′-deoxyguanosines, O^2-alkyl-2′-deoxythymidines, O^4-alkyl-2′-deoxythymidines, 3-alkyladenines, 1,N^6-etheno-2′-deoxyadenosine, and 3,N^4-etheno-2′-deoxycytidine (Rajewsky *et al.*, 1980; Adamkiewicz *et al.*, 1985; Eberle, 1989; Eberle *et al.*, 1989, 1990; Glüsenkamp *et al.*, 1993). While *O*- and *N*-alkylation products in cellular DNA result from exposure to members of the large class of *N*-nitroso compounds (carcinogens, mutagens, cancer therapeutic agents), etheno adducts result from the reaction of DNA with metabolites of vinyl halides.

Based on the use of MAbs, various immunoanalytical methods have been developed that enable us (1) specifically to detect and quantify MAb-binding DNA lesions in the femtomole to attomole range and (2) to measure the kinetics of elimination (repair) of those lesions from DNA (Müller and Rajewsky, 1978, 1980; Müller, 1983; Huh and Rajewsky, 1986; Nehls *et al.*, 1988; Rajewsky, 1989; Nehls and Rajewsky, 1990; Thomale *et al.*, 1990; Satoh *et al.*, 1991; Kang *et al.*, 1992). In this chapter, we describe four of these methods: (1) competitive radioimmunoassay (RIA) following separation by HPLC of modified 2′-deoxynucleosides from DNA digests (Adamkiewicz *et al.*, 1982); (2) immuno-slot-blot (ISB) for quantification of lesions in genomic DNA (Nehls *et al.*, 1984); (3) immunoassay for quantification of DNA lesions in known sequences/genes (immunoaffinity-quantitative PCR) (Hochleitner *et al.*, 1991; Thomale *et al.*, 1994a); and (4) immunocytological assay (ICA) for visualization and quantification of DNA lesions in individual cells (digital imaging of intensified nuclear fluorescence) (Adamkiewicz *et al.*, 1985; Seiler *et al.*, 1993; Thomale *et al.*, 1994b).

2. IMMUNOANALYTICAL METHODOLOGY

2.1. HPLC-Competitive Radioimmunoassay (HPLC-RIA)

(Quantification of specific alkyl-2′-deoxynucleosides in DNA hydrolysates)

2.1.1. Principle

Chromatographic separation of specific alkyl-2′-deoxynucleosides from enzymatically digested DNA and quantification by competitive RIA (see Fig. 7.1).

2.1.2. Materials

Enzymes (Boehringer-Mannheim): DNase I (EC 3.1.21.1), alkaline phosphate (EC 3.1.3.1), snake venom phosphodiesterase (EC 3.1.15.1).

HPLC apparatus (gradient system), integrating UV-detector, fraction collector, reversed-phase column (e.g., Radial-Pak BNVC18 4μ, Waters-Millipore).

Normal and modified 2′-deoxynucleosides as calibration standards and RIA inhibitors, ^{3}H-labeled alkyl-2′-deoxynucleosides as HPLC standards and RIA tracers.

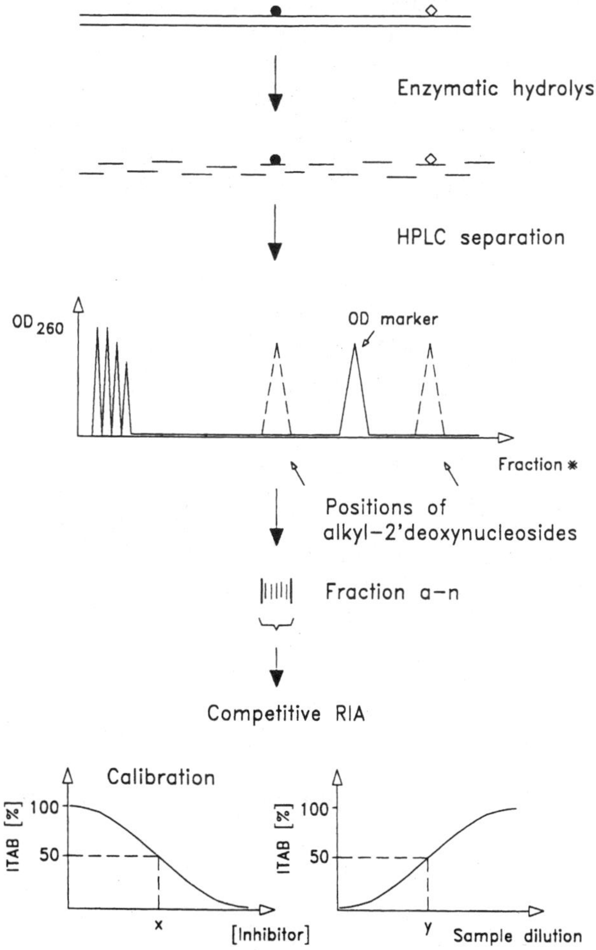

Figure 7.1. Steps used in the HPLC-RIA method to quantify specific DNA alkylation products. HPLC, high-performance liquid chromatography; RIA, radioimmunoassay employing monoclonal antibodies; ITAB, inhibition of tracer–antibody binding; x, concentration of inhibitor resulting in 50% reduction of tracer–antibody binding; y, dilution of pooled and reconstituted fractions a–n leading to 50% ITAB.

2.1.3. Procedure

1. Digest isolated genomic DNA (1 mg/ml TE buffer) to 2′-deoxynucleosides by successive addition of DNase I (per ml of DNA solution: 20 μl of a 10 mg/ml stock DNase I solution prepared in 10 mM Tris-HCl, pH 7.2, and 5 mM $MgCl_2$); incubation at 37°C for 30 min, followed by phosphodiesterase (100 μl/ml from a 1 mg/ml stock solution in 50 mM Tris-HCl, 50 mM $MgSO_4$, 50 mM $MgCl_2$)/alkaline phosphatase (20 μl/ml from a 400 U/ml stock solution in H_2O); incubation for 2 hr at 37°C.
2. Check for complete digestion, the absence of RNA-derived contamination, and precise quantification of normal 2′-deoxynucleosides using HPLC: Apply small aliquots of hydrolysates (equivalent to 10 μg of DNA) to a reversed-phase (C_{18}) HPLC column and separate the constituents using a linear gradient with increasing concentrations of methanol in buffer A (0.1 M ammonium formate, pH 5.0). Use standard mixtures of 2′-deoxynucleosides (ATCG) for identification and calibration in a separate HPLC run.
3. As an internal control for elution times, add an OD marker (e.g., 0.5 nmole of caffeine) to the residual part of DNA hydrolysates and fractionate the samples on a reversed-phase HPLC column (loading in buffer A; see step 2; flow: 1 ml/min, temp.: 40°C). Elute normal and alkylated nucleosides with linearly increasing concentrations of methanol (2–25% within 25 min) in buffer A. Collect fractions around the position (± one fraction) of nucleosides to be measured. These elution positions are predetermined in a separate calibration run using trace amounts of alkyl-2′-deoxynucleosides (see Notes). The reproducibility of elution times is controlled by the position of the OD marker.
4. Pool relevant fractions, dry the sample in a speed-vac concentrator, and reconstitute in a total volume of 100 μl H_2O.
5. Determine the content of specific alkyl-2′-deoxynucleosides by competitive RIA (Müller and Rajewsky, 1978, 1980) using high-affinity MAbs (e.g., MAb ER-6 for O^6-ethyl-2′-deoxyguanosine; Rajewsky *et al.*, 1980) and radiolabeled nucleosides (e.g., [^{3}H]-EtdGuo, specific activity > 20 Ci/mmole) as tracers: Prepare serial dilutions of reconstituted HPLC fractions (see 4) in TBS (10 mM Tris-HCl, pH 7.5; 140 mM NaCl; 0.02% NaN_3), mix 50 μl of each dilution (inhibitor) with 25 μl of tracer (adjusted to 100 cpm/μl with TBS) and 25 μl of MAb solution. The MAb solution is diluted with RIA buffer (TBS containing 2% w/v BSA and 0.2% w/v rabbit IgG) to a concentration binding 50% of the tracer molecules (control by titration).
6. Incubate the RIA mixture at 4°C for 2 hr and separate MAb-bound tracer from free tracer by precipitation: Add 100 μl of saturated ammonium sulfate solution to the mix, incubate for a further 10 min at room temperature (RT), and spin down the precipitate in a microfuge (5 min; 10,000 rpm).
7. Remove 150 μl of the supernatant and determine ^{3}H dpm in this aliquot by liquid scintillation counting.
8. Calculate ITAB (inhibition of tracer–antibody binding) values as described by Müller and Rajewsky (1980), and read the molar concentration of alkyl-2′-deoxynucleosides from a standard calibration curve established in parallel using authentic samples of known concentrations.

2.1.4. Notes

Step 3: To avoid contamination of the HPLC system with alkylated nucleosides, the elution positions must be determined using radiolabeled standards rather than the authentic compounds detected by UV absorption.

2.2. Immuno-Slot-Blot (ISB)

(Quantification of specific alkylation products in genomic DNA).

2.2.1. Principle

Reaction of lesion-specific MAbs with single-stranded DNA slot-blotted onto nitrocellulose or nylon membranes; detection of DNA bound MAbs with radiolabeled or enzyme-coupled second antibodies.

2.2.2. Materials

(Genomic) DNA to be analyzed (minimal amount, 10 μg)

DNA standards containing known concentrations of specific alkylation products (determined by HPLC-RIA)

Slot-blot apparatus

Vacuum oven or UV cross-linker

Suitable, purified MAb, binding to a given lesion in single-stranded (ss) DNA with high specificity

Second antibody to be detected/quantitated either by radioactive label (^{125}I; ^{35}S) or by chemiluminescence (e.g., ECL, Pharmacia, or RadFree, Schleicher & Schuell)

Scanning densitometer for X-ray films or luminescence imager.

2.2.3. Procedure

1. Ensure that DNA samples are free of protein (by UV spectrum analysis) and RNA (e.g., by enzymatic hydrolysis-HPLC analysis, see Method 1)
2. Fractionate DNA by sonication or restriction to a size of 1–10 kb (this facilitates exact pipetting) and dilute samples to a final 30 μg DNA/ml in concentration of TE buffer.
3. Prepare serial dilutions of DNA samples and of standard DNA using nonadducted DNA of the same origin and concentration (30 μg/ml).
4. Denature DNA by heating samples for 10 min at 100°C (water bath or heating block), transfer tubes to ice water and add an equal volume of ice-cold ammonium acetate (2 M).
5. Slot-blot 200 μl of each dilution (total amount of DNA, 3 μg) to a suitable membrane (nitrocellulose or nylon) presoaked in 1 M ammonium acetate. Wash slots carefully with ammonium acetate (1 M), remove membrane, soak in 5× SSC for 5 min, air-dry, and fix DNA by vacuum heating (80°C; 2 hr) or UV cross-linking.
6. Presoak membrane in incubation buffer [PBS containing 0.5% (w/v) casein hydrolysate and 0.1% (w/v) deoxycholate) for 2 hr and incubate with lesion-specific MAb with constant agitation for 1 hr at RT [e.g., MAb EM-21 (Eberle, 1989) for O^6-alkylguanines, concentration 0.1 μg/ml incubation buffer; titrate for each batch of MAb].
7. Wash membrane 3 × 5 min in wash buffer (PBS with the addition of 160 mM NACl and 0.1% Triton X-100); incubate with radiolabeled second antibody (e.g., ^{125}I-labeled anti-mouse Ig $F(ab)_2$, 1 hr, RT) or follow manufacturer's instructions for nonradioactive detection systems (e.g., chemiluminescence).
8. Wash membrane 3 × 5 min as above; wrap in cellophane, expose to X-ray film, and quantitatively evaluate the resulting autoradiographs using a scanning densitometer.

2.2.4. Notes

Blot sample dilutions and calibration standards including untreated control DNA on the same membrane.

Make sure that DNA fragments of the size to be analyzed are efficiently (>90%) trapped as single strands on the chosen type of membrane; control by slot-blotting 3 μg of labeled (e.g., with ^{32}P or biotin) ssDNA fragments.

2.3. Sequence-Specific ALISS Assay

(Quantification of "Adduct Levels In Specific Sequences")

2.3.1. Principle

Immunoaffinity isolation of DNA fragments containing a specific MAb-binding lesion, and subsequent amplification of known (gene) sequences in retained DNA fragments by quantitative PCR (see Fig. 7.2).

2.3.2. Materials

External standard DNA (ESD) for titration of antibody binding: plasmid DNA (5–10 kb) containing, on average, one specific DNA adduct/molecule (see Additional steps, A).

Internal heterologous standard DNA (ISD): plasmid DNA (5–10 kb) carrying, e.g., a bacterial or yeast gene and containing, on average, one specific DNA adduct/molecule (see Additional steps, B). ISD and ESD should be different to avoid cross-contamination.

Suitable lesion-specific antibodies, preferably highly purified MAbs or other high-affinity lesion-specific DNA binding proteins (see Notes, A).

Oligonucleotides as PCR primers for amplification of gene sequences to be analyzed and for hybridization to internal sequences of PCR-amplified genes.

Filter device; liquid scintillation counter; DNA thermal cycler; scanning densitometer or phosphor imager; slot-blot apparatus; speed-vac concentrator.

All plastic materials must be siliconized and autoclaved. To prevent contamination throughout the procedure, only filter tips should be used for pipetting.

Preparation of ISD, preparation of genomic DNA samples, preparation of PCR reactions, and handling of PCR products must each be carried out in separate rooms.

2.3.3. Procedure

Preparation of DNA samples

1. High-molecular-weight genomic DNA (>25 kb) is isolated from exposed cells or tissues. Isolates must be free of protein and RNA contamination (controlled, e.g., by enzymatic hydrolysis to mononucleosides and HPLC analysis).
2. DNA is cleaved with suitable restriction endonuclease(s) to create specific gene fragments of desired length (minimum ~200 bp). The length of specific gene fragments is controlled by Southern analysis.
3. DNA is purified by phenol/chloroform extraction and quantitated by UV spectroscopy.
4. The molar content of specific lesions (e.g., O^6-EtdGuo) in restricted genomic DNA is measured by ISB analysis (see Method 2).

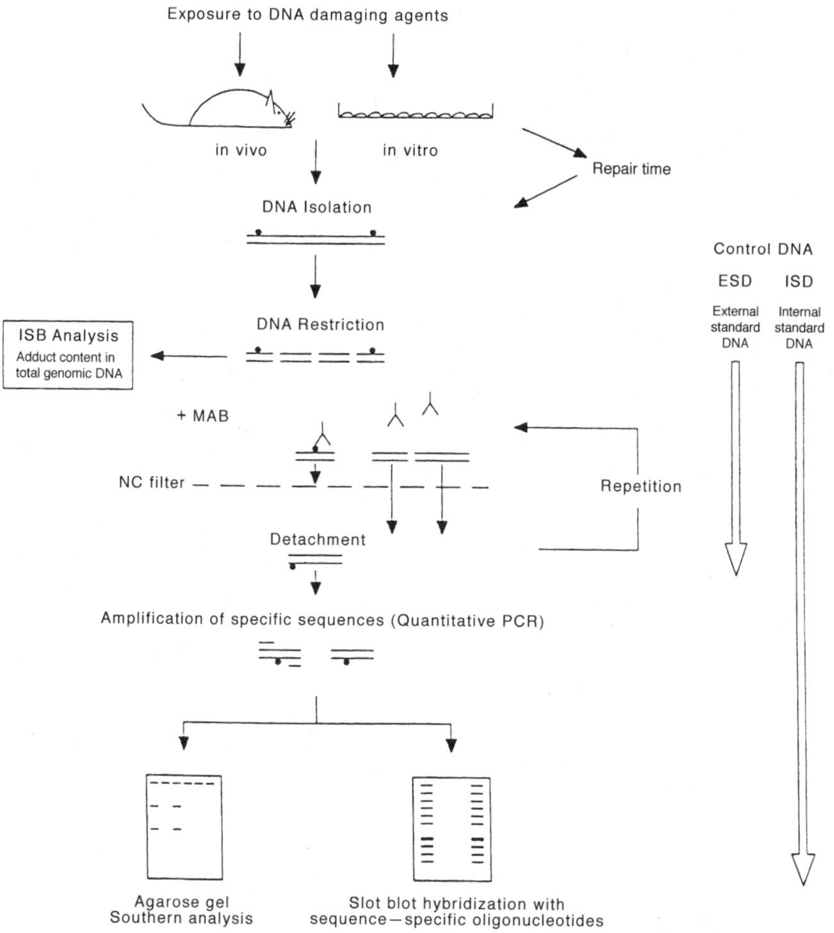

Figure 7.2. Flow diagram of the sequence (gene)-specific assay to measure defined DNA lesions in known gene sequences. ISB, immuno-slot-blot, see Method 2; MAb, monoclonal antibodies binding to defined lesions in dsDNA with high affinity and specificity; NC, nitrocellulose membrane.

Isolation of DNA fragments containing a specific DNA lesion

5. A fixed amount of ISD is added to each sample of restricted DNA (e.g., 50 fg of ISD per 5 μg of genomic DNA). Samples are prepared in duplicate.
6. DNA is incubated with lesion-specific MAb [e.g., 30 μg of Mab ER-6 (Rajewsky *et al.*, 1980) for O^6-EtdGuo] in 100 μl of STE buffer (50 mM Tris-HCl, pH 7.5; 100 mM NaCl, 1 mM EDTA) for 45 min at RT (see Notes, A).
7. MAb-complexed DNA fragments are separated from bulk DNA by passing samples through nitrocellulose (NC) membranes (BA 85, Schleicher & Schuell; 2.5 × 2.5 cm; presoaked in STE buffer) under slight vacuum. Membranes are washed subsequently with 2 × 1 ml STE buffer.
8. To recover DNA from retained MAb–DNA complexes, wet membranes are transferred to 2-ml reaction tubes (DNA side facing inside). The DNA is eluted by adding 300 μl

BTE buffer (5 mM Tris-HCl, pH 7.5; 0.1 mM EDTA, 5% v/v *n*-butanol) and constant rolling agitation for 90 min at RT. The wash is then transferred to a new 2-ml tube.

9. DNA extraction (step 8) is repeated twice with new buffer. Membranes are finally rinsed with 100 μl BTE buffer and the combined filter washes of each sample are dried in a speed-vac concentrator.
10. The eluted DNA is reconstituted in H_2O, steps 6–9 are repeated twice, and DNA is finally dissolved in 70 μl H_2O.

Purification of isolated DNA fragments

11. To eliminate PCR inhibitors (e.g., membrane components, detergents) the DNA is applied to a small phenyl Sepharose column (Pharmacia; bed volume 300 μl; equilibrated with H_2O) and eluted with 80 μl H_2O.
12. The complete eluates are dried in a speed-vac concentrator and finally redissolved in 50 μl H_2O.

Amplification of specific gene sequences in immunoaffinity-isolated DNA fragments by multiplex quantitative PCR (qPCR)

The amount of specific gene sequences in MAb-isolated DNA fragments is quantitated after amplification by qPCR. Appropriate conditions for quantitative coamplification have to be optimized for each specific set of gene sequences including ISD.

13. Aliquots (25 μl) of each purified DNA sample and of standard genomic DNA used for external calibration (see Additional steps, D) are denatured for 10 min at 95°C (the remaining DNA is stored at −20°C for further assays).
14. PCR is performed in 50 μl of PCR buffer containing 50 mM KCl; 10 mM Tris-HCl, pH 8.3; 1.5 mM $MgCl_2$; 0.25 mM dNTP mixture; 0.4 μM each of the upstream and downstream primers (for target sequences and ISD sequence), 1 unit of Taq DNA polymerase (e.g., Ampli Taq, recombinant Taq DNA polymerase, Perkin–Elmer), and the denatured DNA sample.
15. The thermal amplification profile is 0.5 min at 95°C; primer annealing at optimized hybridization temperature for 1 min; primer extension at 72°C for 2 min. The PCR reaction is stopped in the exponential phase of DNA amplification (approximately 25 cycles must be determined for each coamplification). A final cycle with 10 min extension time is included.

Analysis of PCR products

16. Aliquots (e.g., 10 μl) of PCR samples are denatured by addition of 190 μl NaOH (400 mM containing 2.15 mM EDTA) and transferred onto nylon membranes (e.g., NY 13 N, Schleicher & Schuell; presoaked in 5 × SSC buffer) by slot-blot procedure. Slots are washed with 200 μl of 5 × SSC buffer. Membranes are incubated in 5 × SSC buffer for 5 min at RT, air-dried, and the DNA is UV cross-linked to the membrane.

    Identical blot membranes are prepared for hybridization with ISD sequence and for each genomic sequence to be analyzed.
17. Blot membranes are incubated in 6 × SSC, 5 × Denhardt's solution, 0.1% w/v SDS, 0.1 mg/ml BSA, for > 2 hr at 56°C and hybridized with 6 pmole of the appropriate ^{32}P-labeled 20-mer oligonucleotides specific for each target gene sequence(s) and the ISD sequence, respectively, for > 12 hr at 56°C.
18. Membranes are washed once in 6 × SSC, 0.1% w/v SDS, for 10 min at 56°C, once in 2 × SSC, 0.1% w/v SDS, for 10 min at 56°C, and exposed to X-ray films at −70°C.

19. Autoradiographs are evaluated quantitatively by scanning densitometry. An external calibration curve is established by plotting the input amount of genomic external calibration DNA against the corresponding hybridization signals of the PCR products of target gene sequence(s) normalized to the signals of the ISD sequence.
    The signals obtained from target gene sequence(s) in MAb-isolated DNA fragments are normalized to the signals of the coisolated and coamplified ISD sequence. Corresponding amounts of template DNA (= amounts of specific gene fragments trapped by MAbs) are quantitated using the external calibration curve. The adducted nucleotide/normal counterpart molar ratios (e.g., O^6-EtdGuo/dGuo) in specific gene fragments are calculated for each DNA sample.

Additional steps

A. Preparation of ESD

1. The plasmid DNA is linearized with a suitable restriction enzyme to generate recessive 3′-ends, and purified by phenol/chloroform extraction.
2. The plasmid is treated with a DNA reactive agent to introduce, on average, one specific adduct per molecule (for O^6-EtdGuo, e.g., by *in vitro* exposure to EtNU, see Hochleitner *et al.*, 1991). The molar content of the respective modified deoxynucleotide is determined by competitive RIA (see Method 1).
3. Recessive 3′-ends are filled with [^{35}S]-2′-deoxyadenosine-5′-(α-thio)-triphosphate (specific radioactivity, > 1000 Ci/mmole) and dNTPs using the Klenow fragment of DNA polymerase I.
4. The reaction mixture is resuspended in 200 μl of 200 mM NaCl/10 mM Tris-Cl, pH 8, and applied to a DE-52 column (300-μl bed volume). The column is washed with 200 mM NaCl in 10 mM Tris-Cl, pH 8, to remove nonincorporated radiolabeled, and ^{35}S-labeled DNA is eluted with 1 M NaCl in 10 mM Tris-Cl, pH 8.

B. Preparation of ISD

ISD is prepared as described above (additional steps A1 and A2). It should only be handled in *very low* concentrations (e.g., 10 fg/μl solution) to avoid air contamination.

C. Monitoring the isolation procedure of lesion-containing DNA fragments.

The efficiency MAb binding to lesion-containing DNA fragments, the separation of MAb–DNA complexes on NC membranes, and the reelution of trapped DNA are monitored in parallel by using ^{35}S-labeled ESD (~8 ng/sample, ~5×10^3 cpm). Lesion-free ^{35}S-labeled ESD is used to control for nonspecific binding of MAbs to DNA. Samples without MAb are used as a control for nonspecific binding of ^{35}S-labeled adducted or nonadducted ESD to the membrane.

D. Preparation of external calibration standards of genomic DNA for qPCR.

1. High-molecular-weight genomic DNA (isolated from the same type of cells as the DNA to be analyzed) is cleaved by the same restriction enzyme(s) as used in Procedure, step 2.
2. A fixed amount of ISD (50 fg) is added to each sample of a serial dilution of restricted DNA (e.g., 0.1–200 ng of DNA; the dilution range should be correlated to the amount of DNA expected in MAb-isolated fragments).

2.3.4. Notes

A. Selective isolation of DNA fragments containing a specific MAb-binding DNA lesion

The specificity of the ALISS assay is based on MAbs with a high affinity and specificity for a given modified component of dsDNA and very low cross-reactivity with untreated DNA. This requires extensive purification of MAbs from other DNA-binding proteins contained in hybridoma culture supernatants by ion-exchange FPLC, immunoaffinity chromatography, or other suitable methods.

The specificity of antibody binding to lesion-containing DNA fragments must be optimized for each MAb by varying buffer conditions (ionic strength, pH), antibody concentration, or incubation time. It can be monitored by a filter binding assay using radiolabeled ESD (see Additional steps, C).

Quantitative immunoprecipitation of MAb–DNA complexes instead of membrane-trapping and -elution of DNA has also been used to isolate DNA fragments containing specific lesions (Denissenko *et al.*, 1994).

B. Calibration

The validity of the method can be ascertained by analyzing serial dilutions of genomic DNA with known concentrations of a given lesion. O^6-alkylguanine levels are determined using restricted DNA, isolated from cells exposed to an alkylating *N*-nitroso compound, by ISB. This DNA is serially diluted with lesion-free DNA of the same origin and concentration. Samples containing the lesion at molar ratios between 10^{-5} and 10^{-8} are supplemented with a fixed amount of ISD. After isolation of lesion-containing DNA fragments by complexing with the respective MAb target gene/ISD sequences are coamplified by PCR. In this "artificial repair kinetic" a linear correlation should result between the molar lesion concentration in genomic DNA and the amount of amplified PCR products normalized to the ISD signal.

2.4. Immunocytological Assay on Individual Cells (ICA)

(Detection and quantification of specific alkylation products in the nuclear DNA of individual cells)

2.4.1. Principle

Reaction of lesion-specific MAbs with nuclear DNA of fixed single-cell preparations or tissue sections; detection of bound MAbs by fluorescence-labeled second antibodies and quantification by digital imaging of electronically intensified fluorescence signals (see Fig. 7.3).

2.4.2 Materials

Suitable purified lesion-specific MAbs

Anti-(lesion-specific MAb) antibodies, preferentially Fab or $F(ab)_2$ fragments, labeled with a fluorescent dye, preferentially a rhodamine derivative (e.g., TRITC, Texas Red, LRSC) because of its good fading stability

High-performance fluorescence microscope equipped with a 100-W mercury arc lamp

High-sensitivity video camera system

Image analysis system with a minimum two fluorescence parameters (DNA stain and antibody fluorescence)

Fluorescence calibration standard (e.g., fluorescence-labeled beads, as used for flow cytometry, or uranyl glass slides; Corning Glass Works)

2.4.3 Procedure

Staining of DNA adducts

1. Prepare cryostat tissue sections or single-cell preparations (e.g., smears or cytospins) on slides and air-dry at 4°C. If staining is not performed immediately, cell preparations may be stored at <-60°C for up to 1 year.

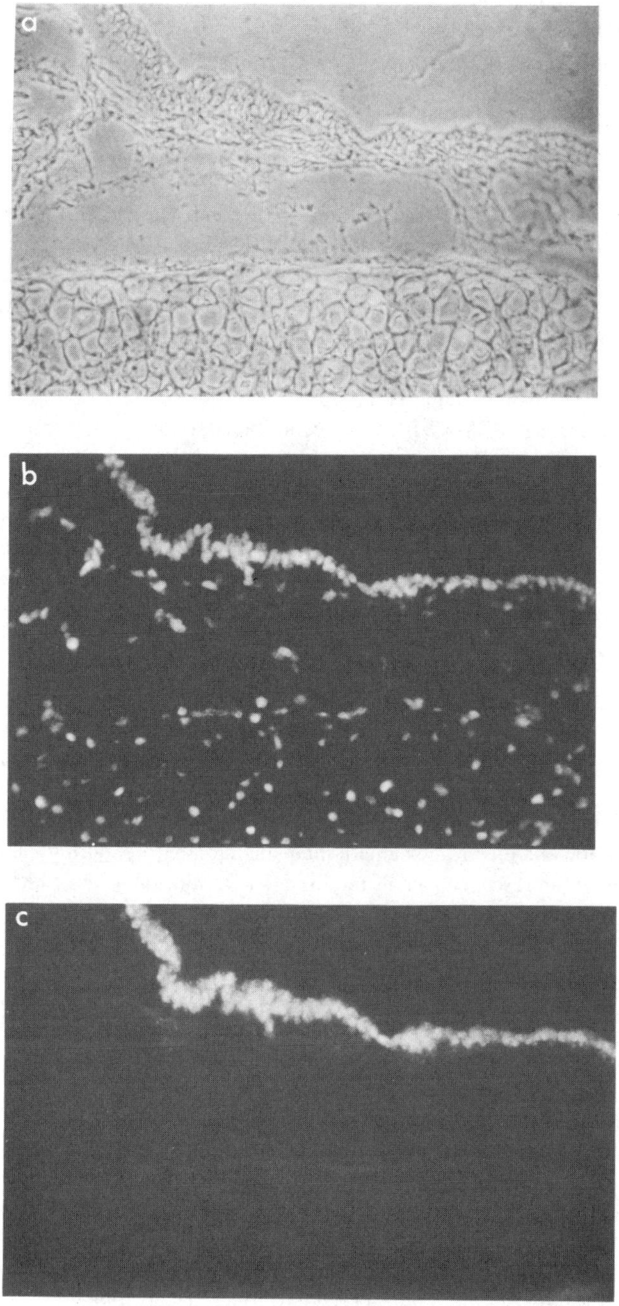

Figure 7.3. Immunocytological staining of O^6-ethyldeoxyguanine in nuclear DNA of hamster tracheal epithelial cells. Hamsters were treated with a single application of DEN (100 μg/g body wt). Cryosections of tracheal tissue were stained according to method 4 (ICA). (a) Phase contrast; (b) DNA staining with DAPI (blue fluorescence); (c) O^6-EtdGuo staining with MAb ER-17 and a second antibody labeled with TRITC (red fluorescence).

2. Treat cells with absolute methanol for 15 min at 4°C (to fix and permeabilize cellular membranes), and rehydrate cells in 2× SSC (0.3 M NaCl, 0.03 M trisodium citrate, pH 7.2) for 15 min at RT.
3. Overlay slides with RNA digestion solution (200 μg/ml RNase A, Sigma; 50 U/ml RNase T1, Boehringer-Mannheim, in 2× SSC) and incubate for 1 hr at 37°C in a moist atmosphere.
4. Preincubate slides in NaCl solution (0.14 M) for 15 min at 0°C and denature cellular DNA by alkali treatment (70 mM NaOH, 0.14 M NaCl, 40% ethanol) for 5 min at 0°C (precise incubation time and temperature are critical).
5. Neutralize and wash with PBS (containing Ca^{2+} and Mg^{2+})/1% BSA for 5 min at 0°C, and overlay slides with blocking solution (PBS/20% BSA) for 20 min at RT.
6. Discard blocking solution and overlay slides with DNA lesion-specific MAb solution [e.g., MAb ER-17 (Eberle, 1989; Seiler *et al.*, 1993) for O^6-EtdGuo; MAb concentration must be titrated for each new batch]. High-affinity MAbs can be diluted down to 0.2 μg/ml PBS/1% BSA. Incubate slides for 16 hr at 4°C in a moist atmosphere.
7. Wash slides 3 × 10 min with 150 ml of PBS/1% BSA, overlay with fluorochromed second antibody solution in PBS/1% BSA (concentration depends on first antibody, and must be titrated); incubate for 45 min at 37°C in a moist atmosphere.
8. Wash slides 3 × 10 min with 150 ml of PBS and incubate with a DNA-staining solution [e.g., 0.3 mM DAPI (4,6-diamidino-2-phenylindole) in PBS for 10 min at RT].
9. Wash slides for 2 min in PBS and cover with mounting medium [PBS; 20% glycerol; 10% polyvinyl alcohol, MS: 30,000; 0.03 M DTE (1,4-dithioerythreitol, nontoxic) or other water-soluble, antifading mounting medium].

Quantification of DNA adducts

In order to obtain reliable averages of homogeneous cell populations and statistical evaluation, at least 100 individual nuclei should be measured for both DNA content and antibody fluorescence signals using an image analysis system. Antibody-derived signals are corrected for DNA content.

To achieve comparable results from different samples, calibration standards must be used to adjust lamp excitation and electronic camera intensification before, during, and after measurements. Calibration of fluorescence signals required to determine absolute adduct concentrations in nuclear DNA is achieved by performing ISB or RIA analyses (see Methods 2 or 1) of DNA isolated from aliquots of the same cell samples, in parallel to ICA.

2.4.4. Notes

Screening for lesion-specific MAbs suitable for ICA is only possible by using the ICA procedure itself.

For multiparameter immunostaining, direct labeling of the primary antibodies with fluorescent dyes with different emission spectra may be necessary to avoid crossbinding of second antibodies.

Specific attention should be given to the optimization of denaturation steps for different types of cells/tissues and different MAbs.

Do not use cross-linking fixatives such as formaldehyde for tissue preservation.

Avoid drying throughout the entire procedure following fixation.

3. DISCUSSION

We have described four MAb-based analytical techniques for the detection and quantification of specific structural modifications of genomic DNA at the level of bulk DNA, known

Table I
Comparison of Different Immunoanalytical Quantification Methods for DNA Alkylation Products (values for O^6-EtGua are presented as examples)

Method	Sensitivity	Detection limit[a] (O^6-EtG/G in DNA)	Specificity in samples with known/unknown exposure		Some typical applications
HPLC-RIA	50 fmole	1×10^{-7}	High	High	Molecular epidemiology; quantification of absolute damage concentrations
ISB	1 fmole	1×10^{-7}	High	Moderate[b]	Screening of DNA samples (epidemiological studies); analysis of formation and repair in experimental cell systems
ALISS	—	0.5×10^{-7}	High	Moderate[b]	Formation of specific DNA lesions and their repair in known DNA sequences; screening for lesion "hotspots"
ICA	1×10^{-21} mole	1×10^{-7}	High	Moderate[b]	Analysis of DNA adduct formation and repair in specific types of cells, tissue sections, and in small bioptic cell samples

[a]Calculation of the detection limit is based on sample sizes of: (1) HPLC-RIA, 1 mg DNA; (2) ISB, 5 μg DNA; (3) ALISS assay, 5 μg DNA; (4) ICA, 100 cells. The input of DNA can be augmented for RIA and ALISS analysis.
[b]The selectivity of the assay is based on one criterion (MAb binding) only; this restricts the specificity if structurally similar, cross-reactive DNA modifications are present in the samples. The parallel application of lesion-specific MAbs exhibiting different cross-reactivity spectra (Eberle, 1989) can further improve the specificity.

sequences (genes), and the nuclear DNA of individual cells (see Table I). This immunoanalytical methodology successfully exploits the exceptional power of antibodies to recognize subtle alterations of molecular structure, and thus to distinguish different modified DNA components from each other and from their normal, unaltered counterparts.

The high affinity and specificity of the used MAbs, together with their individual characteristics of epitope recognition, have made it possible to quantify specific DNA lesions at concentrations as low as 1 in 10^7–10^8 of its normal counterpart molecules. This high degree of sensitivity is a prerequisite for the analysis of specific DNA damage related to environmental or endogenous exposure, for target cell dosimetry of DNA-reactive anticancer agents, and for determining the kinetics of elimination (repair) of specific DNA lesions at the level of individual cells or known genes.

It appears necessary to expand the present repertoire of MAbs by developing antibodies specific to other DNA lesions relevant to carcinogenesis and cancer chemotherapy, and to screen for further MAbs that are suitable for ICA and the sequence/gene- specific immunoaffinity assay. Moreover, efforts may be directed toward combining MAb technology with high-resolution methods for mapping DNA adducts in defined gene sequences at the single nucleotide level (Tornaletti *et al.*, 1993; Pfeifer *et al.*, 1993; Tornaletti and Pfeifer, 1994).

Acknowledgments. We gratefully acknowledge the contributions to this work made by the Laboratory Unit "Monoclonal Antibodies" of this institute (Dr. P. Lorenz), by Dr. U. Kirstein, Dr. P. Nehls, Dr. K.-H. Glüsenkamp, Dipl.-Ing. W. Drosdziok, Mrs. A. Kopplin, and Mrs. I. Spratte, and by our former colleagues Drs. G. Adamkiewicz-Eberle, J. Adamkiewicz, K. Hochleitner, R. Müller, and N.-H. Huh. This work was supported by the Dr. Mildred Scheel Stiftung für Krebsforschung, the Deutsche Forschungsgemeinschaft, and the Fritz Thyssen Stiftung.

REFERENCES

Adamkiewicz, J., Drosdziok, W., Eberhardt, W., Langenberg, U., and Rajewsky, M. F. (1982). High-affinity monoclonal antibodies specific for DNA components structurally modified by alkylating agents, in: *Indicators of Genotoxic Exposure*, Banbury Report 13 (B. A. Bridges, B. E. Butterworth, and I. B. Weinstein, eds.), Cold Spring Harbor Laboratory, Cold Spring Harbor, NY, pp. 265–276.

Adamkiewicz, J., Nehls, P., and Rajewsky, M. F. (1984). Immunological methods for detection of carcinogen–DNA adducts, in: *Monitoring Human Exposure to Carcinogenic and Mutagenic Agents* (A. Berlin, M. Draper, K. Hemminki, and H. Vainio, eds.), IARC Scientific Publ. No. 59, International Agency for Research on Cancer, Lyon, pp. 199–215.

Adamkiewicz, J., Eberle, G., Huh, N., Nehls, P., and Rajewsky, M. F. (1985). Quantification and visualization of alkyl deoxynucleosides in the DNA of mammalian cells by monoclonal antibodies. *Environ. Health Perspect.* **62**:49–55.

Adamkiewicz, J., Ahrens, O., Eberle, G., Nehls, P., and Rajewsky, M. F. (1986). Monoclonal antibody-based immunoanalytical methods for detection of carcinogen-modified DNA components, in: *The Role of Cyclic Nucleic Acid Adducts in Carcinogenesis and Mutagenesis* (B. Singer and H. Bartsch, eds.), IARC Scientific Publ. No. 70, International Agency for Research on Cancer, Lyon, pp. 403–411.

Basu, A. K., and Essigmann, J. M. (1988). Site specifically modified oligodeoxynucleotides as probes for the structural and biological effects of DNA damaging agents. *Chem. Res. Toxicol.* **1**:1–18.

Bohr, V. A., Evans, M. K., and Fornace, A. J., Jr. (1989). Biology of disease. DNA repair and its pathogenetic implications. *Lab. Invest.* **61**:143–161.

Denissenko, M. F., Venkatachalam, S., Yamasaki, E. F., and Wani, A. A. (1994). Assessment of DNA damage and repair in specific genomic regions by quantitative immuno-coupled PCR. *Nucleic Acids Res.* **22**:2351–2359.

Eberle, G. (1989). Monoklonale Antikörper gegen Kanzerogen-DNA Addukte, Ph.D. dissertation, University of Essen, Essen, Germany.

Eberle, G., Barbain, A., Laib, R. J., Ciroussel, F., Thomale, J., Bartsch, H., and Rajewsky, M. F. (1989). 1,N^6-etheno-2′-deoxyadenosine and 3,N^4-etheno-2′-deoxycytidine detected by monoclonal antibodies in lung and liver DNA of rats exposed to vinyl chloride. *Carcinogenesis* **10**:209–212.

Eberle, G., Glüsenkamp, K.-H., Drosdziok, W., and Rajewsky, M. F. (1990). Monoclonal antibodies for the specific detection of 3-alkyladenines in nucleic acids and body fluids. *Carcinogenesis* **11**:1753–1759.

Friedberg, E. C., Walker, G. C., and Siede, W. (1995). *DNA Repair and Mutagenesis*, ASM Press, Washington, DC.

Glüsenkamp, K.-H., Krüger, K., Eberle, G., Drosdziok, W., Jähde, E., Gründel, O., Neuhaus, A., Boese, R., Stellberg, P., and Rajewsky, M. F. (1993). Tautomer-specific anti-(N-3 substituted)-adenine antibodies: New tools in molecular dosimetry and epidemiology. *Angew. Chem. Int. Ed. Engl.* **32**:1640–1643.

Groopman, J. D., Sabbioni, G., and Wild, C. P. (1991). Molecular dosimetry of human aflatoxin exposures, in: *Molecular Dosimetry and Human Cancer* (J. D. Groopman and P. L. Skipper, eds.), CRC Press, Boca Raton, FL, pp. 303–324.

Hemminki, K., and Ludlam, D. B. (1984). Covalent modifications of DNA by antineoplastic agents. *J. Natl. Cancer Inst.* **73**:1021–1023.

Hochleitner, K., Thomale, J., Nikitin, A. Y., and Rajewsky, M. F. (1991). Monoclonal antibody-based, selective isolation of DNA fragments containing an alkylated base to be quantified in defined gene sequences. *Nucleic Acids Res.* **19**:4467–4472.

Huh, N., and Rajewsky, M. F. (1986). Enzymatic elimination of O^6-ethylguanine and stability of O^4-ethylthymine in the DNA of malignant neural cell lines exposed to N-ethyl-N-nitrosourea in culture. *Carcinogenesis* **7**:435–439.

Huh, N., Satoh, M. S., Shiga, J., Rajewsky, M. F., and Kuroki, T. (1989). Immunoanalytical detection of O^4-ethylthymine in liver DNA of individuals with or without malignant tumors. *Cancer Res.* **49**:93–97.

Kang, H. I., Konishi, C., Eberle, G., Rajewsky, M. F., Kuroki, T., and Huh, N.-H. (1992). Highly sensitive, specific detection of O^6-methylguanine, O^4-methylthymine, and O^4-ethylthymine by the combination of high-performance liquid chromatography prefractionation, ^{32}P-postlabeling, and immunoprecipitation. *Cancer Res.* **52**:5307–5312.

Loeb, L. A. (1989). Endogenous carcinogenesis: Molecular oncology into the twenty-first century—residential address. *Cancer Res.* **49**:5489–5496.

Müller, M. R., Seiler, F., Thomale, J., Buschfort, C., Rajewsky, M. F., and Seeber, S. (1994). Capacity of individual chronic lymphatic leukemia lymphocytes and leukemic blast cells for repair of O^6-ethylguanine in DNA: Relation to chemosensitivity *in vitro* and treatment outcome. *Cancer Res.* **54**:4524–4531.

Müller, R. (1983). Determination of affinity and specificity of anti-hapten antibodies by competitive radioimmunoassay. *Methods Enzymol.* **92**:589–601.

Müller, R., and Rajewsky, M. F. (1978). Sensitive radioimmunoassay for detection of O^6-ethyldeoxyguanosine in DNA exposed to the carcinogen ethylnitrosourea *in vivo* or *in vitro*. *Z. Naturforsch.* **33c**:897–901.

Müller, R., and Rajewsky, M. F. (1980). Immunological quantification by high-affinity antibodies of O^6-ethyldeoxyguanosine in DNA exposed to N-ethyl-N-nitrosourea. *Cancer Res.* **40**:887–896.

Müller, R., and Rajewsky, M. F. (1981). Antibodies specific for DNA components structurally modified by chemical carcinogens. *J. Cancer Res. Clin. Oncol.* **102**:99–113.

Nehls, P., and Rajewsky, M. F. (1985a). Differential formation of O^6-ethylguanine in the DNA of rat brain chromatin fibers of different folding levels exposed to N-ethyl-N-nitrosourea in vitro. *Cancer Res.* **45**:1378–1383.

Nehls, P., and Rajewsky, M. F. (1985b). Ethylation of nucleophilic sites in DNA by N-ethyl-N-nitrosourea depends on chromatin structure and ionic strength. *Mut. Res.* **150**:13–21.

Nehls, P., and Rajewsky, M. F. (1990). Monoclonal antibody-based immunoassay for the determination of cellular enzymatic activity for repair of specific carcinogen–DNA adducts (O^6-alkylguanine). *Carcinogenesis* **11**: 81–87.

Nehls, P., Adamkiewicz, J., and Rajewsky, M. F. (1984). Immuno-slot-blot: A highly sensitive immunoassay for the quantitation of carcinogen-modified nucleosides in DNA. *J. Cancer Res. Clin. Oncol.* **108**:23–29.

Nehls, P., Spiess, E., Weber, E., Berger, J., and Rajewsky, M. F. (1988). Distribution of O^6-ethylguanine in DNA exposed to ethylnitrosourea *in vitro* as visualized by electron microscopy using a monoclonal antibody. *Mutat. Res.* **198**:179–189.

Pfeifer, G. P., Drouin, R., and Holmquist, G. H. (1993). Detection of DNA adducts at the DNA sequence level by ligation-mediated PCR. *Mutat. Res.* **288**:39–46.

Rajewsky, M. F. (1989). Formation, distribution, and enzymatic repair of specific carcinogen adducts in genomic DNA: Relevance for malignant transformation, in: *Accomplishments in Cancer Research 1988* (J. G. Fortner and J. E. Rhoads, eds.), Lippincott, Philadelphia, pp. 273–283.

Rajewsky, M. F., Müller, R., Adamkiewicz, J., and Drosdziok, W. (1980). Immunological detection and quantification of DNA components structurally modified by alkylating carcinogens (ethylnitrosourea), in: *Carcinogenesis: Fundamental Mechanisms and Environmental Effects* (B. Pullman, P. O. P. Ts'o, and H. Gelboin, eds.), Reidel, Dordrecht, pp. 207–218.

Satoh, M. S., Moriyama, C., Asai, A., Handa, H., Rajewsky, M. F., Kuroki, T., and Huh, N. (1991). Monoclonal antibody-mediated solid-phase assay for mammalian O^6-alkylguanine DNA alkyltransferase activity. *Anal. Biochem.* **196**:403–409.

Seiler, F., Kirstein, U., Eberle, G., Hochleitner, K., and Rajewsky, M. F. (1993). Quantification of specific DNA O-alkylation products in individual cells by monoclonal antibodies and digital imaging of intensified nuclear fluorescence. *Carcinogenesis* **14**:1907–1913.

Sendowski, K., and Rajewsky, M. F. (1991). DNA sequence dependence of guanine-O^6 alkylation by the N-nitroso carcinogens N-methyl- and N-ethyl-N-nitrosourea. *Mutat. Res.* **250**:153–160.

Singer, B., and Grunberger, D. (1983). *Molecular Biology of Mutagens and Carcinogens*, Plenum Press, New York.

Thomale, J., Huh, N., Nehls, P., Eberle, G., and Rajewsky, M. F. (1990). Repair of O^6-ethylguanine in DNA protects rat 208F cells from tumorigenic conversion by N-ethyl-N-nitrosourea. *Proc. Natl. Acad. Sci. USA* **87**:9883–9887.

Thomale, J., Hochleitner, K., and Rajewsky, M. F. (1994a). Differential formation and repair of the mutagenic DNA alkylation product O^6-ethylguanine in transcribed and non-transcribed genes of the rat. *J. Biol. Chem.* **269**:1681–1686.

Thomale, J., Seiler, F., Müller, M. R., Seeber, S., and Rajewsky, M. F. (1994b). Repair of O^6-alkylguanines in the nuclear DNA of human lymphocytes and leukaemic cells: Analysis at the single-cell level. *Br. J. Cancer* **69**:698–705.

Tornaletti, S., and Pfeifer, G. P. (1994). Slow repair of pyrimidine dimers at p53 mutation hotspots in skin cancer. *Science* **263**:1436–1438.

Tornaletti, S., Rozek, D., and Pfeifer, G. P. (1993). The distribution of UV photoproducts along the human p53 gene and its relation to mutation in skin cancer. *Oncogene* **8**:2051–2057.

Umbenhauer, D., Wild, C. P., Montesano, R., Saffhill, R., Boyle, J. M., Huh, N., Kirstein, U., Thomale, J., Rajewsky, M. F., and Lu, S. H. (1985). O^6-methyldeoxyguanosine in oesophageal DNA among individuals at high risk of oesophageal cancer. *Int. J. Cancer* **36**:661–665.

Chapter 8

Detection of Oxidative DNA Base Damages

Immunochemical and Electrochemical Approaches

Robert J. Melamede, Yoke Wah Kow, Ivan A. Bespalov, and Susan S. Wallace

I. IMMUNOCHEMICAL DETECTION OF OXIDATIVE DNA DAMGES

1.1. Introduction

Antibodies to a variety of oxidized DNA bases have been generated in a number of laboratories including our own (see Table I). Most of these antibodies have been elicited using protein-conjugated haptens of interest. In general, the antibodies have reasonable affinity such that appropriate sensitivity in the various assays can be achieved. The difficulty with antibodies that recognize oxidized DNA bases is that the oxidized bases do not differ largely from their unoxidized derivatives. Thus, the specificity for detecting lesions in DNA must be high especially when one considers the low level of damaged compared to undamaged bases. Even if the sensitivity of the assay can be amplified, cross-reactivity of the antibody with the unoxidized base in DNA remains an obstacle to successful detection of low levels of the oxidized base. This particular problem is not seen as often when antibodies are elicited to the chemical adducts that have features that vary quite dramatically from the unadducted base. An additional consideration is that the lesion must be stable to the procedures used during preparation of the immunogen and during DNA denaturation. The latter is usually necessary since the antibodies often do not recognize the lesion as well in duplex DNA. Despite these shortcomings, antibodies to oxidized bases have been effectively utilized to detect these lesions in oxidized or ionizing radiation-treated DNA *in vitro*.

Table I summarizes the properties of antibodies, reported to date, that react with oxidized DNA bases. The IC_{50} values, the concentration of hapten that inhibits antibody binding by 50%,

Robert J. Melamede, Yoke Wah Kow, Ivan A. Bespalov, and Susan S. Wallace • Department of Microbiology and Molecular Genetics, The Markey Center for Molecular Genetics, The University of Vermont, Burlington, Vermont 05405.

Technologies for Detection of DNA Damage and Mutations, edited by Gerd P. Pfeifer, Plenum Press, New York, 1996.

Table I
Properties of Antibodies to Oxidized DNA Bases

Antibody base specificity	Eliciting antigen	Assay used	IC_{50} hapten	Level of lesion detection DNA	IC_{50} corresponding unmodified hapten (d)	Specificity in DNA
7,8-Dihydro-8-oxoguanine						
Polyclonal[1]	8-Oxoguanosine-α-casein	Competitive RIA	4.2 nM[a]	63 fmole	0.29 mM[a] (7×10^4)	1/100,000 G
Monoclonal[2]	8-Oxoguanosine-BSA	Competitive RIA	5.0 nM[a]	ND	35.5 μM[a] (7×10^3)	ND
Polyclonal[3]	8-Oxoguanosine-BSA	ELISA	0.15 μM[b]	5 fmole	>100 μM[b] ($>7 \times 10^2$)	1/250 G
7,8-Dihydro-8-oxoadenine						
Polyclonal[4]	8-Oxoadenosine-BSA	Competitive RIA	20 nM[b]	40 fmole	40 μM[b] (2×10^3)	—
Polyclonal[3]	8-Oxoadenosine-BSA	ELISA	8 μM[b]	1 fmole	>1.25 mM[b] ($>1.5 \times 10^2$)	1/1,250 A
8,5′-Cycloadenosine						
Polyclonal[5,6]	8,5′-CycloAMP-hemocyanin	Competitive ELISA	28 nM[b]	2 μmole	19 μM[b] (7×10^2)	—
5-Hydroxymethyldeoxyuridine						
Polyclonal[7]	5-Hydroxymethyluridine-BSA	Phage neutralization	ND	0.1 nmole	ND	1/4,000 T
5,6-Dihydroxy-5,6-dihydrothymine						
Polyclonal[8]	OsO_4-DNA	Competitive RIA	ND	4 fmole	ND	1/30,000 T
Polyclonal[9,10]	Thymidine glycol monophosphate-BSA	ELISA	0.75 μM[c]	1 fmole	>5 mM[c] ($>7 \times 10^3$)	1/10,000 T
Monoclonal[11]	OsO_4-polydT	Competitive ELISA	ND	2 fmole	ND	1/200,000 T
Monoclonal[12]	Thymidine glycol monophosphate-BSA	ELISA	5 nM[c]	4 fmole	2.8 μM[c] (5×10^2)	1/5,000 T
5,6-Dihydrothymine						
Polyclonal[13]	Dihydrothymidine monophosphate-BSA	ELISA	0.12 μM[c]	20 fmole	0.73 μM[c] (6×10^3)	1/25,000 T
5-Hydroxycytosine						
Polyclonal[14]	5-Hydroxycytidine monophosphate-BSA	ELISA	0.012 μM[b]	ND	>100 μM[b] ($>8 \times 10^2$)	ND

5-Hydroxyuracil						
Polyclonal[14]	5-Hydroxyuridine monophosphate-BSA	ELISA	0.012 μM[b]	ND	>100 μM[b] (>8 × 10^2)	ND
Abasic site						
Monoclonal[14]	*O*-4-Nitrobenzyl hydroxylamine-BSA	ELISA	0.3 μM[c]	10 fmole	>1 mM[c] (>3 × 10^3)	1/20,000 bases

[1]Degan *et al.* (1991).
[2]Park *et al.* (1992).
[3]Ide *et al.* (1996).
[4]West *et al.* (1982a).
[5]Fuciarelli *et al.* (1985).
[6]Fuciarelli *et al.* (1987).
[7]Lewis *et al.* (1978).
[8]West *et al.* (1982b).
[9]Rajagopalan *et al.* (1984).
[10]Hubbard *et al.* (1989a).
[11]Leadon and Hanawalt (1983).
[12]Chen *et al.* (1990).
[13]Hubbard *et al.* (1989b).
[14]B.-X. Chen, B. F. Erlanger, and S. S. Wallace (unpublished observations).
[15]Chen *et al.* (1992).
[a]Determined by inhibition of hapten–antibody binding.
[b]Determined by inhibition of conjugate–antibody binding.
[c]Determined by inhibition of DNA lesion–antibody binding.
[d]Number in parentheses is the relative inhibition of the unmodified to the modified hapten.

are a rough measure of the affinity of the antibody for the lesion. The IC_{50} values in Table I cannot be directly compared to each other, however, since the binding reactions were done with different "antigens": the free hapten, the immunizing conjugate, or DNA containing the hapten. Some antibodies have a higher affinity for the immunizing conjugate because of stabilizing protein binding sites while others react with higher affinity to haptens in the context of DNA. With respect to lesion detection in DNA, the relative affinity for the modified to the normal base is also very important (Table I, IC_{50} corresponding unmodified hapten). It is this relationship that influences the detection of the lesion in the background or normal bases. Again, relative IC_{50} values determined by antibody binding to free hapten or conjugate do not always reflect this relationship in DNA. The referenced papers should be referred to for details. To summarize, it is a combination of the affinity of the antibody for the lesion in DNA, together with its specificity for the lesion compared to the normal base, that determines the usefulness of the antibody for measuring damage in DNA.

Several of the antibodies to oxidized bases listed in Table I have been successfully used for *in vivo* studies. Both polyclonal (Degan *et al.*, 1991) and monoclonal (Park *et al.*, 1992) antibodies to 7-hydro-8-oxoguanine (8-oxoG) have been used for immunoaffinity columns to detect these adducts in urine and have enabled an *in vivo* analysis of the presumptive repair of 8-oxoG in DNA from both eukaryotes and prokaryotes. Both the monoclonal and polyclonal antibodies to 8-oxoG exhibit a strong affinity for the modified deoxynucleoside; however, the affinity of the monoclonal antibody for the oxidized base is greater. Despite this advantage, the monoclonal antibody cross-reacts significantly with normal guanine, thus making it less suitable for assaying 8-oxoG in DNA. In contrast, a monoclonal antibody to thymine glycol raised by immunization with osmium tetroxide-oxidized poly(dT) has an extremely high specificity for thymine glycol in the background of normal DNA thymine (Leadon and Hanawalt, 1983). Although the affinity of this antibody to thymine glycol appears to be similar to many of the other antibodies to oxidized DNA bases (see references in Table I), it can detect 1 thymine glycol in 200,000 DNA thymines. Because of its specificity, this antibody has been very useful for measuring thymine glycol in oxidized or irradiated cellular DNA from both mammalian (Kaneko and Leadon, 1986; Leadon *et al.*, 1988) and yeast cells (Kaneko *et al.*, 1988). In fact, the detection of strand-selective repair of thymine glycols in yeast (Leadon and Lawrence, 1992) has been established using this antibody. Reviews of the techniques involved in assaying 8-oxoguanine by immunoaffinity (Shigenaga *et al.*, 1994) and thymine glycol by enzyme-linked immunosorbant assay (ELISA) (Leadon, 1988) have been published.

There is no question that the immunochemical detection of oxidative DNA damages has many advantages, in particular, the methodologies are simple and inexpensive. The availability of DNA damage-specific immunochemical reagents, however, remains the constraining factor preventing widespread use of antibodies for detecting DNA damage. In an effort to overcome this limitation, we are currently eliciting in mice, antibodies to modified deoxynucleoside monophosphates conjugated to BSA (7-hydro-8-oxodAMP, 8-oxodGMP, 5-hydroxydCMP, 5-hydroxydUMP, thymidine glycol monophosphate, and deoxyuridine glycol monophosphate) for *in vitro* repertoire cloning and phage display. The phage display system (Scott and Smith, 1990; Devlin *et al.*, 1990; Cwirla *et al.*, 1990) uses the pCOMB3 combinatorial vector to produce monoclonal Fabs in *E. coli* (Barbas *et al.*, 1991; Barbas and Lerner, 1991). Starting with RNA extracted from the spleens of immunized mice, their immunological repertoire is cloned into the filamentous bacteriophage, f1. The Fabs are displayed on the tips of the phage as a fusion product of the Fab heavy chain with the phage Gene 3 coat protein. Light chains associate with the fusion protein in the periplasm to produce functional Fab fragments bound to phage particles that contain the genetic information of the displayed Fab. Antibody selection is carried out via repeated rounds of binding and growth (panning) prior to clonal isolation and characterization.

Soluble Fabs can also be obtained using this same system. We have thus far produced *in vitro*, from rabbit or mouse repertoires, Fabs that bind to hapten–protein conjugates and are specific for oxidized pyrimidines or oxidized purines (Bespalov *et al.* 1996a,b). After further characterization and modification of these antibodies, they will be available for widespread use.

The simplest immunochemical procedure suitable for detecting oxidative base damages in DNA is the direct ELISA, although competitive ELISA and competitive radioimmunoassays (RIA) have been extensively used. For the direct ELISA, DNA is first bound to a 96-well microtiter plate; the plate is washed, blocked, and then washed again. The primary antibody is then bound and the plate is washed; the enzyme-linked secondary antibody is then bound and the plate is washed. Finally, a colorimetric or fluorometric substrate for the enzyme is added and the solution in the wells is spectrophotometrically quantitated.

Figure 8.1 shows the immunochemical detection of 8-oxoG and 8-oxoA (A) and thymine glycol (B) in X-irradiated calf thymus DNA using polyclonal sera from this laboratory (Rajagopalan *et al.*, 1984; Ide *et al.*, 1996). Initially, the slope of the line representing the production of damages as a function of dose is linear but it tapers off at the higher doses. This nonlinear response is an artifact of differential binding of the DNA to the microtiter plate that results because the DNA is fragmented by high doses of X-irradiation as shown in Fig. 8.2B. As can be

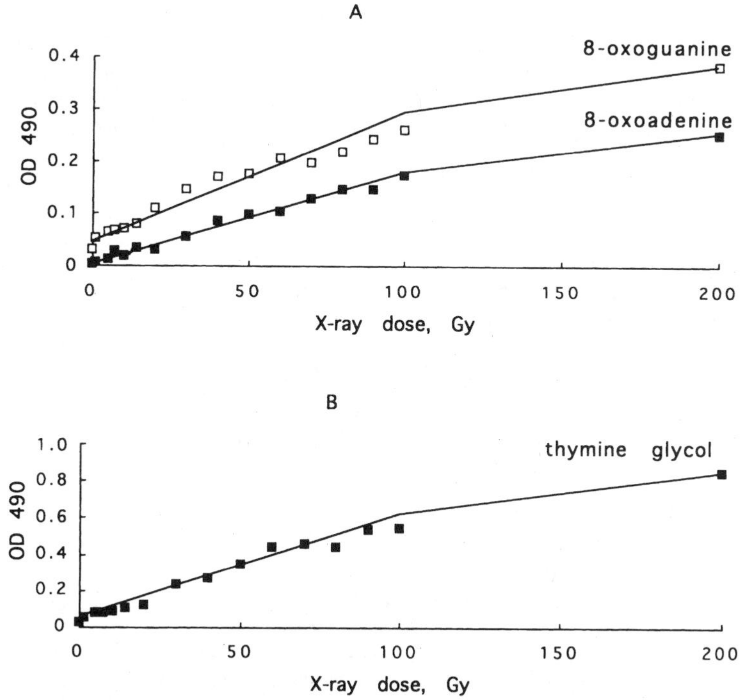

Figure 8.1. ELISA detection of 8-oxoG and 8-oxoA (A) and thymine glycol (B) in X-irradiated calf thymus DNA. Native calf thymus DNA (100–200 μg/ml in PBS) was X-irradiated using a Phillips X-ray generator set at 50 kVp and 2 mA through a beryllium window at a dose rate of 100 Gy/min. The DNA was denatured by boiling for 5 min and 1 μg of DNA in 50 μl was applied to the wells of a microtiter plate (Costar plate A/2, flat bottom, high binding). Polyclonal rabbit serum was used for the ELISA reactions which were performed as described in the text.

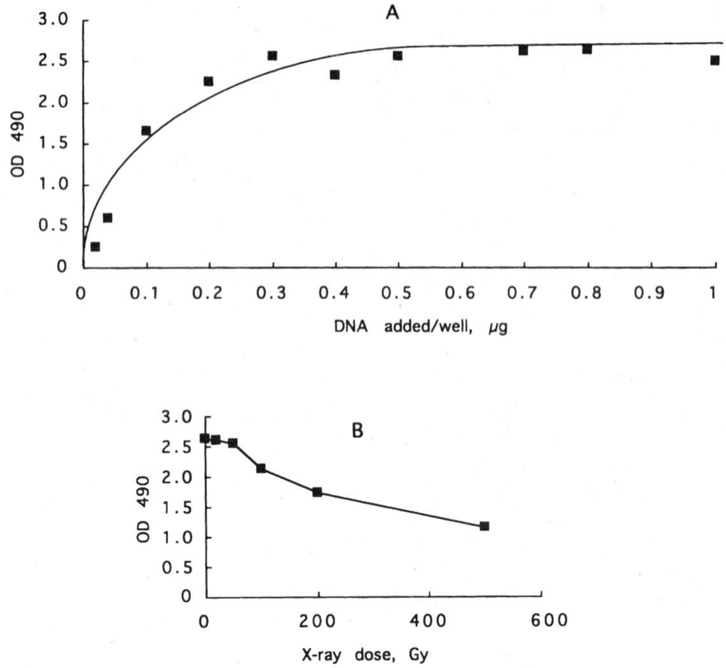

Figure 8.2. ELISA detection of calf thymus DNA bound to microtiter wells. Supernatants (50 μl) from 5× freeze-thawed pellets of *E. coli* expressing anti-DNA (non-lesion-specific) monoclonal Fabs (clone d1290) (Bespalov *et al.*, 1996a) were used for ELISA that was performed as described in the text. (A) DNA binding as a function of the amount of undamaged DNA added to each well. (B) DNA binding to microtiter wells as a function of X-ray dose to the DNA. DNA was X-irradiated as described for Fig. 8.1 and 1 μg of DNA was added to each well.

seen, the amount of DNA bound to the plate, as measured by ELISA using an anti-DNA monoclonal Fab (Bespalov *et al.* 1996a), begins to decrease at 100 Gy. This is approximately the dose at which nonlinearity in dose response begins as shown in Fig. 8.1. These data point out the need to determine the binding efficiencies of the DNA samples being tested. Figure 8.2A shows that the amount of DNA bound to the wells, as measured with the same anti-DNA Fab, increases with increasing amounts of DNA added, saturating at about 0.3 μg. We generally add 1 μg/well (see below).

1.2. Methods

1.2.1. Binding of the DNA to Microtiter Plates

There are a number of methods that have been reported for binding DNA to plates including the use of (1) UV-irradiated polystyrene plates, (2) polyanions such as polylysine or spermidine, (3) high-salt alkali buffer, and (4) overnight drying in phosphate-buffered saline (PBS). We prefer method (4) because it is reliable, reproducible, works without harsh treatment, and does not result in high backgrounds. DNA (50 μl, 20 μg/ml) in PBS (0.14 M NaCl, 1 mM KH_2PO_4, 20 mM Na_2HPO_4, 3 mM KCl, pH 7.4) is added to each well. The uncovered wells are air-dried overnight at 37°C. The wells are washed 3× with H_2O (in-house, polished reverse-osmosis H_2O from the faucet is allowed to flow over the wells). The plates are inverted and smacked onto a few paper towels to empty the wells prior to blocking for 1 hr with 100 μl 3% BSA in PBS or I-block (Tropix) at 37°C.

The quantity of DNA that binds to a microtiter well varies greatly depending on the immobilization procedure used as well as the type of plate used. We find that Costar plate A/2, flat-bottom, high-binding plates work reproducibly well. By overnight drying the DNA solution in the wells, approximately 10–100 ng of DNA can be bound.

1.2.2. Washing Plates

The blocking solution is shaken out of the wells and they are washed 3 times with H_2O. Prior to again washing the plates with H_2O after incubating them with primary or secondary antibody, they are first washed 2× with TPBS (PBS plus 0.2% Tween) from a squeeze bottle, and then 3× with H_2O in order to minimize nonspecific binding to the wells. Again, the plates are inverted and smacked onto paper towels to empty the wells prior to the addition of primary or secondary antibody, or development reagent.

1.2.3. Primary Antibody

Primary antibodies should be titrated to give maximal response without creating unnecessarily high backgrounds and wasting of reagents. Polyclonal sera or monoclonal antibodies are diluted into 1% BSA in PBS or I-block and 50 μl is added to each well and incubated for 1 hr at 37°C.

Competitive assays are slightly more elaborate and time consuming than the direct ELISA. In the competitive ELISA the primary antibody must be added in subsaturating quantities with respect to the bound antigen, therefore a titration must be done prior to the competitive assay. The specifics that should be followed for an analytical competitive ELISA, as would be used for determining an inhibition constant, are detailed in a paper by Friguet *et al.* (1985), as well as in the references in Table I. An advantage of a competitive assay is that the quantity of soluble competing species is limited by the concentration and volume that can be used in the assay and not by the amount of ligand that binds to the microtiter well.

1.2.4. Secondary Antibody

Secondary antibodies are species-specific anti-immunoglobulin molecules that are linked to a reporter enzyme such as peroxidase or alkaline phosphatase. Rabbit Fabs are detected using goat anti-rabbit antibodies labeled with alkaline phosphatase (Jackson Immune Research) or horseradish peroxidase (Bio-Rad). Secondary antibodies should be saturating to detect all bound primary antibodies, but not so excessive as to create high backgrounds. Commercial secondary antibodies are typically diluted from 1:1000 to 1:10,000-fold into 1% BSA in PBS or I-block. Fifty microliters of this is added to each well and incubation carried out for 1 hr at 37°C.

1.2.5. Substrate Development

The phosphatase substrate is made by adding a 15-mg pellet (*p*-nitrophenyl phosphate, Sigma) to 15 ml diethanolamine buffer (10 mM diethanolamine pH 9.5) plus 0.5 mM $MgCl_2$. Prior to adding the phosphatase substrate to the microtiter wells, they are first given a final wash with the diethylamine buffer. Fifty microliters of substrate is added per well and color development is visually monitored prior to reading the optical density at 405 nm. Peroxidase substrate is made by adding 5 ml 0.2 M dibasic Na_2HPO_4 and 5 ml 0.1 M citric acid to 10 ml H_2O and dissolving a 15-mg pellet of *o*-phenylenediamine (Sigma) in the solution. Hydrogen peroxide (10 μl) is added to the substrate mix immediately prior to use. Substrate (50 μl) is added to each well and color development is visually monitored. Prior to reading the optical density at 490 nm, the reactions are stopped with 50 μl of 1 M H_2SO_4. We prefer to use the peroxidase system since it has a much shorter development time than the phosphatase-based one. However, when additional sensitivity is required, secondary antibodies labeled with alkaline phosphatase can be used with chemoluminescent substrates.

2. HPLC/ELECTROCHEMICAL DETECTION OF OXIDIZED BASES

2.1. Introduction

Although HPLC/electrochemical detection is only useful for electrochemically active modified bases, several of the most common oxidative DNA lesions such as 8-oxoG, 7-hydro-8-oxoadenosine (8-oxoA), 5-hydroxycytosine (5-OHC), and 5-hydroxyuracil (5-OHU) fall into this category. Recently, the use of this methodology, developed initially by Floyd and his group (Floyd *et al.*, 1986a,b), has become increasingly widespread for measuring oxidative damage and detailed accounts of these procedures have been published (Shigenaga and Ames, 1991; Shigenaga *et al.*, 1994).

8-Oxoguanine has become the benchmark lesion for oxidative damage and has been measured by HPLC/EC in urine as a free nucleoside (Shigenaga *et al.*, 1989; Fraga *et al.*, 1990; Loft *et al.*, 1994) and from the DNA of cells exposed to a wide variety of oxidizing agents (Kasai *et al.*, 1986, 1989; Fraga *et al.*, 1990; Kiyosawa *et al.*, 1990; Takeuchi *et al.*, 1994). 5-Hydroxycytosine and 5-hydroxyuracil have also been measured by HPLC/EC in DNA oxidized *in vitro* and from DNA extracted from rat tissues (Wagner *et al.*, 1992).

Figure 8.3 shows the production of the 8-oxopurines (A), 8-oxodG and 8-oxodA, and the 5-hydroxypyrimidines (B), 5-OHdC and 5-OHdU, in X-irradiated DNA using HPLC/EC as described below. As can be seen, the production of these lesions is linear with dose and they are readily detectable in the picomolar range.

2.2. Methods

2.2.1. Preparation of DNA Samples

Human cells (1×10^7) are pelleted at 1200 rpm (Dynac Clinical centrifuge) for 2 min. The cell pellet is washed twice with 1 ml of PBS buffer and resuspended in 1 ml of PBS buffer. Similarly, bacterial cells (1×10^{10}) are pelleted at 5000 rpm for 10 min and washed twice with 1 ml of PBS and resuspended in 1 ml of PBS buffer. Three milliliters of 4 M guanidinium hydrochloride is added to the cell suspension, which is left on ice for at least 10 min. At this stage, the cell suspension should become clear. Alternatively, the mammalian cells can be lysed by resuspending the cell pellet into 2 ml of proteinase K buffer at 50°C for 2 hr; 0.6 ml of 3 M sodium acetate is then added to the cell lysate. The DNA is precipitated by adding an equal volume of cold isopropanol (−20°C) and left at −80°C for 20 min. The precipitated DNA is then redissolved in 1 ml of TE buffer, and 200 μl of DMSO is added. The DNA solution is extracted with 2 ml of TE-saturated phenol by vigorous mixing, and 1 ml of chloroform:isoamyl alcohol (24:1) is added to the phenol mix and vortexed for 14 sec. The sample is then centrifuged at 8000 rpm for 10 min. The top layer containing the DNA is removed and placed into a new tube. Another 2 ml of chloroform:isoamyl alcohol is added to extract the excess phenol. Again the top

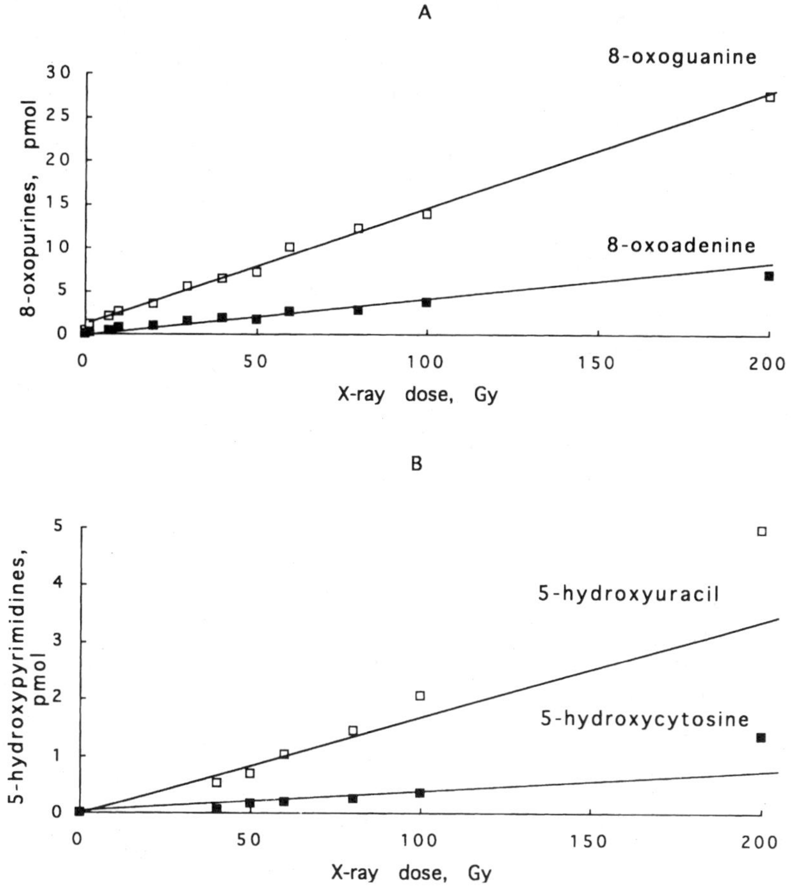

Figure 8.3. HPLC/EC detection of 8-oxopurines (A), 8-oxodG and 8-oxodA, and 5-hydroxypyrimidines (B), 5-OHdC and 5-OHdU, in X-irradiated DNA. Samples were irradiated as described for Fig. 8.1. HPLC/EC was performed isocratically on C-18 columns and detected electrochemically as described in the text.

layer containing the DNA is saved. The process is repeated again. The DNA is then ethanol precipitated with 3 vol of 100% cold ethanol (−20°C) and 0.3 vol of 3 M sodium acetate, and the solution is left at −80°C for 30 min. The precipitated DNA is washed twice with 70% cold ethanol (−20°C) and air-dried. The DNA pellet is redissolved in 1 ml of RNase A buffer and incubated at 37°C for 2 hr. Then 200 μl of DMSO is added and the DNA sample reextracted with TE-saturated phenol as described above. Excess phenol is extracted with 1 ml of chloroform: isoamyl alcohol (24:1) as described above. The DNA is then ethanol precipitated, washed twice with 70% ethanol, and air-dried. This preparation yields approximately 30 and 50 μg of mammalian and bacterial DNA, respectively. The DNA is then dissolved in 50 μl of TE buffer.

2.2.2. Digestion of DNA for HPLC Analysis

Fifty microliters of DNA in TE buffer is incubated at 90°C for 5 min and quick cooled in an ice bath. Then 5 μl of 0.5 M sodium acetate, pH 5.1, and 5 units of nuclease Pl is added and the mixture incubated for 1 hr at 37°C. This is followed by the addition of 20 μl of 0.4 M Tris-HCl, pH 8.0, and 3 units of shrimp alkaline phosphatase and the mixture is incubated for an additional 1 hr at 37°C. The resultant hydrolysate is centrifuged at full speed in a microfuge and the supernatant stored on ice for analysis by HPLC-EC. Samples can be frozen at −80°C for analysis at later times.

2.2.3. HPLC-EC Detection of DNA Lesions

Currently, four DNA base lesions can be easily measured by electrochemical detection: 8-oxodG, 8-oxodA, 5-OHdC, and 5-OHdU. For HPLC-EC analysis, the modified deoxyribonucleosides needed for calibration can be prepared according to published procedures (see Wagner *et al.*, 1992, and references therein for synthesis of 5-OHdC and 5-OHdU; Shigenaga *et al.*, 1989, for 8-oxodG; Berger *et al.*, 1990, for 8-oxodA). C-18 Absorbsphere (Whatman, 4 × 250 mm, 5 μm) or C-18 Microsorb-MV (Rainin, 4 × 250 mm, 5 μm) are the columns routinely used in this laboratory and both columns give comparable elution profiles. One advantage of the Rainin column is that it is roughly three times cheaper than the Whatman column. The columns are equilibrated at 1 ml/min with the HPLC elution buffer with both the UV and electrochemical detector in line. For a new column, this should give a back pressure of about 1500 psi. The electrochemical detector (BAS LC-4C equipped with the glass electrode) is set at an oxidation potential of 0.65 V for 5-OHdU, 5-OHdC, and 8-oxodG, and 0.80 V for 8-oxodA. The signal from both detectors is connected to Rainin's Control/Interface Module (Model 720-127, hardware component of the McIntegrator I marketed by Rainin), which is an 18-bit analog-to-digital converter, allowing data sampling rates up to 20 times per second. The acquired data can then be analyzed with a Macintosh computer installed with the McIntegrator I analysis software. Twenty-five microliters of standard mixtures containing 1 pmole of modified nucleoside and 100 nmole of each of the four normal deoxyribonucleosides is injected after the column is equilibrated. Normal elution requires roughly 45 min under these conditions, where dA is eluted at approximately 38 min; the elution time for dC = 5.5 min, dG = 13.5 min, dT = 15.5 min, 8-oxodG = 21 min, 5-OHdC = 8.5 min, 5-OHdU = 10 min, and 8-oxodA = 41 min. Depending on the stability of the electrochemical detector, the detection limit we have routinely observed is approximately 0.1 pmole for 8-oxodG, 1 pmole for 5-OHdC, 5-OHdU, and 8-oxodA. The DNA hydrolysate, 25 μl, is injected, and eluted isocratically with the HPLC elution buffer. Based on the elution time and peak area, the amount of the lesion of interest can be calculated referencing the peak area of the injected standards.

2.2.4. Storage of the Glass Electrode

One problem we encounter with the use of an electrochemical detector is that it takes a considerable amount of time to get to a stable baseline at high sensitivity (for detection in the low picomole range). Once a stable baseline is established, we routinely keep the HPLC unit running with the elution buffer continuously at a flow rate of 0.1 ml/min. If the unit is not to be used for a short period of time, e.g., 1 week, the system is stored in 50% methanol by exchanging the HPLC unit with the column and detector in line with 50% methanol. When needed, the HPLC system is equilibrated with the HPLC elution buffer, and is ready to use within a few hours. However, for prolonged storage, the glass electrode is deassembled from the detector, and the reference electrode is stored in saturated KCl solution. The HPLC unit is then flushed with 50% methanol.

2.2.5. Buffers

TE buffer: 10 mM Tris-HCl, 1 mM EDTA, pH 8.0.

Proteinase K buffer: Proteinase K, 100 μg/ml, 5 mM EDTA, 0.5% Sarkosyl, 10 mM Tris-HCl, pH 8.0.

TE-saturated phenol: Phenol is removed from the freezer and warmed to room temperature by placing in a 50°C water bath. The phenol is left in the dark bottle and the above thawing is done in the hood. To every 500 ml of phenol, 0.5 g of hydroxyquinoline is added and stirred slowly to dissolve the hydroxyquinoline. Then 200 ml of TE buffer is added and stirred for 30 min to saturate the phenol solution. Excess TE buffer is removed.

RNase A buffer: RNase A, 100 μg/ml, 50 ml Tris-HCl, 10 mM EDTA, 10 mM NaCl, pH 8.0. The solution is heated at 90°C for 5 min to inactivate any contaminating DNases.

HPLC elution buffer: 12.5 mM citric acid, 25 mM sodium acetate, 5% methanol, pH 5.1.

ACKNOWLEDGMENTS. The work from this laboratory was supported by U.S. Department of Energy Grant DE-FG02-88ER60742. We are grateful to Dr. Bernard Erlanger and his collaborators at Columbia University College of Physicians and Surgeons and to many former students and postdoctoral associates for their contributions.

REFERENCES

Barbas, C. F., III, and Lerner, R. A. (1991). Combinatorial immunoglobulin libraries on the surface of phage (Phabs): Rapid selection of antigen-specific Fabs, in: *Methods: A Companion to Methods in Enzymology*, Vol. 2 (R. A. Lerner and D. R. Burton, eds.), Academic Press, Orlando, pp. 119–124.

Barbas, C. F., Kang, A. S., Lerner, R. A., and Benkovic, S. J. (1991). Assembly of combinatorial antibody libraries on phage surfaces: The gene III site. *Proc. Natl. Acad. Sci. USA* **88**:7978–7982.

Berger, M., Anselmino, C., Mouret, J.-F., and Cadet, J. (1990). High performance liquid chromatography-electrochemical assay for monitoring the formation of 8-oxo-7,8-dihydroadenine and its related 2′-deoxyribonucleoside. *J. Liquid Chromatogr.* **13**:929–940.

Bespalov, I. A., Purmal, A. A., Erlanger, B. F., Wallace, S. S., and Melamede, R. J. (1996a). Antibodies to oxidized pyrimidines selected from combinatorial phage display libraries. Submitted for publication.

Bespalov, I. A., Purmal, A. A., Wallace, S. S., and Melamede, R. J. (1996b). Recombinant Phabs reactive with 7-hydro-8-oxoguanine, a major oxidative DNA lesion. *Biochemistry*, **35**:2067–2078.

Chen, B. X., Hubbard, K., Ide, H., Wallace, S. S., and Erlanger, B. F. (1990). Characterization of a monoclonal antibody to thymidine glycol monophosphate. *Radiat. Res.* **124**:131–136.

Chen, B. X., Kubo, K., Ide, H., Erlanger, B. F., Wallace, S. S., and Kow, Y. W. (1992). Properties of a monoclonal antibody for the detection of abasic sites, a common DNA lesion. *Mutat. Res.* **273**:253–261.

Cwirla, S. E., Peters, E. H., Barrett, R. W., and Dower, W. S. (1990). Peptides on phage: A vast library of peptides for identifying ligands. *Proc. Natl. Acad. Sci. USA* **87**:6378–6382.

Degan, P., Shigenaga, M. K., Park, E. M., Alperin, P. E., and Ames, B. N. (1991). Immunoaffinity isolation of urinary 8-hydroxy-2′-deoxyguanosine and 8-hydroxyguanine and quantitation of 8-hydroxy-2′-deoxyguanosine in DNA by polyclonal antibodies. *Carcinogenesis* **12**:865–871.

Devlin, J. J., Panganiban, L. C., and Devlin, P. E. (1990). Random peptide libraries: A source of specific protein binding molecules. *Science* **249**:404–406.

Floyd, R. A., Watson, J. J., Wong, P. K., Altmiller, D. H., and Richard, R. C. (1986a). Hydroxyl free radical adduct of deoxyguanosine: Sensitive detection and mechanisms of formation. *Free Rad. Res. Commun.* **1**:163–172.

Floyd, R. A., Watson, J. J., Harris, J., West, M. and Wong, P. K. (1986b). Formation of 8-hydroxydeoxyguanosine, hydroxyl free radical adduct of DNA in granulocytes exposed to the tumor promoter, tetradecanoylphorbol-acetate. *Biochem. Biophys. Res. Commun.* **137**:841–846.

Fraga, C. G., Shigenaga, M. K., Park, J.-W., Degan, P., and Ames, B. N. (1990). Oxidative damage to DNA during aging: 8-hdroxy-2′-deoxyguanosine in rat organ DNA and urine. *Proc. Natl. Acad. Sci. USA* **87**:4533–4537.

Friguet, B., Chaffotte, A. F., Djavadi-Ohaniance, L., and Goldberg, M. E. (1985). Measurements of the true affinity constant in solution of antigen–antibody complexes by enzyme-linked immunosorbent assay. *J. Immunol. Methods* **77**:305–319.

Fuciarelli, A. F., Miller, G. G., and Raleigh, J. A. (1985). An immunochemical probe for 8,5′-cycloadenosine-5′-monophosphate and its deoxy analog in irradiated nucleic acids. *Radiat. Res.* **104**:272–284.

Fuciarelli, A. F., Shum, F. Y., and Raleigh, J. A. (1987). Intramolecular cyclization in irradiated nucleic acids: Correlation between high-performance liquid chromatography and an immunochemical assay for 8,5′-cycloadenosine in irradiated poly(A). *Radiat. Res.* **110**:35–44.

Hubbard, K., Huang, H., Laspia, M. F., Ide, H. Erlanger, B. F., and Wallace, S. S. (1989a). Immunochemical quantitation of thymine glycol in oxidized and X-irradiated DNA. *Radiat. Res.* **118**:257–268.

Hubbard, K., Ide, H., Erlanger, B. F., and Wallace, S. S. (1989b). Characterization of antibodies to dihydrothymine, a radiolysis product of DNA. *Biochemistry* **28**:4382–4387.

Ide, H., Kow, Y. W., Chen, B.-X., Erlanger, B. F., and Wallace, S.S. (1996). Properties of antibodies elicited to 7-hydro-8-oxopurines and their reaction with X-irradiated DNA. Submitted for publication.

Kaneko, M., and Leadon, S. A. (1986). Production of thymine glycols in DNA by *N*-hydroxy-2-naphthylamine as detected by a monoclonal antibody. *Cancer Res.* **46**:71–75.

Kaneko, M., Leadon, S. A., and Ito, T. (1988). Relationship between the induction of mitotic gene conversion and the formation of thymine glycols in yeast *S. cerevisiae* treated with hydrogen peroxide. *Mutat. Res.* **207**:17–22.

Kasai, H., Crain, P. F., Kuchino, Y., Nishimura, S., Ootsuyama, A., and Tanooka, H. (1986). Formation of 8-hydroxyguanine moiety in cellular DNA by agents producing oxygen radicals and evidence for its repair. *Carcinogenesis* **7**:1849–1851.

Kasai, H., Okada, Y., Nishimura, S., Rao, M. S., and Reddy, J. K. (1989). Formation of 8-hydroxydeoxyguanosine in liver DNA of rats following long-term exposure to a peroxisome proliferator. *Cancer Res.* **49**:2603–2605.

Kiyosawa, H., Suko, M., Okudaira, H., Murata, K., Miyamoto, T., Chung, M.-H., Kasai, H., and Nishimura, S. (1990). Cigarette smoking induces formation of 8-hydroxyguanosine, one of the oxidative DNA damages in human peripheral leukocytes. *Free Rad. Res. Commun.* **11**:1–3.

Leadon, S. A. (1988). Immunological probes for lesions and repair patches in DNA, in: *DNA Repair. A Laboratory Manual of Research Procedures* (E. C. Friedberg and P.C. Hanawalt, eds.), Dekker, New York, pp. 311–326.

Leadon, S. A., and Hanawalt, P. C. (1983). Monoclonal antibody to DNA containing thymine glycol. *Mutat. Res.* **112**:191–200.

Leadon, S. A., and Lawrence, D. A. (1992). Strand-selective repair of DNA damage in the yeast GAL7 gene requires RNA polymerase II. *J. Biol. Chem.* **267**:23175–23182.

Leadon, S. A., Stampfer, M. R., and Bartley, J. (1988). Production of oxidative DNA damage during the metabolic activation of benzo[a]pyrene in human mammary epithelial cells correlates with cell killing. *Proc. Natl. Acad. Sci. USA* **85**:4365–4368.

Lewis, H. L., Muhleman, D. R., and Ward, J. F. (1978). Serologic assay of DNA base damage. I. 5-Hydroxymethyldeoxyuridine, a radiation product of thymidine. *Radiat. Res.* **75**:305–316.

Loft, S., Astrup, A., Buemann, B., and Poulsen, H. E. (1994). Oxidative DNA damage correlates with oxygen consumption in humans. *FASEB J.* **536**:534–537.

Park, E. M., Shigenaga, M. K., Degan, P., Korn, T. S., Kitzler, J. W., Wehr, C. M., Kolachana, P., and Ames, B. N. (1992). Assay of oxidized DNA lesions: Isolation of 8-oxoguanine and its nucleoside derivatives from biological fluids with a monoclonal antibody column. *Proc. Natl. Acad. Sci. USA* **89**:3375–3378.

Rajagopalan, R., Melamede, R. J., Laspia, M. F., Erlanger, B. F., and Wallace, S. S. (1984). Properties of antibodies to thymine glycol, a product of the radiolysis of DNA. *Radiat. Res.* **97**:499–510.

Scott, J. K., and Smith, G. P. (1990). Searching for peptide ligands with an epitope library. *Science* **249**:386–390.

Shigenaga, M. K., and Ames, B. N. (1991). Assays for 8-hydroxy-2′-deoxyguanosine: A biomarker of *in vivo* oxidative DNA damage. *Free Rad. Biol. Med.***10**:211–216.

Shigenaga, M. K., Gimeno, C. J., and Ames, B. N. (1989). Urinary 8-hydroxy-2′-deoxyguanosine as a biological marker of *in vivo* oxidative DNA damage. *Proc. Natl. Acad. Sci. USA* **86**:9697–9701.

Shigenaga, M. K., Aboujaoude, E. N., Chen, Q., and Ames, B. N. (1994). Assays of oxidative DNA damage biomarkers 8-oxo-2′-deoxyguanosine and 8-oxoguanine in nuclear DNA and biological fluids by high-performance liquid chromatography with electrochemical detection, in: *Methods in Enzymology: Oxygen Radicals in Biological Systems, Part D* (L. Packer, ed.), Academic Press, San Diego, CA, pp. 16–33.

Takeuchi, T., Nakajima, M., Ohta, Y., Kanae, M., Takeshita, T., and Morimoto, K. (1994). Evaluation of 8-hydroxydeoxyguanosine, a typical oxidative DNA damage, in human leukocytes. *Carcinogenesis* **15**:1519–1523.

Wagner, J. R., Hu, C. C., and Ames, B. N. (1992). Endogenous oxidative damage of deoxycytidine in DNA. *Proc. Natl. Acad. Sci. USA* **89**:3380–3384.

West, G. J., West, I. W.-L., and Ward, J. E. (1982a). Radioimmunoassay of 7,8-dihydro-8-oxadenine (8-hydroxy adenine). *Int. J. Radiat. Biol.* **42**:481–490.

West, G. J., West, I. W.-L., and Ward, J. F. (1982b). Radioimmunoassay of a thymine glycol. *Radiat. Res.* **90**: 595–608.

Chapter 9

Strategies for Measuring Damage and Repair in Gene-Sized Specific DNA Sequences

Charles A. Smith and Philip C. Hanawalt

1. INTRODUCTION

Interest in how the efficiency of DNA repair might vary among specific categories of cellular DNA dates almost to the origin of the "repair replication" technique, which quantifies the short stretches of DNA synthesized during excision repair (Pettijohn and Hanawalt, 1964). It has always been clear that the biological consequences of DNA damage to the cell or organism would depend strongly on the functional role of the particular segment of DNA suffering the damage. Early studies were confined to comparing repair in classes of DNA that were in relative abundance and could be physically isolated for analysis such as chloroplast and mitochondrial DNA, and genomic satellite DNA. Later, the repair of the highly repetitive alpha DNA sequences in African green monkey cells was investigated in detail using a variety of techniques. This was made possible by the abundance of this alpha DNA species; 8% of the DNA can be easily isolated as pure 172-base-pair fragments by digestion by *Hin*dIII and gel electrophoresis (Zolan *et al.*, 1982). These investigations (reviewed in Smith, 1987) demonstrated complex differences in the repair of this nontranscribed sequence as compared to the remaining, bulk DNA, and gave impetus to efforts to develop methods for studying repair in active genes.

Even if methods were available for isolating in pure form those genomic sequences found in only a few copies per cell, the amount of DNA required for direct analysis of lesion frequency would probably require experiments on a prohibitively large scale. Thus, the methods that have been devised are indirect ones. All make use of the specificity of nucleic acid hybridization to obtain information about unique DNA sequences within the genome. Some more recently devised methods use hybridization to position specific primers for the polymerase chain reaction, which is then used either to amplify lesion-free DNA or to map the sites of lesions. These techniques, discussed elsewhere in this volume, facilitate measurement of DNA damage at the nucleotide level. We will discuss here another class of methods, in which radioactive hybridiza-

---

Charles A. Smith and Philip C. Hanawalt • Department of Biological Sciences, Stanford University, Stanford, California 94305.

Technologies for Detection of DNA Damage and Mutations, edited by Gerd P. Pfeifer, Plenum Press, New York, 1996.

tion probes are used to detect and quantitate specific DNA sequences of interest in the vast collection of DNA isolated from cells. This detection step is carried out after the isolated DNA has been subjected to a procedure that physically partitions the molecules based on their content of DNA lesions, or tracts of repair synthesis.

The techniques we will discuss differ primarily in the means used to alter the behavior of damaged (or repaired) DNA molecules and the fractionation scheme applied to partition them, and they fall into three general categories. In the first, an agent is used to break the DNA quantitatively at the sites of lesions, and the DNA is then fractionated according to its size. The agent may be an enzyme, chemical, or photochemical treatment and the success of the technique depends in large part on the specificity and efficiency of the cleavage reaction. In the second category, the binding of ligands to lesions or repair tracts in the DNA is used to separate those molecules containing such sites from those that do not. The third category is the special case of DNA interstrand cross-links, which protect the DNA from irreversible denaturation. The DNA is subjected to a denaturing treatment, followed by conditions that permit renaturation. Different methods can then be used to separate the cross-link-containing (native) DNA molecules from cross-link-free (single-stranded) molecules. Each of these general methods may be applied to genomic DNA as isolated or after treatment with a suitable restriction endonuclease to obtain unique fragments of interest.

It is not our intent to provide an exposition of detailed protocols for all of the variations in these methods. Rather we wish to discuss the advantages and disadvantages of some of these methods and the considerations that are important in designing and carrying out studies using them.

2. REMOVAL OF CYCLOBUTANE PYRIMIDINE DIMERS—T4 ENDONUCLEASE ASSAYS

The first method devised to measure well-characterized lesions in gene sequences of low abundance (Bohr *et al.*, 1985) has been most widely used and forms the prototype from which other methods have been developed. We will treat this method in some detail and then discuss the special features of other methods derived from this one. A small "uv-endonuclease," encoded by the v gene of bacteriophage T4, termed T4 endonuclease V (TEV), is used to make single-strand scissions at the sites of UV-induced cyclobutane pyrimidine dimers (CPD) in DNA. The enzyme has been extensively characterized and contains both a glycosylase activity directed against the 5′ pyrimidine in a CPD and an AP lyase activity. Acting sequentially, these activities incise only the DNA strand containing the dimer.

2.1. General Procedure

Prior to treatment with TEV, DNA isolated from UV-irradiated cells is cleaved with a suitable restriction endonuclease to produce fragments of the gene or DNA region of interest whose size and location are known and for which specific hybridization probes are available. Equal amounts of restricted DNA are treated or mock-treated with TEV, denatured, and subjected to electrophoresis in alkaline agarose gels to separate the DNA molecules according to their lengths. The DNA is then Southern transferred from the gel onto a suitable support medium such as a nylon membrane and hybridized with ^{32}P-labeled probes to quantify the restriction fragments of interest. The analysis focuses solely on the amount of full-length restriction fragments in each of the two samples, as determined from the strength of hybridization. The ratio of the amount of hybridization to DNA in the appropriate band from the TEV-treated sample to that in its untreated counterpart yields the fraction of molecules in the population that contained *no* TEV-sensitive sites, often termed the "zero class" fraction. With the assumption that introduction of TEV-sensitive sites by irradiation and their removal by repair systems are random processes (discussed below), Poisson analysis is used to calculate the actual frequency of lesions using only the measured zero class fraction.

A detailed description of the basic method may be found in Bohr and Okumoto (1988) and in research articles (e.g., Mellon *et al.*, 1987) that include the use of strand-specific RNA probes. A "cookbook"-level treatment of this method for mammalian cells, yeast, and bacteria has recently been published (Spivak and Hanawalt, 1995). We now discuss factors to be considered in planning a study using the method and interpreting data obtained with it.

2.2. Choice of Fragments and Lesion Frequencies

A number of features of this assay place constraints on its use. To use the Poisson analysis, the zero class must be rather accurately determined. In most cases, the background-to-noise ratio for the autoradiographic analysis limits accurate assessment of the zero class to values greater than about 0.1. Thus, accurate measurement of the initial lesion frequency requires an irradiation dose sufficient to place the zero class fraction for the fragment analyzed into this range. However, as the initial value becomes too high, the changes in its magnitude brought about by repair become difficult to measure with confidence; one is then attempting to determine small differences between two relatively large numbers, and thereby increasing the uncertainty of the measurement. The ideal situation is an initial zero class that is small but accurately measured, so that increases in its value that reflect repair can be measured with confidence.

The initial zero class fraction is a product of the lesion frequency in the DNA, governed by the dose of damaging agent used, and the size of the fragment under study, determined by the available restriction sites and probes for the region of interest. These factors tend to place a lower limit on the size of fragment that can be analyzed. A UV-C dose of 20 J/m^2 would place about 0.6 CPD in a 5-kb DNA fragment, but above this dose repair in the overall mammalian genome is saturated and very few cells survive. In addition, at high UV doses, one must consider the effects of the introduction of significant quantities of the 6–4 photoproducts, the other major lesion introduced by this agent. Bacteria and yeast survive much higher UV doses than do mammalian cells, and experiments have been conducted at doses as high as 60 J/m^2, making measurements possible in appropriately smaller fragments. At the other end of the scale, the maximum size of the restriction fragment chosen is determined primarily by the success with which DNA can be isolated and purified without introducing single-strand breaks. At present, these considerations place an upper limit on useful fragment size of about 30 kb. In practice, this is generally not a serious constraint as most genes of interest are of this size or smaller, and larger restriction fragments are rarely available. However, it would appear to prevent application of the method to studies at doses much below about 5 J/m^2.

Besides the absolute size of the chosen restriction fragment, its size and location with respect to the functional unit under study are also critical. If possible, the restriction fragment chosen should reside within and represent a sizable fraction of the functional unit whose repair is to be studied. The use of a fragment several times the length of a gene of interest, for example, would pose several problems for interpretation of the data. At a frequency of 1–2 lesions per fragment, the majority of lesions would be outside the sequence of interest and efficient repair of lesions within it might be masked by inefficient repair in the surrounding sequences. In addition, a large fragment containing a small gene, for example, may not have been entirely characterized; it may contain other functional units whose repair properties could be very different from those of the gene under study, possibly leading to erroneous conclusions. On the other hand, the location of a fragment within a much larger functional unit can also be important. For example, the kinetics of transcription-coupled repair in a fragment located far downstream from a gene's promoter may be influenced by the presence of lesions induced between the promoter and the fragment under study.

2.3. Complications Related to DNA Replication

All lesion-free fragments present after repair incubation will contribute to the zero class fraction, whether generated by repair of preexisting DNA or by the process of DNA replication, synthesizing new (and therefore lesion-free) stretches of DNA after irradiation. Replication of the CPD-containing DNA has been demonstrated at the sequence level in CHO cells; 24 hr after irradiation the frequency of CPD in replicated parental strands was the same as in unreplicated DNA, and no transfer of CPD to daughter strands was detected (Spivak and Hanawalt, 1992) Thus, in these cells there appears to be no discrimination against replication of DNA strands containing CPD. For actively growing cells, newly replicated DNA is usually removed from analysis by including a density label in the posttreatment incubation medium and separating the newly synthesized DNA from unreplicated DNA in isopycnic CsCl density gradients. Depending on the cell type and dose used, a considerable fraction of the cellular DNA may in fact be removed from the repair analysis as a consequence of replication. For example, actively growing repair-proficient CHO cells irradiated with 10 J/m^2 UV replicate 60–80% of their DNA in 24 hr. This has two consequences. From a practical standpoint, more than twice as many cells are needed for longer incubations as for shorter ones to obtain enough unreplicated DNA for analysis. More importantly, the assay will measure repair only in a minor fraction of the DNA in the population, namely, that which did not successfully replicate. Repair in this fraction may not be characteristic of the rest of the DNA. In those cases in which growing cells and cells not actively synthesizing DNA (such as stationary phase cells or cells treated with hydroxyurea) have been compared, the results have been similar. At higher doses, and with human cells, which replicate much less of their DNA in comparable times after UV irradiation than do CHO cells, these considerations are less important.

2.4. Probes

Obviously, a prerequisite for using this method is that an adequate system for making probes is available. The ideal probe will be (1) free of repetitive DNA sequences, (2) completely homologous to the genomic fragment under study, (3) not homologous to any restriction fragments *larger* than the one under study, and (4) relatively large. However, probes not meeting all of these specifications have been successfully used.

Repetitive DNA in the probe sequence leads to background smears in the lanes of electrophoresed DNA, which can make analysis of hybridization to the fragment of interest difficult or impossible. In a few cases this background has been reduced to a manageable level by prehybridization of the membranes with nonradioactive genomic DNA from which all but repetitive sequences have been removed. The amount and nature of repetitive DNA in the probe determine whether this procedure will be successful. Its success has been mostly confined to situations in which cells amplified for the genomic region under study were used. Subcloning probes toremove repetitive DNA probably requires less effort than using this prehybridization technique.

Often cDNA probes for a particular gene from one species contain sufficient homology to be used successfully for the same gene in another species. However, large stretches of much-less-homologous 3′ untranslated sequences in cDNA probes can lessen their usefulness. Retroviral probes have been used to study the cellular homologues of the viral oncogenes (Madhani *et al.*, 1986).

It is of course essential that the fragment under study be well resolved from any other fragments that hybridize to the probe sequence, and it is also highly desirable that the fragment of interest be the largest one, since analysis of the smaller of two fragments is confounded by the background smear originating from the nuclease digestion products of the larger one. Usually the selection of a large restriction fragment ensures that these conditions are met. However, occasionally the positioning of fragments is such that a cDNA probe or a genomic probe spanning the restriction site defining the fragments precludes analysis without further subcloning of the probe.

Initially, probes were made by nick translation of plasmids into which had been cloned segments of genomic DNA or cDNA. This results in simultaneously measuring lesion frequencies in both strands of the fragment, and led to the discovery that CPD are generally removed from active genes more rapidly than from inactive sequences or the genome as a whole (Bohr *et al.*, 1985). By constructing probes specific for one or the other of the complementary strands, Mellon *et al.* (1987) were able to show that in human and CHO cells it is only the transcription template strand that exhibits more rapid repair. With the subsequent demonstration that this is also true for genes in yeast and bacteria, it has become clearly important to use strand-specific probes for studies of repair in specific sequences. The most common method for producing such probes is to use plasmid systems in which the probe DNA is situated between promoters specific for two different RNA polymerases, and produce radiolabeled RNA probes from one or the other promoter. The vector systems for this application are commercially available and moving DNA sequences from other cloning vectors into them has proven to be relatively straightforward. In many cases the specificity of the probes has been effectively demonstrated by analyses showing radically different rates of repair observed when using probes for the different strands.

Two other methods for preparing strand-specific probes have been used in repair analyses. In one case, each complementary sequence is cloned into a specialized single-stranded vector, constructed to allow excision and subsequent end-labeling of the insert (Venema *et al.*, 1991). A disadvantage of this method is the low specific activity obtained by end-labeling and the effort required to construct the specialized vectors. In the other case, linear PCR is used with one or another primer to generate probes specific for one strand or the other. This is formally analogous to making RNA probes, in that a copy of only one strand is produced from the cloned DNA sequence, but either probe can be produced from the starting material. However, it appears that conditions for PCR must be carefully defined to avoid generating probes that hybridize to both strands, presumably as a result of nonspecific priming (Ruven *et al.*, 1994).

2.5. Amplified Genes

Initial studies used CHO and human cells containing large stable intrachromosomal amplifications of a segment of the genome containing the DHFR gene, which increased the copy number of the sequences analyzed 50-fold (CHO) or 300-fold (human). This produced very strong hybridization signals from amounts of DNA well below the loading limit (10–20 μg) for gels used. It also allowed the use of large genomic probes that contained some middle-repetitive DNA sequences (see above). While it is unlikely that a gene of interest will generally be available for study in an amplified state, it is important to realize that much of the fine-structure analysis of repair, requiring accurate measurement of differences in repair, was greatly facilitated by their availability in amplified form. Although in most cases results obtained with cells not containing amplified genes have been similar to those with amplification, Tang *et al.* (1994a) reported that repair of certain adducts may be less efficient in the amplified DHFR gene in CHO cells than in cells not containing the amplification.

2.6. Measuring Hybridization Intensity

In the initial use of the method, the amount of hybridization was measured by excising the portions of the support membrane containing bands and equal-sized portions immediately above the bands and assaying the radioactivity with scintillation counting. This was made possible by the large signal resulting from the use of amplified genes. Subsequent studies have used autoradiography to measure radioactivity, using various types of densitometers or flat-bed scanners. This makes experiments with single-copy genes feasible and also allows sequential rehybridization with probes for different sequences or different strands from the same gene. This indirect quantitation carries with it a number of technical difficulties, however, which are beyond the scope of this chapter. Many of these can be alleviated with the use of direct assay devices such as phosphorimagers.

2.7. Calculations

The Poisson analysis requires only a measurement of the fraction of molecules in each sample that were resistant to TEV. The measurements are a comparison of hybridization intensity of the single band of interest in the enzyme-treated and mock-treated samples, and thus depend critically on the assumption that identical amounts of DNA have been processed and loaded on the gel. It is for this reason that for each sample, two equal portions are taken, treated and mock-treated, and all of the DNA from each is loaded in adjacent wells in the gel. Knowledge of the absolute amount of DNA loaded is not required, nor is it necessary that all time or dose points contain the same amount of DNA, although when this is possible this makes for a better visual presentation of the results. In some cases a small quantity of plasmid DNA containing a sequence that hybridizes to the probe to be used has been added to the samples at some point prior to the final analysis to serve as a control for pipetting, loading DNA onto the gel and transfer to the support membranes.

As mentioned above, the Poisson analysis assumes that the lesions are removed in a random fashion within the sequence under analysis. The results demonstrating different rates of repair between the transcribed and nontranscribed strands of numerous genes have shown clearly the effects of one type of nonrandom removal. In the case of CHO and human XP-C cells, CPD on the nontranscribed strand are removed very poorly or not at all, while efficient removal occurs from the transcribed strand. The effect on the calculations of having two populations repaired so differently was determined directly by probing sequentially with strand-specific probes and with a probe made by nick-translation, detecting both strands simultaneously. The value obtained with the latter probe was slightly higher than the average for the two strands, determined separately, as predicted from the properties of the Poisson distribution. These experiments were done at initial frequencies of about one lesion per fragment; it can be shown that this overestimate should increase dramatically with increased initial frequency.

Another kind of nonrandom removal that might invalidate the use of the Poisson analysis would be highly processive repair. If repair rapidly proceeded to completion in a particular gene after it began, regardless of the number of lesions in the gene, the fraction of lesion-free fragments would then directly reflect the fraction of lesions removed. With random removal, this relationship is approximated only at very low lesion frequencies; as lesion frequencies increase beyond one per fragment, the removal of a substantial fraction of the lesions is required before lesion-free fragments appear in any quantity. At lesion frequencies much greater than one per fragment, such a processive mechanism would cause an overestimation of the amount of repair, if expressed as the fraction of lesions removed. To what extent this may be a real factor is at present unknown, but should not substantially affect results of studies with frequencies below one lesion per fragment.

These considerations suggest that for accurate measurements of the fraction of lesions removed, initial lesion frequency should be roughly one per restriction fragment. Higher lesion frequencies would seem to be acceptable only for distinguishing generally proficient from deficient repair.

As with most biological measurements, this method is subject to variability from several sources. The number of manipulations required is considerable and determining the source of difficulties with hybridization analysis is not usually straightforward. Nonetheless, results demonstrating significant differences in repair rates or extents between sequences or cell types have been reproduced in a number of different laboratories. Attempts to distinguish small differences in repair, especially where repair efficiencies are low, entail considerable effort, requiring multiple experiments and replicate analysis of DNA samples (Lommel and Hanawalt, 1991).

3. REMOVAL OF BULKY ADDUCTS—UVRABC NUCLEASE ASSAY

Substituting the repair enzyme UVRABC nuclease of *E. coli* for the bacteriophage T4 nuclease has been used to extend assays described above to other bulky lesions, and is discussed in detail in this volume by Tang. Unlike TEV, the UVRABC nuclease is not specific for a single DNA lesion, but incises DNA at the sites of many different lesions, including CPD and 6–4 photoproducts, many bulky chemical adducts, thymine glycols, and abasic sites. Although this modification appears simple in principle, it is complicated by the nature of the UVRABC nuclease itself. TEV is a small, well-characterized, and easily purified enzyme whose activity is relatively stable and easy to measure. It functions catalytically in simple buffers, with no cofactor requirements, allowing the reaction buffer to contain EDTA to suppress action of any contaminating nucleases that do require divalent cation. Conditions can easily be found under which the DNA is incised at most or all of the CPD. It can be prepared with a minimum of biochemical manipulations using cells that can be induced to overproduce the protein, or can often be obtained as a gift from other investigators or obtained commercially. The same statements cannot be made for the UVRABC nuclease. To function as a nuclease, three separately prepared polypeptides must assemble on the DNA. These components must be highly purified and are not as stable as the small PD glycosylases. The complex usually makes two scissions on the DNA strand containing the lesion, one on each side of it. The reaction as a whole requires ATP and divalent cation. However, these components alone are insufficient for release of the complex and the reaction is stoichiometric rather than catalytic.

It is clear that to give a meaningful estimate of the lesion frequency, the system must be capable of reacting with a constant fraction of lesions, over a broad range of actual lesion frequencies. However, Thomas *et al.* (1988) found that for UV photoproducts and adducts of psoralen, *cis*-platinum, and 4-nitroquinoline oxide, their preparations made incisions at only approximately 20% of the adducts. Adduct frequencies in these studies ranged from 1 to 10 adducts per 6.6-kb fragment. It was shown later that about 50% of the adducts were found to be substrates, and after repurification of the DNA 50% of the remaining adducts were susceptible to incision. The favored explanation for this behavior is that some fraction of one or more of the individual components form nonproductive complexes, which nonetheless remain bound to a lesion and prevent the assembly of a productive complex at that site. This could be related to partial inactivation or modification of one or more of the components during preparation. Indeed, workers using other preparations have reported achieving complete or near-complete incision of lesions with different preparations of UVRABC (e.g., van Hoffen *et al.*, 1995; Tang *et al.*, 1994a; Chen *et al.*, 1992). It is not clear whether the differences in efficiency reported arise only from differences in the protein preparations and reaction conditions or if the different affinities of the proteins for different adducts may be partly responsible. If different adducts formed by a single agent were incised to different extents, this would certainly complicate the analysis.

The UVRABC nuclease has been particularly useful in studying repair of the 6–4 photoproduct, induced along with CPD by UV light. Purified DNA can be treated with CPD-specific photolyase prior to reaction with UVRABC. This effectively removes the CPD from the DNA without introducing single-strand breaks and facilitates assay of the remaining UVRABC-sensitive sites, primarily 6–4 photoproducts. It is important to appreciate, however, that generally in samples containing more than one type of lesion it is not possible to determine repair of each separately.

This system promises to open the study of repair in defined sequences to a large number of interesting DNA lesions. However, at present its use will probably be confined to those laboratories with experience in purifying and using the polypeptides that can function, under the proper conditions, as a lesion-specific nuclease.

An important consideration in the study of the repair of chemically induced damage is the degree to which not only the repair itself but also the introduction of lesions is influenced by the particular sequence under study. Heterogeneity can result both from the effects of DNA base sequence or composition on the formation of lesions as well as from differences in accessibility of DNA to the active damaging agent brought about by other DNA ligands and chromatin structure. For UV the immediate base sequence around a potential site for photoproduct formation is important and chromatin structure appears to modulate lesion induction differently for the major photoproducts. While positioning of nucleosomes does not significantly influence the distribution of CPD along the DNA, it has a radical effect on formation of 6–4 photoproducts, which appear to form predominately in the linker DNA regions as opposed to DNA in nucleosome cores. For stretches of DNA of several kilobases with average base composition, these effects would not be expected to lead to significantly different lesion frequencies; indeed to date the reported dose response for T4 endonuclease sites for all sequences studied have been similar, with the exception of ribosomal genes, whose high G+C content appears to lead to a lower efficiency of photoproduct formation. For chemical damaging agents, it has been known for some time that "active" DNA may accumulate more lesions than average; that different genes may have different sensitivities to such agents. This has been observed in studies of formation of psoralen adducts in particular sequences (see below).

4. REMOVAL OF BULKY ADDUCTS—OTHER ADDUCT-SPECIFIC ASSAYS

Any treatment that leads to specific DNA scissions specifically at the sites of adducts can obviously be used to substitute for the enzymatic action described above. Such an approach has been reported for the bulky adducts formed by the cancer-chemotherapeutic agent CC1065 (Tang *et al.*, 1994b). As is the case with smaller *N*-alkylations on purines (see next section), the bases alkylated by CC1065 are labile when heated and apurinic sites are formed, which can then be converted to single-strand breaks in the DNA by raising the pH.

It has been known for some time that adducts formed by benz(*a*)pyrene treatment of DNA sensitize the DNA to cleavage by irradiation by long-wavelength UV light. Laser irradiation with 355-nm light has been used to map the formation of adducts in short sequences and in nucleosomal DNA. We have initiated development of a method to use 355-nm irradiation to study induction and repair of adducts in fragments of genomic DNA (Baird *et al.*, 1994).

The specificity of the treatment used to cleave DNA would appear to be more critical in assays that do not use enzymes. The degree to which the control DNA is susceptible to cleavage must be carefully investigated and such controls should be run with each experiment. For chemical or photochemical treatment, mechanisms for cleavage at sites other than the lesions under study can exist which are independent of the mechanism of cleavage at the lesion, and different preparations of DNA could vary in their sensitivity to such "nonspecific" cleavage.

5. REMOVAL OF ALKYLATED BASES

The simple methylating and ethylating agents have received intense study as model chemical carcinogens. They alkylate DNA at a number of positions, producing *N*-alkylated bases like *N*-7 alkyl guanine and 3-alkyl adenine and *O*-alkylated bases (mainly *O*-6 alkyl guanine) in ratios that vary according to the nature of the agent used. These small adducts are not generally subject to the NER pathway for bulky lesions, but instead are removed from DNA by other mechanisms. Some, such as 7-methylguanine, increase the rate of spontaneous release of the base to yield an AP site; specific removal of both *N*-alkylated products occurs through the action of glycosylases, and some alkyl groups, such as those at the *O*-6 position of guanine, are subject to direct removal by alkyltransferases.

Scicchitano and Hanawalt (1989) developed a method to quantify methylated purines and their removal based on their chemical lability. Restricted DNA from cells treated with the relevant damaging agent is prepared and then incubated at 55°C to convert remaining *N*-methyl purines to AP sites. The DNA is then treated with alkali to cleave the DNA at positions of AP sites and the proportion of fragments free of alkali-labile sites is determined by Southern analysis. Because of the considerable chemical manipulations used to cleave DNA at sites of lesions, it is desirable to carry out a more complex "untreated control" for each sample. An equal portion of each sample is also heated in the presence of methoxyamine, which reacts with the AP sites formed to make them stable to the alkaline treatment. This provides a control that, importantly, is still subject to any lesion-independent action of the heat and alkali.

More recently, Wang *et al.* (1995) have adapted the enzymatic cleavage method to the study of small alkylations by using a DNA glycosylase specific for 3-methyladenine in DNA to cleave at the sites of these lesions. They used AP lyase activity of TEV to convert the AP sites resulting from glycosylase action to single-strand breaks. This made possible the study of the repair of 3-methyladenine, formed in minor amounts by methylating agents.

6. USE OF ANTIBODIES TO LESIONS

A second general method for modifying the behavior of DNA molecules containing lesions is the use of lesion-specific ligands, such as antibodies. A great deal of effort has been expended in recent years to prepare and characterize antibodies to specific DNA lesions. A highly specific antibody combined with a method for separating antibody-bound DNA molecules from unbound molecules should provide a very sensitive assay. This has been attained for thymine glycol, a prominent DNA oxidation product (Leadon and Lawrence, 1992). Besides their specificity, the antibodies used bind their substrate tightly enough to allow separation of the antibody-bound DNA from unbound DNA by a simple precipitation procedure. Samples of the bound and free DNA are then electrophoresed in parallel and their content of a specific fragment is measured by hybridization to radioactive probe. Unlike the cleavage method, this procedure detects the presence of lesion-containing molecules and potentially is more sensitive, because even if the fraction of fragments containing lesions is low, changes in the positive signal it generates can be monitored. Thus, in principle, smaller DNA fragments can be used with this method. This may be of great importance if the ability to separate a restriction fragment bound to a single antibody is strongly dependent on the fragment length. Ideally, the antibody should bind lesion sites in single-stranded rather than double-stranded DNA, as this would allow direct examination of lesion frequency on the individual complementary strands by using appropriate probes. Unfortunately, the lesion-specific antibodies prepared to date do not generally appear to allow separation of appropriate quantities of antibody-bound and free DNA by precipitation to use this technique. In two cases, the limited DNA recoverable by other separation techniques has been quantitated by PCR amplification to allow repair measurements (Denissenko *et al.*, 1994; Thomale *et al.*, 1994).

7. USE OF ANTIBODIES TO BROMOURACIL IN REPAIR PATCHES

A different method, using antibodies that bind to bromouracil in DNA, has proven very useful for study of fine structure of repair. Since this analogue can be specifically incorporated at the sites of repair synthesis in DNA, these antibodies can be used to separate DNA molecules that contain one or more repair patches from the remainder of the DNA (Leadon, 1988). Southern analysis is then used to determine the relative content of specific gene fragments in the repaired versus unrepaired fractions. This allows the simultaneous assay of multiple species detected by a single probe, such as the case for related gene families. This assay does not involve degradation of the restriction fragments, so that resolved multiple bands do not interfere with each other in the analysis.

This analysis measures the proportion of DNA fragments containing the probed sequence that carry enough BrUra to place them in the antibody-bound fraction. The question of how many BrUra-containing repair patches per DNA molecule are required for its precipitation is difficult to address; control experiments demonstrated that the system behaves more or less as if this number is small, at least for the size range of DNA analyzed. The method is used to make *comparisons* of the content of repair patches in different sequences rather than actual measurements of lesion frequencies. Thus, it is possible that differences observed could result from differing initial lesion frequencies, which are not determined by the method.

By using DNA prelabeled with tritium, repair in the sequences of interest can be directly compared to that in the overall genome. The method is applicable to any lesion whose removal is accompanied by insertion of BrUra at the repaired site, although it obviously cannot distinguish repair of different lesions caused by the same agent. It has advantages in situations where an efficiently repaired sequence, e.g., a small gene, is flanked by poorly repaired DNA sequences. With the Southern analysis, lesion-free fragments may not be formed even with efficient repair in the gene owing to poor repair in the flanking sequences. With the antibody method, fragments containing any repair patches are scored, so that differences in repair between different sequences or the same sequence in different cells or functional states can be observed even if the total amount of repair in the fragment containing the sequence is relatively low. This method was exploited to demonstrate transcription-coupled repair of adducts induced in DNA by ionizing radiation of human cells, and its absence in cells from patients with Cockayne's syndrome (Leadon and Cooper, 1993).

8. REMOVAL OF DNA CROSS-LINKING

DNA lesions in which covalent bonds are made at a single site to both of the complementary strands are made by a number of therapeutically important halogenated alkylating agents and bifunctional platinum compounds, mitomycin C, and the family of angular furocoumarins known as psoralens. Cross-links would pose greater difficulties to cells than adducts in only one DNA strand, both because strand separation would be needed for DNA transactions and because they compromise the information redundance in the structure of DNA, normally used by excision repair to restore information lost at the site of a lesion.

A unique feature of the cross-link itself has been used to study its repair. When DNA is subjected to denaturing treatments and quickly returned to conditions under which the rate of reannealing of separate molecules is low, only those molecules containing cross-links return to the duplex state. After such treatment the proportions of single-stranded and duplex DNA can be measured by specific nuclease digestion of single-stranded DNA or by physically separating the two classes of molecules by affinity chromatography, density gradient sedimentation, or, in the case of monodisperse populations of molecules, by using sedimentation or gel electrophoresis. Using the fraction of DNA containing cross-links and an estimate of the size of the DNA fragments, one can calculate the frequency of cross-links. Some of the methods used to gather information about processing of cross-links in the overall genome (Smith, 1988) have been adapted and combined with hybridization detection to examine cross-linking in specific sequences. In one method (Vos, 1988) restricted DNA is denatured and immediately electrophoresed in neutral agarose gels; the increased molecular weight and altered conformation of duplex DNA give it much lower mobility than single-stranded DNA of the same length. Standard Southern analysis is then used to quantitate the fraction of molecules containing cross-links. If the average number of cross-links per fragment is sufficiently low, Poisson analysis can be used to calculate the cross-link frequency. This method detects the appearance of a species (the duplex DNA) *not found* in the control (untreated DNA), rather than measuring changes in a species present at greatest abundance in the control DNA, as is the case for the nuclease method. This gives the cross-link assay considerably greater sensitivity than the cleavage method.

Another method analyzes unrestricted DNA and can be used to compare directly the cross-linking levels in DNA containing a given sequence to that in the genome overall. Cellular DNA is prelabeled by growth of the cells in radioactive DNA precursors prior to DNA damage. After damage and repair incubation, the isolated DNA is sheared under controlled conditions, denatured, and centrifuged to equilibrium in CsCl density gradients at a pH of 10.8, to maximize the difference in buoyant density between single-stranded and duplex DNA. Portions of the fractions obtained from the gradients are then assayed for radioactivity and analyzed by hybridization using a slot-blot apparatus. Sequential hybridizations may be used to obtain data for a variety of sequences. The cross-linked fraction may then be compared for the genome overall and the specific sequences. Since the DNA is not restricted, the size distribution of the DNA must be determined before an estimate of the actual cross-link frequency can be made. This can be done with standard sucrose gradient sedimentation, using the distribution of radioactivity to measure the average size of the genomic DNA from untreated cells (to avoid the complications of the altered sedimentation of cross-linked DNA). This depends on the assumption that the treatment does not alter the sensitivity of cross-linked DNA to shearing and handling. If the cross-link frequency per fragment is relatively small, a reasonable estimate of its actual value can be made; however, since the DNA is a heterogeneous population, the estimate is not as precise as that obtained with restricted DNA. The possibility must be considered that the average size of fragments containing a particular sequence is in fact not the same as the size determined for untreated cells, because of general toxic effects of the damage or because of the repair process itself. Although this is a difficult question to address, one can assess whether the DNA molecules containing the sequence of interest are grossly degraded by analyzing the fractions of the sucrose gradients by hybridization. From these complications, this method seems best suited to a comparative analysis of the initial cross-link frequencies in molecules bearing a number of different sequences of interest (Islas *et al.*, 1991).

For both of these methods, it must be emphasized that it is the amount of cross-linked DNA rather than the frequency of cross-links themselves that is measured. The loss of cross-linked DNA with time does not necessarily indicate complete repair, as an initial DNA strand scission at or near a cross-link would remove that DNA molecule from the cross-linked population.

In the case of the psoralens, these methods can also be extended to measure formation and removal of certain of the monoadducts. Psoralen cross-links are formed in a two-step process, whereby in an initial light-dependent reaction, intercalated psoralen molecules form different monoadduct species. Some of these are capable of a second photoreaction to form interstrand cross-links. The ratio of monoadducts to cross-links is dependent on the particular psoralen used, its concentration, and the irradiation dose. With the more soluble compounds, a combination of high concentration and low doses of light can be used to produce a considerable frequency of monoadducts with only a few percent cross-links. About 60–70% of the monoadducts made this way can be converted to cross-links by reirradiation of the isolated DNA in the absence of unbound psoralen. By reirradiating a portion of each sample DNA just before analysis, these cross-linkable monoadducts can be quantitated.

ACKNOWLEDGMENT. Research in the authors' laboratory is supported by an Outstanding Investigator Grant from the National Cancer Institute (CA44349) to P.C.H.

REFERENCES

Baird, W. M., Smith, C. A., Spivak, G., Mauthe, R. J., and Hanawalt, P. C. (1994). Analysis of the fine structure of the repair of anti-benzo[a]pyrene-7,8-diol-9,10-epoxide-DNA adducts in mammalian cells by laser-induced strand cleavage. *Polycyclic Aromatic Compounds* **6**:169–176.

Bohr, V. A., and Okumoto, D. S. (1988). Analysis of pyrimidine dimer repair in defined genes, in: *DNA Repair: A Laboratory Manual of Research Procedures*, Volume III (E. C. Friedberg and P. C. Hanawalt, eds.), Dekker, New York, pp. 347–366.

Bohr, V. A., Smith, C. A., Okumoto, D. S., and Hanawalt, P. C. (1985). DNA repair in an active gene: Removal of pyrimidine dimers from the DHFR gene of CHO cells is much more efficient than in the genome overall. *Cell* **40**:359–369.

Chen, R. H., Maher, V. M., Brouwer, J., van de Putte, P., and McCormick, J. J. (1992). Preferential repair and strand-specific repair of benzo[a]pyrene diol epoxide adducts in the HPRT gene of diploid human fibroblasts. *Proc. Natl. Acad. Sci. USA* **89**:5413–5417.

Denissenko, M. F., Venkatachalam, S., Yamasaki, E. F., and Wani, A. A. (1994). Assessment of DNA damage and repair in specific genomic regions by quantitative immuno-coupled PCR. *Nucleic Acids Res.* **22**:2351–2359.

Islas, A. L., Vos, J.-M., and Hanawalt, P. C. (1991). Differential introduction and repair of psoralen-DNA interstrand crosslinking in specific human genes. *Cancer Res.* **51**:2867–2873.

Leadon, S. A. (1988). Immunological probes for lesions and repair patches in DNA, in: *DNA Repair: A Laboratory Manual of Research Procedures*, Volume III (E. C. Friedberg and P. C. Hanawalt, eds.), Dekker, New York, pp. 311–326.

Leadon, S. A., and Cooper, P. K. (1993). Preferential repair of ionizing radiation-induced damage in the transcribed strand of an active human gene is defective in Cockayne syndrome. *Proc. Natl. Acad. Sci. USA* **90**:10499–10503.

Leadon, S. A., and Lawrence, D. A. (1992). Strand-selective repair of DNA damage in the yeast GAL7 gene requires RNA polymerase II. *J. Biol. Chem.* **267**:23175–23182.

Lommel, L., and Hanawalt, P.C. (1991). The genetic defect in the Chinese hamster ovary cell mutant UV61 permits moderate selective repair of cyclobutane pyrimidine dimers in an expressed gene. *Mutat. Res.* **255**:183–191.

Madhani, H. D., Bohr, V. A., and Hanawalt, P.C. (1986). Differential DNA repair in a transcriptionally active and inactive proto-oncogene: c-*abl* and c-*mos*. *Cell* **45**:417–423.

Mellon, I., Spivak, G., and Hanawalt, P. C. (1987). Selective removal of transcription-blocking DNA damage from the transcribed strand of the mammalian DHFR gene. *Cell* **51**:241–249.

Pettijohn, D., and Hanawalt, P. C. (1964). Evidence for repair-replication of ultraviolet damaged DNA in bacteria. *J. Mol. Biol.* **9**:395–410.

Ruven, H. J., Seelen, C. M., Lohman, P. H., Mullenders, L. H., and van Zeeland, A. A. (1994). Efficient synthesis of ^{32}P-labeled single-stranded DNA probes using linear PCR, application of the method for analysis of strand-specific DNA repair. *Mutat. Res.* **315**:189–195.

Scicchitano, D., and Hanawalt, P. C. (1989). Repair of N-methylpurines in specific DNA sequences in Chinese hamster ovary cells: Absence of strand specificity in the dihydrofolate reductase gene. *Proc. Natl. Acad. Sci. USA* **86**:3050–3054.

Smith, C. A. (1987). DNA repair in specific sequences in mammalian cells. *J. Cell Sci.* Suppl. **6**:225–241.

Smith, C. A. (1988). Repair of DNA containing furocoumarin adducts, in: *Psoralen DNA Photobiology*, Volume II (F. Gasparro, ed.), CRC Press, Boca Raton, FL, pp. 87–116.

Spivak, G., and Hanawalt, P. C. (1992). Translesion DNA synthesis in the DHFR domain of UV-irradiated CHO cells. *Biochememistry* **31**:6794–6800.

Spivak, G., and Hanawalt, P. C. (1995). Determination of damage and repair in specific DNA sequences, in: *Methods: A Companion to Methods in Enzymology*, Vol. 7, Academic Press, London, pp. 147–161.

Tang, M. S., Pao, A., and Zhang, X. S. (1994a). Repair of benzo(a)pyrene diol epoxide- and UV-induced DNA damage in dihydrofolate reductase and adenine phosphoribosyltransferase genes of CHO cells. *J. Biol. Chem.* **269**:12749–12754.

Tang, M. S., Qian, M., and Pao, A. (1994b). Formation and repair of antitumor antibiotic CC-1065-induced DNA adducts in the adenine phosphoribosyltransferase and amplified dihydrofolate reductase genes of Chinese hamster ovary cells. *Biochemistry* **33**:2726–2732.

Thomale, J., Hochleitner, K., and Rajewsky, M. F. (1994). Differential formation and repair of the mutagenic DNA alkylation product O^6-ethylguanine in transcribed and nontranscribed genes of the rat. *J. Biol. Chem.* **269**: 1681–1686.

Thomas, D. C., Morton, A. G., Bohr, V. A., and Sancar, A. (1988). General method for quantifying base adducts in specific mammalian genes. *Proc. Natl. Acad. Sci. USA* **85**:3723–3727.

van Hoffen, A., Venema, J., Meschini, R., van Zeeland, A. A., and Mullenders, L. H. (1995). Transcription-coupled repair removes both cyclobutane pyrimidine dimers and 6–4 photoproducts with equal efficiency and in a sequential way from transcribed DNA in xeroderma pigmentosum group C fibroblasts. *EMBO J.* **14**:360–367.

Venema, J., van Hoffen, A., Karcagi, V., Natarajan, A. T., van Zeeland, A. A., and Mullenders, L. H. (1991). Xeroderma pigmentosum complementation group C cells remove pyrimidine dimers selectively from the transcribed strand of active genes. *Mol. Cell. Biol.* **11**:4128–4134.

Vos, J.-M. (1988). Analysis of psoralen monoadducts and interstrand crosslinks in defined genomic sequences, in: *DNA Repair: A Laboratory Manual of Research Procedures*, Volume III (E. C. Friedberg and P. C. Hanawalt, eds.), Dekker, New York, pp. 367–398.

Wang, W., Sitaram, A., and Scicchitano, D. A. (1995). 3-Methyladenine and 7-methylguanine exhibit no preferential removal from the transcribed strand of the dihydrofolate reductase gene in Chinese hamster ovary B11 cells. *Biochemistry* **34**:1798–1804.

Zolan, M. E., Cortopassi, G. A., Smith, C. A., and Hanawalt, P. C. (1982). Deficient repair of chemical adducts in alpha DNA of monkey cells. *Cell* **28**:613–619.

Chapter 10

Methods to Measure the Repair of Genes

Vilhelm A. Bohr

1. INTRODUCTION

Recently, there has been a lot of interest and many studies in the area of gene-specific repair. Demonstrations of DNA repair heterogeneity within the genome date back more than a decade. An early example of repair heterogeneity was the demonstration that repair incorporation in the first hours after UV damage was preferentially associated with the nuclear matrix (Mullenders *et al.*, 1984). Other studies had shown DNA damage and repair heterogeneity in different subfractions of the cellular nucleus, within the genome, or in repetitive DNA sequences (reviewed in Bohr *et al.*, 1987). A critical advance in this area of research was the development of a general technique to study the DNA repair process in individual genes (Bohr *et al.*, 1985). This assay employed quantitative Southern blot analysis and it has at times been called the Bohr–Hanawalt technique. It is described below. The principal idea in this approach is to use a DNA repair enzyme to generate a strand break at the site of a lesion, and then to measure the frequency of DNA lesions within specific restriction fragments located within specific genes. Early experiments using this approach showed that genes could be repaired much more efficiently than the bulk of the DNA, and this led to the introduction of the concept of "preferential gene repair." In the past years, numerous experiments have examined this technique and the use of strand-specific DNA probes in this same assay (Mellon *et al.*, 1987) has led to new insight into the molecular link between DNA repair and transcription.

2. METHODS

2.1. Southern Blot Assay

The principles involved are shown in Fig. 10.1. After exposure of cells in culture, or of human tissue, to a DNA-damaging agent, cells are incubated for repair in the presence of bromodeoxyuridine (BrdUrd) in order to label semiconservative replication. The DNA is then extracted and purified, and treated with an appropriate restriction endonuclease. The choice of

Vilhelm A. Bohr • Laboratory of Molecular Genetics, National Institute on Aging, National Institutes of Health, Baltimore, Maryland 21224.

Technologies for Detection of DNA Damage and Mutations, edited by Gerd P. Pfeifer, Plenum Press, New York, 1996.

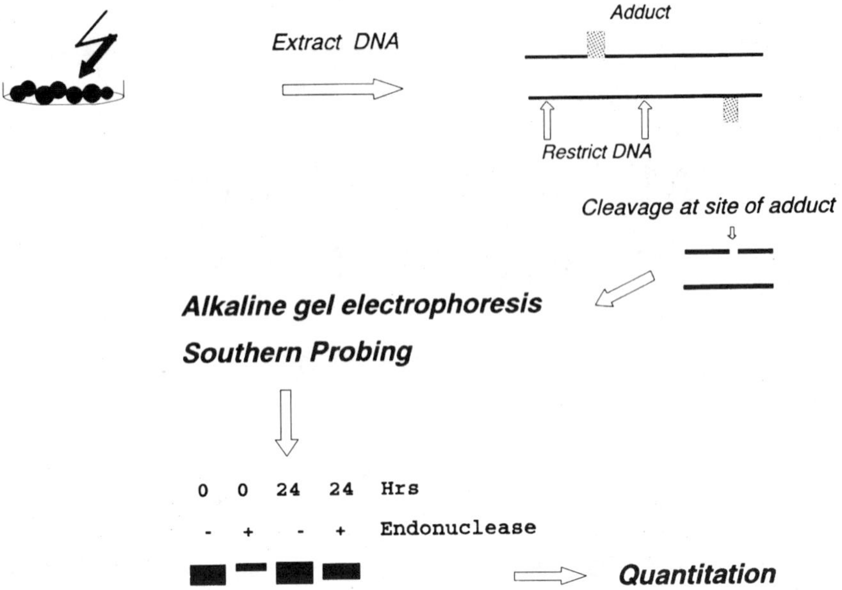

Figure 10.1. General principles involved in gene-specific repair measurements using the Southern blot assay.

endonuclease depends on the restriction map of the respective gene, and on the particular site to be analyzed. Parental DNA is separated by isopycnic CsCl centrifugation, and then treated with a repair enzyme or a chemical agent to generate a single strand break at the site of the lesion. The use of this cutting agent is a key element in the assay. Single-stranded DNA is then resolved on alkaline agarose gels, blotted and probed for the gene and strand of interest. Uncut full-length DNA fragments represent the zero class of the lesion frequency, and the lesion frequency per fragment can then be calculated. A control lane is run where the cutting agent is omitted. The entire procedure is described in detail in Bohr and Okumoto (1988) for further reference.

This procedure has some distinct advantages: one is that once the DNA is fixed on a membrane it can be probed for a number of genes and for the individual DNA strands of genes. Thus, the DNA damage and repair of various genes can be compared in the same biological sample, i.e., in one experiment. Whereas there is some interexperimental variability in this rather laborious, multistep approach, the variability is significantly reduced when comparing the repair of different genes within one experiment. Another distinction of this assay is that the initial DNA damage frequency is inherently measured. There are a number of DNA repair assays where this is not the case, and it is often very important to assess the level of DNA damage introduced as a basis for the DNA repair results; heterogeneity in DNA damage can lead to adverse DNA repair results.

The Southern blot assay has been modified to be used with a number of different types of DNA damage, and most of them are shown in Table I. The method used for cutting lesion-containing DNA is shown there, and the appropriate reference for the procedure is listed. The assays and results shown in Table I are all limited to results obtained in the hamster dihydrofolate reductase (DHFR) gene, which has been used as a model system by us and others. Repair in a 14-kb fragment located at the 5′ end of the DHFR gene has been compared to the repair in

Table I
Hamster DHFR System

Lesion	Probe	Preferential repair?	Reference
UV			
Pyrimidine dimer	T4 endonuclease V	Yes	Bohr *et al.* (1985)
6–4 photoproduct	Photolyase plus ABC excinuclease	Yes	Thomas *et al.* (1989)
Cisplatin			
Intrastrand adduct	ABC excinuclease	Yes	Jones *et al.* (1991)
Interstrand cross-link	Denaturation and neutral gel	Yes[a]	Jones *et al.* (1991)
Bulky adducts			
NAAAF	ABC excinuclease	No	Tang *et al.* (1989)
4NQO	ABC excinuclease	No	Snyderwine and Bohr (1992)
BPDE	ABC excinuclease	Yes	Tang *et al.* (1994)
BcPHDE	ABC excinuclease	Yes	Carothers *et al.* (1992)
Oxidative stress			
8-OH guanosine	FaPy glycosylase		Bohr *et al.* (1995)
Alkylation			
N7 methylpurines	Neutral depurination, alkaline hydrolysis	No	Wassermann *et al.* (1990)
Interstrand cross-links	Neutral gel	Yes	Larminat *et al.* (1993)
X-irradiation	Alkaline conditions		
Psoralen adducts	Neutral gel		Vos and Hanawalt (1987)

[a]Some studies have not found preferential repair of interstrand cross-links.

a downstream inactive DNA region (cs14DO). Also, the strand bias of repair has been measured with these protocols in that same 5′ fragment of the hamster DHFR gene.

A lot of important information has been obtained using these assays. For recent reviews, see Bohr (1991, 1993, 1995), Hanawalt and Mellon (1993), and Hanawalt (1994). In the comparative studies of the gene- and strand-specific repair of different lesions, the initial damage introduced was very similar, about 1–2 adducts per 14 kb. The data (some of which are listed in Table I) show that some lesions are repaired preferentially while others are not. In general, lesions removed by base excision repair are not preferentially repaired, but not all lesions that are repaired by nucleotide excision repair (NER) are preferentially repaired. An interesting example is the difference in repair of pyrimidine dimers and adducts formed by the carcinogen 4-nitroquinoline (4NQO). Both of these lesions are thought to be repaired by NER, but whereas the dimer is preferentially repaired, 4NQO adducts are not. The reason for this discrepancy is not known, and rather than any difference in transcription blockage, there may be a difference in the degree of distortion of chromatin structure that the lesions introduce.

An additional useful feature of the Southern blot method is that it can easily be employed to measure DNA damage and repair in mitochondria. These organelles have been very inaccessible to this type of study because it had previously required the isolation of mitochondrial DNA, a difficult and involved procedure. Now this gene specific repair assay is providing data that have dramatically changed our understanding about mitochondrial DNA repair (LeDoux *et al.*, 1992). Also, nuclear and mitochondrial DNA damage and repair can be compared in the same cells in the same experiment.

Whereas the strand bias of the repair is linked to transcription, preferential gene repair can be dependent or independent of transcription (Beecham *et al.*, 1994). Other factors, possibly including chromatin structure and organization, may play roles in the regulation of preferential DNA repair, and there seems to be a repair hierarchy within cells. Some genes, such as the p53 tumor suppressor gene, are very rapidly repaired (Evans *et al.*, 1993) while others are repaired with much less efficiency. Further support for the notion that gene-specific repair can be independent of transcription is the observation that there are different levels of repair in inactive genes (Evans and Bohr, 1994). Currently, it is an area of very active investigation to clarify aspects of the regulation of preferential and strand-specific DNA repair.

2.2. T4 Polymerase Assay

This approach was developed in order to provide a more sensitive assay for the measurement of gene- and strand-specific DNA repair (Rampino and Bohr, 1994). Single-stranded DNA, capable of hybridizing to gene-specific probes, is generated enzymatically by T4 DNA polymerase. In the absence of dNTPs, the T4 DNA polymerase functions as a 3′-5′ exonuclease that, starting at a free 3′ end, will hydrolyze away one strand until it encounters a blocking lesion. Some lesions, such as UV photoproducts and psoralen interstrand cross-links, have been shown to block the exonucleolytic progression of the T4 polymerase. On incubation with T4 polymerase, the DNA, single-stranded in the ends that are free of lesions, is slot-blotted onto a membrane and hybridized against a labeled gene probe. The degree of probe binding is a direct measure of the proportion of molecules that are free of lesions in the region between the cut site and the distal end of the gene probe sequence. The lack of lesions is related either to a low level of initial damage to the region or to repair activity. To establish this assay, several experiments were performed on a plasmid containing the DHFR gene, and cisplatin was used as the DNA-damaging agent. The kinetic relationship between probe hybridization and cisplatin content was measured. Such experiments also indicated that cisplatin lesions affected a block to the T4 polymerase.

In contrast to other polymerase-based assays (such as several of the ones using PCR, described below), this approach inherently permits strand-specific repair measurements and it can in principle be used with a wide range of DNA-damaging agents that block progression of the T4 polymerase.

The procedure for these experiments has been described (Rampino and Bohr, 1994), and will be detailed here: To stop repair and extract the DNA, a lysis buffer [0.5 M Tris pH 8.0, 20 mM EDTA, 10 mM NaCl, 1% SDS, and 0.5 mg/ml proteinase K (Boehringer-Mannheim, Indianapolis, IN)] is added at 0, 2, 3, or 4 hr after exposure of the cells to the DNA-damaging agent, and the cells are incubated in the lysis buffer for 24 hr at 37°C. The DNA is isolated from the cellular debris by addition of 1/4 vol of saturated (6 M) NaCl solution and centrifugation (500*g* for 30 min). Next the DNA is ethanol precipitated, resuspended in TE buffer (10 mM Tris pH 8,1 mM EDTA), and treated with 100 μg/ml RNase (Boehringer-Mannheim) for 3 hr at 37°C, then reprecipitated and resuspended.

T4 polymerase reaction and slot blotting: Genomic DNA (2 μg) is incubated at 37°C with 10 units of restriction endonuclease (for the DHFR gene we use *Hin*dIII) for 30 min, 25 μl of T4 polymerase buffer, and then 1 unit of T4 polymerase is added. The reactions are stopped with 25 μl of 0.2 mM EDTA. The conditions for the use of T4 polymerase are well described in the manual by Sambrook *et al.* (1989, Chapter 5.38). It is important to maximize the exonuclease activity by omission of dNTPs, and to test the buffer conditions for the individual T4 polymerase batch. The double-stranded DNA molecules, partially digested by T4 polymerase, contain a single-stranded component after the incubation with the enzyme. The DNA enzyme digests are slot-blotted (Hoefer Scientific, San Francisco CA) onto a charged nylon membrane that has been presoaked in H_2O and then 20× SSC buffer (Sambrook *et al.*, 1989). After transfer of the sample DNA the membranes are dried under vacuum at 80°C for 2 hr. Charged nylon membrane binds both double- and single-stranded DNA (see The Genius System User's Guide for Filter Hybridization, Boehringer-Mannheim Corp., Indianapolis, IN), and we have confirmed that the binding of both species was highly efficient for the membranes used in these experiments (Sure Blot™, Oncor, Gaithersburg, MD), by detecting levels of binding that are statistically indistinguishable.

Hybridization and chemiluminescent detection: The Genius chemiluminescent filter hybridization protocols are followed (Boehringer-Mannheim). Prehybridization at 65°C is conducted for 20 hr prior to adding the denatured digoxigenin labeled probe (2.5 ng/ml). Hybridization is then conducted for an additional 20 hr at 65°C. After conjugation with alkaline phosphatase, the membranes are treated with Lumi-Phos™ 530 (Boehringer-Mannheim) and exposed on XMAT Kodak film.

Densitometry and data analysis: Film negatives are digitized on a Molecular Dynamics Personal Densitometer™ (Sunnyvale, CA) using a 100-μm pixel size with 4096 (12-bit) levels of gray scale resolution. Background-corrected band densities are computed and normalized to give band intensities measured in units of optical density. Statistical tests of significance on the data sets are carried out in the Excel (Microsoft Corp., Redmond, WA) and JMP (SAS, Cary, NC) software packages. We have used nonradioactive probing, but the assay appears to work also with radioactive probing.

To assay for gene-specific repair of cisplatin lesions, synchronized human cells were treated with 100 μM cisplatin for 1 hr and then allowed to grow in drug-free media before the DNA was isolated. Two hours of cellular DNA repair yielded DNA to which a DHFR gene probe hybridized intensely (equal to that of unplatinated control DNA), and to which a δ-globin gene probe hybridized weakly. We conclude that, in early G_1 phase, the constitutively active DHFR gene experiences very rapid preferential repair, while the nontranscribed δ-globin gene shows little repair during this phase of the cell cycle (Rampino and Bohr, 1994).

This polymerase assay has been used to detect relatively low levels of DNA damage. Cisplatin concentrations at least several-fold lower than those used in other fine-structure repair assays have been used. Given that the dose is important in gene-specific repair measurements, the increased sensitivity is a useful aspect of this assay. It also may be useful for a variety of DNA lesions; several have already been shown to block the progression of the T4 polymerase, but the range of lesions that can be detected needs to be determined. DNA lesions introduced by oxidative stress can be detected, although it is not clear which of the many possible oxidative lesions actually blocks the enzyme. For oxidative stress, this approach has been used to detect DNA damage after exposure of cells to 50 μM H_2O_2 (Rampino, 1995). It thus may prove useful in repair studies involving oxidative stress.

2.3. Precipitation with Monoclonal Antibody

A method was developed where DNA is incubated with BrdUrd during the repair. Parental DNA is reacted with an antibody toward BrdUrd and precipitated. DNA containing repair patches will precipitate whereas unrepaired DNA remains in the supernatant. Total genomic DNA repair can be calculated based on DNA content (prelabeled DNA) and gene-specific repair can be determined by slot-blot analysis and probing for the gene of interest (Leadon and Cooper, 1993). In principle, this method works for a variety of DNA lesions, and it can also be converted to an assay for DNA damage rather than repair by the use of an antibody to a DNA adduct and then performing a similar quantitation. Recently this approach has been taken using a monoclonal antibody to thymine glycols (Leadon and Snowdon, 1988). A limitation in the method using BrdUrd antibody is that the level of initial DNA damage is not being assessed; it is thus only a measure of the repair without knowing the initial damage level. Another method employing antibody binding plus quantitative PCR technique to detect gene-specific repair (of alkylating agents) has recently been developed (Thomale *et al.*, 1994).

2.4. Quantitative PCR

As PCR technology rapidly develops and improves, it is obviously attractive to use this approach to detect gene-specific DNA repair. It has been demonstrated in various ways that the progression of the (taq) polymerase is blocked at certain DNA lesions and that the lesions can be detected as a decrease in generated PCR product. A previous limitation was that PCR products could not exceed a few kilobases in length, and in order to get enough DNA damage into a small region, it was necessary to expose the cells to vast amounts of DNA damage. This would thus render the experiments nonphysiological. The first of these studies was by Govan *et al.* (1990), and although it was possible to demonstrate repair, the doses were much higher than what other investigators have used in similar studies. Recently, the PCR-based repair techniques have improved and have also become more quantitative (Jennerwein and Eastman, 1991; Kalinowski *et al.*, 1992), and it is now possible to extend to larger DNA fragments by PCR and thus to use lower, more physiological doses of DNA damage for the exposure (Denissenko *et al.*, 1994).

The requirement to use physiological doses in gene-specific repair experiments reflects not only physiological considerations, but also that it is now evident that preferential DNA repair is dose dependent (Larminat *et al.*, 1993). The main problems to overcome remain the challenge to generate quantitative PCR products, and the need to have strand-specific assays since repair can be strand specific. The related assays for ligation-mediated PCR (LMPCR) have been used with considerable success, and this is dealt with in Chapters 15 and 16 of this book.

3. PERSPECTIVES

With the recent very extensive interest in DNA repair, many new approaches and methods are emerging, and the field is undergoing an extensive metamorphosis at this time. The *in vitro* assays for DNA repair have gained much interest and proven extremely useful in the understanding of molecular biochemistry of the incision process of DNA repair (Aboussekhra and Wood, 1994). For mammalian cells, *in vitro* assays are, however, still limited to detection of the repair process in inactive DNA. To understand the molecular interactions involved in transcription coupled DNA repair, it will be necessary to establish *in vitro* assays that permit detection of transcription coupling on plasmid DNA containing single lesions. This type of assay is under development in several laboratories at the time of this writing, and it will undoubtedly be available at the time of publication of this book. When available, such technique should be of wide interest as an incisive way of studying the mechanism of gene-specific DNA repair.

REFERENCES

Aboussekhra, A., and Wood, R. D. (1994). Repair of UV-damaged DNA by mammalian cells and Saccharomyces cerevisiae. *Curr. Opin. Genet. Dev.* **4**:212–220.

Beecham, E. J., Jones, G. M., Link, C., Huppi, K., Potter, M., Mushinski, J. F., and Bohr, V. A. (1994). DNA repair defects associated with chromosomal translocation breaksite regions. *Mol. Cell. Biol.* **14**:1204–1212.

Bohr, V. A. (1991). Gene specific DNA repair. *Carcinogenesis* **12**:1983–1992.

Bohr, V. A. (1993). Gene specific DNA repair. Relation to mutagenesis and genomic instability, in: *Alfred Benzon Symposium 35: DNA Repair Mechanisms* (V. A. Bohr, K. Wassermann, and K. H. Kraemer, eds.), Munksgaard, Copenhagen, pp. 217–228.

Bohr, V. A. (1995). DNA repair fine structure and its relation to genomic instability. *Carcinogenesis*, **16**:2885–2892.

Bohr, V. A., and Okumoto, D. S. (1988). Analysis of pyrimidine dimers in defined genes, in: *DNA Repair: A Laboratory Manual of Research Procedures* (E. C. Friedberg and P. C. Hanawalt, eds.), Dekker, New York, pp. 347–366.

Bohr, V. A., Smith, C. A., Okumoto, D. S., and Hanawalt, P. C. (1985). DNA repair in an active gene: Removal of pyrimidine dimers from the DHFR gene of CHO cells is much more efficient than in the genome overall. *Cell* **40**:359–369.

Bohr, V. A., Phillips, D. H., and Hanawalt, P. C. (1987). Heterogenous DNA damage and repair in the mammalian genome. *Cancer Res.* **47**: 6426–6436.

Bohr, V. A., Taffe, B., and Larminat, F. (1995). DNA repair, oxidative stress, and ageing, in: *Oxidative Stress and Ageing* (R. Cutler and L. Packer, eds.), Birkhauser Verlag, Berlin, pp. 101–111.

Carothers, A. M., Zhen, W., Mucha, J., Zhang, Y., Santella, R. M., Grunberger, D., and Bohr, V. A. (1992). DNA strand-specific repair of $(+/-)-3\alpha,4\beta$-dihydroxy-$1\alpha,2\alpha$-epoxy-1,2,3,4-tetrahydro-benzo[*c*]phenanthrene adducts in the hamster dihydrofolate reductase gene. *Proc. Natl. Acad. Sci. USA* **89**:11925–11929.

Denissenko, M. F., Venkatachalam, S., Yamasaki, E. F., and Wani, A. A. (1994). Assessment of DNA damage and repair in specific genomic regions by quantitative immuno-coupled PCR. *Nucleic Acids Res.* **22**:2351–2359.

Evans, M. K., and Bohr, V. A. (1994). Gene-specific DNA repair of UV-induced cyclobutane pyrimidine dimers in some cancer-prone and premature-aging human syndromes. *Mutat Res.* **314**:221–231.

Evans, M. K., Taffe, B. G., Harris, C. C., and Bohr, V. A. (1993). DNA strand bias in the repair of the p53 gene in normal human and xeroderma pigmentosum group C fibroblasts. *Cancer Res.* **53**:5377–5381.

Govan, H. L., 3rd, Valles-Ayoub, Y., and Braun, J. (1990). Fine-mapping of DNA damage and repair in specific genomic segments. *Nucleic Acids Res.* **18**:3823–3830.

Hanawalt, P. C. (1994). Transcription-coupled repair and human disease. *Science* **266**:1957–1958.

Hanawalt, P. C., and Mellon, I. (1993). Stranded in an active gene. *Curr. Biol.* **3**:67–69.

Jennerwein, M. M., and Eastman, A. (1991). A polymerase chain reaction based method to detect cisplatin adducts in specific genes. *Nucleic Acids Res.* **19**:6209–6214.

Jones, J., Zhen, W., Reed, E., Parker, R. J., Sancar, A., and Bohr, V. A. (1991). Gene specific formation and repair of cisplatin intrastrand adducts and interstrand crosslinks in CHO cells. *J. Biol. Chem.* **266**:7101–7107.

Kalinowski, D. P., Illenye, S., and Van Houten, B. (1992). Analysis of DNA damage and repair in murine leukemia L1210 cells using a quantitative polymerase chain reaction assay. *Nucleic Acids Res.* **20**:3485–3494 (Abstract).

Larminat, F., Zhen, W., and Bohr, V. A. (1993). Gene-specific DNA repair of interstrand cross-links induced by chemotherapeutic agents can be preferential. *J. Biol. Chem.* **268**:2649–2654.

Leadon, S. A., and Cooper, P. K. (1993). Preferential repair of ionizing radiation-induced damage in the transcribed strand of an active human gene is defective in Cockayne's syndrome. *Proc. Natl. Acad. Sci. USA* **90**:10499–10503.

Leadon, S. A., and Snowdon, M. M. (1988). Differential repair of DNA damage in the human metallothionein gene family. *Mol. Cell. Biol.* **8**:5331–5338.

LeDoux, S. P., Wilson, G. L., Beecham, E. J., Stevnsner, T., Wassermann, K., and Bohr, V. A. (1992). Repair of mitochondrial DNA after various types of DNA damage in Chinese hamster ovary cells. *Carcinogenesis* **13**:1967–1973.

Mellon, I., Spivak, G., and Hanawalt, P. C. (1987). Selective removal of transcription-blocking DNA damage from the transcribed strand of the mammalian DHFR gene. *Cell* **51**:241–249.

Mullenders, L. H. F., van Kesteren, A. C., Bussman, C. J. M., van Zeeland, A. A., and Natarajan, A. T. (1984). Preferential DNA repair of nuclear matrix associated DNA in xeroderma pigmentosum group C. *Mutat. Res.* **141**:75–82.

Rampino, N. J. (1995). Oxidative DNA damage and preferential repair. *J. Cell. Biochem.* **294** (Abstract).

Rampino, N., and Bohr, V. A. (1994). Rapid preferential repair of cisplatin lesions at the DUG/DHFR locus comprising the divergent upstream gene and DHFR gene during early G1 phase of the cell cycle assayed by using the exonucleolytic activity of the T4 DNA polymerase. *Proc. Natl. Acad. Sci. USA* **91**:10977–10981.

Sambrook, J., Fritsch, E. F., and Maniatis, T. (1989). *Molecular Cloning: A Laboratory Manual*, 2nd ed., Cold Spring Harbor Laboratory Press, Cold Spring Harbor, NY.

Snyderwine, E. G., and Bohr, V. A. (1992). Gene- and strand-specific damage and repair in Chinese hamster ovary cell treated with 4-nitroquinoline 1-oxide. *Cancer Res.* **52**:4183–4189.

Tang, M. S., Bohr, V. A., Zhang, X. S., Pierce, J., and Hanawalt, P. C. (1989). Quantitation of aminofluorene adduct formation and repair in defined DNA sequences in mammalian cells using UVrABC nuclease. *J. Biol. Chem.* **264**:14455–14462.

Tang, M., Pao, A., and Zhang, X. (1994). Repair of benzo(a)pyrene diol epoxide- and UV-induced DNA damage in dihydrofolate reductase and adenine phosphoribosyltransferase genes of CHO cells. *J. Biol. Chem.* **269**: 12749–12754.

Thomale, J., Hochleitner, K., and Rajewsky, M. F. (1994). Differential formation and repair of the mutagen DNA alkylation product O6 ethylguanine in transcribed and nontranscribed genes of the rat. *J. Biol. Chem.* **269**:1681–1686.

Thomas, D. C., Okumoto, D. S., Sancar, A., and Bohr, V. A. (1989). Preferential repair of (6–4) photoproducts in the dihydrofolate reductase gene of Chinese hamster ovary cells. *J. Biol. Chem.* **264**:18005–18010.

Vos, J., and Hanawalt, P. C. (1987). Processing of psoralen adducts in an active gene: Repair and replication of DNA containing monoadducts and interstrand crosslinks. *Cell* **50**:1789–1799.

Wassermann, K., Kohn, K. W., and Bohr, V. A. (1990). Heterogeneity of nitrogen mustard-induced DNA damage and repair at the level of the gene in Chinese hamster ovary cells. *J. Biol. Chem.* **265**:13906–13913.

Chapter 11

Mapping and Quantification of Bulky-Chemical-Induced DNA Damage Using UvrABC Nucleases

Moon-shong Tang

1. INTRODUCTION

Nucleotide excision repair (NER) is one of the most versatile and conservative repair systems in the biological kingdom. This repair pathway repairs DNA damage caused by a variety of agents, including ultraviolet light (UV), benzo[*a*]pyrene diol epoxide (BPDE), *N*-acetoxy-2-acetyl-aminofluorene (NAAAF), *N*-hydroxy-2-aminofluorene (N-OH-AF), dimethylbenzanthracene diol epoxide (DMBA-DE), and therapeutic drugs (*cis*-platinum, CC-1065, anthramycin, mitomycin C, and psoralen) in both prokaryotes and eukaryotes (for a review see Friedberg *et al.*, 1995; Sancar and Tang, 1993; van Houten, 1990). The initial step of NER involves the recognition of the damaged bases and dual incisions at the 5′ and 3′ sides of the damaged base(s). In *E. coli* cells these steps are controlled by three gene products—the UvrA, UvrB, and UvrC proteins. The *uvrA*, *uvrB*, and *uvrC* genes have been cloned into expression vectors allowing relatively large quantities of these proteins to be readily purified without elaborate procedures. Consequently, the biochemical nature of the recognition of DNA damage and of the dual incisions 5′ and 3′ of the damaged base(s) by the coordinative mechanisms of these three proteins have been extensively studied. The details of these reactions can be found in several review articles (Sancar, 1994; Sancar and Tang, 1993; Grossman and Yeung, 1990; van Houten, 1990), and the following four steps represent a brief synopsis of the reactions of the Uvr proteins:

1. 2 UvrA $\rightarrow$ $(\mathrm{UvrA})_2$
   In solution two UvrA proteins may form a dimer.
2. $(\mathrm{UvrA})_2$ + UvrB $\rightarrow$ $(\mathrm{UvrA})_2 \cdot \mathrm{UvrB}$
   The dimeric form of UvrA may bind to damaged DNA or form a complex with UvrB.

Moon-shong Tang • Department of Carcinogenesis, University of Texas M. D. Anderson Cancer Center, Smithville, Texas 78957.

Technologies for Detection of DNA Damage and Mutations, edited by Gerd P. Pfeifer, Plenum Press, New York, 1996.

3. $(UvrA)_2 \cdot UvrB$ locates and binds to the damaged base(s) and UvrA is released from the complex.
4. UvrC joins the UvrB-damaged base(s) complex, resulting in 5′ and 3′ incisions.

For the sake of simplicity the collective function of these three Uvr proteins in the recognition and incision of DNA damage has been termed UvrABC nuclease or exinuclease even though the three proteins appear to work sequentially rather than as subunits of a complex. Although the UvrABC nuclease can incise a wide range of DNA damage, it does not incise undamaged DNA, and under proper conditions it can incise DNA damage quantitatively (Tang *et al.*, 1989, 1992, 1994a). This unique property has been successfully applied to identify a variety of DNA adducts formed at the nucleotide level (Li *et al.*, 1995; Kohn *et al.*, 1992; Pierce *et al.*, 1989, 1993). Using the UvrABC incision method in combination with the Southern DNA transfer–hybridization technique, several laboratories have successfully quantified several kinds of DNA adduct formation and repair at the gene level (Tang *et al.*, 1989. 1994a; Vreeswijk *et al.*, 1994; Chen *et al.*, 1992; Thomas *et al.*, 1988). This chapter is designed to describe in detail the method using UvrABC nuclease incision analysis for quantification of DNA adduct formation and repair at both the nucleotide and gene levels.

For a successful application of the UvrABC nuclease incision method to quantify DNA adducts there are several precautions that are crucial: (1) having pure Uvr proteins, (2) having pure substrate DNA, and (3) having proper UvrABC–substrate DNA reaction conditions. The following describes in detail how to achieve these goals.

2. METHODS

2.1. Purification of UvrA, UvrB, and UvrC Proteins

Although nucleotide excision repair is one of the major repair mechanisms in *E. coli* cells, Uvr proteins are surprisingly scarce in cells (van Houten, 1990). Fortunately, the *uvrA*, *uvrB*, and *uvrC* genes have been cloned into expression vectors which make the purification of large quantities of Uvr proteins practical (Yeung *et al.*, 1986; Thomas *et al.*, 1985). Uvr proteins isolated from these overexpressed cells seem to have their expected functions: working together, they are able to incise damaged DNA specifically and efficiently.

Sancar's laboratory has constructed three plasmids—pUNC221, pUNC45, and pDR3274—which have the *uvrA*, *uvrB*, and *uvrC* structural genes, respectively, linked to a *tac* promoter; the expression of the *uvr* gene, therefore, can be induced by isopropyl β-D-thiogalactoside (IPTG). These three plasmids have been introduced into cells with an *endA* background, and these cells greatly facilitate the Uvr protein purification process because they do not produce endonuclease I (Thomas *et al.*, 1985). The following are the methods for Uvr protein purification from these cells.

2.1.1. UvrA Purification

UvrA is a DNA binding protein with an isoelectric point (PI) of 6.5; all of the different purification protocols for UvrA take advantage of this property and share a common feature: they separate UvrA proteins by using phosphocellulose- and single-stranded (ss) DNA cellulose columns (Yeung *et al.*, 1986; Thomas *et al.*, 1985; Sancar and Rupp, 1983). Although there are many proteins, including some with nuclease activity in *E. coli* cells, that are able to bind to phosphocellulose and ssDNA cellulose columns, these nucleases appear to be "squeezed out" from binding to both column materials by increasing the amount of UvrA protein in the cell lysates. Slight contamination of these nucleases can then be further diluted out because of the high yield of UvrA. The first step in purifying Uvr proteins therefore is to select for cells containing the *uvr* plasmid which are able to produce a high yield of Uvr proteins by induction; this can be achieved by electrophoretic separation of cell lysates from different colonies induced by IPTG.

A typical purification of UvrA proteins usually starts by growing the cells containing the pUNC45 plasmid in 2 liters of LB (10 g tryptone, 5 g yeast extract, and 5 g NaCl in 1 liter H_2O) containing 20 μg/ml of tetracycline. At $A_{600} = 0.7$, the expression of UvrA can be induced by adding IPTG (0.2–1 mM) for several hours or overnight.

The cells are harvested by repeated centrifugations and first resuspended in a solution of 10% sucrose in buffer P [100 mM KH_2PO_4/K_2HPO_4, pH 7.5, 1 mM EDTA, and 0.1 mM dithiothreitol (DTT)] and then in buffer P. The cells are then sonicated at 0°C by immersion in an ice bath. The cell debris is removed by centrifugation (20,000*g* at 4°C for 60 min), the supernatants loaded onto a preequilibrated phosphocellulose column and eluted with buffer A (50 mM Tris, pH 7.5, 1 mM EDTA, and 0.1 mM DTT) with a KCl gradient ranging from 100 mM to 500 mM. UvrA proteins elute out as a single peak at 200 mM KCl. After adjusting the KCl concentration of the UvrA eluent to 100 mM in buffer A by dilution or dialysis, the solutions are then loaded onto a preequilibrated ssDNA cellulose column and eluted with buffer A containing a KCl gradient of 100 mM to 1 M. UvrA proteins are eluted out at 400 mM KCl. A typical SDS-polyacrylamide gel electrophoresis (PAGE) of purified UvrA is shown in Fig. 11.1. The molecular weight of UvrA, as determined by SDS–PAGE, is 104,000. UvrA stored at −20 to −70°C in 50% glycerol is stable for years.

2.1.2. UvrB Purification

One piece of knowledge crucial for UvrB purification is that a unique outer membrane-bound protease will cleave 40 amino acids away from the C-terminusl of UvrB, converting it to

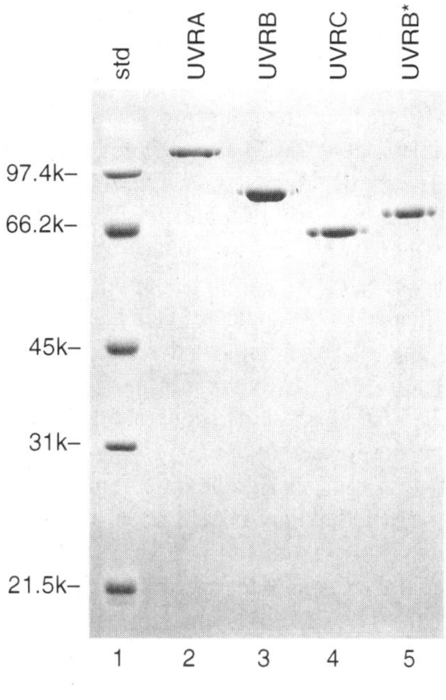

Figure 11.1. SDS–PAGE of purified UvrA, UvrB, UvrB*, and UvrC proteins. The gel was stained with Coomassie blue R-250. k = 1000 daltons, std, standards.

UvrB*, which is relatively inactive in incising DNA damage (Nazimiec *et al.*, 1992; Grossman and Yeung, 1990; Carson and Grossman, 1988). To obtain active UvrB a precautionary step must be taken to reduce this conversion—this can be achieved by adding 1 M $(NH_4)_2SO_4$ to the cell suspension before sonication. The conditions for growing cells containing the *uvrB* plasmid pUNC211, the induction of *uvrB* expression, and the harvesting of cells are the same as those for the UvrA except that the pelleted cells are resuspended in buffer A with 1 M $(NH_4)_2SO_4$. After sonication and removal of the cell debris by centrifugation the cell lysates are dialyzed against buffer A with 100 mM KCl, followed by chromatography in a DEAE Bio-Gel-A column with buffer A containing KCl gradients ranging from 100 mM to 500 mM; UvrB elutes at 0.15 M KCl and UvrB* elutes at 0.2 M KCl. The UvrB proteins are then precipitated by 55% $(NH_4)_2SO_4$, redissolved in buffer A with 20% $(NH_4)_2SO_4$, and chromatographed in a phenyl-agarose column with buffer A containing gradients of 20% $(NH_4)_2SO_4$ to 60% ethylene glycol; UvrB elutes at 40% ethylene glycol and UvrB* elutes at 20% ethylene glycol. The UvrB fractions are then dialyzed against buffer A with 100 mM KCl and chromatographed in an Affi-Gel-Blue column with buffer A containing KCl gradients ranging from 100 mM to 1.5 M; UvrB elutes at 0.5 M KCl. The molecular weight of UvrB as determined by SDS–PAGE is 78,000; a typical SDS–PAGE of UvrB and UvrB* is shown in Fig. 11.1. UvrB stored in buffer A with 100 mM KCl and 50% glycerol at −70°C is stable for years.

2.1.3. UvrC Purification

Among the three Uvr proteins, UvrC is the least stable and the least understood. However, similar to UvrA, it binds to single-stranded DNA and its PI is 7.4 (Selby and Sancar, 1990; Ye and Tang, unpublished results) and thus the purification process is similar to that of UvrA. Typically, the purification starts by growing the cells containing the pDR3274 plasmid in 2 liters of LB with 20 μg/ml of tetracycline. The induction of the expression of the cloned *uvrC* gene, cell harvesting, and sonication of the cells are the same as for UvrA purification except that the pelleted cells are resuspended in buffer A with 10% sucrose and 300 mM KCl. The cell debris is removed by centrifugation, the supernatants are loaded onto a preequilibrated phosphocellulose column, and the chromatography is developed by eluting with buffer A containing KCl gradients ranging from 200 mM to 500 mM. UvrC proteins will elute out as a single peak at 400 mM KCl. After adjusting the KCl concentration of the UvrC eluent to 250 mM by dilution or by dialysis, the solutions are loaded onto a preequilibrated, single-stranded DNA column and eluted with buffer A containing KCl gradients ranging from 100 mM to 1 M. There are two forms of UvrC proteins: one form elutes out at 0.4 M KCl (UvrCI) and the other elutes out at 0.6 M KCl (UvrCII). UvrCI is a monomeric protein while UvrCII is an oligomeric protein. Both forms of UvrC have the tendency to aggregate at high concentrations and low salt conditions. The mechanism behind the formation of these two forms of proteins and their functional differences *in vivo* are not fully understood; however, we have found that both forms are active in DNA damage incision. UvrCII is a double-stranded DNA binding protein and at high protein/DNA ratios it interferes with UvrABC incision (Tang *et al.*, unpublished results); this interference may, through its double-stranded DNA binding activity, prevent proper $(UvrA)_2$UvrB-damaged DNA complex formation. In contrast, UvrCI does not bind to double-stranded DNA, is active at relatively high protein/DNA ratios, and the incision activity of UvrCI is modestly stimulated by a decrease in the protein/DNA ratio. We have found no significant differences in the mode of incising UV-damaged DNA between UvrABCI and UvrABCII under optimal reaction conditions; therefore, for practical purposes, UvrCI is superior to UvrCII for the UvrABC incision assay. It is imperative to separate UvrCI from UvrCII in order to eliminate the possibility of UvrCII interfering with the proper formation of the $(UvrA)_2$UvrB-damaged DNA complex. UvrCI and UvrCII have indistinguishable molecular weights of 68,000, as determined by gradient SDS–PAGE (Ye and Tang, unpublished results); but in 8–10% SDS–PAGE both forms of UvrC move faster than albumin (molecular weight 67,000). A typical SDS–PAGE of purified UvrCI is shown in Fig. 11.1. UvrCI stored at −70°C in 50% glycerol is stable for years. UvrC is unstable at 4°C, and the purification procedures should be completed as quickly as possible.

2.2. Quantification of Bulky-DNA Adducts at Defined Sequences Using UvrABC Incision Analysis and DNA Transfer-Hybridization

2.2.1. Treatment of Cultured Cells with Chemical Carcinogens or Drugs

There are three factors that must be considered in the treatment of cultured mammalian cells with chemicals or drugs: (1) the confluency of the cultured cells, (2) the duration of the treatment, and (3) the possible interaction of the DNA damaging agent with the medium. Most of the DNA binding agents, such as activated bulky chemical carcinogens (BPDE, NAAAF, and N-OH-AF), are electrophilic binding not only to DNA but also to proteins and other substrates. Therefore, removing growth medium and treating the cells with chemicals in buffer usually render reproducible interactions of these compounds with cellular DNA. It is likely that the chemicals may not interact evenly with cells in an overconfluent culture and it thus may be important to use a culture with 50 to 80% confluency for treatment with DNA damaging agents to be certain that the entire cell population will have an equal opportunity to interact with the agents.

Since some DNA adducts are repaired with fast kinetics, it is conceivable that for these types of DNA adducts a significant portion of the adducts formed will be removed during treatment; for example, more than 50% of the antitumor antibiotic CC-1065-induced DNA adducts are repaired within 1 hr in transcriptionally active genes in cultured Chinese hamster ovary (CHO) cells (Tang *et al.*, 1994b). Therefore, it is crucial to determine the optimal duration of chemical treatment by trial and error. The treatment time should be as short as possible but sufficiently extensive to yield enough adduct formation for biochemical analysis. Several factors determine the time period of treatment: the affinity of the agent to cellular DNA, the stability of the agent, and the repair kinetics of the adducts. After a fixed time period of treatment the unbound agents should be removed by washing with buffer.

2.2.2. DNA Purification and Preparation

Impurities in DNA preparations, most often protein contamination, reduce the efficiency of the UvrABC nuclease incision, and it is consequently crucial to have pure DNA for the UvrABC nuclease reaction. DNA purification can be achieved by the regular method. The purified DNA should be dissolved in a low-salt solution, usually TE (10 mM Tris, pH 7.5–8.0, 1 mM EDTA). Salt contamination can also affect the efficiency of the UvrABC incision. We recommend that the protein and salt contamination of DNA be checked by measuring the absorption spectrum of DNA, using wavelengths from 220 to 300 nm.

For detecting DNA damage at the gene level, high-molecular-weight DNA is desirable for two reasons: (1) Since the frequency of DNA damage induced by most DNA damaging agents within a biologically meaningful range is low, it is thus necessary to detect DNA damage in a large size of DNA. (2) High-molecular-weight DNA will contain more intact genes than low-molecular-weight DNA; therefore, high-molecular-weight DNA will have better sensitivity for detection.

The following is a typical procedure for isolating DNA from cultured mammalian cells. To lyse the cells the culture medium must first be removed and the residual medium washed away with DPBS buffer (0.94 mM $MgCl_2$, 1.7 mM $CaCl_2$, 138 mM NaCl 3.9 mM KCl, 2.1 mM

KH_2PO_4, and 12.3 mM Na_2HPO_4, pH 7.4). The cells can then be lysed with lysing buffer (0.5% SDS in 10 mM Tris pH 7.6–8.0, 10 mM EDTA, and 10 mM NaCl) in the presence of proteinase K (100 μg/ml). For complete protein digestion the mixture is incubated for 14 hr at room temperature. Cellular DNA is then precipitated by adding sodium acetate (0.33 M), 2.5 vol of cold 95% ethanol, and the precipitated DNA can be spooled out with a glass rod. The DNA is then dissolved in TE, precipitated by ethanol, and redissolved in TE. The RNA contamination is digested by adding RNase A (10 μg/ml) and incubating 2 hr at 37°C. The RNase and the residual protein contaminations are removed by multiple phenol extractions followed by diethyl ether extractions. The DNA is then precipitated by ethanol, dried, and redissolved in TE.

The purified DNA is digested with restriction enzymes to tailor the size of the DNA. If 5-bromo-2′-deoxyuridine (10 μM) and 5-fluorodeoxyuridine (1 μM) are used for labeling the replicated DNA, then the purified DNA, after phenol/ether extractions to remove the restriction enzymes, is separated by CsCl density (6 g CsCl and 4.6 ml of DNA in TE; refractive index 1.401–1.402) gradient centrifugation; the DNA with "light" density is the nonreplicated DNA and the DNA with hybrid density is the replicated DNA (Fig. 11.2). After removing the CsCl by dialysis the DNA is ready for the UvrABC incision reaction.

2.2.3. UvrABC Incision Reaction

The reaction is performed by adding UvrA, UvrB, and UvrC proteins to the DNA substrate, usually in TE solution, in 50 mM Tris pH 7.5, 100 mM KCl, 1 mM ATP, 10 mM $MgCl_2$, and 0.1 mM DTT, and incubating the mixture at 37°C for 60 min. Mixing the Uvr proteins before addition to substrate DNA may reduce the efficiency of UvrABC incision. One of the crucial factors for obtaining optimal UvrABC activity in this reaction is for the final concentration of KCl to be 100 mM. Since Uvr proteins are stored in different KCl solutions, it is necessary to adjust the final concentration of KCl according to the amount of Uvr proteins used in the reaction. The UvrABC incision should be completed within 60 min; further incubation may increase the nonspecific cutting. If nonspecific cutting is observed, it is usually caused by nuclease contamination either in the substrate DNA or in the Uvr proteins; for the former, phenol extractions should eliminate the problem. An internal standard should be added to the DNA solution before the UvrABC reaction in order to monitor the nonspecific cutting and the variations in recovery for different samples for the following steps. The internal standard should render a similar hybridization band intensity as the sample band and the amount of the internal standard to be used should be determined by trial and error.

Because UvrB and UvrC remain tightly bound to the damaged DNA region in the absence of repair synthesis and helicase II activity after incision in the UvrABC incision reaction, the reaction is a stoichiometric rather than a catalytic reaction (Sancar and Tang, 1993). In order to obtain quantitative UvrABC incision of the damaged DNA, it is important to have a proper protein/DNA damage ratio. Furthermore, excessive amounts of UvrA and UvrC in the reaction mixture very often interfere with proper $(UvrA)_2$UvrB-damaged DNA complex formation, consequently reducing the efficiency of UvrABC incision. Although different ratios of UvrA:UvrB:UvrC have been used, we have found that 1:1:1 works as well as 1:2:1. The optimal protein/DNA ratio which renders maximal UvrABC incision and minimal nonspecific cutting is also dependent on the quality of the Uvr proteins, and we suggest that it should be obtained by trial and error; in our laboratory the molar ratio of protein/DNA(14 kb) in the UvrABC incision reactions has varied from 6 to 14.

After the UvrABC incision reaction, the Uvr protein can be removed by phenol/ether extractions and ethanol precipitation. Carrier tRNA (15 μg) can be added to the DNA solution to increase the recovery. The DNA pellet should be washed with 75% cold ethanol to remove

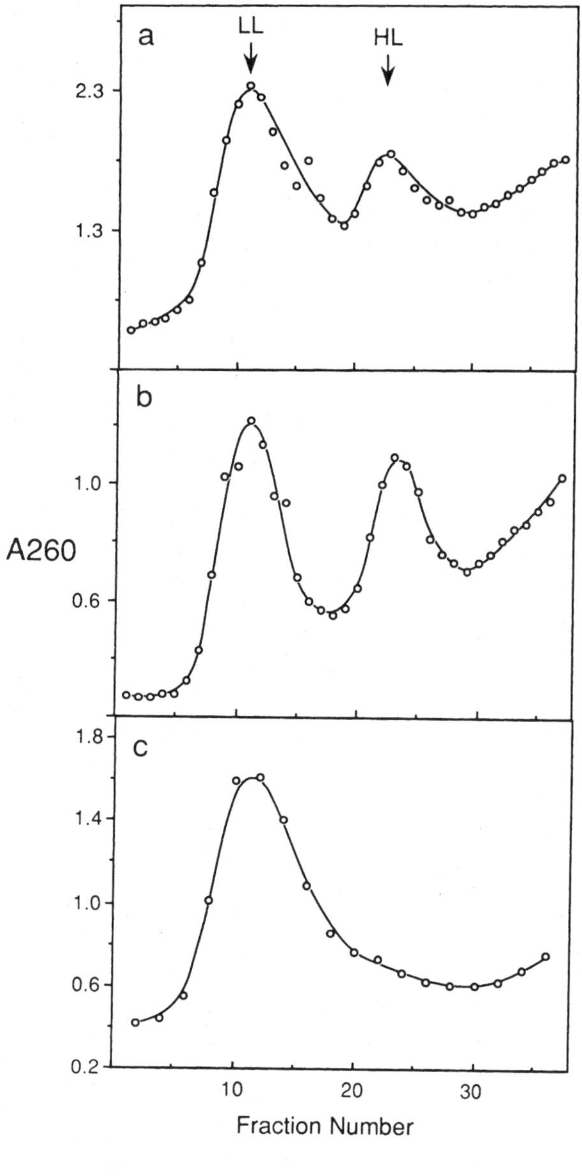

Figure 11.2. Separation of density-labeled DNA by CsCl density gradient sedimentation. Cultured CHO cells were labeled with 5-bromo-2′-deoxyuridine (10 μM) and 5-fluorodeoxyuridine (1 μM), treated with UV (20 J/m²), CC-1065 (60 nM), or BPDE (4 μM). After 24 hr incubation the cellular DNA was isolated and digested with *Asp718* and separated by neutral CsCl gradient sedimentation. Y axis represents the relative amount of DNA, X axis the fraction number. LL, light band; HL, heavy–light hybrid band.

residual salt contamination. The DNA pellet is then dried and redissolved in TE. Because the DNA must be dissolved in a small volume for the next step (denaturing the DNA in 90% formamide), it is necessary to vortex well and dissolve for a long time (overnight).

2.2.4. DNA Denaturation

In order to calculate the number of UvrABC incisions (which should represent the amount of DNA damage), it is necessary to denature the DNA. Although DNA can be denatured by heat treatment, we have found that this treatment *per se* generates DNA breaks. Mild alkaline treatment (30 mM NaOH) has been successfully used for denaturing DNA; however, for detecting alkali-labile DNA adducts such as *N*-(deoxyguanosin-C8-yl)-2-acetylaminofluorene (dG-C8-AAF), *N*-(deoxyguanosin-C8-yl)-2-aminofluorene (dG-C8-AF), and BPDE–DNA adducts, it is impractical to denature the DNA with the alkaline treatment. To overcome these problems we have developed a neutral formamide DNA denaturing method (Tang *et al.*, 1989, 1994a,b). We have found that DNA in a low salt solution such as TE can be denatured by incubation in 90% formamide at 37°C for 60–90 min. This treatment will also dissociate the Uvr protein–DNA complex. The formamide should be freshly deionized with ion exchange resins (AG 501-X8 D resin, Bio-Rad) and the pH of the formamide should be maintained at about 8.

2.2.5. Separation of Single-Stranded DNA by Electrophoresis

After denaturation the size of the denatured DNA is usually separated by electrophoresis. In order to maintain the DNA in denatured form during electrophoresis, alkaline gel electrophoresis has been used, but we have found that the denatured DNA can be separated using regular nondenaturing gel electrophoresis without the DNA being renatured (Tang *et al.*, 1989, 1994a,b). Therefore, nondenaturing gel electrophoresis is our choice for detecting alkali-labile DNA adducts. The gel concentration is dependent on the size of DNA to be separated; for a size of 10 to 20 kb DNA, a 0.5% agarose gel is recommended. The gel is made in TBE buffer (50 mM Tris, 50 mM boric acid, and 1 mM EDTA, and final pH 8.3) with 0.5 μg/ml of ethidium bromide, and the capacity of the wells should be large enough to accommodate a 100-μl sample. We recommend preparing the gel 14 hr before use, removing the comb just before use and checking for leakage with dyes. The electrophoresis is usually conducted at 5 V/cm for a period of time (3 hr) that should result in an adequate separation of DNA according to their size.

2.2.6. Southern Blotting and DNA Hybridization

After electrophoresis the overall DNA separation in the agarose gel can be visualized by a UV transilluminator and recorded by photography. The gel should then be trimmed to proper size for transferal to a nylon membrane. In order to achieve a complete DNA transfer, the gel should be treated with 0.25 N HCl for 30 min to induce depurination, washed with distilled water, and then treated with 0.5 N NaOH/0.6 M NaCl solution for 45 min to induce DNA strand breaks at the depurinated sites. The DNA is transferred in a 0.5 N NaOH/ 0.6 M NaCl solution. For detection of a single-copy gene a charged membrane is recommended, and for detection of multiple-copy genes a regular nylon membrane is recommended. To check the completion of the Southern transfer, the gel should be stained with ethidium bromide (0.5 μg/ml) and visually inspected. The membrane should be washed in distilled water for a few minutes and then dried in a vacuum oven at 80°C for 2 hr. After drying the membrane should be further washed with a solution containing 0.25% SDS and 0.25 × SSPE (1 × SSPE: 149 mM NaCl, 8.5 mM NaH_2PO_4, 10 mM EDTA, pH 7.4) for 30 to 60 min. The membrane can then be inserted into a hybridization tube and incubated with 10 ml of prehybridization solution (50% formamide, 10% dextran, 1 M NaCl, 1% SDS, and 0.5 mg/ml ssDNA) at 42°C for 3 hr to overnight with constant rotation. To hybridize the membrane the volume of the prehybridization solution should be reduced to 5 ml, ^{32}P-labeled probes added, and the membrane should be further incubated at 42°C for 14 to 48 hr. After removing the prehybridization solution containing ^{32}P, the membrane should be washed with washing buffer containing different concentrations of SDS and SSPE. To obtain proper results the washing conditions should follow the vendor's recommendation; however, whether the washing removes the background can be checked by using a Geiger counter.

To detect the hybridized band the membrane can be exposed to an X-ray film, a phosphor screen, or a Betascope. The latter two devices can provide direct quantitative results and also have a higher sensitivity than X-ray film. If X-ray film is used, multiple exposures are recommended to obtain a proper intensity for densitometer scanning.

2.2.7. Quantification

Using UvrABC-sensitive sites to quantify the number of DNA adducts formed in a defined fragment is based on two assumptions: first, the distribution of the DNA adducts is random, and second, the UvrABC incision is quantitative or proportional to the amount of DNA adducts. The first assumption is reasonable since in most cases the size of the defined fragment for detection of DNA adducts is relatively large in comparison with the small number of near-neighbor nucleotides which determine the sequence-dependent adduct formation. Therefore, the formation of adducts in the defined fragment is probably random. The validity of the second assumption, however, must be experimentally tested. If the DNA damaging agent is radioactively labeled, then the quantitative relationship between UvrABC incision and the adduct can be easily established; for example, we have found that the UvrABC nuclease can incise dG-C8-AAF, dG-C8-AF, BPDE–DNA, and DMBA-DE–DNA adducts quantitatively (Chen *et al.*, 1996; Tang *et al.*, 1989, 1992; Pierce *et al.*, 1988). It is difficult to be certain about the relationship between UvrABC incision and the adduct formation for those compounds that are not radioactively labeled; in order to use the number of UvrABC incisions to represent the amount of adduct formation, it is necessary to establish that the UvrABC incision is proportional to the adduct formation.

Although the UvrABC nuclease incises at both the 5′ and 3′ sides of the damaged base(s), the size of the incised or excised fragment including the damaged base is 12–13 nucleotides. This distance of the dual incision is relatively small in comparison with the size of the DNA fragment (a few to 20 kb) to be quantified; therefore, the method described above will register each UvrABC dual incision as a single incision event.

The number of UvrABC incisions in a defined fragment is calculated based on the Poisson distribution equation $P(0) = e^{-n}$, where $P(0)$ represents the fraction of DNA in full length after UvrABC incision and n represents the average number of adducts per DNA fragment. For studying the repair of the DNA adduct the n formed at time zero can be compared with the remaining n at any time point. By the nature of this equation the sensitivity of this method of calculation $\Delta n/\Delta P(0)$ is n dependent; when n is within 0.25 to 2.5 a reasonably reliable $\Delta n/\Delta P(0)$ value can be obtained by the current experimental method. A typical autoradiograph of using UvrABC incision to detect BPDE–DNA adduct in DHFR gene and its 3′ downstream region of CHO cells is shown in Fig. 11.3A and the quantification result is shown in Fig. 11.3B.

2.3. Analysis of Sequence Specificity of Chemical Carcinogen– and Drug–DNA Binding by Using the UvrABC Nuclease Incision Method

We have shown that under carefully controlled conditions the UvrABC nuclease can quantitatively incise monoalkylated DNA adducts induced by bulky chemical carcinogens such as BPDE, NAAAF, N-OH-AF, and DMBA-DE, and antitumor antibiotics such as anthramycin and mitomycin C (Chen *et al.*, 1996; Li *et al.*, 1995; Pierce *et al.*, 1989, 1993; Tang *et al.*, 1992; Nazimiec *et al.*, 1992; Kohn *et al.*, 1992), and we have also successfully used this method to identify the sequence specificity of the DNA binding of these chemicals. There are several crucial factors that determine the successfulness of this application: the purity of the UvrABC, a proper ratio of UvrABC to DNA or DNA adducts, and the number of DNA modifications per

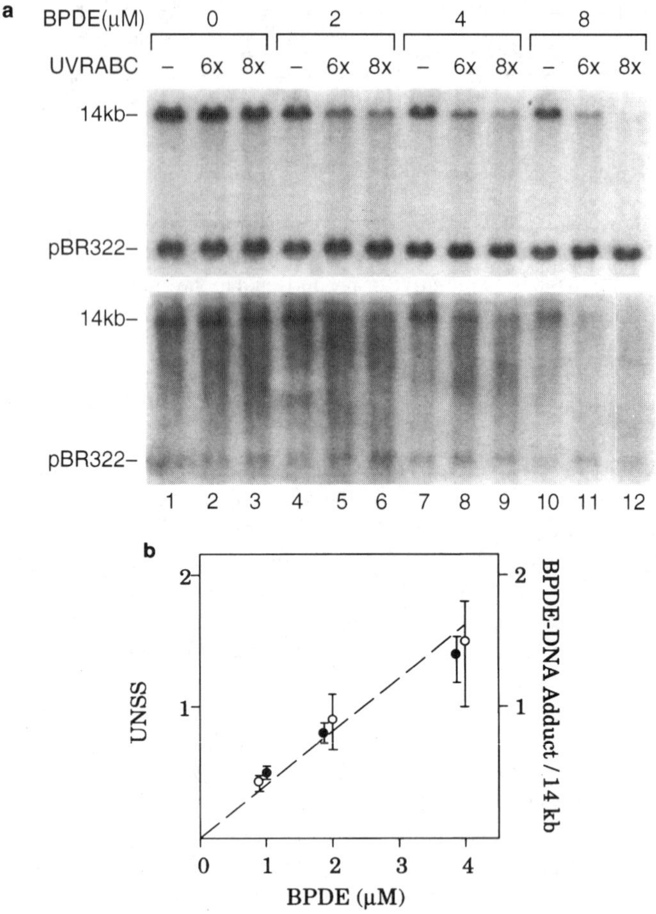

Figure 11.3. Formation of UvrABC nuclease-sensitive sites (UNSS) in the coding region (●) and 3′-downstream noncoding region (○) of DHFR gene domain in CHO cells as a function of BPDE concentration. (a) A typical autoradiograph; (b) the quantifications. (From Tang *et al.*, 1994b.)

substrate DNA fragment. The first two points have been discussed in detail, and the reason why the number of DNA modifications is crucial becomes obvious once the methodology is understood: The method involves modifying a defined DNA fragment of a known sequence and using UvrABC nuclease incision to identify the chemical modification site and the relative degree of modification at different sequences. The DNA fragments must be either 5′- or 3′-end ^{32}P labeled so that the DNA modification-induced UvrABC cut per DNA fragment can be identified and quantified by separating the resulting DNA fragments by denaturing polyacrylamide gel electrophoresis. Since the DNA fragments are single-end labeled, only one UvrABC cut can be identified by autoradiograph; furthermore, a DNA fragment with two DNA modifications within a short distance of each other is very often a poor substrate for the UvrABC nuclease incision. Therefore, in order to determine the sequence specificity of chemical–DNA binding in a defined DNA fragment which is either 5′- or 3′-end labeled with ^{32}P, the number of average chemical modifications per DNA fragment has to be one or less.

2.3.1. 5′- and 3′-End Labeling the DNA Fragment

Because UvrABC nuclease incises DNA adducts only in double-stranded DNA, in order to use UvrABC incision for identification of DNA modifications, the substrate DNA has to be double-stranded. Usually the single-end labeling of a defined double-stranded DNA fragment can be achieved by labeling a restriction-digested fragment at either the 5′ or 3′ ends, followed by a second restriction enzyme digestion to trim away one of the labeled ends. We have found that in order to obtain a high efficiency of labeling, DNA fragments, purified from an agarose or polyacrylamide gel, must be cleaned by multiple phenol/ether extractions before end labeling. For 5′-end labeling the DNA fragments are first dephosphorylated by incubating the DNA fragments with bacterial or calf intestinal alkaline phosphatase; the phosphatases are then removed by multiple phenol/ether extractions and followed by ethanol precipitation. The DNA fragments with both 5′ ends labeled are then digested with a second restriction enzyme to trim away one end labeling and the resulting DNA is separated by polyacrylamide gel electrophoresis. The 3′-end labeling can be achieved by using DNA polymerase I. The same approach as described above can be applied to obtain DNA fragments with a single 3′-end labeling.

2.3.2. Modification of Radioactively Labeled DNA Fragments with Carcinogens or Antibiotics

The method for DNA modification depends on the compound used. In general, low pH conditions should be avoided to reduce depurination of DNA. After DNA modification it is necessary to remove the unreacted drugs or chemicals, and this can usually be achieved by repeated extractions with organic solvents or by ethanol precipitation. Ideally, the number of average DNA modifications should be one or less per DNA fragment.

2.3.3. UvrABC Nuclease Incision

The UvrABC incision reaction for small end-labeled DNA fragments is essentially the same as the reaction described above for cellular DNA. To obtain a quantitative UvrABC incision, it is crucial to have a proper protein/DNA ratio; this can be a problem for DNA fragments with high specific activity since the activity of the diluted Uvr proteins is unstable and a high protein/DNA ratio reduces the enzyme incision efficiency. In order to maintain both an optimal concentration of the Uvr protein in the reaction mixture and an optimal protein/DNA ratio, a small quantity of the undamaged DNA (10 to 200 ng) can be added in the reaction mixture.

There are two important questions in using the UvrABC incision to determine the sequence specificity of chemical carcinogen– or antitumor drug–DNA binding: one, whether the UvrABC incision is quantitative, and two, whether the UvrABC incision has a sequence preference. We have addressed the first question in Section 2.2, and the second question can be addressed as follows: For determining the sequence specificity of chemical–DNA binding the UvrABC incision reaction should be allowed to proceed to completion to compensate for the possible kinetic differences of UvrABC incision at different sequences. Thus, the extent of UvrABC incision at different sequences should represent the extent of chemical modification at those sequences.

2.3.4. Denaturing Polyacrylamide Gel Electrophoresis

After the incision reaction the UvrABC proteins remain tightly bound to the damaged DNA and it is important to remove the proteins from the protein–DNA complexes in order to separate the resulting DNA fragments according to size by gel electrophoresis . This can be achieved by phenol followed by diethyl ether extractions and ethanol precipitation. The DNA pellets should be washed with cold 75% ethanol before redissolving in TE buffer or in denaturing buffer-dye (88% formamide, 0.1% xylene cyanol, and 0.1% bromophenol blue) and then separated in 8–10% denaturing polyacrylamide gel (with 50% urea in 50 mM Tris, pH 7.9, 50 mM borate, 5 mM EDTA) along with DNA from the Maxam and Gilbert sequencing reactions (Maxam and Gilbert, 1980). The gel is then dried and exposed to X-ray film at −70°C or to a phosphor screen for a proper period of time. A typical autoradiograph resulting from the aforementioned procedures is shown in Fig. 11.4.

2.3.5. Densitometer Scanning

The intensity or the radioactivity of the DNA bands separated by gel electrophoresis can be quantified by densitometer scanning, direct scintillation, or phosphorescent counting. One limitation in using densitometer scanning is its limited linear gray scale range (one to two orders of magnitude). Direct scintillation and phosphorescent counting have much wider linear ranges (four orders of magnitude).

3. DISCUSSION

Several methods have been developed for detecting DNA damage at the gene and nucleotide levels, and the crucial element for evaluating these methods is whether the method can identify DNA modifications quantitatively without sequence preference. If a method is able to do so, then it can be used for quantifying DNA adduct formation and for identifying the sequence effect on the DNA adduct formation. We have demonstrated that UvrABC, under proper conditions, can indeed incise bulky carcinogen (BPDE, NAAAF, N-OH-AF, and DMBA-DE)-induced DNA damage quantitatively. The UvrABC incision method is therefore suitable for quantifying these DNA adducts at the gene level.

One potential problem with using the UvrABC incision method for identification of the sequence specificity of DNA adduct formation is that the positions of the UvrABC nuclease incisions at both the 5′ and 3′ side of a DNA adduct may vary from 6 to 7 nucleotides for the 5′ side incision and from 3 to 5 nucleotides for the 3′ side incision. Caution must be exercised in the extrapolation of the UvrABC incision position as the position of the DNA adduct. For determining the DNA adduct formation in a run of repeat bases using UvrABC incision analysis this problem becomes even more severe; in fact, it is theoretically impossible to precisely resolve the sequence specificity of adduct formation in a run of repeat bases using the UvrABC incision assay (Doisy and Tang, 1995). However, in most sequences and for most DNA adducts, the UvrABC nuclease shows remarkably consistent positions of dual incision, namely, 7 nucleotides from the 5′ and 4 nucleotides from the 3′ side of an adduct (Sancar and Tang, 1993; van Houten, 1990). Based on this assumption, a reasonably reliable sequence specificity of adduct formation can be deduced from the UvrABC incision profile. The methods using exonuclease III and λ exonuclease digestion (Li and Kohn, 1991; Walter *et al.*, 1988) or the primer extension method (Doisy and Tang, 1995; Christner *et al.*, 1994) for identifying the sequence specificity of adduct formation can also have a similar if not more severe problem; the stalling of the proceeding of these enzymes by the DNA adduct may be DNA sequence dependent and the positions of the stall may vary at different sequences.

To date, except for the thermal alkaline methods for identifying CC-1065–N3-adenine (Tang *et al.*, 1988; Hurley *et al.*, 1984) adducts and (6–4) photoproducts (Franklin *et al.*, 1982; Brash and Haseltine, 1982; Tang *et al.*, 1991) and the T4 endonuclease V method of identifying

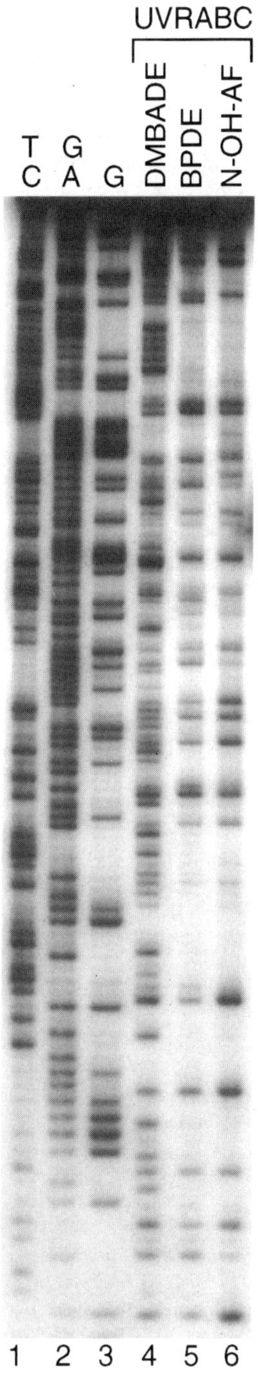

Figure 11.4. Sequence preference of DNA binding of DMBA-DE, BPDE, and N-OH-AF in (*Hin*fI–*Nhe*I) 195-bp fragment of mouse H-*ras* gene detected by UvrABC nuclease. Lanes 1–3, Maxam and Gilbert reactions; lanes 4–6, DNA fragment modified with DMBA-DE, BPDE, and N-OH-AF, respectively, and reacted with UvrABC nuclease.

cyclobutane dipyrimidines (Gordon and Haseltine, 1982; Tang *et al.*, 1991), the UvrABC incision method appears to be the most reliable method for identifying several types of DNA adducts at the gene and sequence levels. Until a new technique is developed, this approach seems to be the most reasonable choice.

ACKNOWLEDGMENTS. I wish to acknowledge Ms. A. Pao, Ms. M. Qian, Ms. X.-S. Zhang, Mr. X. Ye , Mr. Y. Zheng, Mr. Y.-L. Tang, and Drs. J. Chen, M. Nazimeic, V.-S. Li, H. Kohn, and J. Pierce for their significant contributions in developing this method and Ms. Y.-Y. Tang for careful review of this manuscript. I would also like to thank Dr. A. Sancar for kindly providing the *E. coli* cells with plasmids containing the *uvr* genes. This work was supported by an NIH grant (ES03124) and a research grant (3955) from the Council for Tobacco Research.

REFERENCES

Brash, D. E., and Haseltine, W. A. (1982). UV-induced mutation hotspots occur at DNA damage hotspots. *Nature* **298**:189–192.

Carson, P. R., and Grossman, L. (1988). Potential role of proteolysis in the control of UvrABC incision. *Nucleic Acids Res.* **16**:10903–10912.

Chen, J., Pao, A., Yi, Z., Ye, X., Morris, R., Harvey, R., and Tang, M.-s. (1996). Sequence preference of dimethylbenzanthracene diol epoxide-DNA binding in mouse H-ras detected by UvrABC nuclease. Submitted for publication.

Chen, R.-H., Maher, V. M., Bouwer, J., van de Putte, P., and McCormick, J. J. (1992). Preferential repair and strand-specific repair of benzo[a]pyrene diol epoxide adducts in the *HPRT* gene of diploid human fibroblasts. *Proc. Natl. Acad. Sci. USA* **89**:5413–5417.

Christner, D. F., Lakshman, M. K., Sayer, J. M., Jerina, D. M., and Dipple, A. (1994). Primer extension by various polymerase using oligonucleotide templates containing stereoisomeric benzo[a]pyrene–deoxyadenosine adducts. *Biochemistry* **33**:14297–14305.

Doisy, R., and Tang, M.-s. (1995). Effect of aminofluorene and (acetylamino)fluorene adducts on the DNA replication mediated by *Escherichia coli* polymerase I (Klenow fragment) and III. *Biochemistry* **34**:4358–4368.

Franklin, W. A., Lo, K. M., and Haseltine, W. A. (1982). Alkaline lability of fluorescent photoproducts produced in ultraviolet light-irradiated DNA. *J. Biol. Chem.* **257**:13535–13543.

Friedberg, E. C., Walker, G. C., and Siede, W. (1995). *DNA Repair and Mutagenesis*, ASM Press, Washington, DC.

Gordon, L. K., and Haseltine, W. A. (1982). Quantitation of cyclobutane pyrimidine dimer formation in double- and single-strand DNA fragments of defined sequence. *Radiat. Res.* **89**:99–112.

Grossman, L., and Yeung, A. T. (1990). The UvrABC endonuclease system of *Escherichia coli*—A view from Baltimore. *Mutat. Res.* **236**:213–221.

Hurley, L., Reynolds, V. L., Swenson, D. H., Petzold, G. L., and Scahill, T. A. (1984). Reaction of the antitumor antibiotic CC-1065 with DNA: Structure of a DNA adduct with DNA sequence specificity. *Science* **226**: 843–844.

Kohn, H., Li, V.-S., and Tang, M.-s. (1992). Recognition of mitomycin C–DNA monoadducts by UVRABC nuclease. *J. Am. Chem. Soc.* **114**:5501–5509.

Li, V.-S., and Kohn, H. (1991). Studies on the bonding specificity for mitomycin C–DNA monoalkylation processes. *J. Am. Chem. Soc.* **113**:275–283.

Li, V.-S., Choi, D., Tang, M.-s., and Kohn, H. (1995). Structural requirement for mitomycin C DNA bonding. *Biochemisyry* **34**:7120–7126.

Maxam, A. M., and Gilbert, W. (1980). Sequencing end-labeled DNA with base-specific chemical cleavages. *Methods Enzymol.* **65**:499–560.

Nazimiec, M., Grossman, L.,and Tang, M.-s. (1992). A comparison of the rates of reaction and function of UVRB in UVRBC- and UVRAB-mediated anthramycin-N2-guanine-DNA repair. *J. Biol. Chem.* **267**:24716–24724.

Pierce, J. R., Case, R., and Tang, M.-s. (1989). Recognition and repair of 2-aminofluorene- and 2-(acetylamino)-fluorene–DNA adducts by UVRABC nuclease. *Biochemistry* **28**:5821–5826.

Pierce, J. R., Nazimiec, M., and Tang, M-s. (1993). Comparision of sequence preference of tomaymycin- and anthramycin-DNA bonding by exonuclease III and λ exonuclease digestion and UvrABC nuclease incision analysis. *Biochemistry* **32**:7069–7078.

Sancar, A. (1994). Mechanisms of DNA excision repair. *Science* **266**:1954–1956.

Sancar, A., and Rupp, W. D. (1983). A novel repair enzyme: UVRABC excision nuclease of *Escherichia coli*: Cut a DNA strand on both sides of the damaged region. *Cell* **33**:249–260.

Sancar, A., and Tang, M-s. (1993). Nucleotide excision repair. *Photochem. Photobiol.* **57**:905–921.

Selby, C. P., and Sancar, A. (1990). Structure and function of the (A)BC excinuclease of *Escherichia coli. Mutat. Res.* **236**:203–211.

Tang, M.-s., Lee, C.-S., Doisy, R., Ross, L. Needham-VanDevanter, N., and Hurley, L. H. (1988). Recognition and repair of the CC-1065-(N3-adenine)-DNA adduct by the UVRABC nucleases. *Biochem.* **27**:893–901.

Tang, M.-s., Bohr, V. A., Zhang, X.-s., Pierce, J., and Hanawalt, P.C. (1989). Quantification of aminofluorene adduct formation and repair in defined DNA sequences in mammalian cells using the UVRABC nuclease. *J. Biol. Chem.* **264**:14455–14462.

Tang, M.-s., Htun, H., Cheng, Y., and Dahlberg, J. E. (1991). Suppression of cyclobutane and 6–4 dipyrimidines formation in triple-stranded H-DNA. *Biochemistry* **30**:7021–7026.

Tang, M.-s., Pierce, J. R., Doisy, R. P., Nazimiec, M. E., and MacLeod M. C. (1992). Differences and similarities in the repair of two benzo[a]pyrene diol epoxide isomers induced DNA adducts by *uvrA*, *uvrB*, and *uvrC* gene products. *Biochemistry* **31**:8429–8436.

Tang, M.-s., Pao, A., and Zhang, X.-s. (1994a). Repair of benzo(a)pyrene diol epoxide- and UV-induced DNA damage in dihydrofolate reductase and adenine phosphoribosyltransferase genes of CHO cells. *J. Biol. Chem.* **269**:12749–12754.

Tang, M.-s., Qian, M., and Pao, A. (1994b). Formation and repair of antitumor antibiotic CC-1065 induced DNA adducts in the adenine phosphoribosyltransferase and amplified dihydrofolate reductase genes of Chinese hamster ovary cells. *Biochemistry* **33**:2726–2732.

Thomas, D. C., Levy, M., and Sancar, A. (1985). Amplification and purfication of UvrA, UvrB, and UvrC proteins of *Escherichia coli. J. Biol. Chem.* **260**:9875–9883.

Thomas, D. C., Morton, A. G., Bohr, V. A., and Sancar, A. (1988). General method for quantifying base adducts in specific mammalian genes. *Proc. Natl. Acad. Sci. USA* **85**:3723–3727.

van Houten, B. (1990). Nucleotide excision repair in *Escherichia coli. Microbiol. Rev.* **45**:18–51.

Vreeswijk, M. P. G., van Hoffen, A., Westland, B. E., Vrieling, H., van Zeeland, A. A., and Mullenders, H. F. (1994). Analysis of repair of cyclobutane pyrimidine dimers and pyrimidine 6–4 pyrimidone photoproducts in transcriptionally active and inactive genes in Chinese hamster cells. *J. Biol. Chem.* **269**:31858–31863.

Walter, R. B., Pierce, J., Case, R., and Tang, M.-s. (1988). Recognition of the DNA helix stabilizing anthramycin-N2 guanine adduct by UVRABC nuclease. *J. Mol. Biol.* **203**:939–947.

Yeung, A. T., Mattes, W. B., and Grossman, L. (1986). Protein complexes formed during the incision reaction catalyzed by the *E. coli* UvrABC endonuclease. *Nucleic Acids Res.* **14**:2567–2582.

Chapter 12

The Use of DNA Glycosylases to Detect DNA Damage

Timothy R. O'Connor

1. INTRODUCTION

DNA glycosylases, first reported by Lindahl (1974), catalyze the scission of the glycosidic bond releasing damaged or mispaired bases as the first step of the base excision repair pathway (Fig. 12.1) (Dianov and Lindahl, 1994). Removal of damaged bases by a DNA glycosylase is generally associated with a specific type of damage (e.g., uracil-DNA glycosylase excises uracil bases formed by deamination or misincorporation into DNA; Lindahl, 1993). The specificity of DNA glycosylases, however, may also cross over to different types of DNA damage [e.g., AlkA protein, which excises a number of alkylated bases (Table I), also excises formyluracil and hydroxymethyluracil bases formed by oxidation (Bjelland *et al.*, 1994)]. Proteins such as the uracil-DNA glycosylase, the AlkA protein, and the Tag protein leave abasic sites in DNA which are in turn processed by endonucleases cleaving the phosphodiester backbone hydrolytically at these sites (Dianov and Lindahl, 1994; Lloyd and Linn, 1993). In addition to this group of DNA glycosylases, the Fpg, Nth, and MutY proteins of *E. coli* and the UV endonuclease from bacteriophage T4 have physically associated activities incising DNA at abasic sites via β-elimination mechanisms (AP lyases) (Bailly and Verly, 1987; Gerlt, 1993), and as a consequence processing of these lesions may be slightly different than repair of abasic sites (Lloyd and Linn, 1993). The use of these enzymes in the detection of DNA damage is facilitated by the fact that DNA glycosylases are active in the presence of EDTA and function independent of any complex which may form *in vivo*. Table I summarizes several properties and damages recognized by DNA glycosylases.

Despite the advantages of DNA glycosylases in the detection of DNA adducts and modified bases, their use has been restricted by the availability of purified proteins. To address this point, a general protocol for the isolation of the AlkA, Tag, Fpg, and Nth proteins is included. The construction of substrates is critical for the purification of the proteins and the identification of new substrates for DNA glycosylases; therefore, this aspect is addressed first. Other chapters in this volume deal with specific applications of DNA glycosylases.

Timothy R. O'Connor • Groupe 'Réparation des lésions radio- et chimioinduites,' URA147 CNRS, Institut Gustave-Roussy PRII, 94805 Villejuif Cedex, France. *Present address*: Division of Biology, City of Hope National Medical Center, Beckman Research Institute, Duarte, California 91010.

Technologies for Detection of DNA Damage and Mutations, edited by Gerd P. Pfeifer, Plenum Press, New York, 1996.

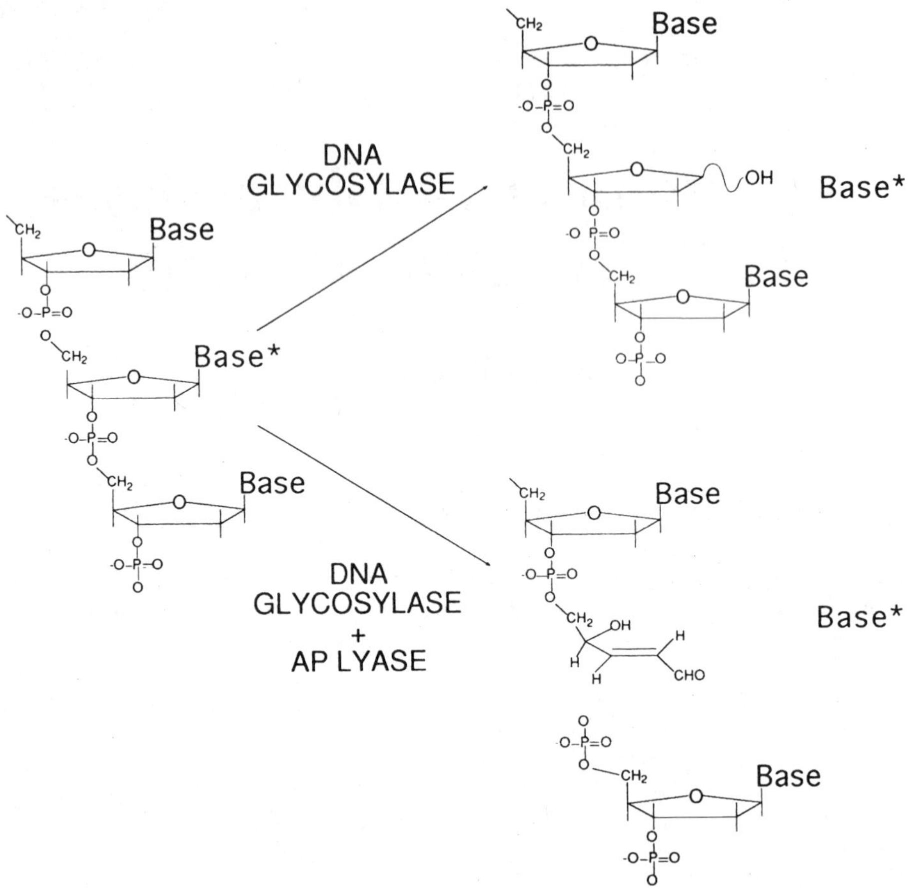

Figure 12.1. Reactions catalyzed by DNA glycosylases. DNA glycosylases catalyze the cleavage of the glycosylic bond and in some cases a b-elimination reaction which breaks the phosphodiester backbone leaving a 5′ phosphate (Dodson *et al.*, 1994). For simplicity a single strand of DNA is shown. DNA glycosylases (except the uracil-DNA glycosylase) generally remove bases from double-stranded DNA. DNA glycosylases useful in the detection of DNA damage and their activities are listed in Table I. In addition to the β-elimination reaction of the AP lyases, the Fpg protein also catalyzes a δ-elimination reaction (not depicted here) which releases the pentenal (an altered deoxyribose) from the DNA.

2. METHODS

2.1. Substrates for DNA Glycosylases

Several different substrates are used to assay DNA glycosylase activity. The best substrates to identify the bases released by a DNA glycosylase provide the possibility to chromatographically separate the bases released. In addition to these substrates which directly identify the bases released, other substrates which indirectly indicate scission of the glycosylic bond of modified bases from plasmids, oligonucleotides, or DNA fragments have been employed. These substrates do not necessarily identify the bases excised, but may strongly suggest that such a reaction has occurred.

2.1.1. Substrates for the Analysis of Modified Bases

Substrates for DNA glycosylases recognizing methylation damage (Tag, AlkA, and Fpg proteins) are prepared by modifying DNA using radiolabeled dimethylsulfate. Similar substrates for DNA damage may be generated by incorporating radiolabeled deoxynucleotides into DNA via nick-translation followed by modification using the damaging agent (Breimer, 1984; Dosanjh *et al.*, 1994; Matijasevic *et al.*, 1992). Although this type of substrate has a higher background than a substrate modified by a radiolabeled damaging agent, such nick-translated substrates are useful when the radiolabeled agent is unavailable or expensive to prepare. Another direct method for identification of modified bases released is the direct modification of the DNA followed by excision of the modified bases by the DNA glycosylase. Prior to use in DNA glycosylase assays the modified DNA is hydrolyzed to release the modified bases which are subsequently analyzed by HPLC. Gas chromatography coupled to selective ion mass spectrometry also has the capacity to identify modified DNA bases, but the use of this technique requires a significant investment in equipment and expertise (Dizdaroglu, 1991). A method modifying calf thymus DNA or poly(dG-dC) using ^3H-labeled dimethylsulfate is presented to generate substrates for the AlkA, Tag, and Fpg proteins. Other substrates which directly identify the bases released may be prepared by modifying nick-translated DNA or nonradiolabeled DNA.

2.1.1a. Preparation of [^3H]-CH_3-Calf Thymus DNA and [^3H]-(Fapy)-Poly(dG-dC) Substrates Using [^3H]-Dimethylsulfate. To prepare a substrate for the AlkA or Tag proteins, purified calf thymus DNA (1 ml, 1 mg/ml) is suspended in 300 mM sodium cacodylate at pH 7. All of the modifications with dimethylsulfate should be performed in a well-ventilated hood. A vial containing 5 mCi (187.5 Bq, 1–5 Ci/mmole) of [3H]dimethylsulfate (NEN) is opened and the solution containing the DNA is used to rinse the vial. If the dimethylsulfate is dissolved in hexane, the two solutions are not miscible and the solution should be agitated to ensure maximal modification of the DNA. The reaction is allowed to proceed overnight at 37°C, and 14 μl of β-mercaptoethanol is added to stop the reaction. The DNA is separated into three 1.5-ml centrifuge tubes, ethanol precipitated, rinsed using 80% ethanol, and the labeled DNA is resuspended in 1 ml of 10 mM Tris-HCl pH 7.0, 1 mM EDTA, and dialyzed at 4°C against several buffer changes until no ^3H is detected in the dialysis solution. The [^3H]-CH_3-calf thymus DNA is reprecipitated, resuspended in 1 ml of the assay buffer, aliquoted, and frozen at −70°C. Care should be taken to avoid cycles of thawing and refreezing this substrate since the glycosylic bonds linking the deoxyribose to *N*7-methylguanine and *N*3-methyladenine, the principal bases modified, are labile compared to the unmodified bases. This protocol modifies approximately 1% of the bases in the DNA. A substrate for the Fpg protein containing ring-opened 7-methylguanines (Fapy) is synthesized by methylating 500 μg (10 O.D.) poly(dG-dC) with [^3H]dimethylsulfate, as for the calf thymus DNA, and subsequently incubating the [^3H]-(CH_3)-poly(dG-dC) in 400 μl of 0.2 M NaOH for 20 min at room temperature to open the imidazole ring of the 7-methylguanine bases. The solution is neutralized with 40 μl of 2.5 M ammonium acetate pH 4.5, and the [^3H]-(Fapy)-poly(dG-dC) formed is precipitated by addition of 1 ml of ethanol and dialysis as for the [^3H]-CH_3-calf thymus DNA substrate (O'Connor *et al.*, 1988).

2.1.1b. Hydrolysis to Release Modified Purines and Pyrimidines from DNA. Following modification, aliquots of the [^3H]-CH_3-calf thymus DNA or [^3H]-(Fapy)-poly(dG-dC) are treated in 75% formic acid for 30 min at 80°C to hydrolyze modified purine bases. To hydrolyze unmodified purines, the DNA is treated in 75% formic acid at 80°C for 1 hr. Modified pyrimidines and purines are released from DNA by resuspending the DNA in 60% formic acid in sealed tubes at 140°C for 30 min (Dizdaroglu *et al.*, 1993; O'Connor *et al.*, 1988). Following hydrolysis the solutions are evaporated to dryness, resuspended in 50–200 ml of water, and injected into the HPLC column.

Table I
Substrates and Assay Conditions for DNA Glycosylases Used to Detect DNA Damage and Mutations

Protein	Molecular mass (kDa)	Bases excised	Origin	Conditions[a]	References
		DNA glycosylases			
AlkA protein	31	3-Methyladenine, 7-methylguanine, 3-methylguanine, *O*2-methylpyrimidines; guanine and adenine adducts formed by ethylating agents; guanine and adenine adducts formed by uracil mustard, mechlorethamine, chlorambucil, and phenylalanine mustard; formyl uracil, hypoxanthine, hydroxymethyl uracil, chlorethylnitrosourea adducts	*E. coli*	A	Bjelland *et al.* (1994), Carter *et al.* (1988), Habraken *et al.* (1991), Matijasevic *et al.* (1992), Mattes *et al.* (1995), Sakumi and Sekiguchi (1990), Saparbaev and Laval (1994)
Tag protein	21	3-Methyladenine, 3-methylguanine	*E. coli*	A	Bjelland and Seeberg (1987), Bjelland *et al.* (1993), Sakumi *et al.* (1986)
3-Methyladenine-DNA glycosylase (human)	32	3-methyladenine, 7-methylguanine, 3-methylguanine, etheno-adenine, etheno-guanine, etheno-cytosine, hypoxanthine, chloroethylnitrosourea adducts[b]	Human cells (recombinant *E. coli*)	B	Chakravarti *et al.* (1991), Dosanjh *et al.* (1994), Mattes *et al.* (1996), O'Connor (1993), O'Connor and Laval (1991), Samson *et al.* (1991), Saparbaev and Laval (1994)
Uracil-DNA glycosylase	26	Uracil, 5-hydroxyuracil	*E. coli*	C	Hatahet *et al.* (1994), Lindahl (1974)
Thymine-DNA glycosylase	55	Removes thymine from guanine–thymine mismatch	Human cells	D	Neddermann and Jiricny (1993)

DNA glycosylases with physically associated AP lyase activities					
Fpg protein	30	8-Oxoguanine, formamidopyrimidines (ring-opened 7-alkylguanines, Fapy-adenine, Fapy-guanine), alkaline-treated AF-guanine (C8 modified), 5-hydroxycytosine, 5-hydroxyuracil	*E. coli*	B	Boiteux (1993), Chetsanga and Lindahl (1978), Hatahet *et al.* (1994)
Nth protein	24	5,6-diHthymine, 5-OH-5-Mehydantoin, 5-OH-5-Mehydantoin, 5-OH-6-uracil, alloxan, 5-OH-6-Huracil, thymine glycol, cytosine glycol, urea residues, pyrimidine hydrates, 5-hydroxycytosine, 5-hydroxyuracil	*E. coli*	B	Boiteux (1993), Boorstein *et al.* (1989), Ganguly *et al.* (1989), Hatahet *et al.* (1994), Kow and Wallace (1987)
T4 uv endonuclease	16	Pyrimidine dimers	Bacteriophage T4	E	Dodson and Lloyd (1989), Lloyd and Linn (1993)
MutY (MicA) protein	39	Removes adenine from guanine–adenine or cytosine–adenine mismatches	*E. coli*	D	Au *et al.* (1989), Tsai-Wu *et al.* (1992)

[a]A: 70 mM Hepes-KOH pH 7.5, 0.5 mM EDTA, 5 mM β-mercaptoethanol, 5% glycerol; B: 100 mM KCl, 70 mM Hepes-KOH pH 7.5, 0.5 mM EDTA, 5 mM β-mercaptoethanol, 5% glycerol; C: 10 mM Tris-HCl pH 7.5, 1 mM EDTA, 5 mM β-mercaptoethanol; D: 25 mM Hepes-KOH pH 7.6, 0.5 mM EDTA, 5 mM β-mercaptoethanol; E: 50 mM NaCl, 50 mM Tris-HCl pH 7.5, 0.5 mM EDTA, 5 mM β-mercaptoethanol.

[b]A recent report (Bessho *et al.*, 1993) suggested that 8-oxoguanine was a substrate for the mouse homologue of the human 3-methyladenine-DNA glycosylase, but in experiments in our laboratory, this activity was not observed using an excess of the homogeneous rat and human 3-meAde-DNA glycosylases. Therefore, the activity is not included in this table.

2.1.2. Site-Specifically Modified Oligonucleotides

An oligonucleotide modified at a specific base is a useful substrate for monitoring DNA glycosylase activity when a substrate with labeled modified bases is not available. Some modified bases such as 8-oxoGua and etheno-Ade are available commercially as phosphoramadites (one source is Glen Research). Other means to obtain site-specifically modified oligonucleotides have also been described including specific incorporation by DNA polymerases, terminal transferase, or ligation of a short oligonucleotide with a modified base into a longer oligonucleotide (Hatahet *et al.*, 1993; Simha *et al.*, 1991).

2.1.2a. 5′ [^{32}P]-End Labeling of Oligonucleotides. Oligonucleotides are labeled in a 25-μl reaction mixture containing 50 pmole primer, 36 pmole [γ-^{32}P]-ATP (4000 Ci/mmole), and 10 units T4 polynucleotide kinase in a buffer consisting of 50 mM Tris-HCl pH 7.5, 10 mM $MgCl_2$, and 2 mM β-mercaptoethanol. The oligonucleotide is separated from unincorporated label using a NENSORB column (New England Nuclear), the oligonucleotide is evaporated to dryness, and resuspended in 50 μl of Tris-HCl pH 7.5, 1 mM EDTA.

2.1.2b. Annealing of Oligonucleotides. A twofold molar excess of the unlabeled complementary strand of the site-specifically labeled strand is heated to 95°C for 5 min in a total volume of 100 μl of 100 mM NaCl, 10 mM Tris-HCl pH 7.5, 1 mM EDTA. The solution with the oligonucleotides is then gradually cooled to room temperature. The labeled oligonucleotide is usually stable for 2–3 weeks.

2.1.3. DNA Fragments Modified by DNA Damaging Agents

Another type of substrate which is useful to show sequence specific excision by DNA glycosylases is a DNA fragment of 200–300 bp which is modified by a DNA damaging agent. This method is best used when the DNA glycosylase is purified to homogeneity or near homogeneity.

2.1.3a. Amplification of a DNA Fragment Using a Labeled Oligonucleotide Primer. A labeled 269-bp fragment from SV40 is prepared by PCR essentially as described (Lion and Haas, 1990; Mattes *et al.*, 1996; Schowalter and Sommer, 1989). An oligonucleotide corresponding to positions 5178–5197 of the SV40 sequence was 5′ end-labeled using [^{32}P]-ATP as described in Section 2.1.2a and purified either by NENSORB column or ethanol precipitation. Amplification was carried out in 100 μl of 50 mM Tris-HCl pH 9.0, 5 mM $MgCl_2$, 200 mM of each dNTP and containing 50 pmole of both the labeled oligonucleotide primer and the unlabeled primer (positions 4929–4948 of SV40), 5×10^8 copies of SV40 DNA (2 ng), 20 nmole of each dNTP, and 2.5 units of AmpliTaq DNA polymerase (Amersham). Amplification was for 16 cycles of 94°C–45 sec, 53°C–30 sec, 72°C–30 sec. After chloroform extraction, the amplified DNA was separated from primers by two centrifugations through a Centricon-100 microconcentrator, and increasing the volume to 2 ml with 50 mM Tris-HCl pH 8.0, 1 mM EDTA before centrifugation. Labeled DNA was recovered by ethanol precipitation and its concentration determined by absorbance at 260 nm.

2.1.3b. Preparation of Damaged DNA. Labeled DNA (0.5–1 μg) is incubated at 25 or 37°C with DNA damaging agents as required in an appropriate buffer in a total volume of 50 μl. After the DNA is damaged, the reaction is stopped. For DNA alkylating agents a convenient method for stopping the reaction is to add β-mercaptoethanol to 200 mM 20 μg of glycogen (Sigma) is added to the stopped reaction followed by ethanol precipitation of the damaged DNA. Markers for damage may be generated using standard chemical sequencing protocols (Maxam and Gilbert, 1980; Sambrook *et al.*, 1989).

2.1.3c. Revealing DNA Damage Chemically. Guanine bases alkylated at the *N*7 position may be revealed by treatment of damaged DNA in 100 μl of 1 M piperidine at 90°C for 30 min. Alkylated adenine bases modified at the *N*3 position may be revealed by heating the modified DNA in 100 μl of 50 mM Hepes-KOH pH 7.5, 0.5 mM EDTA at 80°C for 30 min. The DNA is ethanol precipitated, resuspended in 100 μl of 1 M piperidine, and treated for 30 min at 90°C. Following incubation with piperidine, 10 μl of 3 M sodium acetate (pH 5.2) and 250 μl ethanol are added to the solution and incubated on dry ice for 10 min, centrifuged, and rinsed with 80% ethanol. Remove traces of piperidine by evaporation in a Speed-Vac concentrator (Savant) for 30 min. The DNA is resuspended in a sequencing gel-loading buffer [98% formamide (deionized), 10 mM EDTA, 0.025% xylene cyanol FF, 0.025% bromophenol blue], and the products are separated on a DNA sequencing gel [8–10% PAGE acrylamide:bisacrylamide 29:1, 7 M urea, 90 mM Tris-borate, 1 mM EDTA (TBE)].

2.1.4. Plasmid Substrates for DNA Glycosylases

These sustrates are prepared by treating a DNA substrate with a specific DNA damaging agent such as visible light and methylene blue [generating substrates for the Fpg protein (Floyd *et al.*, 1989; Mûller *et al.*, 1990; O'Connor and Laval, 1989; Pierre and Laval, 1980)]. Assays using damaged plasmids exploit the difference in mobility between superhelical (Form I), nicked (Form II), and linear (Form III) plasmids; therefore, the plasmids should be prepared using CsCl gradients to ensure that 90% or more of the plasmid is in the superhelical form. Since these substrates differ slightly in their preparation, it is usually best to use several different concentrations of DNA damaging agents. At the end of the reaction, the damaging agent is removed and the DNA precipitated. The DNA should be assayed using a small amount of purified protein if available, and analyzed using agarose gel electrophoresis to ensure that the plasmid was not degraded during the modification process.

2.2. DNA Glycosylase Assays

2.2.1. Assays for DNA Substrates Containing Radiolabeled Modified Bases

DNA glycosylases are easily assayed using radiolabeled substrates (Section 2.1.1). In this assay, the modified or excised base remains in the supernatant of an ethanol precipitate. A DNA substrate is incubated with the enzyme or column fraction for 5 min to 1 hr in 50 μl of reaction buffer at 37°C. The assay reactions are performed in 1.5-ml Eppendorf centrifuge tubes. At the end of this time, 50 μl of a solution which contains 1 M NaCl, 1 mg/ml calf thymus DNA, and 0.1 mg/ml of bovine serum albumin is added to the reaction mixture with 300 μl of ethanol at −20°C. The precipitation mixture is incubated at −20 to −40°C for 5–10 min and then centrifuged at 4°C for 15 min. For the purposes of the determination of activity, the excision of the DNA bases should be linear with respect to enzyme concentration. The activity is determined by removal of 300 μl of the ethanol-soluble supernatant, addition of scintillation fluid (3–5 ml), and scintillation spectroscopy.

2.2.1a. Chromatographic Identification of Bases Released by DNA Glycosylases. The use of HPLC detection methods has simplified the unambiguous assignment of bases released by DNA glycosylases. Authentic markers are sometimes available from commercial sources (Sigma, Aldrich). The major advantages in the use of the radiolabeled substrates are the low amount of material needed to detect a base excised and the ability to inject the authentic marker(s) with the radiolabeled bases released. The identification of substrates is performed by reacting the DNA with the enzyme (10–100 ng generally). An excess of substrate should be degraded to provide enough material for several assays. If unlabeled DNA is used as a substrate, buffers with significant ultraviolet absorption may interfere with the identification of modified bases. A reasonable buffer to avoid this problem is phosphate, but the assay should be optimized prior to changing the reaction buffer. Gel filtration may also be used to separate DNA adducts.

2.2.1b. Preparation of the Reaction Products for Chromatographic Analysis. Stop the DNA glycosylase assay by the addition of DNA and bovine serum albumin as above. Precipitate the DNA by addition of 3 vol of ethanol at −20°C as in standard glycosylase assays. The supernatant is then dried using a Speed-Vac. If substrate with radiolabeled modified bases was used, authentic markers may be added. The dried supernatant is resuspended in a small volume (50–100 μl) and injected on the HPLC column.

2.2.1c. HPLC Analysis of Methylated Purines. The C_{18} μBondapak column (Waters) is a versatile column for the analysis of modified and unmodified DNA bases. The separation of methylated purines is presented as an example. In general, the C_{18} μBondapak column may be used in isocratic elution, thus simplifying the use of the HPLC. More complicated base modifications, however, may require the use of gradients to adequately separate the modified bases. Often the retention times may be altered by changing the solvent used in the mobile phase. The parameters for the separation of methylated purines are included as a point of reference (O'Connor *et al.*, 1988): C_{18} μBondapak column, isocratic elution at 1.5 ml/min in 50 mM $NH_4H_2PO_4$ pH 4.5, 2% methanol (v/v) mobile phase. Representative elution times for methylated bases are: *N*1-meAde, 4 min; *N*3-meAde, 7 min; *N*6-meAde, 40 min; *N*7-meAde, 16 min; *N*3-meGua, 9 min; *N*7-meGua, 15 min.

In addition to this method, other detectors for HPLC analysis, such as electrochemistry and fluorescence detection, may also be used to monitor the release of bases.

2.2.1d. Gel Filtration Chromatography Using Bio-gel P2. To prove the identity of different peaks from an initial HPLC analysis, a second chromatographic analysis is generally used. Gel filtration chromatography to identify purine bases using a 40 × 1-cm Bio-gel P2 (Bio-Rad) column in conjunction with the HPLC unambiguously defines the identity of methylated purines. The column is developed at 0.2 ml/min using a potassium tetraborate buffer, pH 10.55. The retention times for this column for methylated purines are (O'Connor *et al.*, 1988): oligonucleotides, 60 min; *N*1-meAde, 140 min; *N*3-meAde, 140 min; *N*7-meAde, 160 min; *N*3-meGua, 90 min; *N*7-meGua, 90 min.

2.2.2. Oligonucleotide-Based DNA Glycosylase Assays

Assays based on removal of modified bases from radiolabeled oligonucleotides may also be used to detect DNA glycosylase activity. The assays are performed identically to those of the plasmid DNA, except DNA sequencing gels or small-format sequencing gels may be used. One microliter of a 20 mg/ml glycogen solution is added to the assay solution following phenol/chloroform extraction to facilitate the ethanol precipitation. In some cases, the purification of the oligonucleotide used for the assays will simplify the interpretation of the results. An aliquot of the damaged DNA (usually $1–10 \times 10^4$ cpm for oligonucleotides and $1–5 \times 10^5$ cpm as measured by Cerenkov counting for fragment DNA) is incubated in the appropriate buffer in the presence of increasing amounts of the DNA glycosylase (usually 10–300 ng of enzyme) for 15–30 min at 37°C. If the DNA glycosylase used has an intrinsic activity cleaving at abasic sites, the DNA is phenol/chloroform extracted, 20 μg of glycogen is added, and the DNA is precipitated with sodium acetate (0.3 M final concentration) and 3 vol of ethanol. The DNA is resuspended in sequence gel-loading buffer, denatured, and loaded on a DNA sequencing gel. If the DNA glycosylase does not have an associated AP lyase activity, another glycosylase which does not recognize the same damage spectrum or endonuclease IV may be used to incise the abasic site. Alternatively, a treatment of the DNA in 100 μl of 1 M piperidine at 37°C for 15 min may be used to reveal the abasic sites generated by the DNA glycosylase. This treatment is mild and does not generate a significant amount of nonspecific background bands. Following this mild piperidine treatment the DNA is precipitated as described above. Fragments are analyzed on a DNA sequencing gel and oligonucleotides are analyzed on small-format 7 M urea–TBE–PAGE gels.

Analysis of Oligonucleotides on Small-Format 7 M Urea–TBE–PAGE Gels. Small denaturing gels, of the format 7 × 10 cm with 0.75-mm spacers, are used to analyze oligonucleotides. A 20% PAGE gel 19:1 acrylamide:bisacrylamide (Bio-Rad) ratio is used with 7 M urea in 1 × TBE. The DNA is reacted with the enzyme as described for the DNA fragments below, precipitated, resuspended in sequence gel-loading buffer, and denatured at 90°C for 3 min prior to loading. The gel is migrated at 400 V for 30 min, covered with Saran Wrap, and autoradiographed.

2.2.3. Plasmid-Based DNA Glycosylase Assays

These DNA glycosylase assays have the advantage that they are inexpensive and do not require the use of radioactivity. The plasmid with a single nick migrates slower on a neutral agarose gel; therefore, the percentage of nicked to supercoiled plasmid reflects the number of lesions removed by a DNA glycosylase. The DNA is incubated with the DNA glycosylase using standard assay conditions and the reaction is stopped by phenol/chloroform extraction and ethanol precipitation to remove proteins. If the glycosylase does not possess an AP lyase activity, another enzyme incubation should be performed to cleave at abasic sites. A protein such as the endonuclease IV of *E. coli* is useful for this, since this endonuclease does not recognize modified bases (Demple and Harrison, 1994). Following enzymatic treatment, the DNA is loaded on a 1% agarose gel in 40 mM Tris-acetate pH 8.0, 1 mM EDTA (TAE buffer). The gel is then stained with ethidium bromide and a photographic negative is obtained. The negative is scanned using a densitometer, and a Poisson formula is used to calculate the number of single-strand breaks (Hegler *et al.*, 1993) with a correction factor for the difference in fluorescence of supercoiled and nicked DNA:

$$\begin{aligned}\text{total number of single-strand breaks} &= -\ln\{1.4(\text{supercoiled})/[1.4(\text{supercoiled}) + \text{nicked}]\}\\ &= \text{enzyme breaks} + \text{damaging agent breaks}\end{aligned}$$

2.3. Purification of AlkA, Tag, Fpg, and Nth Proteins

2.3.1. Production of Recombinant DNA Glycosylases

The first step is the production of the recombinant protein in *E. coli*. The *alk*A [pAlk10 (Clarke *et al.*, 1984; O'Connor and Laval, 1990; Mattes *et al.*, 1996)], *tag* [pTag10 (T. O'Connor, unpublished results; Sakumi *et al.*, 1986; Clarke *et al.*, 1984)], *nth* [pNth10 (Asahara *et al.*, 1989; Graves *et al.*, 1992; Dizdaroglu *et al.*, 1993)], and *fpg* [pFPG230 (Boiteux *et al.*, 1987, 1990)] genes were inserted into expression vectors controlled by the *lac* repressor, and the proteins were produced in *E. coli* JM105. Production of the proteins as a function of time following induction with isopropylthio-β-D-galactoside (IPTG) (induction of the culture at O.D. 1 with 0.5 mM IPTG) is monitored by removing aliquots of a large-scale culture (1 liter) as a function of time, and analyzing the production of the protein by SDS–PAGE or activity. If markers are available for the protein or the protein is produced at high levels in the bacteria, it may be detected using SDS–PAGE, otherwise an activity assay must be used. The length of the induction time should be determined to maximize the amount of protein purified. Cultures from 1–10 liters (producing 10–40 g of bacteria) are the most convenient volumes to use for the purifications on a laboratory scale. Following growth the bacteria are centrifuged and washed with a 100 mM NaCl, 10 mM Tris-HCl pH 8, 1 mM EDTA solution and stored at −20°C for up to

1 year prior to isolation. A small aliquot of the culture (20 ml) should be saved and lysed to determine the level of protein production prior to starting a large-scale purification.

2.3.2. Lysis

The bacteria are resuspended in five times the volume of their wet weight in 250 mM NaCl, 25 mM Tris-HCl, 5 mM EDTA and lysozyme is added (final concentration of lysozyme 1 mg/ml). The lysis suspension is incubated on ice for 20–30 min, then heated to 37°C for 15–20 min. The suspension is then frozen in a dry-ice ethanol bath for 15–20 min, and heated at 37°C for 15–20 min. After this freeze–thaw cycle is repeated two or three times, the lysate is centrifuged for 30 min at 30,000 rpm. The addition of PMSF to a concentration of 0.4 mM to reduce proteolysis by serine proteases is sometimes useful. A kit for the determination of protein concentration by the method of Bradford (Bradford, 1976) is available from Bio-Rad.

2.3.3. Removal of Nucleic Acids—Polyethyleneimine P and/or Ion Exchange Chromatography

Nucleic acids are removed by adding 0.1 vol of 5% PEI-P to the lysate with stirring over a period of 5–10 min (Asahara *et al.*, 1989). Stirring is continued for 25 min at 0°C and the solution is centrifuged at 4°C for 30 min at 30,000 rpm. Alternatively or in conjunction with the PEI-P precipitation, a QMA-Accell (Waters Associates) column may be used (Boiteux *et al.*, 1987). The column is equilibrated at 250 mM NaCl, Buffer A (10 mM Hepes-KOH pH 7.5, 0.5 mM EDTA, 5 mM β-mercaptoethanol, and 5% glycerol). The total proteins are loaded onto the QMA-Accell column and the column is rinsed with 250 mM NaCl, Buffer A until all of the proteins are eluted from the column. Following removal of nucleic acids, the total proteins are dialyzed against Buffer A. This may require two to three changes of the dialysis buffer. To reduce the expense of using a Hepes-KOH buffer for dialysis, the concentration of the Hepes may be reduced for dialysis, or alternatively a Tris-HCl buffer may be substituted.

2.3.4. Phospho Ultrogel A6R Chromatography

Lowering the ionic strength of the lysate and complete removal of nucleic acids is essential for binding of the DNA glycosylases to the Phospho Ultrogel A6R column. The conductimetry of the lysate and the ratio of the OD_{280} nm/OD_{260} nm (conductimetry for the AlkA and Tag A proteins should be less than 1 mS with an $OD_{280\ nm}$/OD_{260} nm of greater than 0.7–0.75). The Phospho Ultrogel A6R (Sepracor, Marlborough, MA) is equilibrated in Buffer A (normally 15–30 ml of resin is sufficient for 10 liters of *E. coli* culture). The total protein, including the protein precipitated during dialysis, is loaded on the Phospho Ultrogel A6R column and rinsed with Buffer A. A gradient of 10 column vol is then performed from Buffer A to 800 mM NaCl, Buffer A (Boiteux *et al.*, 1987; O'Connor *et al.*, 1993; Graves *et al.*, 1992). The fractions are then either assayed using radiolabeled substrates, or analyzed using SDS–PAGE to identify the fractions containing the protein of interest.

2.3.5. Gel Filtration—AcA 54 Chromatography

The Phospho Ultrogel A6R fractions are pooled, and the total proteins are precipitated by adding solid $(NH_4)_2SO_4$ slowly with stirring to a final concentration of 0.5 g/ml of solution over a period of approximately 30 min and the precipitate is centrifuged at 10,000 rpm at 4°C for 20 min. The precipitated proteins are resuspended in a minimal volume of 100 mM NaCl, Buffer A. These proteins are then loaded onto a 1.8-m × 15-mm gel exclusion column of AcA 54 resin (Sepracor) equilibrated in 1 M NaCl, Buffer A.

2.3.6. Hydrophobic Interaction—Phenyl Sepharose CL 4B Chromatography

Following gel filtration chromatography, the fractions are assayed and then pooled. If the protein is homogeneous or nearly so, the hydrophobic interaction column (phenyl Sepharose CL 4B, Pharmacia) may not be useful. If the AcA 54 column contained glycerol, the protein should be precipitated and resuspended in 1.8 M $(NH_4)_2SO_4$. The hydrophobic interaction column is a reversed-phase column which requires that the proteins be loaded at high ionic strength without glycerol. The elution is then performed as a gradient from 100% $(NH_4)_2SO_4$, Buffer A (without glycerol) to Buffer A. The exact conditions for the binding and elution of a particular protein to the phenyl Sepharose resin should be determined on an aliquot of the protein solution.

2.3.7. Concentration of DNA Glycosylases

The fractions containing DNA glycosylase activity are pooled, and dialyzed against 100 mM KCl, 50 mM Hepes-KOH pH 7.5, 0.5 mM EDTA, 10 mM β-mercaptoethanol, 5% glycerol, and 0.4 mM PMSF at 4°C. The protein is concentrated by placing the dialysis tubing in solid polyethylene glycol 6000. Following concentration, an equal volume of glycerol is added to the protein, and the DNA glycosylases may be stored at −20°C.

2.4. Bases Excised by DNA Glycosylases

The cloning of coding sequences for DNA glycosylases has permitted their use as valuable reagents in the detection of DNA damage. The use of these enzymes has provided insight into biological processes including the coupling of transcription and DNA repair (Hanawalt, 1994). Table I lists a variety of DNA glycosylases for the detection of DNA damage. Although a number of other DNA glycosylases have been reported (Sakumi and Sekiguchi, 1990; Lloyd and Linn, 1993), the proteins in Table I have been isolated in sufficient quantities to design experiments. Buffer conditions for all of the DNA glycosylases are also indicated in the table. All of the DNA glycosylases isolated to date are active in the presence of 1–5 mM EDTA. In general, DNA glycosylases are useful for the detection of DNA damage or mutation in several different experimental protocols including detection of global DNA damage and repair, DNA damage at the nucleotide level, gene-specific repair, detection of mutation, and reduction of background in PCR methods.

The first step in the use of DNA glycosylases for the detection of DNA damage is to identify the damaged bases removed by a particular DNA glycosylase. This may be done using the same methods as described for the DNA glycosylase assays and identification of DNA bases.

2.4.1. Coupled Gas Chromatography–Mass Spectrometry

Another method for the detection of bases released by DNA glycosylases is coupled gas chromatography and mass spectrometry with selective ion monitoring (SIM) capability (Boiteux *et al.*, 1992; Dizdaroglu *et al.*, 1993; Dizdaroglu, 1991; Cadet and Weinfeld, 1993). The advantage of this method is that a large number of different modified bases is detected in a single assay. This method was used to determine the substrate specificities of the Nth and Fpg proteins in radiation-damaged DNA (Boiteux *et al.*, 1992; Dizdaroglu *et al.*, 1993). The two major limitations of the technique are the large amount of substrate and enzyme necessary for the assay. In order to perform the assay, the modified DNA bases must be silylated to pass through the GC column. This modification requires the treatment of the bases for 30 min at high temperature (140°C) in the presence of bis(trimethylsilyl)trifluoroacetamide and acetonitrile. Therefore, interpretation of the results must take into account the possibility that bases may undergo rearrangements as a result of the silylation reaction. Nevertheless, the possibility to assay for the release of a large number of bases renders this technique extremely useful.

2.5. Detection of Damage in Genomic or Mitochondrial DNA Using DNA Glycosylases

The details for the isolation of DNA for these experiments are found in Chapters 15 and 16 of this book. In the case of DNA glycosylases with physically associated AP lyase activities, it is necessary to distinguish between the formation of abasic sites in DNA and modified bases. Therefore, comparisons with DNA glycosylase-sensitive sites and mild piperidine treatment-sensitive sites (1 M piperidine, 37°C, 15 min) and/or endonuclease IV-sensitive sites are needed to identify abasic sites generated in DNA by different damaging agents. Prior to treatment of cells with a damaging agent, isolated genomic DNA is damaged with a specific agent which generates a known spectrum of modified bases. The genomic DNA is then digested to completion using the DNA glycosylase recognizing the damaged base. Aliquots of 10 μg of damaged genomic DNA are then subjected to digestion using different enzyme concentrations for the same time period (a working range—10 ng, 50 ng, 100 ng, 500 ng, 1 μg). A control digestion at the highest and lowest enzyme concentrations on undamaged DNA should also be performed to ensure that the glycosylase does not have nonspecific endo- or exonuclease activity. If the DNA glycosylase has an associated activity incising DNA at abasic sites, the sites of the modified bases are automatically cleaved during the enzyme digestion. If the DNA glycosylase does not have a physically associated AP lyase activity, however, further treatment with either piperidine at 37°C for 15 min or endonuclease IV or Nth protein is required to incise the DNA at abasic sites. The modified DNA is treated with increasing amounts of DNA glycosylase, and analyzed on neutral glyoxal agarose gels (see Chapter 3). The DNA glycosylase used in the first step should be removed by extraction with phenol/phenol chloroform/chloroform if a second enzyme is required to incise at abasic sites. This eliminates the possibility of activating contaminating proteins in the different enzyme preparations. Abasic sites may be protected from chemical β-elimination reactions by incubating the DNA in 5 mM methoxyamine (Wang *et al.*, 1995). Once the concentration of enzyme and time parameters have been examined, the digestion conditions should be held constant for a given lot of enzyme. Changing the preparation of enzyme should include a series of experiments to verify that the conditions used are identical to the previous preparation of enzyme. Once the conditions for digestion with a given DNA glycosylase are established, the DNA from cells treated with DNA damaging agents is analyzed using the conditions established in the preliminary experiments. If ligation-mediated PCR (Pfeifer *et al.*, 1993) is to be performed on the DNA, digestion of the DNA with endonuclease IV should not be used to cleave at abasic sites since a ligatable phosphate is not formed. Ligatable 5′-phosphates suitable for ligation-mediated PCR may be generated using AP lyase activity or mild piperidine cleavage.

The covalently closed circular nature of mammalian mitochondrial DNA has allowed the detection of DNA damage in mitochondria following treatment with agents which damage DNA oxidatively (Hegler *et al.*, 1993). The same assay methods for the utilization of plasmid DNAs apply to mitochondrial DNA.

The digestion of damaged DNA by DNA glycosylases has utility in many of the techniques described in this book, including Chapters 1, 4, 5, 7, 9, 10, and 13–17.

2.6. Detection of Mismatched Bases in DNA—Mismatch Repair Enzyme Cleavage (MREC)

In addition to uses in the detection of DNA damage, at least two DNA glycosylases recognize mispaired bases. MutY (MicA) protein and the human thymine-DNA glycosylase involved in mismatch repair have been used in MREC to detect mutations in human tumor cells (Lu and Hsu, 1992; Hsu *et al.*, 1994). The MutY protein excises adenine bases from G/A and C/A mismatches and the thymine-DNA glycosylase excises thymine bases from G/T mismatches (Au *et al.*, 1989; Neddermann and Jiricny, 1993; Tsai-Wu *et al.*, 1992). Thus, these DNA glycosylases may be used to detect certain types of mutations. The potential existence of other types of DNA glycosylases recognizing homomispairs (A/A, G/G, C/C, and T/T) (Yeh *et al.*, 1991; Holmes *et al.*, 1990) suggests that this method may be useful in the general identification of mutations. PCR amplification is performed using standard procedures for 1 mg of DNA from normal cells and potentially mutant cells. The amplified DNA fragment from the normal cells and the mutant cells is mixed in a 1:1 ratio and the annealed fragments are then incubated with the MutY (MicA) protein. This cleaves the DNA at sites of G/A or A/G mismatches by removal and incision of the strand containing the adenine. An oligonucleotide primer is then 5′ end-labeled, annealed to the DNA, and extended using Sequenase (USB), the products are separated on a DNA sequencing gel, and compared with the sequence in the normal cells. This technique may also be used to identify G/T and T/G mismatch using the thymine-DNA glycosylase from human cells.

3. DISCUSSION

DNA glycosylases facilitate the detection of DNA damage and mutations, and therefore their availability should enable researchers to incorporate these proteins into DNA repair experiments. The method of purification of DNA glycosylases presented in this chapter may also be adapted to other DNA glycosylases such as the MutY (MicA) protein (unpublished data), the human 3-methyladenine-DNA glycosylase (O'Connor, 1993), and the Fapy-DNA glycosylase of *S. cerevisiae* (de Oliveira *et al.*, 1994).

ACKNOWLEDGMENTS. Drs. Serge Boiteux and Jacques Laval are thanked for their comments and critical reading of the manuscript. The work described was supported by the CNRS, INSERM, the European Community, and ARC.

REFERENCES

Asahara, H., Wistort, P. M., Bank, J. F., Bakerian, R. H., and Cunningham, R. P. (1989). Purification and characterization of Escherichia coli endonuclease III from the cloned nth gene. *Biochemistry* **28**:4444–4449.

Au, K. G., Clark, S., Miller, J. H., and Modrich, P. (1989). Escherichia coli mutY gene encodes an adenine glycosylase active on G-A mispairs. *Proc. Natl. Acad. Sci USA* **86**:8877–8881.

Bailly, V., and Verly, W. G. (1987). Escherichia coli endonuclease III is not an endonuclease but a b-elimination catalyst. *Biochem. J.* **242**:565–572.

Bessho, T., Roy, R., Yamamoto, K., Kasai, H., Nishimura, S., Tano, K., and Mitra, S. (1993). Repair of 8-hydroxyguanine in DNA by mammalian N-methylpurine-DNA glycosylase. *Proc. Natl. Acad. Sci. USA* **90**:8901–8904.

Bjelland, S., and Seeberg, E. (1987). Purification and characterization of 3-methyladenine-DNA glycosylase I from Escherichia coli. *Nucleic Acids Res.* **15**:2787–2901.

Bjelland, S., Bjoras, M., and Seeberg, E. (1993). Excision of 3-methylguanine from alkylated DNA by 3-methyladenine DNA glycosylase I of *Escherichia coli*. *Nucleic Acids Res.* **21**:2045–2049.

Bjelland, S., Birkeland, N. K., Benneche, T., Volden, G., and Seeberg, E. (1994). DNA glycosylase activities for thymine residues oxidized in the methyl group are functions of the AlkA enzyme in *Escherichia coli*. *J. Biol. Chem.* **269**:30489–30495.

Boiteux, S. (1993). Properties and biological functions of the NTH and FPG proteins of Escherichia coli: Two DNA glycosylases that repair oxidative damage in DNA. *J. Photochem. Photobiol. B: Biol.* **19**:87–96.

Boiteux, S., O'Connor, T. R., and Laval, J. (1987). Formamidopyrimidine-DNA glycosylase of Escherichia coli: Cloning and sequencing of the fpg structural gene and overproduction of the protein. *EMBO J.* **6**:3177–3183.

Boiteux, S., O'Connor, T. R., Lederer, F., Gouyette, A., and Laval, J. (1990). Homogeneous Fpg protein: A DNA glycosylase which excises imidazole ring-opened purines and nicks DNA at apurinic/apyrimidinic sites. *J. Biol. Chem.* **265**:3916–3922.

Boiteux, S., Gajewski, E., Laval, J., and Dizdaroglu, M. (1992). Substrate specificity of the Escherichia coli Fpg protein: Excision of purine lesions in DNA produced by ionizing radiation or photosensitization. *Biochemistry* **31**:106–110.

Boorstein, R. J., Hilber, T. P., Cadet, J., Cunningham, R. P., and Teebor, G. W. (1989). UV-induced pyrimidine hydrates in DNA are repaired by bacterial and mammalian DNA glycosylase activities. *Biochemistry* **28**: 6164–6170.

Bradford, M. (1976). A rapid and sensitive method for the quantitation of microgram quantities of protein utilizing the principle of protein-dye binding. *Anal. Biochem.* **72**:248–254.

Breimer, L. H. (1984). Enzymatic excision from γ-irradiated polydeoxyribonucleotide of adenine residues whose imidazole ring has been ruptured. *Nucleic Acids Res.* **12**:6359–6367.

Cadet, J., and Weinfeld, M. (1993). Detecting DNA damage. *Anal. Chem.* **65**:675A–682A.

Carter, C. A., Habraken, Y., and Ludlum, D.B. (1988). Release of 7-alkylguanines from haloethylnitrosourea treated DNA by E. coli 3-methyladenine-DNA glycosylase II. *Biochem. Biophys. Res. Commun.* **155**:1261–1265.

Chakravarti, D., Ibeanu, G.C., Tano, K., and Mitra, S. (1991). Cloning and expression in Escherichia coli of a human cDNA encoding the DNA repair protein N-methylpurine-DNA glycosylase. *J. Biol. Chem.* **266**: 15710–15715.

Chetsanga, C.J., and Lindahl, T. (1979). Release of 7-methylguanine residues whose imidazole rings have been opened from damaged DNA by a DNA glycosylase from Escherichia coli. *Nucleic Acids Res.* **6**:3673–3683.

Clarke, N. D., Kvaal, M., and Seeberg, E. (1984). Cloning of Escherichia coli genes encoding 3-methyladenine DNA glycosylases I and II. *Mol. Gen. Genet.* **197**:368–372.

de Oliveira, R., Auffret van der Kemp, P., Thomas, D., Geiger, A., Nehls, P., and Boiteux, S. (1994). Formamidopyrimidine-DNA glycosylase in the yeast Saccharomyces cerevisiae. *Nucleic Acids Res.* **22**: 3760–3764.

Demple, B., and Harrison, L. (1994). Repair of oxidative damage to DNA: Enzymology and biology. *Annu. Rev. Biochem.* **63**:915–948.

Dianov, G., and Lindahl, T. (1994). Reconstitution of the DNA base excision-repair pathway. *Curr. Biol.* **4**:1069–1076.

Dizdaroglu, M. (1991). Chemical determination of free radical-induced damage to DNA. *Free Radical Biol. Med.* **10**:225–242.

Dizdaroglu, M., Laval, J., and Boiteux, S. (1993). Substrate specificity of the Escherichia coli endonuclease III: Excision of thymine- and cytosine-derived lesions in DNA produced by radiation-generated free radicals. *Biochemistry* **32**:12105–12111.

Dodson, M. L., and Lloyd, R.S. (1989). Structure–function studies of the T4 endonuclease V repair enzyme. *Mutat. Res.* **218**:49–65.

Dodson, M. L., Michaels, M. L., and Lloyd, R. S. (1994). Unified catalytic mechanism for DNA glycosylases. *J. Biol. Chem.* **269**:32709–32712.

Dosanjh, M. K., Chenna, A., Kim, E., Fraenkel-Conrat, H., Samson, L., and Singer, B. (1994). All four known cyclic adducts formed in DNA by the vinyl chloride metabolite chloracetaldehyde are released by a human DNA glycosylase. *Proc. Natl. Acad. Sci. USA* **91**:1024–1028.

Floyd, R. A., West, M.S., Eneff, K.L., and Schneider, J.E. (1989). Methylene blue plus light mediates 8-hydroxyguanine formation in DNA. *Arch. Biochem. Biophys.* **273**:106–111.

Ganguly, T., Weems, K. M., and Duker, N. J. (1989). Ultraviolet-induced thymine hydrates are excised by bacterial and human DNA glycosylase activity. *Biochemistry* **29**:7222–7228.

Gerlt, J. A. (1993). Mechanistic principles of enzyme-catalyzed cleavage of phosphodiester bonds, in: *Nucleases* (Linn, S., Roberts, R. J., and Lloyd, R. S, eds.), Cold Spring Harbor Laboratory Press, Cold Spring Harbor, NY, pp. 1–34.

Graves, R. J., Felzenswalb, I., Laval, J., and O'Connor, T. R. (1992). Excision of 5′-terminal deoxyribose phosphate from damaged DNA is catalysed by the Fpg protein of Escherichia coli. *J. Biol. Chem.* **267**:14429–14435.

Habraken, Y., Carter, C. A., Sekiguchi, M., and Ludlum, D. B. (1991). Release of N2,3-ethanoguanine from haloethylnitrosourea-treated DNA by Escherichia coli 3-methyladenine DNA glycosylase II. *Carcinogenesis* **12**:1971–1973.

Hanawalt, P. C. (1994). Transcription-coupled repair and human disease. *Science* **266**:1957–1958.

Hatahet, Z., Purmal, A. A., and Wallace, S. S. (1993). A novel method for site specific introduction of single model oxidative DNA lesions into oligodeoxyribonucleotides. *Nucleic Acids Res.* **21**:1563–1568.

Hatahet, Z., Kow, Y. W., Purmal, A. A., Cunningham, R. P., and Wallace, S. S. (1994). New substrates for old enzymes. *J. Biol. Chem.* **269**:18814–18820.

Hegler, J., Bittner, D., Boiteux, S., and Epe, B. (1993). Quantification of oxidative DNA modifications in mitochondria. *Carcinogenesis* **14**:2309–2312.

Holmes, J., Clark, S., and Modrich, P. (1990). Strand-specific mismatch correction in nuclear extracts of human and Drosophila melanogaster cell lines. *Proc. Natl. Acad. Sci. USA* **87**:5837–5841.

Hsu, I.-C., Yang, Q., Kahng, M. W., and Xu, J.-F. (1994). Detection of DNA point mutations with DNA mismatch repair enzymes. *Carcinogenesis* **15**:1657–1662.

Kow, Y. W., and Wallace, S. S. (1987). Mechanism of action of Escherichia coli endonuclease III. *Biochemistry* **26**:8200–8206.

Lindahl, T. (1974). An N-glycosidase from Escherichia coli that releases free uracil from DNA containing deaminated cytosine residues. *Proc. Natl. Acad. Sci. USA* **71**:3649–3653.

Lindahl, T. (1993). Instability and decay of the primary structure of DNA. *Nature* **362**:709–715.

Lion, T., and Haas, O. A. (1990). Nonradioactive labeling of probe with digoxigenin by polymerase chain reaction. *Anal. Biochem.* **188**:335–337.

Lloyd, R. S., and Linn, S. (1993). Nucleases involved in DNA repair, in: *Nucleases*. (Linn, S., Roberts, R. J., and Lloyd, R. S., eds.), Cold Spring Harbor Laboratory Press, Cold Spring Harbor, NY, pp. 263–316.

Lu, A.-L., and Hsu, I.-C. (1992). Detection of single DNA base mutations with mismatch repair enzymes. *Genomics* **14**:249–255.

Matijasevic, Z., Sekiguchi, M., and Ludlum, D. B. (1992). Release of N2,3-ethenoguanine from chloroacetaldehyde-treated DNA by Escherichia coli 3-methyladenine DNA glycosylase II. *Proc. Natl. Acad. Sci. USA* **89**:9331–9334.

Mattes, W. B., Lee, C.-S., Laval, J., and O'Connor, T. R. (1996). Excision of DNA adducts of nitrogen mustards by bacterial and mammalian 3-methyladenine-DNA glycosylases. *Carcinogenesis, in press.*

Maxam, A. M., and Gilbert, W. (1980). Sequencing end-labeled DNA with base-specific chemical cleavage. *Methods Enzymol* **65**:499–559.

Müller, E., Boiteux, S., Cunningham, R. P., and Epe, B. (1990). Enzymatic recognition of DNA modifications induced by singlet oxygen and photosensitizers. *Nucleic Acids Res.* **18**:5969–5973.

Neddermann, P., and Jiricny, J. (1993). Purification of a mismatch-specific thymine-DNA glycosylase from HeLa cells. *J. Biol. Chem.* **268**:21218–21224.

O'Connor, T. (1993). Purification and characterisation of human 3-methyladenine-DNA glycosylase. *Nucleic Acids Res.* **21**:5561–5569.

O'Connor, T. R., and Laval, J. (1989). Physical association of the formamidopyrimidine DNA glycosylase of Escherichia coli and an activity nicking DNA at apurinic/apyrimidinic sites. *Proc. Natl. Acad. Sci. USA* **86**: 5222–5226.

O'Connor, T. R., and Laval, F. (1990). Isolation and structure of a cDNA expressing a mammalian 3-methyladenine-DNA glycosylase. *EMBO J.* **9**:3337–3342.

O'Connor, T. R., and Laval, J. (1991). Human cDNA expressing a functional DNA glycosylase excising 3-methyladenine and 7-methylguanine. *Biochem. Biophys. Res. Commun.* **176**:1170–1177.

O'Connor, T. R., Boiteux, S., and Laval, J. (1988). Ring-opened 7-methylguanine residues are a block to *in vitro* DNA synthesis. *Nucleic Acids Res.* **16**:5879–5894.

O'Connor, T. R., Graves, R. J., de Murcia, G., Castaing, B., and Laval, J. (1993). Fpg protein of Escherichia coli is a zinc finger protein whose cysteine residues have a structural and/or functional role. *J. Biol. Chem.* **268**:9063–9070.

Pfeifer, G. P., Drouin, R., and Holmquist, G.P. (1993). Detection of DNA adducts at the DNA sequence level by ligation-mediated PCR. *Mutat. Res.* **288**:39–46.

Pierre, J., and Laval, J. (1980). Micrococcus luteus endonucleases for apurinic/apyrimidinic sites in deoxyribonucleic acid. 2. Further studies on the substrate specificity and mechanism of action. *Biochemistry* **19**:5024–5029.

Sakumi, K., and Sekiguchi, M. (1990). Structures and functions of DNA glycosylases. *Mutat. Res.* **236**:161–172.

Sakumi, K., Nakabeppu, Y., Yamamoto, Y., Kawabata, S., Iwanga, I., and Sekiguchi, M. (1986). Purification and structure of 3-methyladenine-DNA glycosylase I of Escherichia coli. *J. Biol. Chem.* **261**:15761–15766.

Sambrook, J., Fritsch, E. F., and Maniatis, T. (1989). *Molecular Cloning: A Laboratory Manual*, 2nd ed., Cold Spring Harbor Laboratory Press, Cold Spring Harbor, NY.

Samson, L., Derfler, B., Boosalis, M., and Call, K. (1991). Cloning and characterization of a 3-methyladenine DNA glycosylase cDNA from human cells whose gene maps to chromosome 16. *Proc. Natl. Acad. Sci. USA* **88**: 9127–9131.

Saparbaev, M., and Laval, J. (1994). Excision of hypoxanthine from DNA containing dIMP residues by the *Escherichia coli*, yeast, rat, and human alkylpurine DNA glycosylases. *Proc. Natl. Acad. Sci. USA* **91**:5873–5877.

Schowalter, D. B., and Sommer, S. S. (1989). The generation of radiolabeled DNA and RNA probes with polymerase chain reaction. *Anal. Biochem.* **177**:90–94.

Simha, D., Palejwala, V. A., and Humayun, M. Z. (1991). Mechanisms of mutagenesis by exocyclic DNA adducts. Construction and in vitro template characteristics of an oligonucleotide bearing a single site-specific ethenocytosine. *Biochemistry* **30**:8727–8735.

Tsai-Wu, J. J., Liu, H. F., and Lu, A.-L. (1992). Escherichia coli MutY protein has both N-glycosylase and apurinic/apyrimidinic endonuclease activities on A–C and A–G mispairs. *Proc. Natl. Acad. Sci. USA* **89**:8779–8783.

Wang, W., Sitaram, A., and Scicchitano, D. A. (1995). 3-Methyladenine and 7-methylguanine exhibit no preferential removal from the transcribed strand of the dihydrofolate reductase gene in Chinese hamster ovary B11 cells. *Biochemistry* **34**:1798–1804.

Yeh, Y. C., Chang, D. Y., Masin, J., and Lu, A.-L. (1991). Two nicking enzyme systems specific for mismatch-containing DNA in nuclear extracts from human cells. *J. Biol. Chem.* **266**:6480–6484.

Chapter 13

PCR-Based Assays for the Detection and Quantitation of DNA Damage and Repair

F. Michael Yakes, Yiming Chen, and Bennett Van Houten

1. INTRODUCTION

Exposure to genotoxic agents from both environmental and endogenous sources which result in damage to cellular DNA poses a significant health risk to the individual if such damage is left unrepaired. Several human diseases, including Cockayne's syndrome and xeroderma pigmentosum, have been associated with defects in the repair of DNA damage (Friedberg *et al.*, 1995). An overall decrease in DNA repair capacity is observed in the latter, whereas the former is associated with a defect in a transcription-coupling factor which facilitates rapid repair in the transcribed strand of an active gene. One general method for the analysis of gene- and strand-specific repair is based on Southern analysis of DNA strand breaks induced by a damage-specific endonuclease (Smith and Mellon, 1990; Bohr and Okumoto, 1988). This endonuclease-sensitive site (ESS) technique employs the use of T4 endonuclease V which cleaves DNA at pyrimidine dimers (Ganesan *et al.*, 1980), eliminating its ability to hybridize to a radioactive probe on alkaline Southern analysis. Although this methodology has been pivotal in the elucidation of gene-specific repair in single-copy genes (Bohr *et al.*, 1987; Mellon *et al.*, 1987; Mellon and Hanawalt, 1989), there are certain limitations. It requires some information regarding restriction sequence information flanking the gene or genomic segment of interest, a lesion-specific endonuclease to incise near the damaged base, and perhaps most significant is the requirement for large quantities of DNA (5–10 μg) generally used in Southern assays.

Our laboratory (Chandrasekhar and Van Houten, 1994; Van Houten *et al.*, 1993; Kalinowski *et al.*, 1992) has been involved in the development of a polymerase chain reaction (PCR) for the detection and quantitation of DNA damage, first reported by Govan *et al.* (1990). This assay is based on the premise that several DNA lesions are able to block the progression of *Taq* polymerase (Murray *et al.*, 1992; Ponti *et al.*, 1991), which results in a decrease in amplification. Only those DNA templates, which do not contain *Taq*-blocking lesions, will be amplified. Thus,

---

F. Michael Yakes, Yiming Chen, and Bennett Van Houten • Sealy Center for Molecular Science, University of Texas Medical Branch, Galveston, Texas 77555.

Technologies for Detection of DNA Damage and Mutations, edited by Gerd P. Pfeifer, Plenum Press, New York, 1996.

the PCR assay effectively measures the fraction of template molecules that contain no damage. Recently, quantitative PCR (QPCR) was used to measure the formation and repair of UV-induced photoproducts in a 1.2-kb fragment of the lacI gene and a 3.2-kb fragment of the the *lacZ* gene from *Escherichia coli* (Chandrasekhar and Van Houten, 1994). In addition, studies have been performed on a 2.6-kb fragment of the *dhfr* gene and a 2.3-kb mitochondrial fragment of murine leukemic L1210 cells following exposure to UV and cisplatin (Kalinowski *et al.*, 1992). Jennerwein and Eastman (1992) have investigated the cisplatin lesion frequency in a series of fragments ranging from 150 to 2000 bp of the hamster *aprt* gene. These early studies demonstrated the ability to detect and monitor gene-specific DNA damage and repair with PCR. Although the amount of template DNA used in the PCR reactions was significantly lower than that used in ESS/Southern assays, the high doses required to detect 1 lesion per amplification target could be considered outside of a physiologically relevant range. This problem is directly related to the size of the amplification target. Therefore, in order to increase the sensitivity of the PCR assay for the detection of gene-specific damage and repair, it is essential to amplify larger fragments of DNA. Techniques have recently become available for the amplification of DNA fragments in the range of 10–20 kb from genomic DNA (Cheng *et al.*, 1994a).

QPCR is dependent on several technical procedures which are common to many laboratories. Although the isolation of intact, high-molecular-weight DNA is very important and will be discussed in further detail (see Methods), subsequent procedures will also determine the success or failure of QXLPCR which include: DNA quantitation, defining conditions for QPCR, quantitation of amplification products, and calculation of lesion frequencies (Fig. 13.1). Perhaps one of the most essential components of this assay is determining a precise DNA concentration. Any associated errors in this process will invalidate subsequent QXLPCR results due to differences in the amount of template utilized during the first round of PCR. As outlined in Fig. 13.1, the analysis of lesions that do not block a thermostable polymerase is also applicable with the use of damage-specific endonucleases. Independent of the specific application, both protocols converge at the QPCR step. The most significant advancement in this assay is the ability to perform extended-length PCR (XLPCR), which is then fine-tuned to occur under quantitative conditions. The QPCR box in Fig. 13.1 represents a significant portion of this developing technology and some effort must be spent to establish quantitative conditions. Following QPCR, amplification products are resolved on agarose gels, exposed to a Phosphor screen and quantitated using ImageQuant™ from Molecular Dynamics (Sunnyvale, CA). To calculate the lesion frequency, the amount of amplification from the damaged template (A_D) is divided by the amount of amplification for the control template (A_O). Assuming a random distribution of lesions, and using the Poisson equation [$f(x) = e^{-\lambda}\lambda^x/x!$ where λ = the average lesion frequency] for the nondamaged templates (i.e., zero class; $x = 0$), the lesion frequency per genomic strand can be determined: $\lambda = -\ln A_D/A_O$. Therefore, an average of one lesion per strand would be expected to decrease the amplification to 37%.

Although the individual steps outlined in Fig. 13.1 may appear labor-intensive, when compared directly, the sensitivity of this assay is equivalent to or greater than the ESS/Southern assay (Kalinowski *et al.*, 1992). The requirement for large amounts of DNA is significantly reduced and amplification of large fragments allows for use of physiological relevant doses of a particular genotoxic agent. This chapter outlines conditions which have been established for quantitative extended-length PCR (QXLPCR) in the detection of gene-specific adducts. In addition, we demonstrate the ability to detect and quantify low-fluence UV-induced photoproducts in a 17.7-kb fragment of the human β-*globin* gene cluster with nanogram quantities of DNA. We also discuss the use of this technique to investigate other genomic regions of various target sizes.

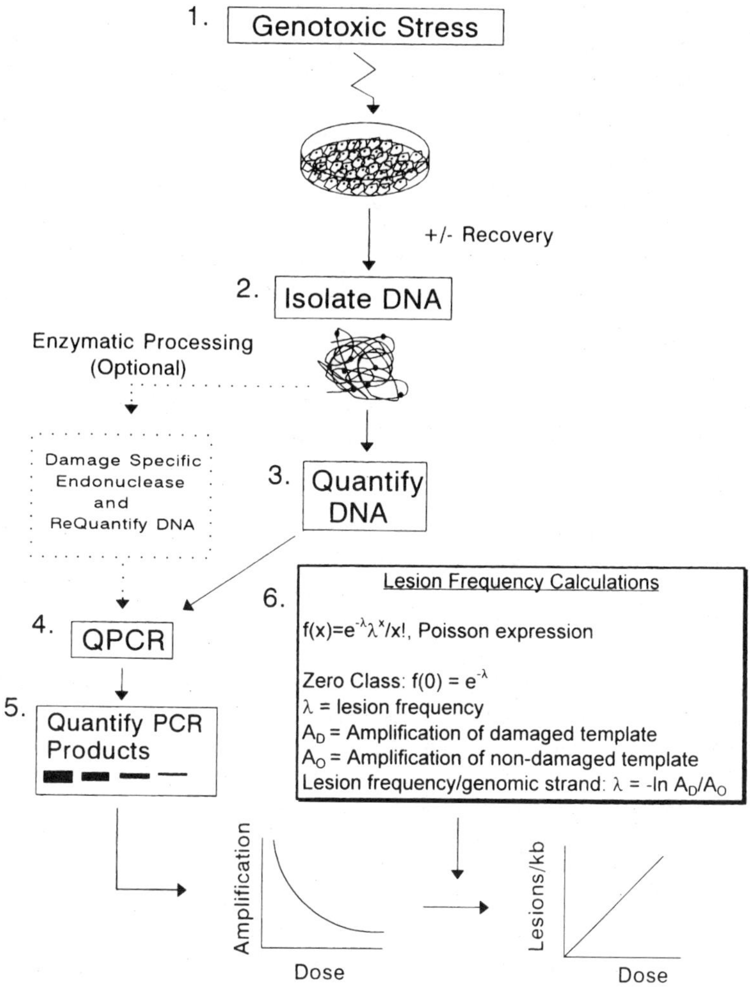

Figure 13.1. Strategy for the detection of gene-specific damage by QPCR. This methodology is dependent on the careful execution of the following steps: DNA isolation, DNA quantitation, QPCR, quantitation of amplification products, and calculation of lesion frequencies. Each step is described in detail in the text. For the detection of lesions which do not effectively block the progression of thermostable polymerase, it is necessary to add the optional enzymatic processing step indicated by the dashed arrows and boxes.

2. METHODS

2.1. Cell Culture and Treatment

The experiments described in this chapter utilize SV40-transformed fibroblasts, GM00637E, (NIGMS Human Genetic Mutant Cell Repository, Coriell Institute, Camden, NJ) maintained at 37°C, 5% CO_2 in MEM Earle's medium supplemented with 15% fetal bovine serum (Hyclone Laboratories, Logan, UT), 2× MEM amino acids (Sigma, St. Louis, MO), 2× MEM nonessential amino acids (Gibco-BRL), 2× MEM vitamin solution (Gibco-BRL), and 2.0 mM L-glutamine

(Gibco-BRL). Cells are routinely split 1:6 at confluency. To reduce the amount of error that may be associated with the analysis of one sample, all experimental doses are represented by at least three 60-mm dishes each containing 1×10^6 cells. Cells are seeded from confluent flasks and allowed to reach logarithmic growth for 20–24 hr prior to treatment. Following removal of the growth media, the cells are washed with room-temperature PBS (6.5 mM Na_2HPO_4, 1.5 mM KH_2PO_4, 2.6 mM KCl, and 137 mM NaCl) and the monolayer irradiated in the absence of PBS or media with a Sylvania G8T5 germicidal lamp (254 nm) at an incident dose rate of 0.5 J/m^2 per sec. The cells are harvested immediately by a brief trypsin (0.25%) treatment, washed with ice-cold PBS, pelleted, and either stored frozen or used immediately for DNA isolation.

2.2. DNA Isolation and Quantitation

Previously our laboratory, in collaboration with Dr. Suzanne Cheng (Roche Molecular Systems), has demonstrated the significance of DNA template integrity on successful amplification of long PCR targets (Cheng *et al.*, 1995). Comparison of several isolation protocols revealed that preparation of DNA by high-salt methods was comparable to DNA that was prepared with commercially available kits. However, the time involved with such kits is significantly lower when compared to the high-salt methods, especially when processing multiple samples. In an attempt to reduce the number of exogenous lesions that would render unusable templates and to optimize the time required to go from cell pellet to purified DNA, we routinely isolate DNA with the QIAamp® Tissue Kit from QIAGEN Inc. (Chatsworth, CA). DNA is isolated according to the manufacturer's protocol with the following modifications: frozen cell pellets or freshly harvested cells are washed once with ice-cold PBS, lysed with the supplied buffer for 10 min at 50°C in the presence of proteinase K (2.5 μg/ml), and precipitated with 2 vol of absolute ethanol as indicated by the manufacturer (QIAGEN Genomic DNA Handbook, 1994). The QIAGEN protocol recommends cell lysis at 70°C; however, lysis at such a high temperature may generate heat-sensitive lesions that would hinder PCR. The entire mixture is applied to a provided spin column, centrifuged at 8000 rpm for 2 min, washed with the supplied buffer, and eluted with prewarmed (50°C) 1× TE (10 mM Tris, 1 mM EDTA), pH 8.0. Elution with distilled water or other low-salt buffers such as 1× TEN (10 mM Tris, 10 mM EDTA, 100 mM NaCl), pH 8.0 has also been used with equal success. The purified DNA is further dialyzed in a Microdialyzer System 100 (Pierce, Rockford, IL) against 1× TE, pH 8.0 for 14–16 hr at room temperature which has been shown to improve amplification efficiency (Van Houten *et al.*, 1993). Verification of intact, high-molecular-weight DNA suitable for long PCR can be determined by agarose gel electrophoresis (data not shown).

QPCR requires an exact determination of DNA concentration. Since the QPCR assay is absolutely dependent on the initial template concentration, detailed attention to the determination of DNA concentration is essential. Our laboratory has adopted an alkaline fluorimetry assay (Kowalski, 1979) which is rapid and extremely sensitive. An initial DNA concentration for each sample is determined by adding 2.0–5.0 μl of the stock DNA solution to 1.2 ml of an alkaline phosphate buffer (20 mM KH_2PO_4, 0.5 mM EDTA, 0.5 μg/ml ethidium bromide, pH 11.8), vortexed, and analyzed with an A4-Filter Fluorimeter (Optical Technology Devices Inc., Elmsford, NY) with excitation at 365 nm and emission at 600 nm. It should be noted that the pH of this buffer is critical for quantitative results. As a result of the initial fluorimeter reading, an approximate DNA concentration is assigned. The original DNA stock solution is then diluted, on average two- to fivefold, to 10 ng/μl followed by a more precise fluorimetry analysis. For example, we routinely analyze 20 μl of the diluted DNA sample in increments of 5 μl, recording the fluorimeter reading at each volume. It is important to point out that the fluorimeter value should increase linearly with each added aliquot, i.e., the fluorimeter reading observed after the addition of the second 5 μl should be twice that of the first reading. A standard curve ranging from 0 to 160 ng is generated with λ/*Hin*dIII-digested DNA (Gibco-BRL) and the concentration of the DNA is determined using the slope of the curve. Typical yields of DNA from 1×10^6 cells are in the range of 1–2 μg in 90–100 μl of buffer. As a practice, DNA concentrations calculated for each fluorimeter reading from a single sample that differ by more than 5% require redilution and requantitation. In our experience, inconsistent fluorimeter readings are most likely indicative of a nonhomogeneous sample, pipet/dilution error(s), or changes in the pH of the alkaline phosphate buffer. As previously indicated, preparations of genomic DNA by high salt methods are suitable for long PCR (Cheng *et al.*, 1994, 1995); however, obtaining a homogeneous DNA solution that results in reproducible fluorimeter readings and more consistent PCR results is difficult. We have found that the use of the QIAGEN kit has reduced this problem.

2.3. PCR Amplification

2.3.1. PCR Components and Reagents

For our quantitative PCR amplifications we routinely use the GenAmp® XL PCR kit (Perkin–Elmer/Roche, Alameda, CA) which includes a 3.3× XL PCR buffer, 2.5 mM $Mg(OAc)_2$, and *r*Tth DNA polymerase (400 units; 2 units/μl). Although the 3.3× buffer is suitable for many long PCR applications, refinement of the buffer system may be necessary for specific PCR targets. In most instances, this will be highly dependent on the specific fragment of interest and may not be directly transferable to other fragments. The development of this buffer is based on a report by Cheng *et al.* (1994) in which several PCR components were changed to allow amplification of long fragments. One important change was from a Tris-HCl-based reaction to a Tricine-based reaction. Tris buffers change pH according to temperature such that a pH 8.0 buffer at room temperature can decrease to pH 6.8 at 94°C. Lower pH's and higher temperatures favor depurination and production of DNA strand breaks. The addition of the cosolvents, DMSO and glycerol, also appears to be essential for the production of long PCR fragments. Both reagents lead to a decrease in the effective DNA melting temperature, while the former increases the thermostability of the polymerase. In addition, we have found that amplification of some target fragments demonstrates an extreme sensitivity to the magnesium concentration in the reaction. The optimal concentration is dependent not only on the target fragment, but also on the DNA extraction procedure and is most easily determined by conducting amplification reactions with varying magnesium concentrations. Therefore, in designing quantitative PCR experiments, attention to buffer composition, and both magnesium and enzyme concentrations can affect the outcome of long PCR.

We have found that with increased target sizes, the need for high-quality reagents also increases. For example, the storage and handling of the deoxynucleotide triphosphates (dNTPs; Pharmacia Biotech) is crucial to success. Newly purchased dNTPs are mixed to a final concentration of 2.5 mM with respect to each deoxynucleotide, aliquoted into working volumes, and stored at −80°C. The use of small volumes eliminates the need for repeated freeze–thaw cycles which may have an adverse effect on amplification. Routinely, quality control experiments particularly with new lots of *rTth* DNA polymerase and dNTPs are performed to confirm that the same level of quantitative PCR is maintained between lots. Although this may be considered a technical drawback, we believe that quality control minimizes the number of areas to analyze in possible future troubleshooting episodes.

Cross-contamination of PCR reactions with products from previous reactions is a potentially serious hazard. We highly recommend the use of individual pipets with aerosol-resistant tips for pre- and post-PCR procedures. Ideally, maximal separation of each preparative stage, i.e., DNA preparation and quantitation, PCR setup, and electrophoresis, should be observed. However, this may not always be possible and efforts should be practiced to minimize possible sources of cross-contamination.

2.3.2. Primer Selection

One important consideration in QXLPCR is the selection of primer sets which are specific to a particular gene of interest. The selection of primers for the amplification of small target fragments is not a significant obstacle. Unfortunately, the selection of primers that span thousands of bases that will amplify under quantitative conditions can be a difficult task. For example, we are able to independently amplify two adjacent regions of the human *hprt* gene, although attempts to amplify across the entire span with the distal primers have been unsuccessful (data not shown). Generally, primers should be 20–24 nucleotides in length and have a G+C content of over 50% to yield a T_M of approximately 68°C. Primers should not end in complementary sequences, contain large repeats or palindromic sequences. DNA-analysis software such as Oligo™ 4.0 (National Biosciences Inc., Plymouth, MN) or other equivalent programs can be used to evaluate selected primers. When selecting primers for large fragments, it is strongly suggested that several primer sets be chosen. Once such a primer set(s) has been found, quantitative conditions must be established (see Section 2.4). Oligonucleotide primers are used directly as supplied by the manufacturer without further purification.

2.3.3. Reaction Conditions and Agarose Gel Electrophoresis

All PCR amplification reactions (50 μl) are performed in either a GeneAmp™ PCR System 2400 or 9600 using 2.5–50 ng template DNA in 1× Tricine buffer (20 mM Tricine, pH 8.7/ 85 mM potassium acetate; Cheng *et al.*, 1994) with 1.1 mM $Mg(OAc)_2$, 100 μg/ml nonacetylated BSA, 0.2 mM dNTPs, 4% DMSO (vol/vol), 8% glycerol (vol/vol), and 0.2 μM primers (*hprt*, β-*globin*, and mitochondria). The primer sequences corresponding to the fragments used in this study are listed in Table I. For quantitative PCR, 0.2 μCi [α-^{32}P]dATP (NEN DuPont) per reaction is included. A master mix omitting the DNA and *rTth* polymerase was prepared for multiple samples and aliquoted into 0.5-ml PCR reaction tubes (Perkin–Elmer) which is then followed by the addition of DNA. We have found that prior dilution of the DNA samples to the same concentration, generally 3 ng/μl, minimizes potential errors that might be associated with pipetting different volumes from stock solutions. "Hot Start" reactions are performed by the addition of 1 unit of *rTth* DNA polymerase when the samples are at 75°C. The standard thermocycler profile for nonquantitative conditions consists of a Hot Start hold at 75°C for 2 min, an initial denaturation at 94°C for 2 min followed by 36 cycles of denaturation at 94°C for 30 sec and 68°C primer extension for 12 min. On completion of the 36 cycles, the extension reaction is continued for an additional 10 min at 72°C. As shown in Fig. 13.2, fragments from the β-*globin* gene ranging in size from 2.7 to 17.7 kb can be successfully amplified with this protocol. In addition, note that lane 6 contains the amplification product of the entire mitochondrial genome.

2.4. Defining QPCR Conditions

Once amplification of a particular fragment has been accomplished, quantitative conditions can be established. This process begins with cycle and DNA template tests. These particular assays not only satisfy reagent quality control and concomitantly measure the integrity of the DNA template, but they also are the most accurate and reliable test of quantitative conditions. As described by Kalinowski *et al.* (1992), it is essential that the only limiting factor in the

Table I
QXLPCR Target Fragments and Respective Primer Sets

Gene	Target length (kb)	Primer set	
		Forward	Reverse
hprt[a]	2.7	b	e
	6.2	a	e
	10.4	b	g
β-*globin*[b]	13.5	RH 1022	RH 1053
	17.7	RH 1024	RH 1053
Mitochondria[c]	16.2	H408	L15996

[a]Van Houten *et al.* (1993).
[b]Accession number: J00179.
[c]Cheng *et al.* (1994).
Primer sequences (5′–3′):
a: TGTGGCAGAAGCAGTGAGTAACTG
b: TGGGATTACACGTGTGAACCAACC
e: TGTGACACAGGCAGACTGTGGATC
g: GCTCTACCCTCTCCTCTACCGTCC
RH 1022: CGAGTAAGAGACCATTGTGGCAG
RH 1024: TTGAGACGCATGAGACGTGCAG
RH 1053: GCACTGGCTTAGGAGTTGGACT
H408: CTGTTAAAAGTGCATACCGCCA
L15996: CTCCACCATTAGCACCCAAAGC

amplification of a target sequence be the amount of nondamaged template. For a given amount of template, therefore, the amplification of a specific fragment should be performed at a cycle number that is within the exponential component of the PCR (Chandrasekhar and Van Houten, 1994; Van Houten *et al.*, 1993; Kalinowski *et al.*, 1992) It should be stressed that *quantitative experiments involving damaged DNA cannot be performed without first demonstrating exponential amplification of the nondamaged template*. Exponential amplification may occur only within a certain number of cycles, after which the amplification signal may begin to plateau. PCR experiments performed at cycles outside of this range are not quantitative. Figure 13.3 is

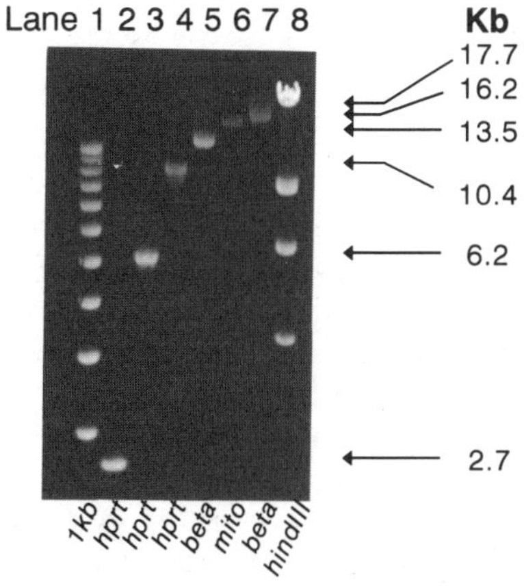

Figure 13.2. Amplification of target fragments from the *hprt*, β-*globin* genes, and the entire mitochondrial genome. Following purification of genomic DNA, each fragment of interest was independently amplified from 50 ng of starting template according to the conditions outlined in the text. XLPCR products were electrophoresed on a 0.4% horizontal SeaKem GTG (FMC Bioproducts) agarose gel in 1× TBE, stained for 1 hr with 1 μg/μl ethidium bromide, destained, and photographed. Lanes: 1, 1-kb molecular size standard; 2, 2.7-kb *hprt*; 3, 6.2-kb *hprt*; 4, 10.4-kb *hprt*; 5, 13.5-kb β-*globin*; 6, 16.2-kb mitochondrial genome; 7, 17.7-kb β-*globin*; 8, λ/*Hin*dIII molecular size standard. (See Table I for sequence and source of primers.)

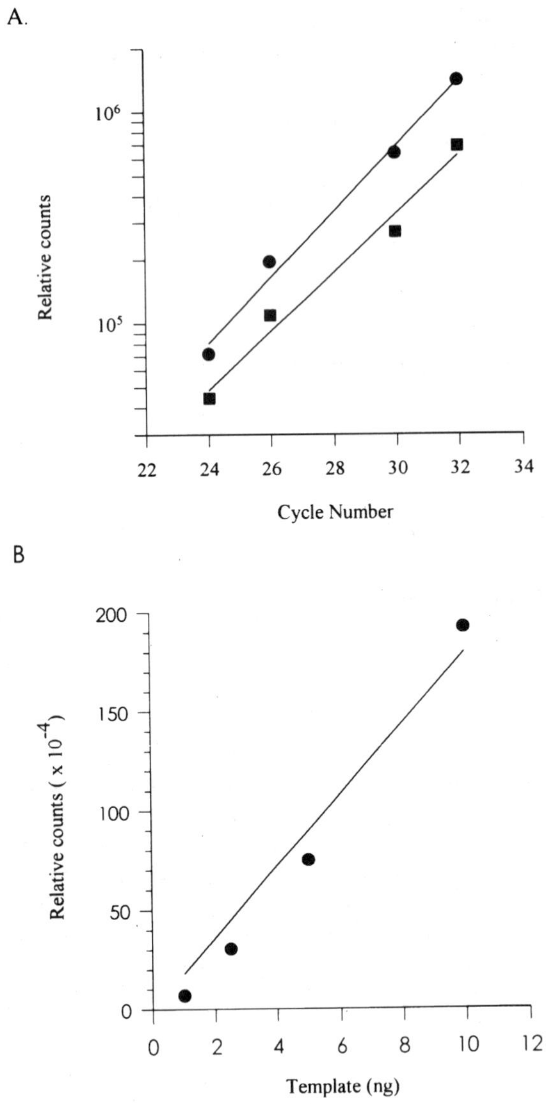

Figure 13.3. Determination of optimal conditions for QXLPCR of the 17.7-kb β-*globin* fragment. (A) Cycle test. Purified genomic DNA, 15.0 ng (●) or 7.5 ng (■), was used to amplify the 17.7-kb β-*globin* fragment that would result in exponential amplification for each template concentration as described in the text. The thermocycler was programmed for a total of 34 cycles, although samples were removed at the completion of 24, 26, 30, or 32 cycles, respectively, during the 94°C ramp for the following cycle. Amplification products were electrophoresed on a 1% vertical agarose gel, dried, and quantitated as described. Each point represents the mean of a duplicate loaded sample. (B) Template titration. Purified genomic DNA was diluted from a stock solution of known concentration to 3 ng/μl which was then further diluted to deliver in 5 μl the following amount of template: 1.0, 2.5, 5, and 10 ng, respectively. QXLPCR was then performed at 27 cycles, the products electrophoresed on a 1% vertical agarose gel and quantitated as described in the text. Each point represents the mean of a duplicate loaded sample.

representative of experiments establishing quantitative conditions for the 17.7-kb β-*globin* fragment. Cycle tests were performed to determine quantitative conditions for the 17.7-kb fragment of the β-*globin* gene using 15.0 and 7.5 ng of template, respectively. Reactions were performed as described earlier. At the completion of 24, 26, 30, and 32 cycles, one tube of each template concentration was removed during the 94°C ramp for the subsequent cycle and stored until completion of the experiment. Figure 13.3A reveals that the cycle range for exponential amplification of the 17.7 kb is between 24 and 32 cycles. Amplification at cycle numbers greater than 32 results in a plateau of synthesis (data not shown). Note that a 50% reduction in the amount of template also results in exponential amplification within the same cycle number range. Therefore, performing quantitative amplifications at 27 cycles ensures that the observed decreases in amplification are due to unavailable template (i.e., damaged) and not some other component of the PCR. We have found with the analysis of other fragments that the addition of an autoextension profile, adding 15 sec per extension per cycle, contributes to quantitative conditions (data not shown). This increase in time allows for the completion of initiated synthesis which may be more significant for fragments containing strong secondary structure.

To ensure that the amplification signal is linear with respect to the amount of added template, "template tests" must be performed with serial titrations of undamaged DNA. DNA, diluted to 3 ng/μl from a stock solution, is further serially diluted to deliver in 5 μl the following template concentrations: 10, 7.5, 5 and 2.5 ng. QXLPCR of the 17.7-kb fragment was performed as described. However, the number of cycles was reduced from 36 to 27 in order to be within the exponential component of the PCR. Results of a representative experiment are shown in Fig. 13.3B. It is expected that a reduction in the amount of template available for PCR will consequently reduce the amplification signal by the same fraction. All amplification products are resolved and quantitated as described below.

Currently we have found that the highest degree of resolution and reproducible quantitation of PCR products is with 1% vertical-agarose gels. An agarose plug (1%) is first poured in the bottom buffer reservoir of the vertical gel apparatus (Gibco-BRL) to act as a support. Subsequently, a 1% agarose gel is cast with 1.5-mm spacers in standard 1× TBE electrophoresis buffer. Equal volumes (10 μl) from each reaction tube are loaded in duplicate in 6× sample loading buffer III (Sambrook *et al.*, 1988) and electrophoresed at 5 V/cm for 4 hr. For quantitative conditions, it is essential that all of the sample is loaded into the well and that nothing remains in either the microfuge tube or pipet tip. Following electrophoresis, the 1% agarose plug containing a significant portion of the unincorporated radioactivity is carefully discarded to radioactive waste. The gels are washed with distilled water to remove any free counts, placed between two sheets of cellophane (Hoefer Scientific) and vacuum-dried at 60°C for 2 hr. Dried gels are exposed to a Phosphor screen (Molecular Dynamics) for 12–18 hr and quantitated using ImageQuant™ (Molecular Dynamics). As shown in Fig. 13.3B, there is a *direct linear relationship* between the amount of template added and the amplification signal, such that a 50% reduction in template concentration reduces the amplification signal by 50%. Once a detailed template test has been performed for the target of interest, a 50% template dilution of the nondamaged experimental control *should always be included* with each PCR as an internal control. We routinely observe that a 50% reduction in template leads to a 50% (± 10%) reduction in amplification. Rare data sets which exceed a margin of 10% error, i.e., < 40% or > 60%, are not used.

2.5. Precision of the QPCR Assay

In the development of this assay we have systematically addressed several potential sources of error. One particular area that is as equally important as the cycle and template tests described above, especially when multiple samples are compared, is examining the amount of variability between wells of the thermocycler. Because quantitative conditions for a specific fragment will only be as reliable as the performance of the well, it is recommended that only wells which result in similar amplification efficiencies be used. We have found that once a well test has been performed for a particular thermocycler, it generally does not require further analysis. Figure 13.4 demonstrates a two-concentration template test which directly measured the performance of 24 wells within the System 2400 thermocycler. DNA templates, 15.0 and 7.5 ng, were used in this experiment. Master QXLPCR mixes were prepared and the DNA aliquoted such that 12 samples of each template concentration were analyzed by QXLPCR. Placement of the 24 samples in the thermocycler was random. As expected the mean (± S.D.) amplification signal for 7.5 ng of

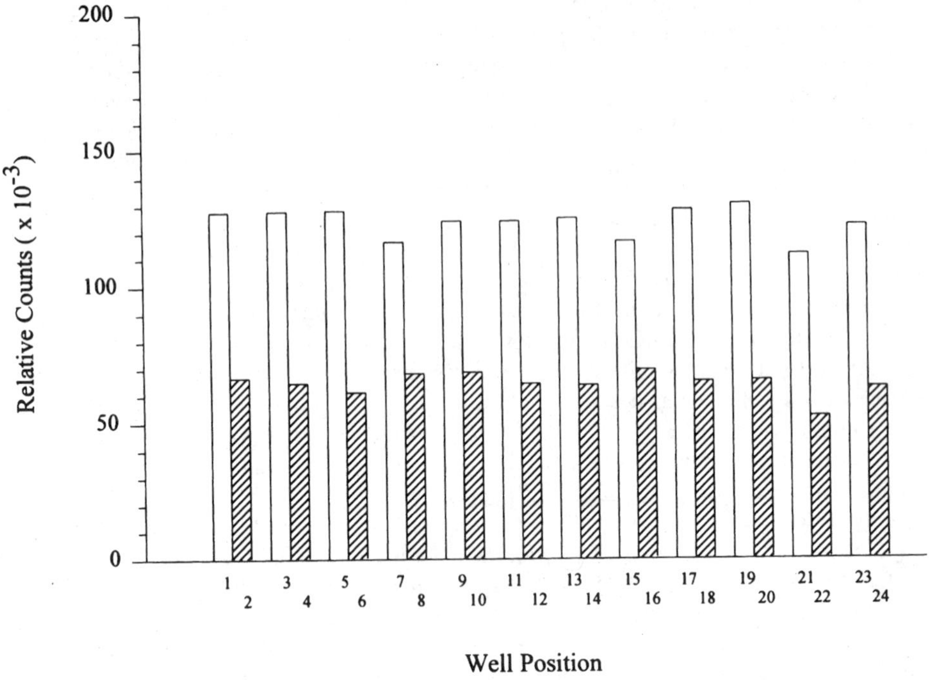

Figure 13.4. Analysis of thermocycler well-to-well variation in the amplification of the 17.7-kb β-*globin* fragment. Purified genomic DNA was diluted to 3 and 1.5 ng/μl, respectively. Aliquots delivering 15.0 and 7.5 ng of template were pipetted into 12 separate reaction tubes and QXLPCR was performed at 27 cycles with a Perkin–Elmer GeneAmp PCR System 2400. Positioning of the reaction tubes in the thermocycler was random. Amplification products were analyzed and quantitated as described in the text.

template, $6.4 \times 10^5 \pm 0.4 \times 10^5$ relative counts, was half that of the 15.0-ng template, $12.3 \times 10^5 \pm 0.5 \times 10^5$ relative counts. The coefficient of variation [(S.D./ mean) × 100] for each template was 7.0 and 4.7%, respectively. This experiment demonstrates that there is minimal variability between the wells and therefore, using the conditions established above, quantitative analysis of damaged template can be performed with confidence.

2.6. Using QXLPCR to Detect DNA Damage following UV Irradiation

Having established quantitative conditions, analysis was performed of the 17.7-kb fragment from the β-*globin* gene following UV irradiation. Amplification reactions utilized 15.0 ng of template at 27 cycles, as described earlier. The relative counts obtained for each damaged-DNA sample corresponding to the specific dose were averaged (A_D) and normalized to the average relative counts for the nondamaged template (A_O). This amplification ratio, (A_D/A_O), was then plotted as a function of dose (Fig. 13.5A). As expected, amplification of the 17.7 kb decreases in a dose-dependent manner. This represents the fraction of DNA molecules within the starting 15 ng that contain no damage. Note that we are able to detect a 31% decrease in amplification following exposure to 2.5 J/m² of UV. Earlier work by Kalinowski *et al.* (1992) showed that a dose nearly tenfold higher was required to see a similar decrease in amplification of a 2.6-kb target from the mouse *dhfr* gene. The increased sensitivity of this assay is clearly demonstrated with the

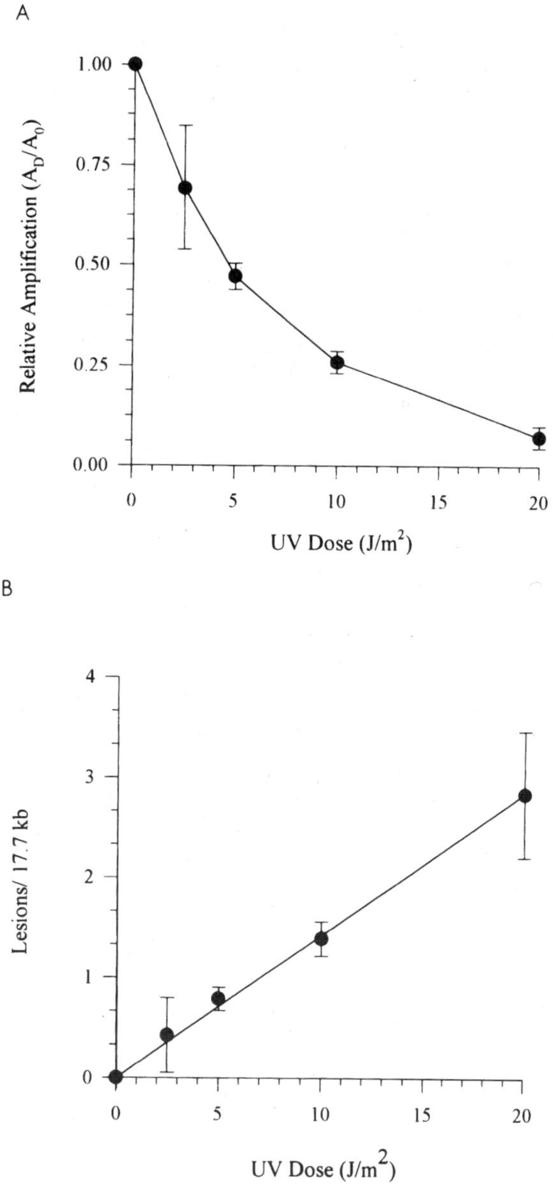

Figure 13.5. Detection and quantitation of DNA damage within the 17.7-kb β-*globin* fragment following exposure to UV. Normal human fibroblasts (1×10^6 cells) were seeded on 60-mm petri dishes in triplicate, allowed to reach logarithmic growth phase, and irradiated (0.5 J/m^2 per sec; 254 nm) with increasing doses of UV. Cells were harvested immediately, the DNA extracted, quantitated, and QXLPCR performed in triplicate at 27 cycles. Each point represents the mean value for three PCR reactions derived from three 60-mm plates and error bars equal ± S.D. (A) Exponential decrease in amplification following UV irradiation. The mean raw counts for damaged samples (A_D) were normalized to the mean raw counts for non-damaged samples (A_O) and plotted as a function of UV dose. (B) UV dose response. The UV-induced lesion frequency for each dose was calculated from the experiment described in panel A using the Poisson expression (Fig. 13.1).

amplification of a longer target fragment. This also permits analysis of lower doses of UV which are less likely to be cytotoxic. Using the Poisson equation as outlined in Figure 13.1, a lesion frequency was calculated for the zero class of molecules. At the lowest UV dose analyzed, 2.5 J/m^2, this assay was able to detect 0.42 photoproduct per 17.7 kb, whereas at 10 J/m^2, approximately 1.5 photoproducts were detected It is important to stress that the QXLPCR assay measures template integrity. Thus, any DNA lesion which disrupts the progression of the thermostable polymerase will decrease the amplification signal. These DNA lesions include not only bulky adducts such as UV-induced 6–4 cyclobutane and pyrimidine photoproducts, but also DNA strand breaks. As an example, high doses of a genotoxic agent which may lead to apoptotic cell death known to generate DNA strand breaks will be measured with this assay.

3. DISCUSSION

3.1. Advantages of the QXLPCR Assay

The experiments discussed in this chapter have described the general utility of a PCR-based methodology for the detection and quantitation of DNA damage in single-copy genes of mammalian cells. In addition, QXLPCR can be used to follow the kinetics of DNA repair. Current limits of detection with this assay are approximately 1 lesion/10^5 nucleotides with 5–30 ng of mammalian DNA (equivalent to approximately 1000–5000 cells). Therefore, in combination with the speed and ease of PCR, the low amount of template DNA required and lack of dependence on restriction enzyme sites make QXLPCR a powerful technique. QXLPCR may be particularly advantageous when cells or tissue samples are a limiting factor in designing experiments. This assay, because of its ease, sensitivity, and need for low numbers of cells, should be amenable to studying DNA damage and/or repair capacity in human populations.

3.2. Technical Limitations

We have continued to increase the overall sensitivity of this assay by optimizing conditions which are associated with two significant factors: (1) the size of the target fragment, since larger regions are more inclined to sustain damage at lower doses than smaller regions, and (2) the ability to accurately detect and quantitate small differences in the amount of amplification. Amplification of longer fragments, greater than 20 kb, would significantly increase the level of sensitivity. However, quantitative amplification of such fragments will no doubt have conditions that will be different from those discussed here, and will require independent optimization.

As indicated earlier, a significant limitation to this developing methodology is the generation of PCR primers. In our experience, producing primers that will amplify sufficiently long targets is an empirical process that can only be evaluated by performing PCR.

Analysis of DNA damage and repair at various points within the cell cycle would provide valuable information regarding DNA repair kinetics. The UV dose curve presented in Fig. 13.5 was generated with exponentially growing cells which were harvested immediately after exposure. Dilution of DNA damage due to DNA replication could lead to a misinterpretation of repair. Therefore, repair course studies involving the analysis of DNA damage beyond one round of replication may require a "damage dilution correction" factor. Otherwise, observed increases in amplification associated with such dilution could be misinterpreted as repaired template. Alternatively, isolation of nonreplicated, parental DNA using BrdU labeling and CsCl gradients could be performed to measure DNA damage and repair independent of DNA replication.

3.3. Future Directions

This PCR assay is dependent on the ability of a particular DNA lesion to act as an absolute block to *rTth* polymerase, such as the case demonstrated here with UV-induced photoproducts. As a result of UV exposure, the strand containing the lesion is removed from the available template pool. However, not every genotoxic agent is expected to produce a lesion that inhibits the progression of *rTth* polymerase. Analysis of DNA templates containing lesions which do not block *rTth* polymerase can be analyzed following enzymatic treatment with a damage-specific endonuclease(s) (see Chapters 12, 16). Cleavage of the DNA would subsequently generate a DNA strand break, and therefore act as a direct block to the polymerase.

Gene-specific repair, for the most part, is a manifestation of strand-specific repair. Development of an assay that results in quantitative, one-sided linear amplification of a DNA strand within a target gene may provide further insight into the heterogeneity and complexity of DNA repair in mammalian cells (Chandrasekhar and Van Houten, 1994). Finally, recent studies have described a gene-specific assay for induced damage which combines the use of lesion-specific antibodies to capture damaged genomic DNA followed by quantitative PCR of the gene fragment of interest (Denissenko *et al.*, 1994; Hochleitner *et al.*, 1991). In these immunocoupled experiments a clear dose response was demonstrated, although extrapolation to a lesion frequency per genome does not appear to be straightforward.

ACKNOWLEDGMENTS. We wish to thank Suzanne Cheng and Russell Higuchi of Roche Molecular Systems for helpful discussions and the generous gift of *rTth* polymerase and PCR reagents. This work is supported by NIEHS grants 1RO1ES07038-02 and 1RO1ES07218-01.

REFERENCES

Bohr, V. A., and Okumoto, D. S. (1988). Analysis of pyrimidine dimers in defined genes, in: *DNA Repair: A Laboratory Manual for Research Procedures*, Volume 3 (E. C. Friedberg and P. C. Hanawalt, eds.), Dekker, New York, pp. 347– 366.

Bohr, V. A., Phillips, D. H., and Hanawalt, P.C. (1987). Heterogeneous DNA damage and repair in the mammalian genome. *Cancer Res.* **47**:6426–6436.

Chandrasekhar, D., and Van Houten, B. (1994). High resolution mapping of UV-induced photoproducts in the *Escherichia coli lacI* gene. Inefficient repair of the non-transcribed strand correlates with high mutation frequency. *J. Mol. Biol.* **238**:319–332.

Cheng, S., Fockler, C., Barnes, W. M., and Higuchi, R. (1994a). Effective amplification of long targets from cloned inserts and human genomic DNA. *Proc. Natl. Acad. Sci. USA* **91**:5695–5699.

Cheng, S., Higuchi, R., and Stoneking, M. (1994b). Complete mitochondrial genome amplification. *Nature Genetics* **7**:350–351.

Cheng, S., Chen, Y., Monforte, J. A., Higuchi, R., and Van Houten, B. (1995). Template integrity is essential for PCR amplification of 20–30 kb sequences from genomic DNA. *PCR Methods Appl.* **4**:294–298.

Denissenko, M. F., Venkatachalam, S., Yamasaki, E. F., and Wani, A. A. (1994). Assessment of DNA damage and repair in specific genomic regions by quantitative immuno-coupled PCR. *Nucl. Acids Res.* **22**:2351–2359.

Friedberg, E. C., Walker, G. C., and Siede, W. (1995). *DNA Repair and Mutagenesis*, ASM Press, Washington, DC.

Ganesan, A. K., Smith, C. A., and van Zeelan, A. A. (1980). Measurement of pyrimidine dimer content of DNA in permeabilized bacterial or mammalian cells with endonuclease of bacteriophage T4, in: *DNA Repair: A Laboratory Manual of Research Procedures*, Volume 1 (E. C. Friedberg and P. C. Hanawalt, eds.), Dekker, New York, pp. 89–97.

Govan, H. L., Valles-Ayoub, Y., and Braun, J. (1990). Fine mapping DNA damage and repair in specific genomic segments. *Nucleic Acids Res.* **18**:3823–3829.

Hochleitner, K., Thomale, J., Nikitin, A. Y., and Rajewsky, M. F. (1991). Monoclonal antibody based, selective isolation of DNA fragments containing an alkylated base to be quantified in defined gene sequences. *Nucleic Acids Res.* **19**:4467–4472.

Jennerwein, M. M., and Eastman, A. (1991). A polymerase chain reaction-based method to detect cisplatin adducts in specific genes. *Nucleic Acids Res.* **19**:6209–6214.

Kalinowski, D. P., Illenye, S., and Van Houten, B. (1992). Analysis of DNA damage and repair in murine leukemia L1210 cells using a quantitative PCR assay. *Nucleic Acids Res.* **20**:3485–3494.

Kowalski, D. (1979). A procedure for the quantitation of relaxed closed circular DNA in the presence of superhelical DNA: An improved fluorometric assay for nicking-closing enzyme. *Anal. Biochem.* **93**: 346–354.

Mellon, I., and Hanawalt, P. C. (1989). Induction of the *Escherichia coli* lactose operon selectively increases repair of its transcribed DNA strand. *Nature* **342**:95–98.

Mellon, I., Spivak, G., and Hanawalt, P. C. (1987). Selective removal of transcription-blocking DNA damage from the transcribed strand of the mammalian DHFR gene. *Cell* **51**:241–249.

Murray, V., Motyka, H., England, P. R., Wickham, G., Lee, H. H., Denny, W. A., and McFadyen, W. D. (1992). The use of *Taq* DNA polymerase to determine the sequence specificity of DNA damage caused by cis-diamminedichloroplatinum (II), acridine-tethered platinum (II) diammine complexes or two analogues. *J. Biol. Chem.* **267**:18805–18809.

Ponti, M., Forrow, S. M., Souhami, R. L., D'Incalci, M., and Hartley, J. A. (1991). Measurement of the sequence specificity of covalent DNA modification by antineoplastic agents using *Taq* DNA polymerase. *Nucleic Acids Res.* **19**:2929–2932.

QIAGEN Genomic DNA Handbook. (1994). QIAGEN, Inc., Chatsworth, CA.

Sambrook, J., Fritsch, E. F., and Maniatis, T. (1989). *Molecular Cloning: A Laboratory Manual*, 2nd ed., Cold Spring Harbor Laboratory Press, Cold Spring Harbor, NY.

Smith, C. A., and Mellon, I. (1990). Clues to the organization of DNA repair systems gained from studies of intragenomic repair heterogeneity, in: *Advances in Mutagenesis Research*, Volume 1 (G. Obe, ed.), Springer-Verlag, Berlin, pp. 153–194.

Van Houten, B., Chandrasekhar, D., Huang, W., and Katz, E. (1993). Mapping DNA lesions at the gene level using quantitative PCR methodology. *Amplifications/Perkin-Elmer* **10**:10–17.

Chapter 14

DNA Damage Analysis Using an Automated DNA Sequencer

Gopaul Kotturi, Wolfgang C. Kusser, and Barry W. Glickman

1. INTRODUCTION

Advances in biotechnology and molecular genetics have made possible a better understanding of the molecular nature of mutation. For example, the discovery of genetically altered proto-oncogenes and tumor suppressor genes in cancerous cells has led to a better understanding of the links between mutation and cancer. Similarly, the ability to study mutation and mutational specificity *in vivo* and *in vitro* has led to an increased appreciation of the mechanisms of mutation and the role that DNA damage and DNA repair play in determining the specificity of mutagenesis. In turn, differences in both the cellular metabolism of exogenous chemicals and DNA repair can at least in part explain tissue, gender, and species specificity of carcinogenesis. We remain, however, a long way off from being able to predict the individual risks implicated with the mutagenic potential of chemical and physical agents. A part of this problem reflects our lack of knowledge of how individual lesions are handled in different tissues and different species against the genetic makeup of an individual.

In the past few years, an increasing amount of sequence data has confirmed that the mutational specificity of mutagenic agents appears specific for each agent, providing a mutational "fingerprint." It is also clear that the mutational fingerprint is a composite image with at least three components: (1) the ability of the genetic target to reveal mutation (the target filter which is related to the product and the mutation selection criteria), (2) the distribution of DNA damage, and (3) the specificity of DNA repair. The importance of the latter two factors can be emphasized by the realization that the initial distribution of DNA damage will contain clues as to the relevance of the different types of lesions to the mutational process; and, finally, that it is the absence of repair or misrepair of lesions that will determine their biological potency. It is from these origins that the distribution of damage in DNA derives its importance. This chapter describes an automated approach to the study of the distribution of DNA damage.

Gopaul Kotturi, Wolfgang C. Kusser, and Barry W. Glickman • Centre for Environmental Health, Department of Biology, University of Victoria, Victoria, British Columbia V8W 2Y2, Canada.

Technologies for Detection of DNA Damage and Mutations, edited by Gerd P. Pfeifer, Plenum Press, New York, 1996.

1.1. Development of Automated DNA Sequencers

Automated DNA sequencers is a term currently used to describe instruments that utilize fluorescent chromophores instead of radioactive isotopes to detect DNA fragments. Fluorescence of the labeled molecules generated during dideoxy-mediated DNA sequencing reactions is recorded in real time during gel electrophoresis (Smith *et al.*, 1985; Sanger *et al.*, 1977). The fluorescent signal is directly converted into a digital signal which can be acquired, stored, and processed into sequence information "automatically" by a computer (Smith *et al.*, 1986). Compared to standard autoradiographic techniques, several steps are eliminated and human handling is greatly reduced. These steps include: (1) precise timing of the electrophoresis, (2) drying the gel, (3) exposing of the film, and waiting one or more days before developing the autoradiogram, and (4) generation of the DNA sequence data. Table I lists the various automated DNA sequencers and the ancillary software required to perform DNA damage experiments. We have no experience with the Licor Biotechnology automated DNA sequencer and do not report on this system here although it can be similarly applied to the study of DNA damage distribution.

1.2. Ancillary Applications of Automated DNA Sequencers

Since highly fluorescent chromophores and laser-induced fluorescence are employed, the detection limits can be extremely low, on the order of 3×10^{-18} to 1×10^{-16} mol per band (Ansorge *et al.*, 1987; Smith *et al.*, 1985). In addition to their high sensitivity, these instruments can be designed to have a wide dynamic range of detection. This represents a dramatic improvement over conventional autoradiograms and should facilitate the adaptation of automated DNA sequencers to nonsequencing applications. These supplementary applications were not initially recognized, particularly the subsequent requirement for ancillary downstream software to perform peak integration, data analysis, and data presentation. An early study relied on the correlation of peak height to DNA quantity (Koehler *et al.*, 1991) largely because of the lack of downstream software. The first software to become available still required extensive input from the user (Shoukry *et al.*, 1991). Current analysis software, listed in Table I, greatly increases the flexibility of automated DNA sequencers.

The use of automated DNA sequencers to determine the sequence specificity of DNA damaging agents was established using approaches such as cleavage techniques (Shoukry *et al.*, 1993; Koehler *et al.*, 1991; Shoukry *et al.*, 1991), or DNA polymerase and DNA polymerase $3' \rightarrow 5'$ exonuclease activity (Sage *et al.*, 1992). Several other techniques that can be adapted are single-strand conformation polymorphism analysis (SSCPA) (Iwahana *et al.*, 1994), mismatch determination (Verpy *et al.*, 1994), T-cell receptor clonality (Segurado and Schendel, 1993), and quantitative polymerase chain reaction (QPCR) analysis (Porcher *et al.*, 1992).

Table I
Various Automated DNA Sequencers
and Associated Integration and Peak Identification Software

Company	Model	Ancillary software
Pharmacia Biotech (Piscataway, NJ)	A.L.F. Blue™ A.L.F. Express™	1. Fragment Manager™ 2. ALF2SMA™ and SMART™ Manager
Perkin/Elmer Applied Biosystems Division (Foster City, CA)	373A™ 377™	1. Genescan™ 2. Genotyper™
Licor Biotechnology Division (Lincoln, NE)	4000™ 4000 LX™	RFLP Scan (from Scanlytics)

2. METHODS

As outlined above, the techniques involved in utilizing an automated DNA sequencer do not differ in any fundamental way from conventional sequencing protocols. There are a few modifications and precautions which appear to be more relevant to fluorescently tagged DNA and these are discussed in detail. In order to illustrate the transition to fluorescence technology, we describe the detection of UVC (254 nm)-induced DNA damage. These techniques have been widely applied using radioactivity-based protocols both *in vitro* (Drobetsky and Sage, 1993; Brash *et al.*, 1987) and *in vivo* (Gao *et al.*, 1994; Tornaletti and Pfeifer, 1994; Kunala and Brash, 1992; Pfeifer *et al.*, 1991). UVC light induces lesions which are predominantly the cyclobutane pyrimidine dimers (CPDs) and 6–4 pyrimidine-pyrimidone photoproducts (6–4 Py<>Py). The sequence specificity of both of these photoproducts is determined by distinct cleavage techniques at the site of the photoproduct which includes: (1) the treatment with hot alkali for cleavage at 6–4 Py<>Py (Lippke *et al.*, 1981) and (2) the enzymatic T_4 endonuclease V cleavage at CPDs (Gordon and Haseltinc, 1982). Thus, the length of the cleaved DNA fragment reveals the position of the damage and quantitating the peak determines the amount of damage at that DNA site. What follows is a description of the procedure to determine the distribution of 6–4 Py<>Py, starting with DNA amplification through to the analysis and presentation of the data.

2.1. Template Generation

It is essential to generate pure full-length DNA template for the DNA damage experiment because after UV irradiation greater than 80% of the DNA fragments remain undamaged (Goodisman and Dabrowiak, 1992). This ensures that, on average, a "single hit" or 1 adduct per DNA molecule occurs. In a 300-bp fragment with approximately 100 possible sites of damage, each peak represents about 0.05 to 3.0% of the total quantity of DNA. Any nonspecific DNA fragments from the amplification severely impact the accuracy of the experiment if they result in a peak.

2.1.1. Amplification

In vitro studies use template DNA generated by PCR which facilitates the incorporation of a fluorescent label since the primers are labeled on the 5′ end with fluorescein phosphoamidite. The first decision is the position and length of the primer sequence. Computer programs such as OLIGO® (National Biosciences, Hamel, MN) greatly facilitate primer design. The PCR conditions can be optimized by adjusting the magnesium concentration, annealing temperatures, and invoking a "touchdown" protocol to minimize the formation of nontarget fragments.

The specific target sequence used in this example was the first 230 bp of the human *hprt* cDNA. Flanking regions of 30–50 bp are added to each end of the target sequence for several reasons. Interference of any undesired ssDNA ends is reduced. Since the primer peak from a sequencing reaction is ~40 bp wide and sequencing reactions are used to determine peak positions, no peak positions can be determined until after the primer peak is past the detector. The length of the fragment is 297 bp and is amplified using the following PCR conditions. In a 100-μl reaction volume the following components are added: 10–100 pg of plasmid DNA (containing the human *hprt* cDNA), 2 μl of 25 pmole/μl of each primer, 1 μl of 5 U/μl of AmpliTaq (Perkin–Elmer, Norwalk, CT), 10 μl of 10× buffer (600 mM KCl, 150 mM Tris·HCl pH 8.9, 27.5 mM $MgCl_2$); 2 μl of 25 mM of each dNTP (Pharmacia Biotech, Piscataway, NJ); and double-distilled H_2O to volume. Depending on the DNA strand being studied, the appropriate primer is 5′-labeled fluorescein phosphoamidite (Pharmacia Biotech). The 297-bp fragment is amplified with a Perkin–Elmer 9600 GeneAmp PCR system using 30 cycles of: (1) 94°C for 1 min, (2) 57.5°C for 30 sec, and (3) 72°C for 1 min. The primer sequences are:

HPRT-21F: 5′-CCTGAGCAGTCAGCCCG-3′
HPRT-255R: 5′-ATCACTATTTCTATTCAGTGC-3′

2.1.2. Template Purification

Our method of choice to reduce the nonspecific fragments generated by PCR is to use a low-melting-point (LMP) agarose gel followed by phenol/chloroform extraction of the excised band. This produces a product of sufficient yield and purity to perform DNA damage experiments. Specifically, the PCR fragment is loaded on a 4.0% LMP agarose gel (FMC BioProducts, Rockland, ME) *without* ethidium bromide (EtBr). Although in the absence of EtBr, some DNA may be lost, this protocol *avoids* the exposure of the DNA to EtBr and UV light which may introduce nicks. All lanes of a 70-ml gel (10 × 15 cm) are loaded with the PCR mixture and electrophoresed at 50 V for 16–18 hr. To localize the desired band a longitudinal slice is cut the from the edge of the gel, soaked in a 0.1 μg/ml EtBr solution for 10 min, and visualized with a UV transilluminator. The gel slice is cut such that its length reveals the position of the predominant band. Using the truncated gel slice repositioned in the gel tray, the remainder of the appropriate DNA band is excised. The excised gel band is melted at 60°C in 2 vol of TE buffer (50 mM Tris·HCl pH 7.6, 1 mM EDTA) and the DNA is extracted using phenol, phenol/chloroform (50/50), and chloroform/isoamyl alcohol (24:1). The product is ethanol precipitated. The PCR product is resuspended in H_2O to yield a final DNA concentration of ~30 ng/μl and stored at −20°C until required. DNA concentrations are determined using Hoechst 33258 intercalation and measuring the fluorescence on a Hoefer Fluorimeter TK100 (Hoefer Scientific Instruments, San Francisco, CA). The fluorescein label of the PCR fragment contributes ~3% of the signal and this minor systematic error is ignored.

To avoid exposing the DNA to phenol, the DNA in the gel band could have been recovered by dialysis, but usually the yields of DNA are much lower compared to the phenol extraction protocol. Control experiments for damage inducing agents such as UVC light, benzo[a]pyrene diol epoxide, and coriandrin reveal no significant background damage (unpublished results).

2.2. Internal Standards and Sequencing Reactions

Internal standards are required to compare the results from separate runs. The design of the automated DNA sequencer determines the number and type of controls required. Our laboratory's earlier efforts relied on the Perkin/Elmer Applied Biosystems Division (PE/ABD) 370/373A system (Foster City, CA) (Sage *et al.*, 1992; Koehler *et al.*, 1991; Shoukry *et al.*, 1991). More recently we acquired Pharmacia Biotech A.L.F.™ DNA sequencers (Piscataway, NJ) and applied them to the study of sequence-specific DNA damage.*

In the case of the Pharmacia A.L.F.™, three fluorescein-labeled internal DNA standards are employed. The first internal standard (IS-1) monitors the degradation of the fluorescein from chemical reactions and the recovery of DNA after DNA precipitations. The IS-1 standard is added to the DNA sample following UVC radiation. The length of IS-1 is 317 bp. Its length is 20 bp longer than the full-length fragment because the width of the undamaged peak is equivalent to about 20 bp of sequence. The second and third standards (IS-2 and IS-3), 322 and 347 bp, respectively, are required specifically for the Pharmacia A.L.F.™ because this instrument has 40 fixed photodiodes and the DNA may not migrate directly in front of the photodiode. The IS-2 and IS-3 standards are required in order to ascertain that each photodiode is responding in a like manner. All three internal standards are prepared and purified as described above.

*Kotturi, G., Erfle, H., Koop, B. F., and Glickman, B. W. Correlation of UV-induced mutational spectra and the *in vitro* damage distribution at the human *hprt* gene, unpublished.

Sequencing reactions are added as external controls which are used to align the damage-related peaks with the corresponding sequence position. Cycle sequencing is usually employed because of modest template requirements compared to T_7 polymerase-based sequencing. For the PE/ABD 373A, cycle sequencing using dye-labeled dideoxyterminator or dye-labeled primer protocols can be used effectively. One of the four nucleotide reactions with the dye most distinct from the labeled template can be added in the same lane to accurately determine the peak position. For the experiments with the Pharmacia A.L.F.™, the following cycle sequencing protocol is used: A master mixture of 2.6 μl of 10× sequencing buffer (300 mM Tris·HCl pH 9.0, 300 mM KCl, 50 mM $MgCl_2$), 1.5 μl of the appropriate labeled primer at 2.5 pmole/μl (Dalton Chemicals, North York, Ontario, Canada), 125 ng of *hprt* cDNA, and 1.2 μl of 5 U/μl AmpliTaq (Perkin–Elmer) is combined with double-distilled H_2O to a volume of 26 μl. The labeled primer is identical to the labeled primer used to amplify the PCR fragment. Six microliters of the master mixture is aliquoted into one of four PCR tubes. Four microliters of one of the termination mixtures (A, C, G, or T) is added into one of 200 μl thin-walled PCR tubes (Perkin–Elmer). The final concentration of the dNTP and ddNTP in each of the termination mixture is: (1) 60 μM for each dNTP and (2) "A" and "T" RXNs: 0.8 mM ddATP or ddTTP, respectively, "C" RXN: 0.4 mM ddCTP, "G" RXN: 0.08 mM. The sequencing reactions are carried out in a Perkin–Elmer Thermocycler 9600 programmed for the following cycles: (1) 94°C–2 min and then (2) 25 cycles of the following: 94°C–10 sec, 50°C–20 sec, 72°C–30 sec. After the reaction is completed, 6 μl of deionized formamide (USB, Cincinnati, OH) and 5 mg/ml Dextran Blue 2000 dye (Pharmacia Biotech) are added. The dye facilitates loading and is excluded from the gel. At this point the samples either are stored at −20°C until required or are denatured for 2 min at 95°C and 8–10 μl is loaded onto the gel. Another practical advantage of utilizing a fluorescein-based assay is that the sequencing reactions can be made up in bulk and stored at −20°C for several months without degradation.

2.3. DNA Damage Induction and Fragment Cleavage

Aliquots of 300 ng of the purified PCR fragment in ~ 10-μl drops are placed in a sterile petri dish. The petri dish is placed in wet ice and the distance from the 8-watt Sylvania germicidal light bulb (UVC, 254 nm) is adjusted such that the dose rate is 8 J/m^2 per sec as calibrated by using chemical actinometry (Murov, 1973). To detect 6–4 Py<>Py a dose of 1.2 kJ/m2 is used. Prior to the addition of piperidine, 1 μl of the first internal standard (IS-1) is added to estimate the recovery of the fluorescently tagged DNA after the cleavage treatments. The DNA is treated with 1 M piperidine at 90°C for 30 min in a total volume of 100 μl, followed by an ethanol precipitation to recover the DNA. The DNA is dried overnight under vacuum without the application of any further heat to remove traces of piperidine. Typically there is a loss of 40% of the fluorescent signal attributed to the degradation of the fluorescein during the piperidine treatment.

2.4. Electrophoresis of DNA Samples

Following the cleavage procedure, the samples consist of fragmented DNA of varying length. The full-length fragment corresponds to the undamaged template while the smaller fragments reveal sites of cleavage and, hence, DNA lesions.

2.4.1. Sample Dilution

Several dilutions of the sample are usually required because the photodiode signal is attenuated. Peaks with attenuated tops cannot be accurately quantitated. Based on an initial starting quantity of 300 ng, four dilutions (1.0, 0.5, 0.1, and 0.05) consistently allowed both the damage peaks to be accurately quantitated as well as bringing the concentration of the undamaged fragment below the level of attenuation. The DNA is lyophilized until it is ready to be examined at which time the sample is resuspended into 16.5 μl of H_2O and aliquoted to make the above dilutions. A final volume of 10 μl is maintained for each dilution. The IS-2/IS-3 mixture is added (1.0 μl), as well as a further 5.0 μl of deionized formamide containing Dextran Blue 2000. The final volume to be loaded is 16 μl. Samples can be stored at −20°C or are heated to 94°C for 2 min and immediately placed onto wet ice to cool. Liquid loss due to condensation is reversed as the water droplets are spun down in a microfuge and the microfuge tubes are replaced into the ice until loaded.

2.4.2. Lane Assignments

Sequence lane assigments of the samples require some preplanning. There are 40 lanes available on the Pharmacia A.L.F.™ In comparison, the PE/ABD 373A may allow either 24 or 36 lanes and may have two samples per lane. Usually there is enough space on the gel to have several different "exposed" samples. In the case of the Pharmacia A.L.F.™, the 40 lanes are arranged into ten clones. Sequencing reactions used to determine the sequence position of the peaks associated with the DNA damage must be loaded in the correct 4 lanes to automatically process the data into sequencing information. Three sequencing reactions (4 lanes each) are loaded and are usually located in clones 3, 5, and 8 or lanes 9–12, 17–20, and 29–32, respectively. Up to 3 lanes are used for internal standards to determine the quantity of IS-1 without loss of yield (lanes 38–40). This leaves 25 lanes or sufficient space for three to five exposed samples plus the control sample based on four dilutions per sample. Some partial sequencing reactions are loaded into lanes to concentrated samples where the best results are expected. This facilitates peak alignment between damage-induced peaks and DNA sequence reactions and minimizes problems due to bleeding by spacing concentrated samples. The dilutions are grouped together and the most dilute samples are loaded into lanes 1–6.

2.4.3. Preparation of the Polyacrylamide Gel

The cleaning and assembly of the gel cassettes can be accomplished by following the detailed manufacturer's instructions. The gel quality is of prime importance when using an automated DNA sequencer. We presently use 6% polyacrylamide, 19:1 polyacrylamide/*bis*acrylamide (5% C) Ready Mix gels (Pharmacia Biotech) which provide good reproducibility. Other gels that provide good separation are the recently developed Hydrolink Mutation Detection Enhancement™ or Longranger™ gels (AT Biochem, Malvern, PA).

2.4.4. Electrophoresis Conditions

Once the gel is warmed, wells are prewashed with electrophoresis buffer from the top reservoir to ensure that any salts such as urea do not cause a stacking effect which shifts the DNA bands. The electrophoresis conditions which can be adjusted are temperature, and electrical settings. In addition, the polyacrylamide concentration can also be altered. Reducing the laser power from the default 3 W to 2 W decreases the background fluorescence which increases the effective dynamic range of the instrument. Figure 14.1 shows the resulting resolution under a variety of gel conditions. Results from the standard PE/ABD 373A configuration are shown in Fig. 14.1a. This configuration has a "well-to-detector" distance of 24 cm. Plates with a "well-to-detector" distance of 34 and 48 cm are also available. As seen in Fig. 14.1c, the resolution under standard conditions used for DNA sequencing is not optimal. The high temperature and short

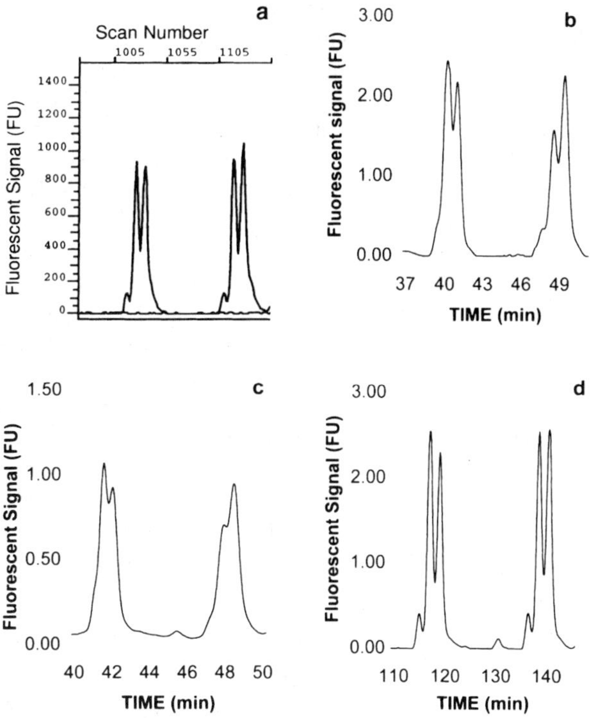

Figure 14.1. Illustration of the resolution from various automated DNA sequencers and electrophoretic conditions. All frames show the same piperidine cleavage products induced by 1.2 kJ/m^2 of UV (254 nm) light in the transcribed strand of the human *hprt* cDNA. The electropherograms are the raw data from automated DNA sequencers. (a) Electropherogram of a sample run on a PE/ABD 373A with 24-cm "well- to-detector" plates and a 6% polyacrylamide gel and other electrophoretic conditions recommended by PE/ABD. (b–d) Data from a Pharmacia A.L.F.™ with the common conditions of: power = 21 W, current = 34 mA, voltage = 1500 V. The electropherograms are runs at (b) 6% polyacrylamide and 25°C, (c) 6% polyacrylamide and 40°C, and (d) 12% polyacrylamide and 40°C.

"well-to-detector" distance of 19 cm are the major reasons. Typical run times are 5–6 hr; in fact, diagnostic information is possible after 2–3 hr. Herein lies another significant advantage of using the automated DNA sequencer. Optimal resolution is obtained with a 12% gel (see Fig. 14.1d) but the electrophoresis time is increased to 13 hr.

2.5. Data Analysis

After the end of the electrophoresis run, the data are stored in a digital format on a hard drive. The analysis of data from the PE/ABD 373A is performed with the 672 Genescan™ software which is different from the DNA sequencing software. Because the DNA sequence of the DNA sequencing reactions cannot be calculated, it is more time-consuming to determine the DNA bases which are damaged.

Before the electrophoresis run using a PE/ABD 373A instrument is started, a sample sheet is completed and this information is used to automatically detect the samples across the width of the gel. Also in the setup, the user can define which of the different dyes will be present. After the run is complete, the software automatically scans for the maximum intensity of the peaks in the width dimension and tracks any "bent" lanes. This is a valuable procedure which can compensate for any defects in the gel. If the lane autodetection misses a lane, or if the lane "bending" algorithm does not accurately track the lane, it is possible to make adjustments. The user can view the electropherogram which is in two dimensions, time and fluorescence intensity. The time units are only displayed in "scans" which represent the number of times the laser has scanned across the gel. Alternatively, if a PE/ABD size standard such as GENESCAN-500™ was run on the gel, the scans can be converted to "base pairs." The assignment of size is only a rough estimate because the DNA fragment may migrate at a slightly different rate. The best alignment is achieved by running partial sequencing reactions in the same lane.

The parameters which define how the Genescan™ program will integrate the data are chosen. The baseline is automatically generated and the area under the peaks is integrated. Unfortunately, the baseline cannot be edited to account for baseline drifts near the primer peak. Raw data from the electropherogram cannot be formatted and readily imported into a spreadsheet program to generate a custom plot. Electropherograms can be saved as graphic files (e.g., PICT files) and imported into a graphics program but it is difficult and time-consuming to properly format and annotate the plot in a manner which displays the data efficiently. Peak tables can be exported directly to a spreadsheet program. The Genotyper™ software has the same limitations for peak quantitation but does allow the user to analyze data from various experimental runs.

In the case of the Pharmacia A.L.F.™, before proceeding with peak quantitation software, the data from sequencing reactions are processed and the DNA sequence is calculated. Any migration anomalies related to the particular DNA sequences such as GC compressions are determined and facilitate peak alignment. For the Pharmacia A.L.F.™ there are two data analysis programs. The first program which is similar to the PE/ABD 672 Genescan™ software is called Fragment Manager™. This Windows™-based program displays electropherograms quickly and is very good for quantitating results from experiments that have less than five peaks and have a narrow size distribution ($<$ 30 bp). The user is somewhat limited by the number of parameters that can be adjusted to calculate the area under the peaks. The most difficult noise to accommodate occurs when the salt front passes the detector because there is a large shift in the baseline. This is significant because at the trailing edge of the salt front, the DNA fragments start to pass the detector. As shown in Fig. 14.2, the fluorescent signal of the salt front ($<$ 50 min) is mostly, but not entirely, compensated to subtract the background fluorescence. If integration is performed using this electropherogram, the area of the peak labeled "TC23" would be too large. Unless these effects on the baseline can be appropriately compensated, errors in integration occur.

Our experience indicates that the most effective software for peak quantitation is a program developed to evaluate data from a Pharmacia SMART™ system. To utilize this software, a Pharmacia A.L.F.™ file is converted to a format which can be imported into the Pharmacia SMART™ Manager software. The conversion software is called ALF2SMA™. The software translates raw data generated by Pharmacia A.L.F.™ Manager up to and including version 2.5. The ALF2SMA™ program is a simple one line command executed from an OS/2 window that selects the lanes and time interval that will be converted. Once the file is converted, the relevant lanes or "curves" are opened in the SMART™ Manager. Of primary interest are the baseline and peak integration calculations. Both have several parameters which allow the user to change the baseline to overcome problems. The parameters which can be chosen are: shortest baseline segment; noise window; maximum baseline level; and highest acceptable slope. While the units of the abscissa remain in minutes, the ordinate units are absorbance (AU) because the SMART™ Manager is primarily designed to run with a UV spectrophotometer. The ordinate is scaled from the ALF™ Manager software from 0 to 100% of full scale to 0 to 4.095 AU in the SMART™ Manager program. Usually to produce the correct baseline for the region of the electropherogram with a rapidly decreasing baseline, the two parameters that are changed are the highest acceptable slope and shortest baseline segment. Values suggested by the program are 0.006–0.008 AU/min and 8–10 min, respectively, while values that are more appropriate are 0.015–0.025 AU/min for the highest slope and 2–5 min for the shortest baseline segment. By changing the parameters too drastically, smaller peaks, especially later in the electrophoretic run when the peaks broaden, may be confused with noise. If this occurs, the baseline can be manually edited by adding or deleting "baseline points" to problem segments of the curve and recalculated. The recalculation

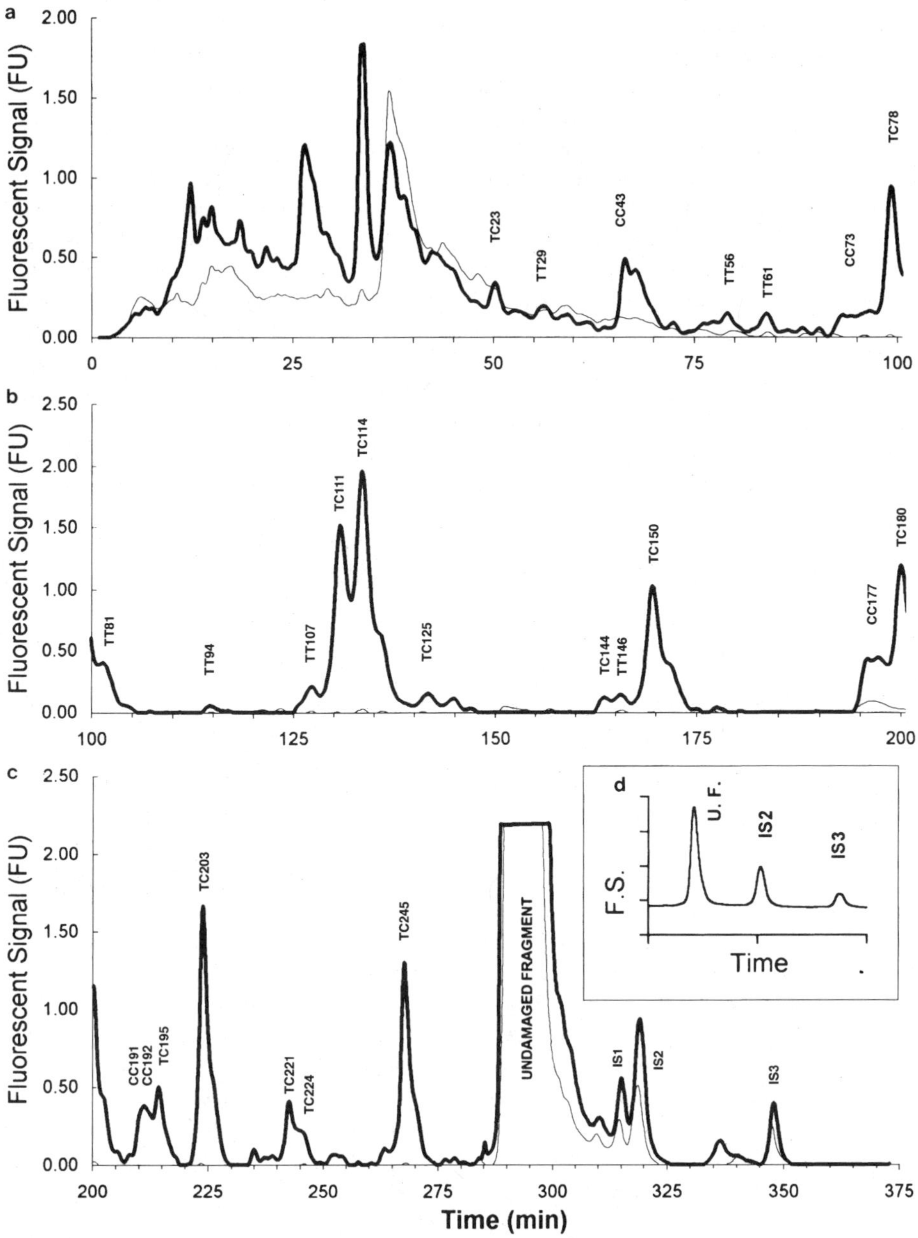

Figure 14.2. Distribution of piperidine cleavage products induced by 1.2 kJ/m^2 of UV (254 nm) light in the nontranscribed strand of the human *hprt* cDNA. The electropherograms are the raw data from a Pharmacia A.L.F.™ automated DNA sequencer. The DNA is resolved on a 6% polyacrylamide gel under the standard electrophoretic conditions. The fluorescent units (FU) are linear arbitrary units. Cleavage products were matched with DNA sequencing termination products in adjacent lanes to determine the base position. The control experiment is shown with the decreased line intensity. Panels a–c show different time ranges in the same experiment. Panel d, as an insert to panel c, shows the dilution series which brings the undamaged fragment (U.F.) into the range that does not have the signal attenuation.

is very "robust" and accurately redraws the baseline consistent with the new baseline points which even cause large discontinuities in the previous baseline. Infrequently, a systematic noise is present in both the control and exposed experimental results which must be taken into consideration before peaks are quantitated. This may be a result of inadequate purification of the full-length fragment. To compensate for this problem, curves can be subtracted. All anomalies are usually accounted for at this point and the area under the curve can be integrated and a peak table is generated. To reject peaks that represent noise, similar to the other software, it is possible to set minimum peak height, width, and area and the maximum number of largest peaks. The peak table from the SMART™ Manager contains useful calculations such as peak width at one-half height, which the user can use to judge the quality of the data. Peak area and retention time are used to correlate a peak to the DNA sequence. At this point it is usually most efficient to export the peak table into a spreadsheet program. In the example using UVC light, each peak is in turn analyzed with exposed DNA sample and the "T" and "C" reaction being displayed to assign sequence position to the DNA damage. If the DNA lesion remains on the 3′-end of the fluorescently labeled DNA fragment, there may be a retention time shift (Comess *et al.*, 1992). But with the aforementioned 30- to 50-bp flanking regions, this effect is minimal. Finally, the electropherogram can be exported into a spreadsheet format. The number of data points selected should be between 1300 and 1800 points. After the electropherogram curve is imported, the data can be formatted and annotated to a much greater degree than is possible in other peak quantitation programs. An example of typical experimental results is shown in Fig. 14.2.

2.6. Relative Mobility of DNA Fragments

One complication using automated DNA sequencers is that DNA molecules are moving past the detector during the real-time collection of data. This manifests itself as shorter DNA fragments elute with sharper peaks and, as the size of the fragments increase, peaks broaden which results in a lower peak height-to-area ratio. Peak spreading is due to the effects of diffusion and the difference in relative velocities. Diffusion does not affect the accuracy of the integrated area but it does impact the peak resolution, or the ability to separate two peaks.

A more important problem is the ability to determine the relative mobilities or velocities of two DNA fragments of different size. Smaller DNA fragments pass the detectors at a faster rate or velocity than larger fragments. If a fluorescent molecule passes the detector slower than another molecule, it will generate a greater fluorescent signal. The average velocity from "well to detector" is not accurate because the velocity of the fragments (< 300 bp) changes during the course of the electrophoretic run. The important place to determine the relative velocity is as the molecules pass the detector. Commercial products such as size standards (e.g., SIZER 50-500™, Pharmacia; GENESCAN-500™, PE/ABD) with equimolar DNA fragments at about every 25–50 bp can be obtained. Integration of data electrophoresed from these products could quickly determine the function of relative mobilities because the area should increase with DNA fragment size. However, our experience is that the actual concentration of each of the components in the SIZER 50-500™ varies considerably. To date, we have not determined the accuracy of the GENESCAN™ products. While the quantity of each of the components in the SIZER 50-500™ is not precisely 5 fmole/μl, the size is accurate. From a plot of retention time (time to reach the detector) as a function of DNA fragment size, an exact linear relationship is evident ($R^2 = 0.9996$). This relationship is only valid for DNA fragments from 50 to 500 bp under the conditions used. Thus, the velocities of the fragments at the detector are a linear function of DNA size within this range. Smaller fragments do not travel linearly until they approach the detector. By using the retention times of larger fragments (e.g., 400, 450, and 500 bp), an accurate relationship for velocity of the DNA fragments at the detector can be determined. Since the large fragments do not travel significantly in the initial part of the electrophoresis, the average velocity equals the velocity at the detector. The velocity at the detector for the shorter fragments is extrapolated with the knowledge of the above linear relationship. The function can be easily applied to the table of integrated areas based on either retention time or DNA size to correct for the different mobilities. The function for the velocity at the detector is

$$\text{Velocity} = 9.75 \times 10^{-2}\ \text{cm/min} - 9.61 \times 10^{-5}\ \text{cm/(min bp)} * [\text{DNA size}]$$

where the DNA size is in base pairs. The velocity of a 50-bp fragment is 0.093 cm/min compared to a 300-bp fragment with a velocity of 0.69 cm/min. The ratio of the velocities is 0.75; therefore, if the area of a 50-bp fragment is used as a basis for calculation, the integrated area for the 300 bp must be reduced by 25% in order to compare the relative amounts of DNA damage. We want to emphasize that this calculation should be repeated periodically if retention times shift from experiment to experiment or if different electrophoretic conditions are used. Peaks integrated from Fig. 14.2 are corrected in a similar manner. The area for the amount of 6–4 Py<>Py at peak "TC78" is 500 FU*min and the area of peak "TC203" is 880 FU*min. The velocities of the fragments are 0.088 and 0.076 for the 100-bp (TC78) and 225-bp (TC203) fragments, respectively. The area of the "TC203" peak is reduced by 14% to 760 FU*min to be able to compare equivalent amounts of damage.

3. DISCUSSION

Associated with any technology there are trade-offs which have to be weighed. Automated DNA sequencers have a demonstrated advantage over radioactivity based apparatus. There are clear benefits: (1) elimination of radioisotopes, (2) data are obtained in 5–7 hr or by the following morning, (3) samples can be stored for months, (4) data are in a digital form, and (5) the possibility of automation for screening and risk analysis. In terms of throughput, the use of automated DNA sequencers only becomes an advantage when ancillary software is available. The main advantage of using the fluorescence based technology is that fluorescence of molecules such as fluorescein and rhodamine have extremely low detection limits and have a potentially wide dynamic range of response which can span several orders of magnitude. This is tempered with the realization that the technical design of the fluorescence detection system must be capable of producing a linear response over a large range.

3.1. Developing Technologies

Another technology that is being applied to DNA sequencing and which offers great potential is use of capillary electrophoresis (CE). This technology uses linear polyacrylamide and replaces urea with formamide to increase the stability of the gel. The linear acrylamide formation allows the replacement of the gel after each run. Higher applied voltages generate better separations and increased speeds (Khrapko *et al.*, 1994; Huang *et al.*, 1992; Chen *et al.*, 1991; Karger *et al.*, 1991). Typical conditions on the CE instrument generate 400 bp of sequence information in 0.5 hr. The main limitation appears to be the number of parallel capillaries of an instrument. For the CE instrument to be a practical improvement over the conventional automated DNA sequencers, a large number of capillaries must be run simultaneously. Naturally, much will depend on the lifetime of a capillary that is used repetitively. Problems such as lane bending will be eliminated since all of the sample passes in front of the detector. However, at present the injection of material into the capillary is done either by applying pressure or electrokinetically. The variability in the quantity of injected DNA is sufficiently small to generate

DNA sequence information. However, our preliminary experience suggests that this is not sufficiently precise to utilize in DNA damage experiments unless an internal standard is introduced.

4. CONCLUSIONS

The application of DNA sequencing technology to other endpoints is being facilitated by the development of ancillary products and appropriate software. Here we have reviewed the application of these technologies for assessing the distribution of DNA damage. This application can be expected to have two major uses. The first is the analysis of DNA damage caused by environmentally relevant agents. The distribution of DNA damage and the analysis of its repair can be expected to have significant impacts on our understanding of the origins of cancer in humans. A second likely application is in the design of drugs targeted to specific DNA sequences. Such drugs may be used to target virus-specific sequences, or human sequences amplified during the course of a disease. The automation of the analysis of DNA sequence specificity of drug binding will be of practical assistance to the development of these types of drugs.

ACKNOWLEDGMENTS. The authors express their gratitude to Christian Ooste for his support in the performance of the experiments on the Beckman P/ACE 5000 CE system, Michael Ashwood-Smith for his many helpful comments regarding the coriandrin experiments, and Paul Romaniuk and Steve Hendy for their cooperation and assistance using the PE/ABD 373A and Genescan™.

REFERENCES

Ansorge, W., Sproat, B., Stegemann, J., Schwager, C., and Zenke, M. (1987). Automated DNA sequencing: Ultrasensitive detection of fluorescent bands during electrophoresis. *Nucleic Acids Res.* **15**:4593–4602.

Brash, D., Seetharam, S., Kraemer, K. H., Seidman, M. M., and Bredberg, A. (1987). Photoproduct frequency is not the major determinant of UV base substitution hot spots or cold spots in human cells. *Proc. Natl. Acad. Sci. USA* **84**:3782–3786.

Chen, D. Y., Swerdlow, H. P., Harke, H. R., Zhang, J. Z., and Dovichi, N. J. (1991). Low-cost, high-sensitivity laser-induced fluorescence detection for DNA sequencing by capillary gel electrophoresis. *J. Chromatogr.* **559**:237–246.

Comess, K. M., Burstyn, J. N., Essigmann, J. M., and Lippard, S. J. (1992). Replication inhibition and translesion synthesis on templates containing site-specifically placed cis-diamminedichloroplatinum(II) DNA adducts. *Biochemistry* **31**:3975–3990.

Drobetsky, E. A., and Sage, E. (1993). UV-induced G:C to A:T transitions at the aprt locus of Chinese hamster ovary cells cluster at frequently damaged 5′-TCC-3′ sequences. *Mutat. Res.* **289**:131–136.

Gao, S., Drouin, R., and Holmquist, G. P. (1994). DNA repair rates mapped along the human PGK1 gene at nucleotide resolution. *Science* **263**:1438–1440.

Goodisman, J., and Dabrowiak, J. C. (1992). Quantitative aspects of DNase I footprinting, in: *Advances in DNA Sequence Specific Agents* (L.H. Hurley, ed.), JAI Press, Greenwich, CT, pp. 25–50.

Gordon, L. K., and Haseltine, W. A. (1982). Quantitation of cyclobutane dimer formation in double and single stranded DNA fragments of defined sequence. *Radiat. Res.* **89**:99–112.

Huang, X. C., Quesada, M. A., and Mathies, R. A. (1992). DNA sequencing using capillary array electrophoresis. *Anal. Chem.* **64**:2149–2154.

Iwahana, H., Yoshimoto, K., Mizusawa, N., Kudo, E., and Itakura, M. (1994). Multiple fluorescence-based PCR-SSCP analysis. *Biotechniques* **16**:296–305.

Karger, A. E., Harris, J. M., and Gesteland, R. F. (1991). Multiwavelength fluorescence detection for DNA sequencing using capillary electrophoresis *Nucleic Acids Res.* **19**:4955–4962.

Khrapko, K., Hanekamp, J. S., Thilly, W. G., Belenkii, A., Foret, F., and Karger, B. L. (1994). Constant denaturant capillary electrophoresis (CDCE): A high resolution approach to mutational analysis. *Nucleic Acids Res.* **22**:364–369.

Koehler, D. R., Awadallah, S. S., and Glickman, B. W. (1991). Sites of preferential induction of cyclobutane pyrimidine dimers in the nontranscribed strand of lacI correspond with sites of UV-induced mutation in Escherichia coli. *J. Biol. Chem.* **266**:11766–11773.

Kunala, S., and Brash, D. E. (1992). Excision repair at individual bases of the Escherichia coli lacI gene: Relation to mutation hot spots and transcription coupling activity. *Proc. Natl. Acad. Sci. USA* **89**:11031–11035.

Lippke, J. A., Gordon, L. K., Brash, D. E., and Haseltine, W. A. (1981). Distribution of UV light-induced damage in a defined sequence of human DNA: Detection of alkali-sensitive lesions at pyrimidine nucleoside-cytosine sequences. *Proc. Natl. Acad. Sci. USA* **78**:3388–3392.

Murov, S. L. (1973). *Handbook of Photochemistry*, Dekker, New York.

Pfeifer, G. P., Drouin, R., Riggs, A. D., and Holmquist, G. P. (1991). In vivo mapping of a DNA adduct at nucleotide resolution: Detection of pyrimidine (6–4) pyrimidine photoproducts by ligation-mediated polymerase chain reaction. *Proc. Natl. Acad. Sci. USA* **88**:1374–1378.

Porcher, C., Malinge, M. C., Picat, C., and Grandchamp, B. (1992). A simplified method for determination of specific DNA or RNA copy number using quantitative PCR and an automated DNA sequencer. *Biotechniques* **13**:106–113.

Sage, E., Cramb, E., and Glickman, B. W. (1992). The distribution of UV damage in the lacI gene of Escherichia coli: Correlation with mutation spectrum. *Mutat. Res.* **269**:285–299.

Sanger, F., Nicklen, S., and Coulson, A. R. (1977). DNA sequencing with chain-terminating inhibitors. *Proc. Natl. Acad. Sci. USA* **74**:5463–5467.

Segurado, O. G., and Schendel, D. J. (1993). Identification of predominant T-cell receptor rearrangements by temperature-gradient gel electrophoresis and automated DNA sequencing. *Electrophoresis* **14**:747–752.

Shoukry, S., Anderson, M. W., and Glickman, B. W. (1991). A new technique for determining the distribution of N7-methyl guanine using an automated DNA sequencer. *Carcinogenesis* **12**:2089–2092.

Shoukry, S., Anderson, M. W., and Glickman, B. W. (1993). Use of fluorescently tagged DNA and an automated DNA sequencer for the comparison of the sequence selectivity of SN1 and SN2 alkylating agents. *Carcinogenesis* **14**:155–157.

Smith, L. M., Fung, S., Hunkapiller, M. W., Hunkapiller, T. J., and Hood, L. E. (1985). The synthesis of oligonucleotides containing an aliphatic amino group at the 5′ terminus: Synthesis of fluorescent DNA primers for use in DNA sequence analysis. *Nucleic Acids Res.* **13**:2399–2412.

Smith, L. M., Sanders, J. Z., Kaiser, R. J., Hughes, P., Dodd, C., Connell, C. R., Heiner, C., Kent, S. B. H., and Hood, L.E. (1986). Fluorescence detection in automated DNA sequence analysis. *Nature* **321**:674–679.

Tornaletti, S., and Pfeifer, G. P., (1994). Slow repair of pyrimidine dimers at p53 mutation hotspots in skin cancer. *Science* **263**:1436–1438.

Verpy, E., Biasotto, M., Meo, T., and Tosi, M. (1994). Efficient detection of point mutations on color-coded strands of target DNA. *Proc. Natl. Acad. Sci. USA* **91**:1873–1877.

Chapter 15

Ligation-Mediated PCR for Analysis of UV Damage

Silvia Tornaletti and Gerd P. Pfeifer

1. INTRODUCTION

UV irradiation of DNA produces two major types of DNA lesions: the cyclobutane pyrimidine dimers (CPDs) and the pyrimidine (6–4) pyrimidone photoproducts [(6–4) photoproducts]. Cyclobutane dimers are formed between the 5,6 bonds of two adjacent pyrimidines, 5′-TpT, 5′-TpC, 5′-CpT, or 5′-CpC. The (6–4) photoproducts have a covalent bond between positions 6 and 4 of two adjacent pyrimidines, and are detected most frequently at 5′-TpC and 5′-CpC sequences. The (6–4) photoproducts are formed at a rate of approximately 30% of that of CPDs and this ratio appears to be DNA sequence-dependent (Mitchell and Nairn, 1989). Both photoproducts are mutagenic in *Escherichia coli* and in mammalian cells (Brash, 1988).

The distribution of UV-induced lesions appears to be influenced by the local chromatin environment. For example, CPDs are distributed with a 10-base periodicity when the DNA is wrapped around nucleosome core particles (Gale *et al.*, 1987). Binding of transcription factors can also modify the lesion spectrum after UV irradiation (Selleck and Majors, 1987; Pfeifer *et al.*, 1992; Tornaletti and Pfeifer, 1995a). Therefore, it is important to identify the type and frequency of UV-induced lesions formed *in vivo* at specific positions within a gene. In addition, repair of UV-induced CPDs is heterogeneous along the mammalian genome. Active genes are repaired faster than inactive loci, the transcribed strand of active genes is repaired preferentially due to transcription–repair coupling (Bohr *et al.*, 1985; Mellon *et al.*, 1987; Selby and Sancar, 1993), and repair of CPDs can vary even between neighboring base positions (Tornaletti and Pfeifer, 1994; Gao *et al.*, 1994).

To analyze the frequency of UV lesions along specific human gene sequences, we have developed techniques based on the ligation-mediated polymerase chain reaction (LMPCR). LMPCR is a single-sided PCR procedure that allows the amplification and detection of gene-specific DNA fragments which have variable ends on one side (defined by the position of the lesion along a sequence) and a fixed end on the other side (defined by the position of a gene-specific primer). This technique has been previously used for *in vivo* footprinting (Mueller and

Silvia Tornaletti and Gerd P. Pfeifer • Department of Biology, Beckman Research Institute of the City of Hope, Duarte, California 91010. *Present address of S.T.*: Department of Biological Sciences, Stanford University, Stanford, California 94305.

Technologies for Detection of DNA Damage and Mutations, edited by Gerd P. Pfeifer, Plenum Press, New York, 1996.

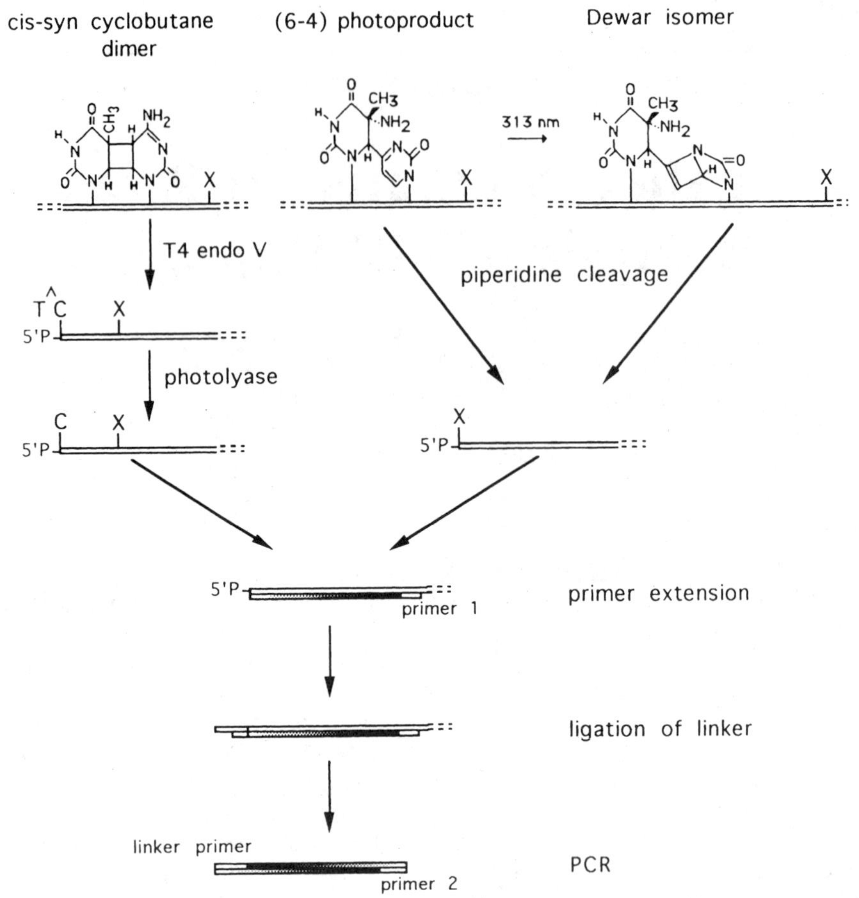

Figure 15.1. Outline of cyclobutane pyrimidine dimer and (6–4) photoproduct detection by ligation-mediated polymerase chain reaction. Photoproducts of the dinucleotide TpC are shown. X indicates an intact DNA base 3′ to the lesion. The open bar below the photoproducts represents the sugar phosphate backbone.

Wold, 1989; Pfeifer and Riggs, 1993; Rozek and Pfeifer, 1993), genomic sequencing (Pfeifer *et al.*, 1989), and determination of DNA methylation patterns (Pfeifer *et al.*, 1989; Tornaletti and Pfeifer, 1995b).

The LMPCR procedure, as applied to the detection of CPDs and (6–4) photoproducts, is outlined in Fig. 15.1. First, the adduct is converted into a DNA single-strand break with a 5′ phosphate group. In the case of cyclobutane dimers, this is achieved by enzymatic cleavage of UV-irradiated DNA with a cyclobutane dimer-specific endonuclease (Haseltine *et al.*, 1980; Gordon and Haseltine, 1980), followed by treatment with *E. coli* photolyase, to cleave the overhanging pyrimidine dimer at the 5′ end of the T4 endonuclease V cleavage products (Pfeifer *et al.*, 1992). The photolyase digestion appears to be an absolute requirement for LMPCR amplification, probably because the ligation requires an intact 5′ base with a 5′ phosphate group. The (6–4) photoproducts are sensitive to alkaline cleavage and can be assayed as piperidine-labile sites (Lippke *et al.*, 1981). Piperidine cleavage at (6–4) photoproducts results in ligatable breaks, allowing direct amplification by LMPCR without further modification. In conjunction with *E. coli* endonuclease III cleavage, LMPCR has been used for mapping of pyrimidine monoadducts which represent a minor class of UV photoproducts (Drouin and Holmquist, 1993). After specific cleavage and denaturation of DNA, primer extension of a gene-specific oligonucleotide (primer 1) produces DNA molecules that have a blunt end on one side. The ligation of an oligonucleotide linker onto the 5′ end of each DNA molecule provides a common sequence at all 5′ ends. The longer oligonucleotide of the linker (linker-primer) can then be used in conjunction with a second gene-specific primer (primer 2) in a PCR reaction for exponential amplification of the intervening sequence. (See Chapter 16 for a more detailed description of the individual steps of the method.) After 18–20 amplification cycles, the DNA fragments are separated on a sequencing gel, electroblotted onto a nylon membrane, and hybridized with a gene-specific probe to visualize the sequence ladder.

LMPCR has been applied to map the distribution and repair of UV photoproducts along single-copy genes at the DNA sequence level (Pfeifer *et al.*, 1991, 1992; Törmänen and Pfeifer, 1992; Tornaletti *et al.*, 1993; Tornaletti and Pfeifer, 1994; Gao *et al.*, 1994; Tornaletti and Pfeifer, 1995a). The methods for detection of UV photoproducts are described in detail below. Some of the steps of the LMPCR protocol are similar to those described in Chapter 16, but there are differences in several details, and for convenience, the complete protocol is given here.

2. MATERIALS AND METHODS

2.1. UV Irradiation

Cells are grown to confluent monolayers and irradiated from a germicidal lamp (254 nm) in petri dishes after removal of the medium and washing the cells with phosphate-buffered saline (PBS). UV doses are determined with a UVX radiometer (Ultraviolet Products, San Gabriel, CA). For *in vitro* treatment, purified DNA (1 μg/μl) is irradiated in water in 5-μl droplets on Parafilm.

2.2. DNA Purification

- Lyse cells (1×10^7 to 1×10^8) after UV irradiation by adding 10 ml of buffer A (0.3 M sucrose, 60 mM potassium chloride, 15 mM sodium chloride, 60 mM Tris-HCl, pH 8.0, 0.5 mM spermidine, 0.15 mM spermine, 2 mM EDTA) containing 0.5% Nonidet-P40.
- Incubate on ice for 5 min.
- Sediment nuclei by centrifugation for 5 min at 4°C.
- Wash the nuclear pellet with 15 ml of buffer A.
- Resuspend nuclei in 2 to 5 ml of buffer B (150 mM NaCl, 5 mM EDTA, pH 8.0), add an equal volume of buffer C (20 mM Tris-HCl, pH 8.0, 20 mM NaCl, 20 mM EDTA, 1% sodium dodecyl sulfate) containing 600 μg/ml of proteinase K.
- Incubate for 3 hr at 37°C.
- Add RNase A to a final concentration of 100 μg/ml.
- Incubate for 1 hr at 37°C.
- Extract DNA twice with phenol–chloroform and once with chloroform.
- Add 0.1 vol of 3 M sodium acetate, pH 5.2, and 2.5 vol of ethanol.
- Incubate 5 min at room temperature.
- Spin briefly to pellet the DNA. Remove supernatant. Wash the pellet with 75% ethanol and air-dry.

- Dissolve DNA in 10 mM Tris-HCl, pH 7.6, 1 mM EDTA (TE buffer) to a concentration of approximately 0.2 μg/μl.

2.3. Cleavage at Cyclobutane Pyrimidine Dimers

- Mix the following: 10 μg of UV-irradiated genomic DNA, 10 μl of 10 × T4 endonuclease V buffer (1 × buffer consists of: 50 mM Tris-HCl, pH 7.6, 1 mM EDTA, 50 mM NaCl, 1 mM dithiothreitol, 100 μg/ml bovine serum albumin), and T4 endonuclease V enzyme (kindly provided by S. Lloyd), in a final volume of 100 μl. The amount of enzyme needs to be determined empirically: Genomic DNA is irradiated with 500 J/m^2 of UV light and is incubated with different amounts of T4 endonuclease V. The completion of the cleavage reaction is monitored by running the DNA on alkaline agarose gels (see Chapter 3) and the optimum amount of enzyme can be determined from this assay. T4 endonuclease V produces very little nonspecific degradation of non-UV-irradiated DNA, and an excess of enzyme (up to 10-fold) can be used.
- Incubate the reaction mix at 37°C for 1 hr.
- Add dithiothreitol to a final concentration of 10 mM.
- Add 5 μg of *E. coli* photolyase (kindly provided by A. Sancar).
- Preincubate the mixture at room temperature under yellow light for 3 min.
- Irradiate the open tubes from two 360-nm black lights (Sylvania 15W F15T8) for 1 hr at room temperature at a distance of 5 cm.
- Extract once with phenol–chloroform and once with chloroform.
- Precipitate the DNA by adding 1/10 vol of 3 M sodium acetate (pH 5.2) and 2.5 vol of ethanol.
- Put samples on dry ice for 20 min.
- Centrifuge samples for 15 min at 14,000 rpm in an Eppendorf centrifuge at 4°C, and remove the supernatant.
- Wash pellets with 1 ml of 75% ethanol.
- Air-dry pellets.
- Dissolve DNA in TE buffer to a concentration of about 1 μg/μl.
- Determine the frequency of cyclobutane dimers by running 1 μg of the samples on a 1.5% alkaline agarose gel.

2.4. Cleavage at (6–4) Photoproducts

- Dissolve 50 μg of UV-irradiated DNA in 100 μl of 1 M piperidine (Fluka, Ronkonkoma, NY), freshly diluted.
- Heat at 90°C for 30 min in a heat block in tightly closed tubes (lead weight on top). During this treatment the Dewar isomers of (6–4) photoproducts are also cleaved (Fig. 15.1).
- Add 10 μl of 3 M sodium acetate (pH 5.2) and 2.5 vol of ethanol.
- Keep on dry ice for 20 min.
- Spin at 14,000 rpm, 15 min; remove the supernatant.
- Wash with 1 ml of 75% ethanol.
- Remove all traces of remaining piperidine by drying the sample overnight in a vacuum concentrator. Dissolve DNA in TE buffer to a concentration of about 1 μg/μl.
- Determine the frequency of (6–4) photoproducts by running 1 μg of the samples on a 1.5% alkaline agarose gel. Because the DNA is already efficiently denatured, (6–4) photoproduct cleavage can also be monitored on a neutral agarose gel.

2.5. Ligation-Mediated Polymerase Chain Reaction (LMPCR)

2.5.1. Primer Design

LMPCR uses three gene-specific primers: primer 1 (the Sequenase primer), primer 2 (the PCR primer), and primer 3 (the primer used to make a hybridization probe). Primer 3 (T_m 60 to 68°C) should hybridize approximately 70 to 80 nucleotides 3′ to a sequence position of interest to allow for good band separation and visibility of this region on the sequencing gel. We use a computer program (Oligo 4.0; National Biosciences Inc., Plymouth, MN; Rychlik and Rhoads, 1989) for design of primers and calculation of T_m. Sequences are extracted from GenBank and are pasted directly into this program. After the position of primer 3 has been fixed, primer 2 (T_m 60 to 68°C) is designed to be 5′ of primer 3 (farther away from the sequence of interest) and can overlap with primer 2 by up to 6–8 bases. Primer 1 (T_m 48 to 54°C) is placed farther 5′ of primer 2 and can overlap a few bases with primer 2. With this strategy, our success rate for designing new primer sets that work well in LMPCR is between 80 and 90%. It is preferable to have LMPCR primers of high purity, i.e., there should not be much *n*-1 material after oligonucleotide synthesis. If necessary, primers can be gel-purified (Sambrook *et al.*, 1989).

2.5.2. Primer Extension

A gene-specific primer (primer 1) is used on piperidine- or T4 endonuclease-cleaved fragments for primer extension with Sequenase.

- Mix the following in a siliconized tube: 1.5 μg of piperidine or T4 endo V/cleaved genomic DNA, 0.6 pmole of primer 1, and 3 μl of 5× Sequenase buffer [250 mM NaCl, 200 mM Tris-HCl (pH 7.7)] in a final volume of 15 μl.
- Denature at 95°C for 3 min, then incubate at 45°C for 30 min.
- Cool on ice, then spin for 5 sec.
- Add 7.5 μl cold, freshly prepared Mg–DTT–dNTP mix (20 mM $MgCl_2$, 20 mM dithiothreitol, 0.25 mM of each dNTP), and 5 units of Sequenase 2.0 (United States Biochemical, Cleveland, OH).
- Incubate at 48°C for 10 min, then cool on ice.
- Add 6 μl of 300 mM Tris-HCl (pH 7.7).
- Heat-inactivate the Sequenase at 67°C for 15 min.
- Cool on ice, then spin 5 sec.

2.5.3. Ligation

The primer-extended molecules that have a 5′-phosphate group after chemical or enzymatic cleavage are ligated to an unphosphorylated synthetic double-stranded linker. Linkers are prepared by annealing in 250 mM Tris-Cl (pH 7.7) a 25-mer (5′-GCGGTGACCCGGGAGATCTGAATTC, final concentration 20 pmole/μl) to an 11-mer (5′-GAATTCAGATC, final concentration 20 pmole/μl) (Mueller and Wold, 1989). This mixture is heated to 95°C for 3 min, then incubated at 70°C for 1 min, and gradually cooled to 4°C over a time period of 1–2 hr. Linkers are stored at −20°C. They are always thawed and kept on ice.

- Add 45 μl of freshly prepared ligation mix [13.33 mM $MgCl_2$, 30 mM DTT, 1.66 mM ATP, 83 μg/ml bovine serum albumin (BSA), 3 units per reaction T4 DNA ligase (Promega, Madison, WI), and 100 pmole linker per reaction].
- Incubate overnight at 18°C.
- Heat-inactivate at 70°C for 10 min.

- Precipitate the DNA by adding 8.4 μl of 3 M sodium acetate (pH 5.2), 10 μg *E. coli* tRNA, and 220 μl ethanol; keep on dry ice for 20 min.
- Centrifuge for 15 min at 4°C at 14,000 rpm.
- Wash pellets with 0.5 to 1 ml of 75% ethanol.
- Air-dry samples.
- Dissolve pellets in 50 μl water.

2.5.4. Polymerase Chain Reaction

Gene-specific fragments are amplified with a second, nested gene-specific primer (primer 2) and the common linker-primer, the longer oligonucleotide of the linker. The primers used in the amplification step are 21- to 28-mers. The annealing temperature in the PCR is chosen to be at the T_m of the gene-specific primer (calculated T_m between 60 and 68°C).

- Add 50 μl of freshly prepared 2× Taq polymerase mix [20 mM Tris-HCl (pH 8.9), 80 mM NaCl, 0.02% gelatin, 4 mM $MgCl_2$, 0.4 mM of each dNTP, containing 10 pmole of primer 2, 10 pmole of the linker primer, and 3 units of Taq polymerase (Boehringer-Mannheim, Indianapolis, IN].
- Cover the samples with 50 μl mineral oil and spin briefly.
- Cycle 20 times at 95°C for 1 min, 60 to 68°C for 2 min, and 74°C for 3 min.
- To extend completely all DNA fragments and uniformly add an extra nucleotide through the terminal transferase activity of Taq polymerase, an additional Taq polymerase extension step is performed. If this step is omitted, there can be double bands for each sequence position and assignment of photoproducts to specific sequences may become difficult. Add 1 unit of fresh Taq polymerase per sample together with 10 μl of 1× Taq polymerase mix. Incubate for 10 min at 74°C.
- Stop the reaction by adding sodium acetate (pH 5.2) to 300 mM, EDTA to 10 mM, and 10 μg *E. coli* tRNA.
- Extract with 70 μl phenol and 120 μl chloroform (premixed).
- Add 2.5 vol of ethanol and put on dry ice for 20 min.
- Centrifuge samples for 15 min at 14,000 rpm at 4°C.
- Wash pellets in 1 ml of 75% ethanol.
- Air-dry pellets.

2.5.5. Gel Electrophoresis

The PCR-amplified fragments are separated on 0.4-mm-thick, 60-cm-long sequencing gels consisting of 8% polyacrylamide and 7 M urea in 0.1 M Tris–borate–EDTA (TBE) buffer, pH 8.3 (Sambrook *et al.*, 1989).

- Dissolve pellets in 1.5 μl water and add 3 μl formamide loading dye [94% formamide, 2 mM EDTA (pH 8.0), 0.05% xylene cyanole, 0.05% bromophenol blue].
- Heat the samples to 95°C for 2 min prior to loading, and load one-half of the sample with an elongated, flat tip (National Scientific, San Rafael, CA).
- Run the samples until the xylene cyanole marker reaches the bottom of the gel.

2.5.6. Electroblotting

Electroblotters for transfer of sequencing gels are available from Hoefer Scientific (San Francisco, CA) and Owl Scientific (Cambridge, MA). Home-made transfer boxes can also be used (Pfeifer and Riggs, 1993).

- After the run, transfer the lower part of the gel (length 40 cm) to Whatman (Clifton, NJ) 3MM paper and cover it with Saran Wrap.
- Pile three layers of Whatman 17 paper, 43 × 19 cm, presoaked in 90 mM TBE, onto the lower electrode. Squeeze the paper with a rolling bottle to remove air bubbles between the paper layers.
- Place the gel piece covered with Saran Wrap onto the paper and remove the air bubbles between the gel and the paper by wiping the Saran Wrap with a soft tissue.
- When all air bubbles are squeezed out, remove the Saran Wrap and cover the gel with a GeneScreen (DuPont, Boston, MA) nylon membrane cut somewhat larger than the gel and presoaked in 90 mM TBE.
- Put three layers of Whatman 17 paper presoaked and cut as above onto the nylon membrane.
- Place the upper electrode onto the paper. Transfer at 1.6 A and 12 V.
- After 45 min, remove the nylon membrane and mark the DNA side.

2.5.7. Hybridization

- Dry the nylon membrane briefly at room temperature, then UV irradiate the membrane at a dose of 1000 J/m^2.
- The hybridization is performed in rotating 250-ml glass cylinders in a hybridization oven (Hoefer). Wet the nylon membrane in 90 mM TBE and roll it into the cylinders so that the membrane sticks completely to the walls of the cylinders. This can be done by rolling the membrane first onto a 25-ml pipette and then unspooling it into the cylinder.
- Prehybridize with 15 ml hybidization buffer [0.25 M sodium phosphate (pH 7.2), 1 mM EDTA, 7% SDS, 1% BSA] for 15 min at 60 to 68°C.
- Remove the prehybridization solution. Dilute the labeled probe into 7 ml hybridization buffer. Hybridize overnight at 60 to 68°C depending on the G+C content of the probe.
- After hybridization, wash the membrane 5 min with 300 ml washing buffer I [20 mM sodium phosphate (pH 7.2), 1 mM EDTA, 0.25% BSA, 2.5% SDS], prewarmed to 60–65°C, and afterwards approximately 30 min with 2 liters prewarmed buffer II [20 mM sodium phosphate (pH 7.2), 1 mM EDTA, 1% SDS] changing buffer II three times.
- Dry the membrane briefly at room temperature, wrap it in Saran Wrap, and expose to Kodak (Rochester, NY) XAR-5 films. Nylon membranes can be rehybridized if several sets of primers have been included in the primer extension and amplification reactions (Pfeifer *et al.*, 1989). Probes can be stripped by soaking the filters in 0.2 M NaOH for 30 min at 45°C.

Exposure times will vary between 1 and 16 hr. If longer exposure times are required, the experiment is usually a failure because there was insufficient amplification. This can have several causes: (1) there was an insufficient amount of DNA as starting material, (2) the lesion frequency was too low, (3) cleavage was inefficient, (4) the primers did not work properly, or (5) an error was made during the procedure. When exposure times need to be higher than 16 hr, there is often a lane-to-lane variability in band patterns for duplicate lanes also indicating that the procedure has not worked correctly.

2.5.8. Preparation of Hybridization Probes

We prepare labeled single-stranded probes by repeated primer extension by Taq polymerase from the third gene-specific primer (primer 3), which overlaps 3′ to the PCR primer, on a cloned, double-stranded template DNA, which can be either a plasmid or a PCR product. PCR products can be made by using genomic DNA as a template and primers 3 from two oppositely oriented

LMPCR primer sets (see Törmänen and Pfeifer, 1992, for an example). The PCR products are separated on low-melting agarose gels and isolated by standard procedures (Sambrook *et al.*, 1989). Either of the two primers used initially for making the PCR product can be used to make a single-stranded probe by linear PCR.

- Mix 5 ng of a PCR product (or 50 ng of the respective restriction-cut plasmid DNA) with 20 pmole of primer 3, 100 μCi of $\alpha[^{32}P]$dCTP, 5–10 μM of the other three dNTPs, 10 mM Tris-HCl (pH 8.9), 40 mM NaCl, 0.01% gelatin, 2 mM $MgCl_2$, and 3 units of Taq polymerase in a volume of 100 μl. Cover with 50 μl mineral oil.
- Cycle 30 cycles at 95°C for 1 min, 60–68°C for 1 min, and 74°C for 2 min.
- Extract once with phenol/chloroform.
- Add ammonium acetate to a concentration of 0.7 M, 10 μg *E. coli* tRNA, and 2.5 vol ethanol.
- Incubate 5 min at room temperature.
- Centrifuge for 5 min at 14,000 rpm at room temperature, remove supernatant.
- Resuspend pellet in 100 μl water and add to hybridization mix.

3. EXAMPLE

Figure 15.2 shows an example of the distribution of UV photoproducts mapped by LM-PCR. The sequences shown are from the upstream promoter region of the human single-copy gene *JUN*. The background of the procedure is visible in the nonirradiated samples (Fig. 15.2, lanes 5 and 10). The background is lower with the enzymatic treatment used to detect cyclobutane dimers (lane 5) when compared to the chemical method used to cleave at (6–4) photoproducts (lane 10). In UV-irradiated samples, UV-specific signals are seen at adjacent pyrimidines (lanes 6–9, 11–14). In irradiated purified DNA, the distribution of UV-photoproducts shows a pattern similar to that obtained when cloned, end-labeled DNA is used for UV irradiation (Gordon and Haseltine, 1980; Brash and Haseltine, 1982; Glickman *et al.*, 1986; Bourre *et al.*, 1987). The cyclobutane dimers are seen at all possible dipyrimidine sequences, 5′-TT, 5′-CT, 5′-TC, and 5′-CC (lanes 6 and 7). The (6–4) photoproducts are seen preferentially at 5′-TC and 5′-CC sequences, and they are almost undetectable at 5′-TT and 5′-CT positions (lanes 11 and 12). In irradiated cells, the distribution of UV photoproducts can show striking differences when compared to the photoproduct pattern obtained after irradiation of purified DNA. In particular, the photoproduct frequency within sequences bound by sequence-specific DNA binding proteins (transcription factors) can be supressed or enhanced relative to naked DNA. For example, a 5′-TC dinucleotide within the AP-1 transcription factor binding site produces a (6–4) photoproduct signal 4 times less in irradiated cells with respect to irradiated DNA (lanes 11–14). Changes in the photoproduct pattern were also detected along sequences that bind an unknown transcription factor (Fig. 15.2, sequences delimited by brackets with a question mark). The complete analysis of the *JUN* promoter showed that changes in the photoproduct pattern between the samples irradiated *in vivo* and the samples irradiated *in vitro* always corresponded to transcription factor binding sequences (Tornaletti and Pfeifer, 1995a).

4. DISCUSSION

We have described novel and sensitive PCR-based methods to map CPDs and (6–4) photoproducts at the sequence level in mammalian cells. The sensitivity of the technique has limitations related to the specificity of the method used to produce single-strand breaks at the damage site. The background of the procedure (no UV irradiation) is relatively low when T4

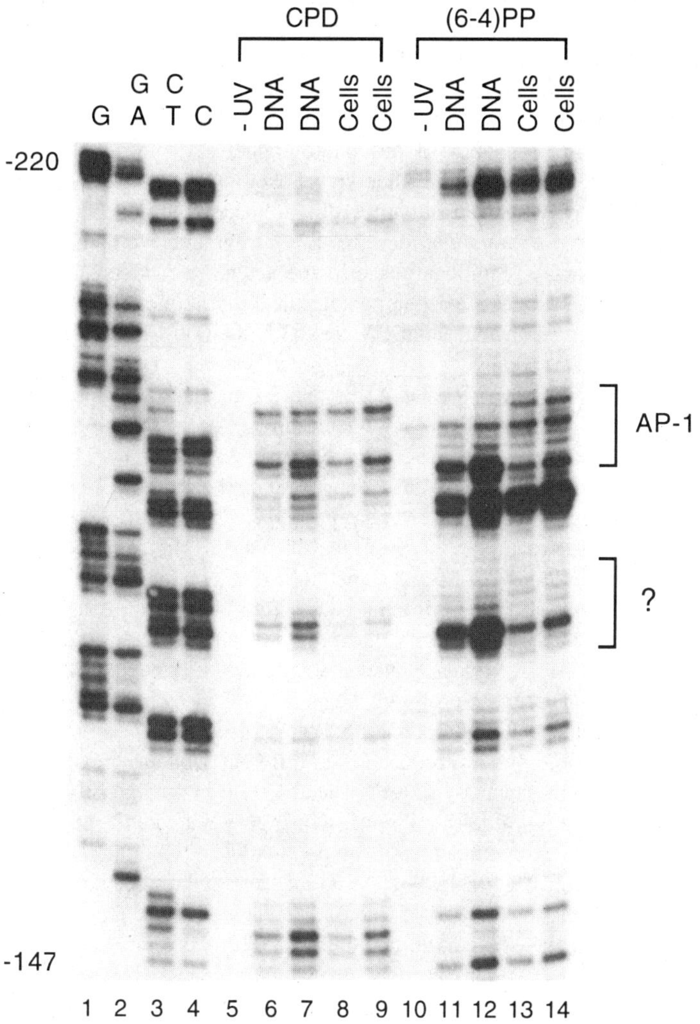

Figure 15.2. Distribution of UV photoproducts along the promoter of the human *JUN* gene in UV-irradiated HeLa cells. Sequences are from the upper strand between nucleotide −147 and −220 relative to the major transcription initiation site. Lanes 1–4 are Maxam–Gilbert sequencing reactions (Maxam and Gilbert, 1980). Lanes 5–14 show the photoproduct frequencies in UV-irradiated DNA and UV-irradiated HeLa cells. Lanes 5–9, cyclobutane pyrimidine dimers (CPDs); lanes 10–14, (6–4) photoproducts [(6–4)PPs]. Lanes 5 and 10, no UV irradiation; lanes 6 and 11, purified DNA irradiated at 1000 J/m^2; lanes 7 and 12, purified DNA irradiated at 2000 J/m^2; lanes 8 and 13, cells irradiated at 1000 J/m^2; lanes 9 and 14, cells irradiated at 2000 J/m^2. On the right, the consensus binding site for transcription factor AP-1 is delimited by brackets. Question mark indicates a "photo-footprint" caused by an unknown transcription factor.

endonuclease V is used to cleave DNA at cyclobutane dimers. This specificity allows mapping of cyclobutane dimers not only at high UV doses (1000 J/m^2) but also at much lower doses (12 J/m^2; Tornaletti and Pfeifer, 1994), making it possible to analyze repair of cyclobutane dimers in human cells at nucleotide resolution and at physiologically relevant levels of UV damage. The chemical treatment used to detect (6–4) photoproducts produces a higher background in the non-UV-treated samples, due to heating in alkaline piperidine. This increased background makes detection of (6–4) photoproducts by LMPCR difficult at UV doses below 80 J/m^2 and has so far prevented us from using this method to study repair of (6–4) photoproducts. Another limitation is related to the minimun amount of DNA molecules required to obtain a signal above LMPCR background. These limitations have been discussed earlier (Pfeifer *et al.*, 1993).

Theoretically, a high frequency of one photoproduct (e.g., cyclobutane dimers) could influence the photoproduct pattern of the other type of photoproduct [e.g., (6–4) photoproducts], by preventing the DNA polymerase from proceeding to a break introduced at sites of the other photoproduct farther downstream along the same DNA fragment. We found that this does not seem to influence the assay significantly, since the sequence ladders obtained are always continuous for both photoproducts and the photoproduct distribution did not show significant sequence-dependent changes even at high UV doses, up to 1000 J/m^2 (Tornaletti *et al.*, 1993; Tornaletti and Pfeifer, 1994). However, for fragments containing cyclobutane dimer hotspots (long runs of pyrimidines, A+T-rich DNA), it may be necessary to include an enzymatic photoreactivation step for removal of cyclobutane dimers before precise mapping of (6–4) photoproducts can be accomplished. Alternatively, lower UV doses may be used.

The LMPCR amplification reactions are dependent on the sequence of the fragments to be amplified. The sequence of a broken end that participates in primer extension and ligation reactions determines, at least to some degree, the intensity of the band seen on the LMPCR autoradiogram. For example, weak LMPCR signals are often seen for stretches of purines that are not interrupted by one or several pyrimidines (see Fig. 15.2 for an example). Fortunately, pyrimidine-containing sequence stretches, the preferred targets for UV damage formation, always give very good LMPCR signals. However, even here the signal intensity varies for bands produced in the Maxam–Gilbert control sequencing reactions which should have similar break frequencies for every base position before LMPCR amplification. In order to obtain a quantitative value for relative photoproduct frequencies at different sequence positions, one needs to determine the relation of the intensity of a UV-specific band to the intensity of the band at the same position in the Maxam–Gilbert pyrimidine sequencing lane (T+C), i.e., normalize for the different LMPCR amplification efficiencies for individual bands (Pfeifer *et al.*, 1991). Note that for a CPD at a 5′-TpC sequence, the LMPCR band migrates at the same position as the T in the Maxam–Gilbert T+C lane, and for a (6–4) photoproduct at the same sequence, the LMPCR band comigrates with the C of the Maxam–Gilbert T+C lane (this is related to the initial cleavage mechanisms; see Fig. 15.1). DNA repair experiments measure relative lesion frequencies over a time course at a defined sequence position. Therefore, LMPCR amplification bias does not play a role in measuring repair rates for individual positions.

ACKNOWLEDGMENTS. We thank A. Sancar for kindly providing *E. coli* photolyase and S. Lloyd for T4 endonuclease V. This work was supported by a grant from the National Institute of Environmental Health Sciences (ES06070 to G.P.P.).

REFERENCES

Bohr, V. A., Smith, C. A., Okumoto, D. S., and Hanawalt, P. C. (1985). DNA repair in an active gene: Removal of pyrimidine dimers from the DHFR gene of CHO cells is much more efficient than in the genome overall. *Cell* **40**:359–369.

Bourre, F., Renault, G., and Sarasin, A. (1987). Sequence effect on alkali-sensitive sites in UV-irradiated SV40 DNA. *Nucleic Acids Res.* **15**:8861–8875.

Brash, D. E. (1988). UV mutagenic photoproducts in Escherichia coli and human cells: A molecular genetics perspective on human skin cancer. *Photochem. Photobiol.* **48**:59–66.

Brash, D. E., and Haseltine, W. A. (1982). UV-induced mutation hotspots occur at DNA damage hotspots. *Nature* **298**:189–192.

Drouin, R., and Holmquist, G. P. (1993). UV-induced pyrimidine monoadducts and their in vivo photofootprints, in: *Chromosomal Aberrations: Origin and Significance* (G. Obe and A.T Natarajan, eds.), Springer Verlag, Berlin, pp. 10–20.

Gale, J. M., Nissen, K. A., and Smerdon, M. J. (1987). UV-induced formation of pyrimidine dimers in nucleosome core DNA is strongly modulated with a period of 10.3 bases. *Proc. Natl. Acad. Sci. USA* **84**:6644–6648.

Gao, S., Drouin, R., and Holmquist, G. P. (1994). DNA repair rates mapped along the human PGK-1 gene at nucleotide resolution. *Science* **263**:1438–1440.

Glickman, B. W., Schaaper, R. M., Haseltine, W. A., Dunn, R. L., and Brash, D. E. (1986). The C-C (6–4) UV photoproduct is mutagenic in *Escherichia coli. Proc. Natl. Acad. Sci. USA* **83**:6945–6949.

Gordon, L. K., and Haseltine, W. A. (1980). Comparison of the cleavage of pyrimidine dimers by the bacteriophage T4 and Micrococcus luteus UV-specific endonucleases. *J. Biol. Chem.* **255**:12047–12050.

Haseltine, W. A., Gordon, L. K., Lindan, C. P., Grafstrom, R. H., Shaper, L. N., and Grossman, L. (1980). Cleavage of pyrimidine dimers in specific DNA sequences by a pyrimidine dimer DNA-glycosylase of M. luteus. *Nature* **285**:634–641.

Lippke, J. A., Gordon, L. K., Brash, D. E., and Haseltine, W. A. (1981). Distribution of UV light-induced damage in a defined sequence of human DNA: Detection of alkaline-sensitive lesions at pyrimidine nucleoside-cytidine sequences. *Proc. Natl. Acad. Sci. USA* **78**:3388–3392.

Maxam, A. M., and Gilbert, W. (1980). Sequencing end-labeled DNA with base-specific chemical cleavages. *Methods Enzymol.* **65**:499–560.

Mellon, I., Spivak, G., and Hanawalt, P. C. (1987). Selective removal of transcription blocking DNA damage from the transcribed strand of the mammalian DHFR gene. *Cell* **51**:241–249.

Mitchell, D. L., and Nairn, R. S. (1989). The biology of the (6–4) photoproduct. *Photochem. Photobiol.* **49**: 805–819.

Mueller, P. R., and Wold, B. (1989). In vivo footprinting of a muscle specific enhancer by ligation mediated PCR. *Science* **246**:780–786.

Pfeifer, G. P., and Riggs, A. D. (1993). Genomic footprinting by ligation mediated polymerase chain reaction, in: *Methods in Molecular Biology*, Volume 15, (B. A. White, ed.), Humana Press, Totowa, NJ, pp. 153–168.

Pfeifer, G. P., Steigerwald, S. D., Mueller, P. R., Wold, B., and Riggs, A. D. (1989). Genomic sequencing and methylation analysis by ligation mediated PCR. *Science* **246**:810–813.

Pfeifer, G. P., Drouin, R., Riggs, A. D., and Holmquist, G. P. (1991). In vivo mapping of a DNA adduct at nucleotide resolution: Detection of pyrimidine (6–4) pyrimidone photoproducts by ligation-mediated polymerase chain reaction. *Proc. Natl. Acad. Sci. USA* **88**:1374–1378.

Pfeifer, G. P., Drouin, R., Riggs, A. D., and Holmquist, G. P. (1992). Binding of transcription factors creates hot spots for UV photoproducts in vivo. *Mol. Cell. Biol.* **12**:1798–1804.

Pfeifer, G. P., Drouin, R., and Holmquist, G. P. (1993). Detection of DNA adducts at the DNA sequence level by ligation-mediated PCR. *Mutat. Res.* **288**:39–46.

Rozek, D., and Pfeifer, G. P. (1993). In vivo protein–DNA interactions at the c-jun promoter: Preformed complexes mediate the UV response. *Mol. Cell. Biol.* **13**:5490–5499.

Rychlik, W., and Rhoads, R. E. (1989). A computer program for choosing optimal oligonucleotides for filter hybridization, sequencing and *in vitro* amplification of DNA. *Nucleic Acids Res.* **17**:8543–8551.

Sambrook, J., Fritsch, E. F., and Maniatis, T. (1989). *Molecular Cloning: A Laboratory Manual*, 2nd ed., Cold Spring Harbor Laboratory Press, Cold Spring Harbor, NY.

Selby, C. P., and Sancar, A. (1993). Molecular mechanism of transcription–repair coupling. *Science* **260**:53–58.

Selleck, S. B., and Majors, J. (1987). Photofootprinting in vivo detects transcription-dependent changes in yeast TATA boxes. *Nature* **325**:173–177.

Törmänen, V. T., and Pfeifer, G. P. (1992). Mapping of UV photoproducts within ras protooncogenes in UV-irradiated cells: Correlation with mutations in human skin cancer. *Oncogene* **7**:1729–1736.

Tornaletti, S., and Pfeifer, G. P. (1994). Slow repair of pyrimidine dimers at p53 mutation hot spots in skin cancer. *Science* **263**:1436–1438.

Tornaletti, S., and Pfeifer, G. P. (1995a). UV-light as a footprinting agent: Modulation of UV-induced DNA damage by transcription factors bound at the promoters of three human genes. *J. Mol. Biol.* **249**:714–728.

Tornaletti, S., and Pfeifer, G. P. (1995b). Complete and tissue-independent methylation of CpG sites in the p53 gene: Implication for mutations in human cancers. *Oncogene* **10**:1493–1499.

Tornaletti, S., Rozek, D., and Pfeifer, G. P. (1993). The distribution of UV photoproducts along the human p53 gene and its relation to mutations in skin cancer. *Oncogene* **8**:2051–2057.

Chapter 16

Ligation-Mediated PCR for Analysis of Oxidative DNA Damage

Régen Drouin, Henry Rodriguez, Gerald P. Holmquist, and Steven A. Akman

1. INTRODUCTION

Reactive oxygen species (ROS), including superoxide anion, hydrogen peroxide (H_2O_2), hydroxyl radical, and singlet oxygen, may play an important role in promoting aging and neoplastic transformation (reviewed in Breimer, 1990; Floyd, 1990; Halliwell and Gutteridge, 1990; Piette, 1991; Ames *et al.*, 1993; Guyton and Kensler, 1993; Nohl, 1993). Part of this role may be mediated by ROS-induced DNA mutations at critical sites. ROS, which are produced by any oxidative stress, are known to cause promutagenic damage due to a direct interaction of hydroxyl radicals and singlet oxygen with DNA (Breimer, 1990). ROS can be produced by a variety of exogenous and intracellular mechanisms, including ionizing radiation, cigarette smoke, air pollutants, toxins, UV light, inflammation, and intracellular metabolism (Guyton and Kensler, 1993). Ames (1987) has estimated that each human cell sustains an average of 10^3 "oxidative hits" each day.

The transition metal ion-catalyzed reduction of H_2O_2 has served as a useful model reaction for generating ROS. H_2O_2 is an oxidizing agent that does not react readily with DNA. However, in the presence of reactive transition metal ions [Fe(II) and Cu(I)], H_2O_2 generates extremely reactive oxidizing radical species (Cheeseman and Slater, 1993). H_2O_2 is generated by aerobic phosphorylation in mitochondria at a rate of approximately 1 mole of H_2O_2 produced per 50 moles of O_2 consumed (Boveris, 1977). Since the H_2O_2 generated in the mitochondria is sufficiently stable to diffuse to the nucleus, transition metal ion/H_2O_2-mediated DNA damage is a candidate for one source of oxidative DNA damage during normal aerobic metabolism.

Régen Drouin and Gerald P. Holmquist • Division of Biology, Beckman Research Institute of the City of Hope, Duarte, California 91010. **Henry Rodriguez and Steven A. Akman** • Department of Medical Oncology and Therapeutics Research, City of Hope National Medical Center, California 91010. *Present address of R.D.*: Research Unit in Human and Molecular Genetics, Department of Pathology, Hospital Saint-François d'Assise, Laval University, Quebec G1L 3L5, Canada. *Present address of S.A.A.*: Department of Cancer Biology, Wake Forest Cancer Center, Winston–Salem, North Carolina 27157.

Technologies for Detection of DNA Damage and Mutations, edited by Gerd P. Pfeifer, Plenum Press, New York, 1996.

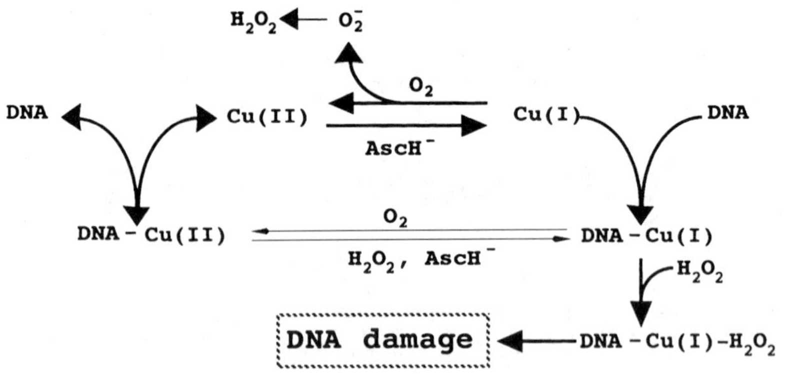

Figure 16.1. Reactions leading to DNA damage in aerobic solutions. The DNA association constant of Cu(I) is 10^9 (Prütz *et al.*, 1990) while that of Cu(II) is 10^4 (Bryan and Frieden, 1967). Consequently, 50 μM Cu(II) distributes almost equally between bound and unbound forms in the presence of 10 μg genomic DNA. In the presence of an excess of ascorbate, soluble Cu(II) becomes reduced and tightly bound to DNA. The DNA–Cu(I) then forms a complex with H_2O_2. During the resolution of this complex, DNA damage occurs by mechanisms still in dispute (Stoewe and Prütz, 1987; Masarwa *et al.*, 1988; Yamamoto and Kawanishi, 1989; Milne *et al.*, 1993), and DNA-bound Cu(I) is oxidized, completing one redox cycle.

Cu(II)/ascorbate/H_2O_2-mediated DNA damage in aerobic aqueous solutions may be induced *in vitro* and *in vivo* through formation of a DNA–Cu(I)–H_2O_2 complex (Fig. 16.1) (Goldstein and Czapski, 1986; Stoewe and Prütz 1987; Chevion, 1988). Ascorbate reduces Cu(II) to Cu(I) which binds tightly to DNA (Fig. 16.1) (Minchekova and Ivanov, 1967; Izatt *et al.*, 1971; Prütz *et al.*, 1990), preferentially at GC-rich DNA sequences (Minchekova and Ivanov, 1967; Izatt *et al.*, 1971; Pezzano and Podo, 1980; Prütz *et al.*, 1990; Geierstanger *et al.*, 1991; Kagawa *et al.*, 1991). Kinetic analysis (Goldstein and Czapski, 1986; Stoewe, and Puütz, 1987; Chevion, 1988; Masarwa *et al.*, 1988), inhibitor studies (Stoewe and Prütz, 1987; Aruoma *et al.*, 1991; Dizdaroglu *et al.*, 1991b), and studies of copper ion-mediated cleavage of small radiolabeled DNA molecules (Kazakov *et al.*, 1988; John and Douglas, 1989) suggest that the reaction of Cu(I)–DNA complexes with H_2O_2 results in the induction of site-specific oxidative DNA damage and oxidation of the Cu(I) by mechanisms still in dispute (Stoewe and Prütz, 1987; Masarwa *et al.*, 1988; Yamamoto and Kawanishi, 1989; Milne *et al.*, 1993).

ROS induce several classes of DNA damage, including single-strand breaks, double-strand breaks, modified bases, abasic sites, and DNA–protein cross-links (Dizdaroglu, 1991, 1992; Halliwell and Aruoma, 1991). DNA base damage caused by reduction of H_2O_2 by Fe(II) or Cu(I) has been identified and quantified *in vitro* (Aruoma et *al.* 1991; Dizdaroglu *et al.*, 1991b) and *in vivo* (Dizdaroglu *et al.*, 1991a) by analytic techniques, such as gas chromatography/mass spectrometry (GC-MS) (Aruoma *et al.*, 1991; Dizdaroglu *et al.*, 1991a,b), high-performance liquid chromatography with electrochemical detection (Floyd *et al.*, 1986, 1988; Kasai *et al.*, 1986), or ^{32}P-postlabeling. Such techniques have revealed the types (Table I) and amounts of modified bases induced by ROS (Aruoma *et al.*, 1991; Dizdaroglu *et al.*, 1991a,b), but not their distribution along DNA sequences. To date, mapping of copper ion/H_2O_2-induced DNA damage at the sequence level of resolution has been limited to piperidine-sensitive damage in small target genes subcloned into plasmids (Sagripanti and Kraemer, 1989; Yamamoto and Kawinishi, 1989, 1992) and the identity of the modified bases cleaved by hot piperdine has not been determined. Hot piperidine cleaves a wide variety of base modifications. Furthermore, all oxidized bases are not equally sensitive to it and the cleavage rate of undamaged bases is high. For these reasons,

Table I
Types of Modified Bases Induced by ROS[a]

Abasic sites[b,c]
Pyrimidine damage
5,6-Dihydroxy-5,6-dihydrocytosine[b] (cytosine glycol)
5-Hydroxyhydantoin[b]
5-Methyl-5-hydroxyhydantoin[b]
5-Hydroxycytosine[b] (5-OHdC)
5-Hydroxyuracil[b] (5-OHdU)
5,6-Dihydroxycytosine
5,6-Dihydroxy-5,6-dihydrothymine[b] (thymine glycol)
β-Ureidoisobutyric acid[b]
Urea[b]
5-Hydroxymethyluracil
Purine damage
7-Hydro-8-oxoguanine[c]
Imidazole ring-opened purines
2,6-Diamino-4-hydroxy-5-formamidopyrimidine[c] (FAPY guanine)
4,6-Diamino-5-formamidopyrimidine[c] (FAPY adenine)
7-Hydro-8-oxoadenine
2-Oxadenine (isoguanine)

[a]Adapted from Auroma *et al.* (1991), Dizdaroglu *et al.* (1991a,b), Wallace (1988), Doetsch and Cunningham (1990), Boiteux (1993), Hatahet *et al.* (1994).
[b]Modified bases recognized by Nth protein.
[c]Modified bases recognized by Fpg protein.

identification and cleavage of modified bases by specific enzymes has replaced chemical cleavage for many nucleotide resolution mapping techniques.

The Nth protein from *Escherichia coli* (also known as endonuclease III) and the Fpg protein from *E. coli* (also known as formamidopyrimidine DNA glycosylase) are capable of recognizing and cleaving most of the ROS-induced modified bases (Table I). Both enzymes combine an *N*-glycosylase activity with a class I abasic endonuclease activity and act more efficiently on duplex DNA. They specifically recognize oxidatively damaged bases, cleave the glycosidic bond to release the modified base, and nick the phosphodiester bond 3′ to the abasic site by a β-elimination (Nth protein) or β-∂-elimination reaction (Fpg protein), leaving ends with a 5′-phosphate suitable for ligation (Wallace, 1988; Doetsch and Cunningham, 1990; Boiteux, 1993). The ligation-mediated polymerase chain reaction (LMPCR), which maps at nucleotide resolution the frequency of breaks with 5′-P_i ends, can be used for mapping of ROS-induced, endonuclease-sensitive, DNA base modifications.

LMPCR has been applied to map the distribution of oxidized bases along single-copy genes at the DNA sequence level (Rodriguez *et al.*, 1995). By utilizing an exponential amplification step, LMPCR is a genomic sequencing method several thousand-fold more sensitive than the original genomic sequencing method of Church and Gilbert (1984). It uses 20-fold less DNA, and allows short autoradiographic exposure times to obtain high-resolution bands in the final autoradiogram. The unique aspect of LMPCR is the blunt-end ligation of an asymmetric (5′-overhang to avoid self-ligation or ligation in the wrong direction) double-stranded linker onto the 5′-end of each nicked DNA molecule (Fig. 16.2). The blunt end is created by the extension of primer 1 until a strand break is reached. The linker provides a common sequence at all 5′-ends. This allows exponential PCR amplification which requires a separate primer at each end of the region of interest, using the longer oligonucleotide of the linker (linker-primer) and a second

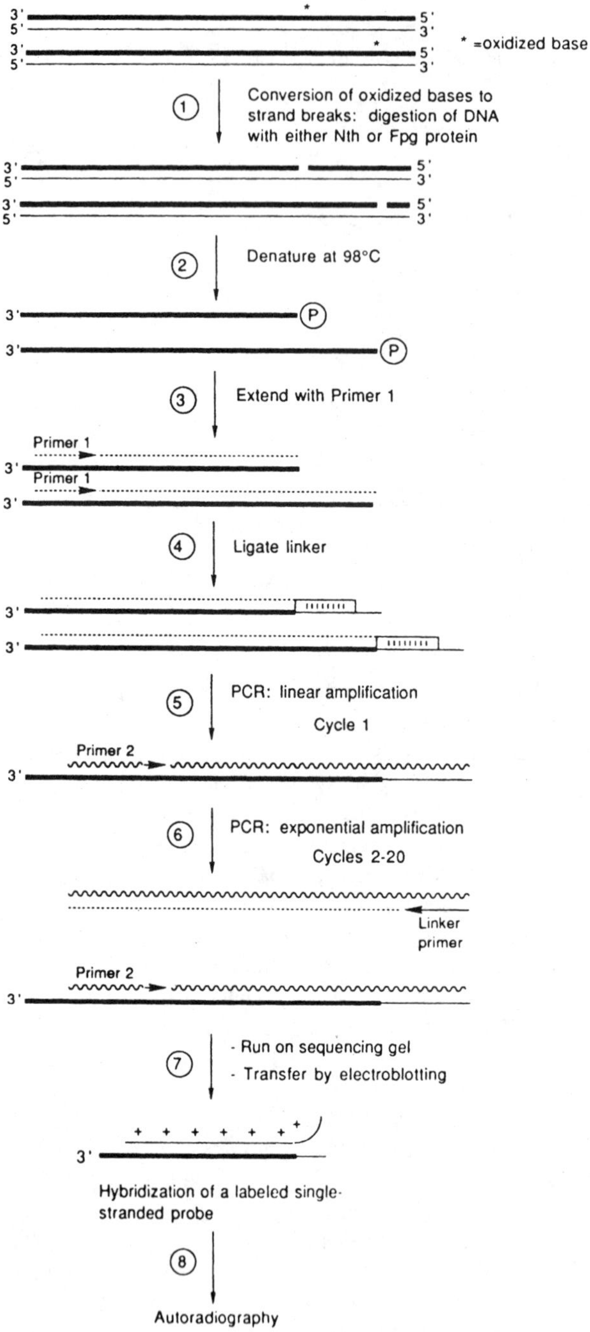

Figure 16.2. Outline of the LMPCR procedure. Step 1, specific conversion of oxidized bases to strand breaks; step 2, denaturation of genomic DNA; step 3, annealing and extension of primer 1 with Sequenase; step 4, ligation of the linker; step 5, first cycle of PCR amplification, this cycle is a linear amplification because only the gene-specific primer 2 can anneal; step 6, cycle 2 to 20 of exponential PCR amplification of gene-specific fragments with primer 2 and the linker primer (the longer oligonucleotide of the linker); step 7, separation of the DNA fragments on a sequencing gel and transfer of the sequence ladder to a nylon membrane by electroblotting; step 8, visualization of the sequence ladder by hybridization with a labeled single-stranded probe which abuts on primer 2.

nested gene-specific primer (primer 2). After 19–21 cycles of PCR, the DNA fragments are size-fractionated on a sequencing gel. LMPCR preserves the quantitative representation of each fragment of the sequence ladder in the original population of cleaved molecules and each sequence ladder can be amplified very reproducibly (Pfeifer *et al.*, 1993a,b).

2. MATERIALS AND METHODS

LMPCR can map H_2O_2-induced modified bases and single-strand breaks *in vitro* and *in vivo*. *In vivo* treatment means that cells are treated with H_2O_2, then the DNA is purified. *In vitro* treatment indicates that the DNA is extracted from cells, stripped of any proteins and dialyzed extensively, then exposed to the Fenton reaction (Cu–ascorbate–H_2O_2).

2.1. Hydrogen Peroxide Treatment of Human Skin Fibroblasts

Human skin fibroblasts are grown in 150 mm^2 dishes to confluent monolayers in Dulbecco's modified Eagle medium (DMEM) with 10% (v/v) fetal bovine serum. The medium is removed and the cells are washed with 0.9% NaCl. Twenty-five milliliters of serum-free medium MEM (minimum essential medium which contains only 1 mM sodium phosphate) containing 1 to 50 mM H_2O_2 (stock solution: 30% or 8.8 M) is then added.

- Incubate at 37°C for 30 min.
- Remove medium containing H_2O_2.
- Wash the cells once with 25 ml 0.9% NaCl.
- Add 7 ml buffer A (0.3 M sucrose, 60 mM KCl, 15 mM NaCl, 60 mM Tris-HCl pH 8, 0.5 mM spermidine, 0.15 mM spermine, 2 mM EDTA) containing 0.5% Nonidet-P40.
- Incubate on ice for 5 min.
- Scrape the cells.
- Wash the dish with 3 ml buffer A with 0.5% NP-40.
- Sediment nuclei by centrifugation at 4000 rpm for 10 min/4°C.
- Continue according to Section 2.2.

2.2. DNA Purification

- For *in vitro* treatment, cells are detached by exposure to trypsin or sedimented by centrifugation if they are grown in suspension.
- Resuspend the cells in 20–25 ml of buffer A.
- Add 20–25 ml of buffer A containing 1% Nonidet P40.
- Incubate on ice for 5 min.
- Sediment nuclei by centrifugation at 4000 rpm for 10 min.
- Wash once with 10 ml of buffer A.
- Resuspend nuclei in 100–500 μl of buffer A.
- Dilute the suspension with 1 ml of buffer B (150 mM NaCl, 5 mM EDTA pH 8) per 10^7 nuclei.
- Add an equivalent volume of buffer C [20 mM Tris-HCl pH 8, 20 mM NaCl, 20 mM EDTA, 1% sodium dodecyl sulfate (SDS)] containing 600 μg/ml of proteinase K.
- Incubate at 37°C for 3 hr.
- Add RNase A to a final concentration of 100 μg/ml.
- Incubate at 37°C for 1 hr.

- Purify DNA by extraction with phenol (1 or 2 times as needed), phenol–chloroform (1 or 2 times as needed), and chloroform.
- Precipitate DNA in 200 mM NaCl and 2 vol of ethanol.
- Recover DNA by spooling the floating DNA filament with a micropipette tip. If DNA is in small pieces or not clearly visible, recover DNA by centrifugation (4K/30 min/4°C), but expect RNA contamination. RNA contamination does not cause any problems for LMPCR.
- Wash DNA once with 70% ethanol.
- Centrifuge the DNA (4000/10 rpm min/4°C).
- Air-dry DNA pellet.
- Dissolve DNA in 10 mM HEPES pH 7.4 and 1 mM EDTA (HE buffer).
- Carefully measure DNA concentration by either spectrophotometry or fluorometry after staining with 4′,6-diamidino-2-phenylindole (DAPI). If there is RNA contamination, the DNA concentration has to be measured with a fluorometer after staining with DAPI.
- Adjust DNA concentration to 66 μ/ml with HE buffer (this last step concerns only the DNA which will be dialyzed and treated *in vitro*).

2.3. DNA Dialysis

- Fill dialysis bags (Spectra-Por, M_r cutoff 6000–8000, pretreated according to Maniatis *et al.*, 1982) with a maximum volume of 1250 μl of a DNA solution at a concentration of 66 μg/ml.
- Dialyze DNA against two exchanges of 800 ml of distilled water at 4°C for 12 to 18 hr. The concentration of DNA after dialysis is usually diluted to 58–64 μg/ml.

2.4. Cu/Ascorbate/H_2O_2 Treatment of Purified DNA

Each reaction contains 10 μg of target DNA in a final volume of 268 μl/ $CuCl_2$ and L-ascorbic acid solutions are freshly prepared, the latter is kept in the dark. Warm DNA solution to room temperature.

- Add 26.8 μl of a 500 μM concentrate of $CuCl_2$ to 10-μg aliquots of dialyzed target DNA (~161 μl). Mix with the pipet.
- Incubate at room temperature for 30 min.
- Add 26.8-μl aliquots of 10 mM Chelex® resin (Bio-Rad, Richmond, CA)-treated potassium phosphate pH 7.5, 1 mM ascorbate, and a 10× concentrate of H_2O_2. Mix gently with a pipet.
- Incubate at room temperature for 30 min with gentle rocking.
- Stop the reaction by adding 82.8 μl of distilled water, 9 μl of 50 mM EDTA pH 8, 40 μl of 3 M sodium acetate pH 7, and 1000 μl of 100% ethanol. Centrifuge.
- Wash DNA pellets with 80% ethanol.
- Air-dry DNA pellets.

2.5. Enzyme Digestion (Step 1, Fig. 16.2)

- Dilute 10-μg aliquots of DNA with 50 μl of water and 50 μl of 2× Nth protein buffer (100 mM Tris-HCl pH 7.6, 200 mM KCl, 2 mM EDTA, 0.2 mM dithiothreitol, 200 μg/ml bovine serum albumin).
- Add enzyme in 5 μl of dilution buffer (50 mM Tris-HCl pH 7.6, 100 mM KCl, 1 mM

EDTA, 0.1 mM dithiothreitol, 500 μg/ml bovine serum albumin, 10% (v/v) glycerol), the total digestion volume is 105 μl. In our conditions, the optimal amount of enzyme per 105 μl for digesting 10 μg of DNA is 400 ng for Fpg protein and 100 ng for Nth protein. For mapping oxidative treatment-induced strand breaks, no enzyme is added to the 5 μl of dilution buffer.

- Incubate at 37°C for 60 min.
- Stop the digestion by adding 250 μl of water, 50 μl of 0.8% SDS, and mix well.
- Extract with phenol, phenol–chloroform, and chloroform.
- Precipitate the DNA by adding 18 μl of 5 M NaCl and 1000 μl of 100% ethanol.
- Wash DNA pellets with 80% ethanol.
- Air-dry DNA pellets.
- Dissolve the pellets in water to a final concentration of 0.2 μg/μl.
- Determine the relative frequency of strand breaks and oxidative damage on a neutral agarose gel electrophoresis after dimethyl sulfoxide (DMSO) denaturation and glyoxal treatment (Fig. 16.3) (see Chapter 3).
- Add Sequenase buffer (final concentration 40 mM Tris-HCl pH 7.7, 50 mM NaCl) to the DNA to obtain a final DNA concentration of 0.16 μg/μl.

2.6. Ligation-Mediated Polymerase Chain Reaction: Primer Extension (Steps 2 and 3, Fig.16.2)

A gene-specific primer (primer 1) is used to initiate primer extension with Sequenase. The primer 1 used in the first-strand synthesis are 17- to 20-mer oligonucleotides and have a calculated melting temperature (T_m) of 50 to 56°C. Optimally, their T_m, as calculated by a computer program (Rychlik and Rhoads, 1989), should be about 10°C lower than that of subsequent primers (Rodriquez *et al.*, 1995).The first-strand synthesis reaction is designed to require very little primer with a lower T_m so that this primer does not interfere with subsequent steps (Mueller and Wold,

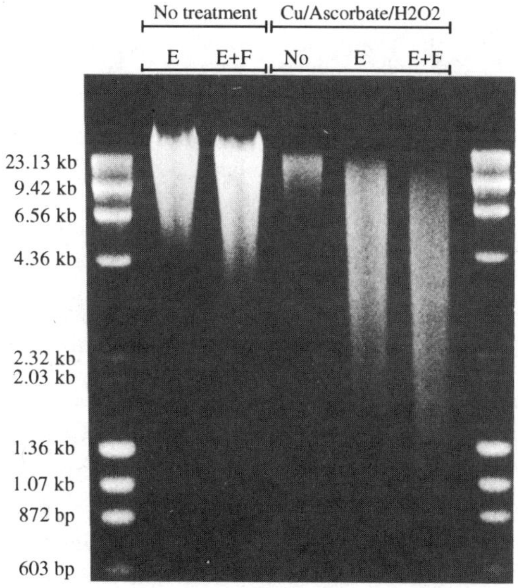

Figure 16.3. Frequency of strand breaks and oxidized bases induced by Cu(II)/ascorbate H_2O_2. Purified human male fibroblast DNA was treated *in vitro* either with buffer alone (lanes 2 and 3) or 50 μM Cu(II), 100 μM ascorbate, 5 mM H_2O_2 for 30 min at 37°C (lanes 4–6). After treatment, DNA was digested with either buffer alone (lane 4), Nth protein (E, lanes 2 and 5), or Fpg + Nth proteins (E + F, lanes 3 and 6). DNA was denatured in 50% (v/v) DMSO at 50°C and renaturation was blocked with glyoxal, then electrophoresed through a neutral 1.5% agarose gel. Lanes 1 and 6 contain *Hin*dIII lambda phage + *Hae*III-digested ΦX174 DNA molecular weight standards.

1991; Pfeifer and Riggs, 1993a,b; Pfeifer *et al.*, 1993b). The primer extension is carried out in siliconized 0.625-ml tubes; a thermocycler (MJ Research Inc.) is used for all incubations.

- Mix 1.28 to 1.6 μg of DNA in Sequenase buffer, 1 pmole of primer 1, 40 mM Tris-HCl pH 7.7, and 50 mM NaCl in a final volume of 15 to 18 μl.
- Denature at 98°C for 3 min.
- Incubate at 45 to 50°C, depending on the T_m of primer 1, for 15 min.
- Cool to 4°C.
- Add 9 μl of the following mix: 7.5 μl of $MgCl_2$–dNTP mix (20 mM $MgCl_2$, 20 mM dithiothreitol, and 0.25 mM of each dNTP), 1.1 μl H_2O, and 0.4 μl of Sequenase^{R}2.0 (13 U/μl, U.S. Biochemicals, Cleveland, OH).
- Incubate at 48 to 49°C for 5 min, 50°C for 1 min, 51°C for 1 min, 52°C for 1 min, 54°C for 1 min, 56°C for 1 min, 58°C for 1 min, and 60°C for 1 min. Then the samples are cooled to 4°C.
- Add 6 μl of ice-cold 310 mM Tris-HCl pH 7.7.
- Incubate at 67°C for 15 min to inactivate the Sequenase, then cool to 4°C.

2.7. LMPCR: Ligation (Step 4, Fig. 16.2)

The DNA molecules that have a 5′-phosphate group and a double-stranded blunt end are suitable for ligation. A DNA linker with a single blunt end is ligated directionally onto the double-stranded blunt end of the extension product using T_4 DNA ligase. This linker has no 5′ phosphate and is staggered to avoid self-ligation and provide directionality. Also, the duplex between the 25-mer (5′-GCGGTGACCCGGGAGATCTGAATTC) and 11-mer (5′-GAATTCAGATC) is stable at the ligation temperature, but denatures easily during subsequent PCR (Mueller and Wold, 1989, 1991). The linker is prepared by annealing in 250 mM Tris-HCl pH 7.7, 20 pmole/μl each of the 25-mer and 11-mer, heating at 95°C for 3 min, transferring quickly at 70°C, and cooling gradually to 4°C over a period of 3 hr. Linkers are stored at −20°C and thawed on ice before use.

- Add 45 μl of the following mix: 13.33 mM $MgCl_2$, 30 mM dithiothreitol, 1 mM ATP, 83.3 μg/ml BSA, 100 pmole of linker, and 6.25 units of T_4 DNA ligase (5 U/μl, Boehringer-Mannheim).
- Incubate at 18°C overnight.
- Stop the reaction and, while kept on ice, precipitate DNA by adding 25 μl of 10 M ammonium acetate, 1 μl of 0.5 M EDTA pH 8, 1 μl of 20 μg/μl glycogen, and 260 μl of ice-cold 100% ethanol.
- Air-dry DNA pellets.
- Dissolve DNA pellets in 50 μl of water.

2.8. LMPCR: PCR (Steps 5 and 6, Fig. 16.2)

At this step, gene-specific fragments can be exponentially amplified because primer sites are available at each end of the target fragments, primer 2 on one end and the longer oligonucleotide of the linker on the other end. Primer 2 may or may not overlap with primer 1. If it does overlap, the overlap should not be more than 7 or 8 bases (Mueller and Wold, 1991; Pfeifer and Riggs 1993a,b; Pfeifer *et al.*, 1993b).

- Add 50 μl of the Taq polymerase mix: 0.02% gelatin, 20 mM Tris-HCl pH 8.9, 4 mM $MgCl_2$, 80 mM KCl, 0.25 mM of each dNTP, 10 pmole of primer 2, 10 pmole of linker primer (LP25), and 3.0 units of Taq DNA polymerase (5 U/μl, Boehringer-Mannheim). The reaction is overlaid with 50 μl of mineral oil.

- Cycle 20 to 22 times at 95°C for 1 min (98°C for 3 min for the first cycle), 61 to 73°C (1 to 2°C below the calculated T_m of primer 2) for 2 min and 74°C for 3 min. The last extension should be done for 10 min to fully extend all DNA fragments.
- Add stop mix under the mineral oil layer: 13 μl of 3 M sodium acetate pH 5.2, 3 μl of 0.5 M EDTA pH 8, and 9 μl of water.
- Extract with 250 μl of premixed phenol–chloroform (92 μl–158 μl), then precipitate with ethanol.
- Air-dry DNA pellets.
- Dissolve DNA pellets in 7 μl of premixed formamide dye [1 part water, 2 parts 94% formamide, 2 mM EDTA pH 7.7, 0.05% xylene cyanole FF, 0.05% bromophenol blue (Pfeifer and Riggs, 1993a,b; Pfeifer *et al.*, 1993b)] in preparation for sequencing gel electrophoresis.

2.9. Gel Electrophoresis and Electroblotting (Step 7, Fig. 16.2)

PCR-amplified fragments are separated by electrophoresis through an 8% polyacrylamide/ 7 M urea gel, 0.4 mm thick and 60 to 65 cm long, then transferred to a nylon membrane by electroblotting (Pfeifer and Riggs, 1993a,b; Pfeifer *et al.*, 1993b).

- Heat the samples at 95°C for 2 to 3 min, then keep them on ice prior to loading.
- Run the gel at the voltage necessary to keep the temperature of the gel between 45 and 50°C.
- Stop the gel when the green dye (xylene cyanole FF) reaches the bottom of the gel.
- Cover the lower part of the gel (~40 to 42 cm) with a clean Whatman (Clifton, NJ) 3MM paper, remove the gel from the glass plate and cover it with Saran Wrap.
- Perform electroblotting using an electroblotting apparatus (HEP3, Owl Scientific Inc.) according to the manufacturer's instructions. Individually layer three sheets of Whatman 17 paper, 43 × 19 cm, presoaked in 90 mM TBE, and squeeze out the air bubbles between the paper layers with a rolling bottle.
- Put 50 ml of 90 mM TBE on the top layer and place the gel quickly on the Whatman papers before TBE is absorbed. Before removing the Saran Wrap, with a soft tissue remove all air bubbles under the gel. Check with a flashlight.
- Cover the gel with a charged nylon membrane (Qiabrane, Qiagen, Chatsworth, CA) presoaked in 90 mM TBE, then cover with three layers of presoaked Whatman 17 paper and squeeze out air bubbles with rolling bottle. Papers can be reused several times except for papers immediately under and above the gel.
- Place the upper electrode of the electroblotting apparatus onto the paper.
- Electrotransfer for 45 min at 2 A and 15 V.
- UV-crosslink (1000 J/m^2) the blotted DNA to the membrane, taking care to expose the DNA side of the membrane. If probe stripping and rehybridization are planned, keep the membrane damp.

2.10. Hybridization (Step 8, Fig. 16.2)

The hybridization is performed in a rolling 80-mm-wide × 22-cm-long borosilicate glass hybridization tube in a hybridization oven (Hoefer, San Francisco, CA). The nylon membrane is soaked in 90 mM TBE and, using a 25-ml pipette, placed in the tube so that the membrane sticks completely to the wall of the hybridization tube.

- Prehybridize with 15 ml of hybridization buffer (250 mM sodium phosphate pH 7.2, 1 mM EDTA, 7% SDS, 1% BSA) at 60 to 68°C for 15 min.
- Decant the prehybridization buffer and add the labeled probe in 6 to 8 ml of hybridization buffer.
- Hybridize at 60 to 68°C (2°C below the calculated T_m of the probe) overnight.
- Wash the membrane with prewarmed washing buffers for 30 to 45 min. The buffers should be kept in an incubator or waterbath set at a temperature of 4°C higher than the hybridization temperature. The membrane is placed in a tray located on an orbital shaker. Wash with buffer I (20 mM sodium phosphate pH 7.2, 1 mM EDTA, 2.5% SDS, 0.25% BSA) for 10 min and with buffer II (20 mM sodium phosphate pH 7.2, 1 mM EDTA, 1% SDS) for about 30 min. Buffer II is changed three times.
- Wrap the membrane in Saran Wrap. Do not let the membrane become dry if stripping and rehybridization are planned after exposure to the film.
- Expose membrane to Kodak XAR-5 X-ray film for 0.5 to 8 hr with intensifying screens at −70°C. Nylon membranes can be rehybridized if few sets of primers have been included in the primer extension and amplification reactions (Pfeifer and Riggs, 1993a,b; Pfeifer *et al.*, 1993b). Probes can be stripped by soaking the membranes in boiling 0.1% SDS solution twice for 5 to 10 min.

2.11. Preparation of Single-Stranded Hybridization Probes

The [^{32}P]dCTP-labeled single-stranded probe is prepared by repeated linear primer extension, between 30 and 35 cycles, with Taq polymerase. Primer 3 is extended on a double-stranded template which can be a plasmid or PCR product. To avoid probes longer than 200 bp, plasmid DNA is cut with an appropriate restriction enzyme (for examples, see Rodriguez *et al.*, 1995). A third primer (primer 3), which is used to make the probe, should be on the same strand as the amplification primer (primer 2), just 5′ to the corresponding primer 2, with no more than 7 or 8 bases of overlap with primer 2, and have a T_m of 60 to 68°C.

- Prepare 150 μl of the following mix: 0.01% gelatin, 10 mM Tris-HCl pH 8.9, 2 mM $MgCl_2$, 40 mM KCl, 0.25 mM of dATP, dGTP, and dTTP, 20 to 40 ng of plasmid or 5 ng of PCR products, 75 pmole of primer 3, 2.5 units of Taq DNA polymerase, and 10 μl of [^{32}P]dCTP (3000 Ci/mmole).
- Cycle 30 to 35 times at 95°C for 1 min, 60 to 68°C for 2 min, and 74°C for 3 min.
- Precipitate the probe by adding 37.5 μl of 10 M ammonium acetate, 20 μg of glycogen, and 420 μl of ice-cold 100% ethanol.
- Resuspend the probe in 100 μl of 10 mM Tris-HCl pH 7.4 and 1 mM EDTA.
- Dilute the probe in 6 to 8 ml of hybridization buffer.

On the final autoradiogram, each band represents a nucleotide position where a break was induced, and the signal intensity of the band reflects the number of DNA molecules with ligatable ends terminating at that position (Fig. 16.4). Figure 16.4 displays sequences from the transcribed strand (bottom strand) around the transcription initiation site of the human single-copy gene phosphoglycerate kinase 1 (*PGK*1) localized on the X chromosome. The bands with high signal intensities indicating numerous DNA molecules were nicked at the corresponding nucleotide position. These bands map most frequently to guanine bases, frequently to cytosine bases, infrequently to thymine bases, and rarely to adenine bases (Rodriguez *et al.*, 1995). *In vitro* and

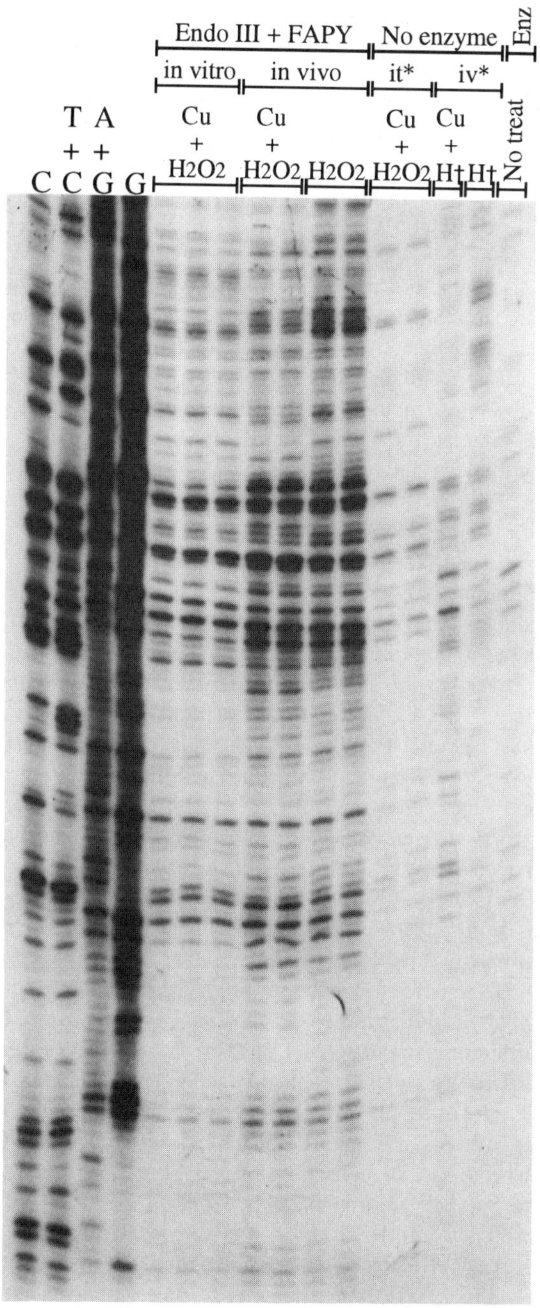

Figure 16.4. LMPCR analysis of oxidative damage in the transcribed strand of the human *PGK*1 gene. Primer set E (Pfeifer and Riggs, 1991; Rodriguez *et al.*, 1995) allows the display of sequences around the transcription initiation site. Lanes 1–4 represent LMPCR of 0.4 μg of HeLa cell DNA treated with the standard chemical cleavage reactions (Maxam and Gilbert, 1980; Pfeifer and Riggs, 1993b). Lanes 5–7 represent LMPCR of purified human male fibroblast DNA treated with 50 μM Cu(II), 100 μM ascorbate, 5 mM H_2O_2, for 30 min at 37°C followed by DNA purification and digestion with Nth protein and Fpg protein. Lanes 8 and 9 represent LMPCR of human male fibroblasts incubated overnight with 50 μM Cu(II) diluted in serum-free MEM, then treated with 37 mM H_2O_2 for 30 min at 37°C followed by DNA isolation and digestion of DNA with the endonucleases. Lanes 10 and 11 represent LMPCR of human male fibroblasts treated with 37 mM H_2O_2 followed by digestion with the same enzymes. Lanes 12 and 13 represent LMPCR of Cu(II)/ascorbate/H_2O_2-treated DNA digested with buffer only. Lanes 14 and 15 represent LMPCR of H_2O_2-treated fibroblasts, with and without Cu(II) preincubation, digested with buffer only. Lane 16 represents LMPCR of DNA treated with buffer only, followed by digestion with Nth and Fpg proteins. Note the similarity between the *in vitro* and *in vivo* band patterns. The *in vitro* lanes were loaded with slightly less DNA. it*, *in vitro*; iv*, *in vivo*.

in vivo, the frequency of oxidative DNA damage is strikingly affected by the local DNA sequence, the 5′ bases of d(pGn) and d(pCn) being damage hotspots as were the most internal guanines of d(pGGGCCC) and d(pCCCGGG) (Rodriguez *et al.*, 1996). *In vitro*, local factors influencing the efficiency of formation of the DNA–Cu(I)–H_2O_2 complex or the efficiency of base oxidation by this complex are important determinants of the damage distribution. The mechanisms of Cu–H_2O_2-induced DNA damage and the sequence specificity will not be discussed further, being beyond the scope of this chapter and having been discussed at length elsewhere (Rodriguez *et al.*, 1995; Drouin *et al.*, 1996).

3. DISCUSSION

LMPCR is an extremely sensitive technique for quantifying and mapping the frequency of rare DNA breaks along single-copy genes at nucleotide resolution. We have adapted the LMPCR technique by combining it with Nth and Fpg proteins to recognize and cleave DNA at oxidized bases to facilitate mapping of oxidative damage distributions. The approach we have described allows us to quantify and segregate DNA strand breaks and DNA base modifications and partly differentiate the relative contribution of each type of base modification, i.e., abasic sites, oxidized pyrmidines, and oxidized purines. The discussion will address the following: (1) technical notes on LMPCR, (2) background signal, (3) efficiency of enzyme digestion, and (4) selective mapping of oxidative DNA damage.

The DNA used for LMPCR must be very clean and its concentration accurately determined. Fluorometry combined with DAPI staining appears to be the most precise way of measuring double-strand DNA. The optimal starting amount of DNA varies between 1 and 2 μg. Problems in the autoradiogram, including high background, spurious bands, and missing bands, can often be traced to poor quality of the DNA, excessive amount of DNA, or inaccurate determination of starting DNA concentration. The primers, particularly primer 2 and the linker oligonucleotides, must be very carefully selected and be either gel or HPLC purified. Special attention should be given to the T_m of each primer and their relative position (Mueller and Wold, 1991; Pfeifer and Riggs, 1993a,b; Pfeifer *et al.*, 1993b).

In vitro and *in vivo* treatment conditions and reagent concentrations have been adjusted to induce oxidative damage at a frequency of 1 oxidized base per 2 to 4 kb (Rodriguez *et al.*, 1995). This is an optimal break frequency for LMPCR mapping studies. Prior to LMPCR, it is important to determine the break frequency by agarose gel electrophoresis (Fig. 16.3) (see Chapter 3). By using 1 to 2 μg of DNA (representing 150,000 to 350,000 cells), there are a sufficient number of DNA molecules in the starting material for any nucleotide position to be amplified reproducibly by LMPCR if the break frequency ranges from 1 break per 2 kb to 1 break per 4 kb. These conditions, amount of DNA and break frequency, are particularly suited to be combined with the electroblotting of the sequence ladder on a nylon membrane and hybridization of a labeled probe (Pfeifer *et al.*, 1989); the method using labeled primer 2 followed by gel electrophoresis (Mueller and Wold, 1989) requires longer exposure time, more DNA, or higher break frequency (unpublished data). The hybridization step adds more specificity and a higher specific activity than do the end-labeled primers, allowing more sensitivity and a sharper sequence ladder (unpublished data).

LMPCR is a complex and long multistep procedure. At any step, there are slightly different conditions between samples (viz., pipetting artifacts) and the efficiency of the reaction (viz., efficiency of the ligation) may be different from one sample to the other. These factors may impact the reproducibility of signal intensity at any particular band. The sample-to-sample variability is, we believe, primarily due to Poisson noise (Pfeifer *et al.*, 1993a), and is inversely proportional to the break frequency. For break frequencies less than 1 break per 2000 bases, we recommend processing of duplicate or triplicate samples to estimate this variability and allow proper interpretation of the signal intensity.

The background is defined by LMPCR signals unrelated to treatment-induced DNA damage. It is produced during the purification process, during storage of the DNA and by nonspecific glycosylase activity (Fig. 16.4, lane 16) (Pfeifer *et al.*, 1993a). Oxidative damage such as strand breaks, abasic sites, and oxidized bases can be generated by phenol extraction and exposure to Tris-HCl. Freshly prepared (>8 weeks) phenol kept at 4°C and equilibrated with HEPES pH 8, and using HEPES instead of Tris-HCl for DNA storage is recommended to limit background oxidative damage.

Most of the ROS-induced damaged bases have been identified by their recognition and cleavage by the DNA-glycosylase activity of Nth and Fpg proteins (Table I). The associated AP lyase activity of these enzymes cleaves the phosphodiester backbone at the abasic site or site of base modification to produce the 5′-phosphoryl groups which are substrates for LMPCR. The enzymatic conditions used in our studies produced terminal cleavage; at least 90% cleavage was achieved at every potentially cleavable position as measured visually from LMPCR autoradiographs. This condition is twofold more extreme than terminal cleavage as measured by glyoxal gel electrophoresis.

For purposes of clarity in the following discussion, we will call the single- and double-stranded nicks in DNA resulting from oxidation of the sugar-phosphate backbone as "frank strand breaks." The vast majority of these breaks have 5′-phosphoryl groups, are ligatable, and can be mapped by LMPCR (Rodriguez *et al.*, 1995). For treated and digested DNA, such as DNA of lanes 5–11 of Fig. 16.4, each band signal comprises background signal, frank strand break signal, abasic sites, and Nth and/or Fpg-sensitive oxidized bases. The abasic sites induced by the oxidative treatment represent a very minor proportion of what is cleaved by the Nth and Fpg proteins (Rodriguez *et al.*, 1995). For each nucleotide position, the background signal and the frank strand break signal can be quantitatively subtracted from the total signal and the signal remaining represents the modified base signal (Fig. 16.4, lane 5 − lane 12 − land 16 = modified base signal) (Rodriguez *et al.*, 1995). In other words, autoradiographic signals from a treated and digested DNA sample can be dissected to determine the relative contribution of the background, frank strand breaks, and oxidized bases. After *in vitro* and *in vivo* oxidative treatment the ratio of modified bases to frank strand breaks is 3.3–9:1 (Rodriguez *et al.*, unpublished). *In vitro*, this ratio can only be reached after extensive dialysis of purified DNA against water. Low ratios can be caused by inadequate purity of the distilled water used for dialysis (Rodriguez *et al.*, 1995).

REFERENCES

Ames, B. N. (1987). Oxidative DNA damage, cancer, and aging. *Ann. Intern. Med.* **107**:526–545.

Ames, B. N., Shigenaga, M. K., and Hagen, T. M. (1993). Oxidants, antioxidants, and the degenerative diseases of aging, *Proc. Natl. Acad. Sci. USA* **90**:7915–7922.

Aruoma, O. I., Halliwell, B., Gajewski, E., and Dizaroglu, M. (1991). Copper-ion-dependent damage to the bases in DNA in the presence of hydrogen peroxide. *Biochem. J.* **273**:601–604.

Boiteux, S. (1993). Properties and biological functions of the NTH and FPG proteins of *Escherichia coli*; two DNA glycosylases that repair oxidative damage in DNA. *J. Photochem. Photobiol. B Biol.* **19**:87–96.

Boveris, A. (1977). Mitochondrial production of superoxide radical and hydrogen peroxide. *Adv. Exp. Med. Biol.* **75**:67–82.

Breimer, L. H. (1990). Molecular mechanisms of oxygen radical carcinogenesis and mutagenesis: The role of DNA base damage. *Mol. Carcinogen.* **3**:188–197.

Bryan, S. E., and Frieden, E. (1967). Interaction of copper(II) with deoxyribunucleic acid below 30 degrees. *Biochemistry* 6:2728–2734.

Cheeseman, K. H., and Slater, T. F. (1993). An introduction to free radical biochemistry. *Br. Med. Bull.* 49:481–493.

Chevion, M. (1988). A site-specific mechanism for free radical induced biological damage: The essential role of redox-active transition metals. *J. Free Radicals Biol. Med.* **5**:27–37.

Church, G. M., and Gilbert, W. (1984). Genomic sequencing. *Proc. Natl. Acad. Sci. USA* **81**:1991–1995.

Dizdaroglu, M. (1991). Chemical determination of free radical-induced damage to DNA. *J. Free Radicals Biol. Med.* **10**:225–242.

Dizdaroglu, M. (1992). Oxidative damage to DNA in mammalian chromatin. *Mutat. Res.* **275**:331–342.

Dizdaroglu, M., Rao, G., Halliwell, B., and Gajewski, E. (1991a). Damage to the DNA bases in mammalian chromatin by hydrogen peroxide in the presence of ferric and cupric ions. *Arch. Biochem. Biophys.* **285**:317–324.

Dizdaroglu, M., Nackerdien, Z., Chao, B.-C., Gajewski, E., and Rao, G. (1991b). Chemical nature of in vivo DNA base damage in hydrogen peroxide-treated mammalian cells. *Arch. Biochem. Biophys.* **268**:388–390.

Doetsch, P. W., and Cunningham, R. P. (1990). The enzymology of apurinic/apyrimidinic endonucleases. *Mutat. Res.* **236**:173–201.

Drouin, R., Rodriguez, H., Gao, S., Gebreyes, Z., O'Connor, T. R., Holmquist, G. P., and Akman, S. A. (1996). Cupric ion/ascorbate/H_2O_2-induced DNA damage: DNA-bound copper ion primarily induces base modifications. *Free Rad. Biol. Med.*, in press.

Floyd, R. A. (1990). Role of oxygen free radicals in carcinogenesis and brain ischemia. *FASEB J.* **4**:2587–2597.

Floyd, R. A., Watson, J. J., Harris, J., West, M., and Wong, P. K. (1986). Formation of 8-hydroxy-deoxyguanosine, hydroxyl free radical adduct of DNA in granulocytes exposed to tumor promoter, tetradeconyl phorbolacetate. *Biochem. Biophys. Res. Commun.* **137**:841–846.

Floyd, R. A., West, M. S., Eneff, K. L., Hogsett, W. E., and Tingey, D. T. (1988). Hydroxyl free radical mediated formation of 8-hydroxyguanine in isolated DNA. *Arch. Biochem. Biophys.* **262**:266–272.

Geierstanger B. H., Kagawa, T. F., Chen, S.-L., Quigley, G. J., and Ho, P. S. (1991). Base-specific binding of copper(II) to Z-DNA. *J. Biol. Chem.* **266**:20185–20191.

Goldstein, S., and Czapski, G. (1986). The role and mechanism of metal ions and their complexes in enhancing damage in biological systems or in protecting these systems from toxicity of O_2^-. *J. Free Radical Biol. Med.* **2**:3–11.

Guyton, K. Z., and Kensler, T. W. (1993). Oxidative mechanisms in carcinogenesis. *Br. Med. Bull.* **49**:523–544.

Halliwell, B., and Aruoma, O. I. (1991). DNA damage by oxygen-derived species. *FEBS Lett.* **281**:9–19.

Halliwell, B., and Gutteridge, J. M. C. (1990). Role of free radicals and catalytic metal ions in human disease: An overview. *Methods Enzymol.* **186**:1–85.

Hatahet, Z., Kow, Y. W., Purmal, A. A., Cunningham, R. P., and Wallace, S. S. (1994). New substrates for old enzymes. 5-hydroxy-2′-deoxycytidine and 5-hydroxy-2′-deoxyuridine are substrates for *Escherichia coli* endonuclease III and formamidopyrimidine DNA N-glycosylase, while 5-hydroxy-2′-deoxyuridine is a substrate for uracil DNA N-glycosylase. *J. Biol. Chem.* **269**:18814–18820.

Izatt, R. M., Christensen, J. J., and Rytting, J. H. (1971). Sites and thermodynamic quantities associated with proton and metal ion interaction with ribonucleic acid, deoxyribonucleic acid, and their constituent bases, nucleosides, and nucleotides. *Chem. Rev.* **71**:439–457.

John, D. C. A., and Douglas, K. T. (1989). Apparent sequence preference in cleavage of linear B-DNA by the Cu(II):thiol system. *Biochem. Biophys. Res. Commun.* **165**:1235–1242.

Kagawa, T. F., Geierstanger, B. H., Wang, H.-J., and Ho, P. S. (1991). Covalent modification of guanine bases in double-stranded DNA. *J. Biol. Chem.* **266**:20175–20184.

Kasai, H., Crain, P. F., Kuchino, Y., Nishimura, S., Ootsuyama, A., and Tanooka, H. (1986). Formation of 8-hydroxyguanine moiety in cellular DNA by agents producing oxygen radicals and evidence for its repair. *Carcinogenesis* **7**:1849–1851.

Kazakov, S. A., Astashkina, T. G., Mamaev, S. V., and Vlassov, V. V. (1988). Site-specific cleavage of single-stranded DNAs at unique sites by a copper-dependent redox reaction. *Nature* **335**:186–188.

Maniatis, T., Fritsch, E. F., and Sambrook, J. (1982). *Molecular Cloning: A Laboratory Manual*, Cold Spring Harbor Laboratory Press, Cold Spring Harbor, NY.

Masarwa, M., Cohen, H., Meyerstein, D., Hickman, D. L., Bakac, A., and Espenson, J. H. (1988). Reactions of low-valent transition-metal complexes with hydrogen peroxide. Are they "Fenton-like" or not? 1. The case of Cu^+ and Cr^{2+}. *J. Am. Chem. Soc.* **110**:4293–4297.

Maxam, A. M., and Gilbert, W. (1980). Sequencing end-labeled DNA with base-specific chemical cleavages. *Methods Enzymol.* **65**:499–560.

Milne, L., Nicotera, P., Orrenius, S., and Burkitt, M. J. (1993). Effects of glutathione and chelating agents on copper-mediated DNA oxidation: Pro-oxidant properties of glutathione. *Arch. Biochem. Biophys.* **304**:102–109.

Minchekova, L. E., and Ivanov, V. I. (1967). Influence of reductants upon optical characteristics of the DNA–Cu^{2+} complex. *Biopolymers* **5**:615–625.

Mueller, P. R., and Wold, B. (1989). In vivo footprinting of a muscle specific enhancer by ligation-mediated PCR. *Science* **246**:780–786.

Mueller, P. R., and Wold, B. (1991). Ligation mediated PCR: Applications to genomic footprinting. *Methods* **2**: 20–31.

Nohl, H. (1993). Involvement of free radicals in ageing: A consequence or cause of senescence. *Br. Med. Bull.* **49**:653–667.

Pezzano, H., and Podo, F. (1980). Structure of binary complexes of mono- and polynucleotides with metal ions of the first transition group. *Chem. Rev.* **80**:366–401.

Pfeifer, G. P., and Riggs, A. A. (1991). Chromatin differences between active and inactive X chromosomes revealed by genomic footprinting of permeabilized cells using DNase I and ligation-mediated PCR. *Genes Dev.* **5**:1102–1113.

Pfeifer, G. P., and Riggs, A. D. (1993a). Genomic footprinting by ligation mediated polymerase chain reaction, in: *PCR Protocols: Current Methods and Applications* (B. White, ed.), Humana Press, Totowa, NJ, pp. 153–168.

Pfeifer, G. P., and Riggs, A. D. (1993b). Genomic sequencing, in: *DNA Sequencing Protocols* (A. Griffin and H. Griffin, eds.), Humana Press, Totowa, NJ, pp. 169–181.

Pfeifer, G. P., Steigerwald, S. D., Mueller, P. R., Wold, B., and Riggs, A. D. (1989). Genomic sequencing and methylation analysis by ligation mediated of PCR. *Science* **246**:810–813.

Pfeifer, G. P., Drouin, R., and Holmquist, G. P. (1993a). Detection of DNA adducts at the DNA sequence level by ligation-mediated PCR. *Mutat. Res.* **288**:39–46.

Pfeifer, G. P., Singer-Sam, J., and Riggs, A. D. (1993b). Analysis of methylation and chromatin structure. *Methods Enzymol.* **225**:567–583.

Piette, J. (1991). Biological consequences associated with DNA oxidation mediated by singlet oxygen. *J. Photochem. Photobiol. B Biol.* **11**:241–260.

Prütz, W. A., Butler, J., and Land, E. J. (1990). Interaction of copper(I) with nucleic acids. *Int. J. Radiat. Biol.* **58**:215–234.

Rodriguez, H., Drouin, R., Holmquist, G. P., O'Connor, T. R., Boiteux, S., Laval, J., Doroshow, J. H., and Akman, S. A. (1995). Mapping of copper/hydrogen peroxide-induced DNA damage at nucleotide resolution in human genomic DNA by ligation-mediated PCR. *J. Biol. Chem.* **270**:17633–17640.

Rodriguez, H., Drouin, R., Holmquist, G. P., and Akman, S. (1996). Repair of a hydrogen-peroxide-induced *in vivo* footprint in the human hypoxia-inducible factor 1 binding site of the PGK1 gene (unpublished).

Rychlik, W., and Rhoads, R.-E. (1989). A computer program for choosing optimal oligonucleotides for filter hybridization, sequencing and *in vitro* amplification of DNA. *Nucleic Acids Res.* **17**:8543–8551.

Sagripanti, J.-L., and Kraemer, K. H. (1989). Site-specific oxidative DNA damage at polyguanosines produced by copper plus hydrogen peroxide. *J. Biol. Chem.* **264**:1729–1734.

Stoewe, R., and Prütz, W. A. (1987). Copper-catalyzed DNA damage by ascorbate and hydrogen peroxide: Kinetics and yield. *J. Free Radicals Biol. Med.* **3**:97–105.

Wallace, S. S. (1988). AP endonucleases and DNA glycosylases that recognize oxidative DNA damage. *Environ. Mol. Mutagen.* **12**:431–477.

Yamamoto, K., and Kawanishi, S. (1989). Hydroxyl free radical is not the main active species in site-specific DNA damage induced by copper(II) ion and hydrogen peroxide. *J. Biol. Chem.* **264**:15435–15440.

Yamamoto, K., and Kawanishi, S. (1992). Site-specific DNA damage by phenylhydrazine and phenelzine in the presence of Cu(II) ion or Fe(III) complexes: Roles of active oxygen species and carbon radicals. *Chem. Res. Toxicol.* **5**:440–446.

Chapter 17

Single-Strand Ligation PCR for Detection of DNA Adducts

Keith A. Grimaldi, Simon R. McAdam, and John A. Hartley

1. INTRODUCTION

Many anticancer drugs, chemical carcinogens, and UV irradiation form covalent adducts on nucleotides in DNA. These adducts can interfere with transcription and replication of DNA and in so doing can trigger cell death or potentially carcinogenic mutations. Studies with anticancer drugs such as cisplatin and the nitrogen mustards have shown that even relatively simple DNA damaging agents show a degree of sequence preference in adduct formation on isolated plasmid DNA (Cullinane *et al.*, 1993; Hartley *et al.*, 1986; Mattes *et al.*, 1986; Ponti *et al.*,1991). Agents with a greater degree of sequence selectivity are under development with the aim of improving the specificity of anticancer therapy. Carcinogens such as benzo[*a*]pyrene also show some selectivity for particular nucleotides when reacted with isolated plasmid DNA (Lobanenkov *et al.*, 1986; Puisieux *et al.*, 1991). These observations illustrate the importance of studying DNA damage at the nucleotide level of a single-copy gene in cells where DNA exists in a highly ordered structure complexed with many proteins, and where other intracellular components could affect the reactivity of drugs and carcinogens. It has also become clear that it is important to be able to study the repair of individual lesions since repair mechanisms play an important part in determining sensitivity of a tumor cell to chemotherapy, and in preventing mutation. Recent work has shown that mutation hot spots in the E. coli *lacI* gene and the human *p53* gene corresponded to sites of slow repair of UV-induced pyrimidine dimers (Kunala and Brash, 1992; Tornaletti and Pfeifer, 1994). It would be interesting to learn how widespread this phenomenon is and whether similar correlations exist for chemical carcinogens.

To begin to explore some of these areas we have developed single-strand ligation PCR (sslig-PCR), a technique that can be used to detect DNA adducts at the nucleotide level in single-copy genes in mammalian cells. The method depends on the blockage of a DNA polymerase by many types of adducts formed by chemical carcinogens, anticancer drugs, and UV irradiation. sslig-PCR can be used for a wide variety of agents as it does not require the generation of a strand

---

Keith A. Grimaldi, Simon R. McAdam, and John A. Hartley • CRC Drug-DNA Interactions Research Group, Department of Oncology, University College London Medical School, London W1P 8BT, England.

Technologies for Detection of DNA Damage and Mutations, edited by Gerd P. Pfeifer, Plenum Press, New York, 1996.

break at the site of the adduct. It can also be used to detect damage to DNA caused by treatments such as X-irradiation which break DNA strands.

To allow the detection of damage at the nucleotide level in a single-copy gene in mammalian cells, the main problem to be overcome is that of sensitivity. The observation that covalent DNA lesions can block *taq* polymerase (or other thermostable polymerases such as Pfu and Vent) has been exploited recently to develop a method that can detect the sequence specificity of adduct formation on plasmid DNA *in vitro* (Ponti *et al.*, 1991). The measurement of the sites of adduct formation requires the use of linear PCR with a single specific oligonucleotide primer, the elongation of which will be blocked at sites of lesions. The products of this reaction can be directly detected on a sequencing gel when highly purified plasmid DNA is used as template (Ponti *et al.*, 1991) but the sensitivity is insufficient to look at single-copy genes. Such sensitivity could be reached by exponentially amplifying the sequence ladder obtained, but only one end will be defined (by the initial primer). This problem has been overcome in sslig-PCR by the ligation of a single-stranded oligonucleotide to the undefined ends of the products of linear amplification to allow subsequent exponential amplification. This makes use of the property of T4 RNA ligase to ligate single-stranded deoxyribo-oligonucleotides to single-stranded DNA (Tessier *et al.*, 1986).

The method of sslig-PCR is outlined in Fig 17.1. It involves a first-round PCR using a single 5′-biotinylated primer (1-B) which defines the area of the gene, and the strand, to be studied. Thirty cycles of linear amplification by PCR generates a family of single-stranded molecules of varying length for which the 5′-end is defined by primer 1-B and for which the 3′-ends are defined by the positions of the DNA adducts. In order to exponentially amplify these molecules, which are captured and isolated by binding to streptavidin-coated magnetic beads, a single-stranded oligonucleotide ("ligation oligonucleotide") is ligated to their 3′-OH ends using T4 RNA ligase (Tessier *et al.*, 1986). The "ligation oligonucleotide" bears a 5′-phosphate which is essential for ligation and a 3′-terminal amine group which blocks self-ligation. With both ends of the DNA molecules defined they can then be exponentially amplified, using primers P2 and "ligation primer," and detected by autoradiography following a final PCR with a single ^{32}P end-labeled primer (P3).

We have found single-stranded ligation PCR to be a sensitive and reproducible technique and we have applied it to study the interaction of several anticancer drugs within a region of a single copy gene (Grimaldi *et al.*, 1994). The importance of studying adduct formation within the cellular environment was highlighted by the discovery of a new site of lesion formation by cisplatin which occurred only in cells and not in naked genomic DNA (Grimaldi *et al.*, 1994). sslig-PCR is particularly useful in the development of sequence-specific anticancer agents and has been used, for example, to confirm that the sequence specificity of AT486, a novel sequence-selective minor groove DNA cross-linking agent, is maintained inside cells (Smellie *et al.*, 1994). sslig-PCR has also been used to map the sites of UV-induced pyrimidine dimers, benzo[*a*]pyrene adducts, and the sites of strand breakage caused by X-irradiation (unpublished observations).

2. METHODS

2.1. Buffers and Reagents

10× Teoa (store at 4°C)

250 mM triethanolamine pH 7.2
10 mM EDTA

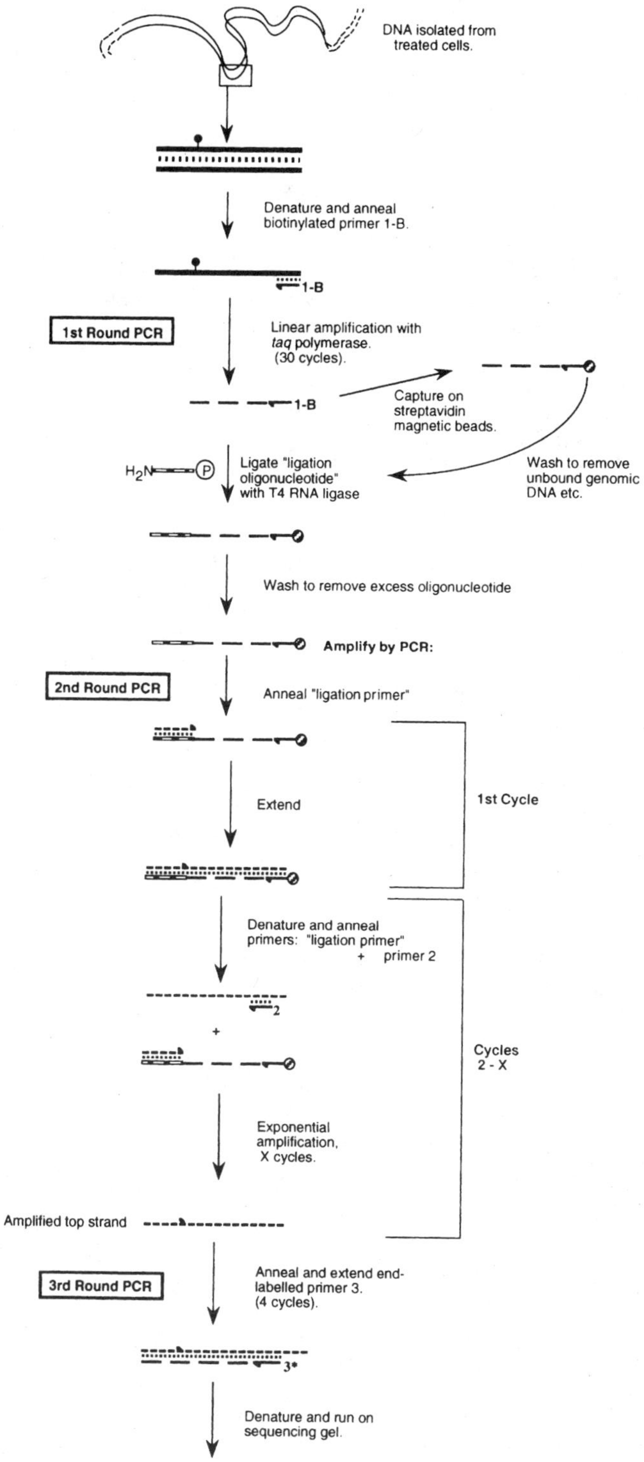

Figure 17.1. sslig-PCR method.

Cell lysis buffer (store at room temp.)
400 mM Tris-Cl pH 8.0
60 mM EDTA
150 mM NaCl
1% (w/v) SDS

Thermostable polymerase
Obtained from Advanced Biotechnologies (UK).

10× PCR buffer
Buffer IV supplied by Advanced Biotechnologies was used and found to be superior in terms of product yield and efficiency compared to the original Perkin–Elmer Cetus formula. Composition:
200 mM $(NH_4)_2SO_4$
750 mM Tris-Cl pH 9.0
0.1% (v/v) Tween

5× washing and binding buffer (WBB) (store at 4°C)
25 mM Tris-Cl pH 7.6
5 mM EDTA
5 M NaCl

TE (pH 7.6) (store at 4°C)
10 mM Tris-Cl pH 7.6
1 mM EDTA

10× ligation buffer (store at −70°C)
0.5 M Tris-Cl pH 8.0
100 mM $MgCl_2$
10 mM hexammine (III) cobalt chloride
100 μg/ml bovine serum albumin
200 μM ATP

T4 RNA ligase
Obtained from New England Biolabs (activity = 20 units/μl)

PEG (store at room temp.)
50% (w/v) PEG 8000

Oligonucleotide primers
A set of three nested primers are required for each strand of the region under study. The first primer must be 5′-biotinylated. As an example the following primers were used to look at exon 1 of the human H-*ras* gene:
To detect adducts on the nontranscribed strand:
1-B: 5′-CAG GGC CAC AGC ACC ATG CA
P2: 5′-AGC ACC ATG CAG GGG ACC A
P3: 5′-GGC GCT AGG CTC ACC TCT AT
To detect adducts on the transcribed strand:
1-B: 5′-GTA GGC ACG CTG CAG TCC TT
P2: 5′-AGT CCT TGC TGC CTG GCG TT
P3: 5′-GGC CTG GGC CTG GCT GAG CA

"Ligation oligonucleotide"
5′-ATC GTA GAT CAT GCA TAG TCA TA
This oligonucleotide should be gel or HPLC purified. It must also be 5′-phosphorylated and

bear a 3′-terminal amine group to block self-ligation (it is best if these modifications are incorporated at synthesis). Alternatively the 3′ end can be blocked by addition of a dideoxy nucleotide using terminal transferase as described in Tessier *et al.*, (1986).

"Ligation primer"

5′-TAT GAC TAT GCA TGA TCT ACG AT

This oligonucleotide, which is complementary to the "ligation oligonucleotide," must be gel or HPLC purified.

Oligonucleotide end-labeling reaction

Oligonucleotides were end-labeled with T4 polynucleotide kinase using Gibco-BRL kits with forward reaction buffer.

5× forward reaction buffer:
300 mM Tris-Cl pH 7.8
75 mM 2-mercaptoethanol
50 mM $MgCl_2$
1.65 μM ATP

Sequencing gel loading dye

96% (v/v) formamide (deionized)
20 mM EDTA
0.03% (w/v) xylene cyanol
0.03% (w/v) bromophenol blue

Sequencing gel

6% sequencing gels (80 cm × 20 cm × 0.4 mm) were prepared with Sequagel (National Diagnostics). Composition:
5.7% acrylamide
0.3% bisacrylamide
8.3 M urea
0.1 M Tris-borate pH 8.3
2 mM EDTA

2.2. DNA Damaging Agent Treatments

2.2.1. Treatment of Isolated DNA

Three-microgram aliquots of genomic DNA are incubated with DNA damaging agent at 37°C, or irradiated with UV, for appropriate times in Teoa buffer in a total volume of 50 μl. DNA is then precipitated with ethanol, washed twice with 70% ethanol, dried, and resuspended in 10 μl H_2O for sslig-PCR.

2.2.2. Treatment of Cells

Cells are treated in tissue culture medium in the presence or absence of fetal calf serum according to the agent used and the length of incubation. Short incubations (e.g., 1–5 hr) may be carried out in serum-free medium if the damaging agent is also likely to react with proteins. Depriving the cell of serum will, however, disrupt the cell's homeostatic environment, and this should be taken into account when designing experiments and interpreting results. We have found that the DNA damaging effect of cisplatin is not significantly diminished when cells are treated in the presence of 5% serum despite the known reactivity of cisplatin with protein. Different agents will no doubt behave in different ways and whether to include serum or not will have to be determined empirically.

2.2.2a. Suspension Cultures. 10^6 cells/well in 24-well plates are treated with the DNA damaging agent in 1 ml of the appropriate tissue culture medium at 37°C. The cells are harvested and washed 3× with 1 ml of tissue culture medium in 1.5-ml microfuge tubes (tubes are spun for 5 min at 270*g*, at 4°C). At this point the cell pellet is either processed immediately for DNA extraction or may be stored frozen. For repair experiments the cells are returned to culture plates for further incubation at 37°C.

2.2.2b. Adherent Cultures. Cells are grown to confluence in 2-cm-diameter wells of six-well plates. After treatment with damaging agent they are gently washed with fresh medium while attached to the plate. The cells are harvested by trypsinization and the cell pellet is then processed for DNA extraction as for cells from suspension cultures. Alternatively the cells may be lysed *in situ* with cell lysis buffer. In practice this is not the preferred method since the cell lysis buffer contains SDS and the small volume used creates a viscous, frothy mixture which makes efficient and reproducible extraction difficult.

2.3. DNA Isolation from Cells

sslig-PCR involves the isolation of DNA from many cell samples treated in parallel. A quick, simple, and reproducible "single-tube" method based on the DNA isolation protocol described by Marmur (1961) was developed in order to avoid the necessity of measuring DNA concentrations of many small samples.

The cell pellet obtained after the drug treatment is resuspended in 340 μl cell lysis buffer to which is added 100 μl 5 M sodium perchlorate. After vortexing, the suspension is incubated with shaking at 37°C for 20 min, then at 65°C for 20 min with occasional agitation. Chloroform (580 μl) precooled at −20°C is added and the mixture is rotated for 20 min at room temperature. After centrifugation for 10 min at 11,600*g* in a microfuge, half of the aqueous DNA-containing upper layer (220 μl, equivalent to 5×10^5 cells from suspension cultures) is removed and the DNA precipitated with 440 μl absolute ethanol, washed twice with 70% ethanol, and dried under vacuum. When DNA is prepared from adherent cultures the amount of the aqueous DNA-containing layer to be removed will depend on cell size since a monolayer culture of large cells will obviously yield less DNA than a culture of smaller cells. With care it is possible to remove up to 400 μl without disturbing the interface.

Having isolated the DNA it is often useful to digest it with a restriction enzyme that cuts at a site a few hundred base pairs upstream of the binding site of the biotinylated primer that is to be used in the first PCR step (primer 1-B). This creates a stop site at which the elongation of the primer will be halted in the absence of downstream adducts. This will be seen as a "full-length" band on the autoradiograph, the intensity of which will decrease in proportion to the extent of DNA damage in the region between the biotinylated PCR primer and the restriction site. The intensity of this band thus allows the amount of damage in this region to be measured which is especially helpful for establishing "single-hit" kinetics.

2.4. Single-Strand Ligation PCR

The following protocol is used for the nontranscribed strand of the human H-*ras* gene; certain conditions such as annealing temperature and $MgCl_2$ concentration will have to be adjusted according to which primers are used.

2.4.1. First-Round PCR

First-round "linear" PCR is carried out, in a volume of 40 μl, using 3 μg of drug-treated isolated DNA, or DNA isolated from the equivalent of 5×10^5 cells. The PCR is carried out using 0.6 pmole of 5′-biotinylated primer 1-B and the reaction mixture is composed of 20 mM $(NH_4)_2SO_4$, 75 mM Tris-HCl pH 9.0, 0.01% (v/v) Tween 20 (PCR buffer), 1.5 mM $MgCl_2$, 0.01% gelatin, 250 μM each dNTP, and 1 unit thermostable polymerase. The DNA was initially denatured at 94°C for 5 min and then subjected to 30 cycles of 94°C–1 min, 62°C–1 min, 72°C–1 min + 1-sec extension per cycle without oil overlay on a PT-100 thermal cycler with hot bonnet (MJ Research, USA). The mixture was finally incubated at 72°C for 5 min and then cooled to 4°C.

2.4.2. Capture and Ligation

To capture and purify the products of biotinylated primer extension, 10 μl of 5× WBB (washing and binding buffer) is added to the PCR mixture which is then transferred to 1.5-ml microfuge tubes containing 5 μl (50 μg) washed streptavidin M-280 Dynabeads (Dynal UK). The suspension is incubated for 30 min at 37°C with occasional agitation. The beads are sedimented in a magnetic rack and washed three times with 200 μl 10 mM TE pH 7.6. The beads are resuspended in 10 μl ligation mixture made up of:

1 μl 10× ligation buffer
5 μl 50% PEG 8000
1 μl "ligation oligonucleotide" (20 pmole/μl)
1 μl T4 RNA ligase (20 units/μl)
2 μl H_2O

The mixture is ligated overnight at 22°C. After ligation the beads are washed three times with 200 μl TE pH 7.6, resuspended in 40 μl H_2O, and transferred to 0.5-ml PCR tubes for the second-round PCR which is carried out with the template attached to the beads.

2.4.3. Second-Round PCR

The second-round PCR mixture, in a final volume of 100 μl, contains 10 pmole each of primer P2 and the gel-purified "ligation primer" that is complementary to the "ligation oligonucleotide." The buffer composition is as for first-round PCR except that 2.5 units of thermostable polymerase is used. The cycling conditions are: an initial denaturation at 94°C for 5 min, then X cycles of 94°C–1 min, 58°C–1 min, 72°C–1 min + 1-sec extension per cycle with a final 5-min step at 72°C. The number of cycles (X) in this step has to be determined empirically for each set of primers. It generally falls between 22–28 cycles (see Fig. 17.2).

2.4.4. Third-Round PCR

The third-round PCR is carried out by adding 10 μl PCR reaction buffer containing 5 pmole of ^{32}P 5′ end-labeled primer P3 and 1 unit thermostable polymerase. The mixture is subjected to four further cycles of 94°C–1 min, 64°C–1 min, and 72°C–1 min with a final 5-min step at 72°C. The reaction mixture is precipitated with ethanol, resuspended in 5 μl formamide loading buffer, denatured at 95°C for 3 min, cooled on ice and electrophoresed at 2500–3000 V for 3 hr in a 80 cm × 20 cm × 0.4 mm, 6% acrylamide sequencing gel. The gel is then dried onto Whatman 3MM paper, supported by a layer of Whatman DE 81 paper to bind the shorter fragments which will pass through the 3MM, and autoradiographed.

2.5. Results

Figure 17.2A shows the adducts present on the nontranscribed strand of exon 1 of the human H-*ras* gene after treatment of DNA with the anticancer drug cisplatin. The experiment was a

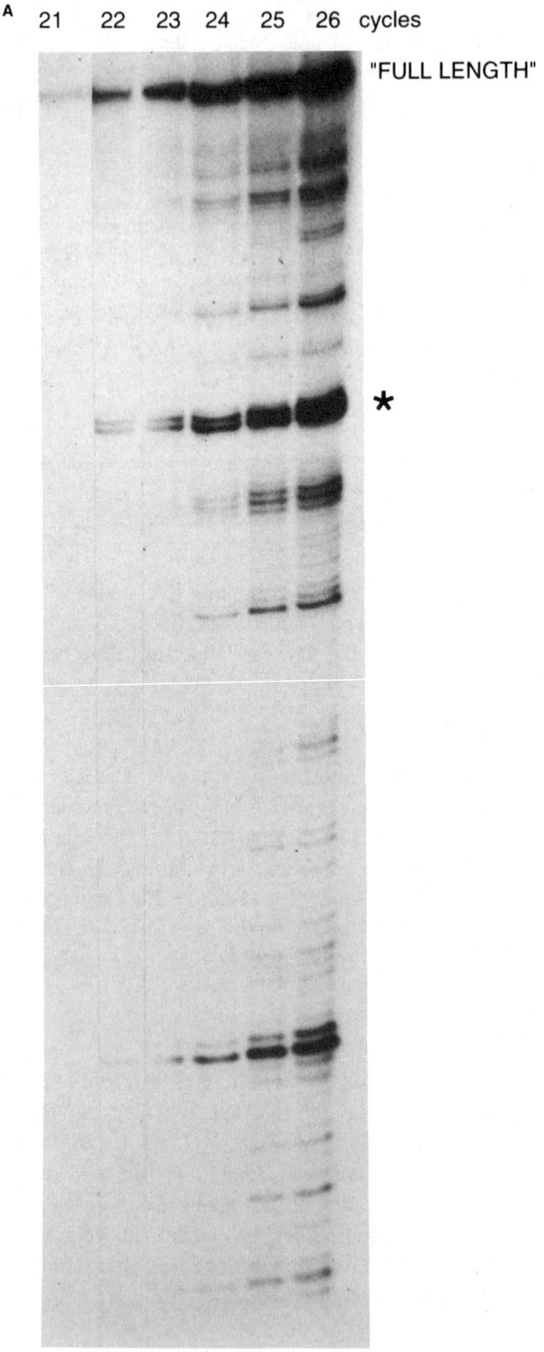

Figure 17.2.

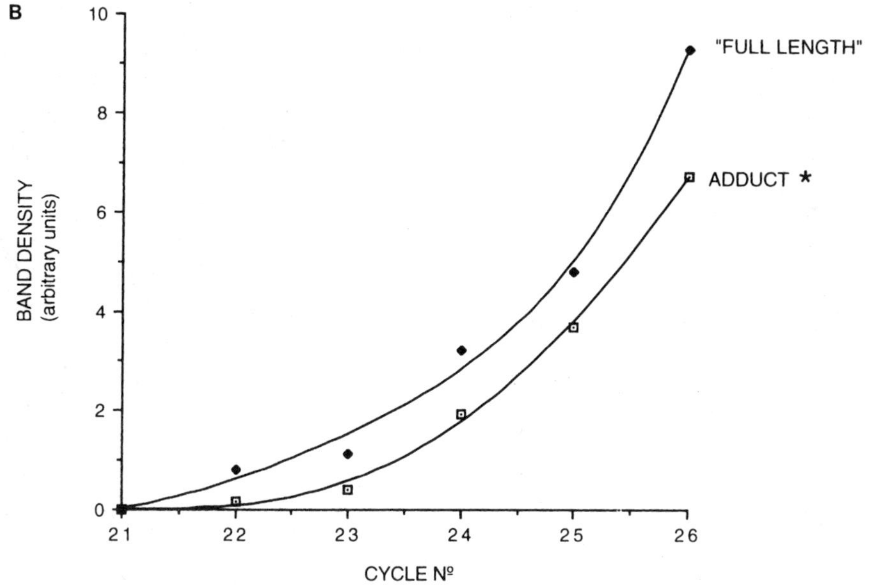

Figure 17.2. sslig-PCR cycle titration. sslig-PCR was carried out on cisplatin (10 μM)-treated DNA and the number of cycles in the second-round, exponential, PCR was varied. (A) Sites of adducts on the nontranscribed strand of exon 1 of the H-*ras* gene. "Full length" is the termination site generated by digestion with restriction endonuclease Sty 1. * represents a cisplatin adduct which was quantitated. (B) Bands corresponding to "full length" and adduct * were quantitated by densitometry and the band intensity was plotted against the number of cycles in the second-round PCR.

cycle titration of the second-round exponential PCR. As can be seen the intensity of the bands increase with increasing cycle number. Quantitation of two of the bands [the "full-length" band of the cisplatin-treated DNA and the band (*) corresponding to the site of a cisplatin adduct] by both phosphor imaging (not shown) and densitometry shows that the intensities increase exponentially (Fig. 17.2B). It is important that the sslig-PCR is within the exponential range when stopped so that the intensities of the individual bands will be proportional to the frequency of adduct formation at particular sites. With H-*ras* for example, 24 cycles were chosen. This is well within the exponential range but with reasonable incorporation of radioactivity into the products, allowing detection by autoradiography after an overnight exposure without intensifying screens.

Figure 17.3 shows the intracellular adducts formed by cisplatin on the nontranscribed strand of exon 1 of the H-*ras* gene. As can be seen there is a marked heterogeneity in the distribution of lesions. Cisplatin forms mainly intrastrand cross-links at 5′-AG and GG sites (Fichtinger-Schepman *et al.*, 1985; Roberts and Friedlos, 1981), but it is clear that the frequency of adduct formation at these sites varies considerably. The region of the H-*ras* gene which contains the sites of activating mutations, codons 12 and 13, contains several GG pairs and a run of three guanines in close proximity which unexpectedly are infrequent sites of cisplatin adduct formation while other regions clearly contain strong binding sites.

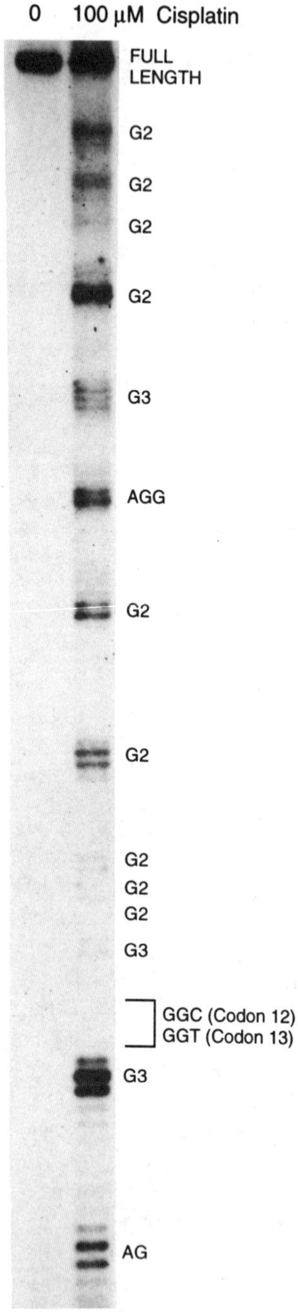

Figure 17.3. Cisplatin treatment of human fibroblasts. Human fibroblasts were treated for 18 hr with 100 μM cisplatin, the DNA was isolated and subjected to sslig-PCR. Adducts shown are on the nontranscribed strand of exon 1 of the H-*ras* gene. Potential cisplatin binding sites are indicated as are the positions of mutation hot spots codons 12 and 13.

3. DISCUSSION

3.1. Optimization and Troubleshooting

Being a multistep method, there are many potential problems. We have encountered many of them—maybe all!—and this section will exploit the benefits of our experience.

3.1.1. Biotinylated Primer and Paramagnetic Beads

It has happened that primers which should have been biotinylated were not. Also we have had one batch of paramagnetic beads which did not bind efficiently. These are rare occurences but nevertheless possible and quite easy to test for:

1. Incubate 10 pmole of biotinylated primer with 5 μl of washed beads, in PCR buffer for 10 min, sediment the beads, and use the supernatant in a conventional PCR. The quantity of product should be significantly reduced (usually by at least 50%) compared to the product obtained with uncaptured primers. It may not be completely reduced unless pure primers are used because the "failure sequence" oligonucleotides present (which would not bear the 5′ biotin), although being shorter, can still participate in the PCR. It is also a good idea to perform this test in parallel on an unbiotinylated primer pair to control for any dilution or loss of primer that may occur when incubating with the beads.
2. Carry out a conventional PCR with 10 pmole of each primer, one of which is biotinylated. Then capture the product on the beads, run the unbound fraction on an agarose gel, and a reduction in product after capture (of around 50%) will be observed if all is functioning correctly. Again it is a good idea to run a parallel test with unbiotinylated primers.

3.1.2. Efficiency of Ligation and Ligase Contamination

We have had one batch of T4 RNA ligase apparently contaminated with exonuclease activity which removes the 3′-terminal amine block on the "ligation oligonucleotide." The result on the autoradiograph is a ladder of intense bands at intervals of 20–23 bp due to serial self-ligation (the "ligation oligonucleotide" is 23 bp).

The efficiency of the ligation step can be tested using donor and acceptor nucleotides as described in Tessier *et al.* (1986). If exonuclease is present, the labeled oligonucleotide will be degraded.

3.1.3. PCR Components

The activity of the thermostable polymerase, efficiency of primers, quality of the genomic DNA template, and appropriate annealing temperatures can be determined by conventional PCR. If the primers are inefficient, the presence of formamide (1–10%) and/or DMSO (1–10%) can often improve both efficiency and specificity.

3.1.4. Resolution and Background

The main reason for poor resolution of bands on the autoradiograph is inadequate drying of the sequencing gel before autoradiography.

A common problem encountered when setting up sslig-PCR is too much background, that is, nonspecific bands on the autoradiograph caused by either spontaneous premature termination of primer extension, or nonspecific primer binding, in the first-round PCR. There are several possible reasons why this may occur:

- Nonoptimal $MgCl_2$ concentration especially in the first-round, linear PCR
- Annealing temperatures too low

- Too much genomic DNA template in the first-round PCR
- Too many cycles in the second-round PCR (As with all PCR assays, a small level of background is inevitable. This level, though, will be exaggerated if too many cycles are used in the exponential, second-round PCR.)

3.2. Summary

sslig-PCR has proved to be a sensitive method for studying the sequence selectivity of adduct formation in single-copy genes in mammalian cells. It has been useful in cancer therapy-related research into the intracellular behavior of established drugs such as cisplatin, as well as novel sequence specific cytotoxic agents.

In mutagenesis studies its ability to detect a wide variety of lesions should make it a useful tool in the investigation of adduct formation and repair mechanisms. Detailed information about these areas will be essential to the further understanding of the processes of mutagenesis and carcinogenesis.

REFERENCES

Cullinane, C., Wickham, G., McFadyen, W., Denny, W., Palmer, B., and Phillips, D. (1993). The use of bi-directional transcription footprinting to detect platinum–DNA crosslinks by acridine-tethered platinum diamine complexes and cisplatin. *Nucleic Acids Res.* **21**:393–400.

Fichtinger-Schepman, A., Van der Veer, J., Den Hartag, J., Loman, P., and Reedijk, J. (1985). Adducts of the antitumor drug cis-diamminedichloroplatinum(II) with DNA—Formation, identification, and quantification. *Biochemistry* **24**:707–713.

Grimaldi, K., McAdam, S., Souhami, R., and Hartley, J. (1994). DNA damage by anti-cancer agents resolved at the nucleotide level of a single copy gene: Evidence for a novel binding site for cisplatin in cells. *Nucleic Acids. Res.* **22**:2311–2317.

Hartley, J., Gibson, N., Kohn, K., and Mattes, W. (1986). DNA sequence selectivity of guanine-N7 alkylation by three antitumor chlorethylating agents. *Cancer Res.* **46**:1943–1947.

Kunala, S., and Brash, D. (1992). Excision repair of individual bases of the Escerichia coli *lacI* gene: Relation to mutation hot spots and transcription coupling activity. *Proc. Natl. Acad. Sci. USA* **89**:11031–11035.

Lobanenkov, V., Plumb, M., Goodwin, G., and Grover, P. (1986). The effect of neighbouring bases on G-specific DNA cleavage mediated by treatment with the anti-diol epoxide of benzo[a]pyrene in vitro. *Carcinogenesis* **7**:1689–1695.

Marmur, J. (1961). A procedure for the isolation of deoxyribonucleic acid from micro-organisms. *J. Mol. Biol.* **3**: 208–218.

Mattes, W., Hartley, J., and Kohn, K. (1986). DNA sequence selectivity of guanine-N7 alkylation by nitrogen mustards. *Nucleic Acids Res.* **14**:2971–2987.

Ponti, M., Forrow, S., Souhami, R., D'Incalci, M., and Hartley, J. (1991). Measurement of the sequence specificity of covalent DNA modification by antineoplastic agents using *taq* DNA polymerase. *Nucleic Acids. Res* **19**: 2929–2933.

Puisieux, A., Lim, S., Groopman, J., and Ozturk, M. (1991). Selective targeting of *p53* gene mutational hotspots in human cancers by etiologically defined carcinogens. *Cancer Res.* **51**:6185–6189.

Roberts, J., and Friedlos, F. (1981). Quantitative aspects of the formation and loss of DNA interstrand crosslinks in Chinese-hamster cells following treatment with cis-diamminedichloroplatinum(II) (cisplatin). 1. Proportion of DNA–platinum reactions involved in DNA crosslinking. *Biochim. Biophys. Acta* **655**:146–151.

Smellie, M., Grimaldi, K., Bingham, J., McAdam, S., Thompson, A., Thurston, D., and Hartley, J. (1994). Cross-linking and sequence specific binding in isolated DNA and intact cells by C8-linked pyrrolobenzodiazepine dimers DSB-120 and AT-486. *Proc. Am. Assoc. Cancer Res.* **35**:534.

Tessier, D., Brousseau, R., and Vernet, T. (1986). Ligation of single stranded oligodeoxyribonucleotides by T4 RNA ligase. *Anal. Biochem* **158**:171–178.

Tornaletti, S., and Pfeifer, G. (1994). Slow repair of pyrimidine dimers of *p53* mutation hotspots in skin cancer. *Science* **263**:1436–1438.

Part II

Technologies for Detection of Mutations

Chapter 18

Heterodupiex Analysis

Damjan Glavač and Michael Dean

1. INTRODUCTION

Many of the current methods used for mutation detection in genomic DNA rely on conformational changes in either double- or single-stranded DNA structure. Conformations can be distinguished by electrophoresis due to their different response to an electrical field. The rate of migration or mobility of DNA molecules through the electrical field depends not only on the shape of molecules, but also on the ionic strength, viscosity, and temperature of the medium in which the molecules are moving. For example, the single-stranded conformation polymorphism (SSCP) technique (Orita *et al.*, 1989) depends on the altered mobility of single-stranded molecules and is a powerful diagnostic tool for detecting sequence variations.

Double-stranded mismatched structures are called heteroduplexes and several techniques have been developed to distinguish hetero- and homoduplexes such as RNase A cleavage (Myers *et al.*, 1985), denaturing gradient gel electrophoresis (DGGE) (Myers *et al.*, 1985; Myers *et al.*, 1987; Sheffield *et al.*, 1989), chemical mismatch cleavage (CMC) (Cotton *et al.*, 1988), conformation-sensitive gel electrophoresis (Ganguly and Prockop, 1990), and cleavage with bacteriophage resolvases (Mashal *et al.*, 1995; Youil *et al.*, 1995). Heteroduplex analysis (HA) is a technique in which heteroduplex structures can be directly detected on polyacrylamide gels because they migrate slower than their corresponding homoduplexes. Nagamine *et al.* (1989) observed "a PCR artifact," the generation of heteroduplexes, when homologous loci differing by an 18-bp deletion were amplified and noted their potential diagnostic use. Subsequently, it was shown that this technique may prove useful for the identification of heterozygotes carrying insertion/deletion mutations (Triggs-Raine and Gravel, 1990, Olds *et al.*, 1993). The first attempts to explore the possibility that single-base mismatches in double-stranded DNA can be detected as differential mobility from homoduplexes were not successful. Bhattacharyya and Lilley (1989) were not able to detect any differences for 12 different single-base mismatches. Subsequently, it was shown that single-base heteroduplexes can also be resolved either on an alternative gel matrix (Keen *et al.*, 1991) or under optimized electrophoretic conditions (White *et al.*, 1992).

Heteroduplex structures are generated during polymerase chain reaction (PCR) amplification in late cycles when two homologous DNA segments, or alleles, which differ in sequence are

Damjan Glavač • Institute of Pathology, Medical Faculty, Ljubljana, Slovenia. **Michael Dean** • Laboratory of Viral Carcinogenesis, National Cancer Institute, Frederick Cancer Research and Development Center, Frederick, Maryland 21702.

Technologies for Detection of DNA Damage and Mutations, edited by Gerd P. Pfeifer, Plenum Press, New York, 1996.

amplified. In HA mismatches are formed purposely by crossmatching with wild-type samples (i.e., mixing of equal amounts of PCR product of mutant and wild-type DNAs together, denaturing, and slowly cooling). In many cases these DNA species are resolvable on polyacrylamide gels because heteroduplex molecules are retarded in the gel (Fig. 18.1).

In recent years several variations of the original protocols were developed to make HA more sensitive and to screen for sequence variations in disease genes. Sensitivity can be increased by mixing samples with a driver molecule containing a deletion or by using mildly denaturing conditions to enhance the tendency of single-base mismatches to produce conformational changes (conformation-sensitive gel electrophoresis). In addition, it has been shown that highly specific cleavage of heteroduplexes can be accomplished with bacteriophage resolvases.

Because of the simplicity and nonisotopic means of detection, HA techniques have been used in various fields of molecular biology and medicine. Heteroduplex analysis has been applied both to the discovery of new mutations in disease genes and to the detection of known alterations (Table I). One of the first genes studied with HA was the cystic fibrosis transmembrane conductance regulator (CFTR) gene. The most common mutation in the CFTR gene is a 3-bp deletion, ΔF508 (Kerem *et al.*, 1989), which is principally detected by HA (Rommens *et al.*, 1990). The ΔF508 mutation is a natural heteroduplex generator, and other mutations in this exon can be detected simultaneously (Highsmith, 1993; Ravnik-Glavac *et al.*, 1994). Heteroduplex analysis alone or combined with SSCP has been used to detect several of the known CF mutations (White *et al.*, 1990, 1992; Cuppens *et al.*, 1992; Leoni *et al.*, 1992; Glavač and Dean, 1993; Claustres *et al.*, 1993; Meitinger *et al.*, 1993; Chevalier-Porst *et al.*, 1993; Ravnik-Glavac *et al.*, 1994). Approaches for rapid screening of CF mutations using HA have also been described (Dodson and Kant, 1991). HA has also been used for the screening for mutations in a number of other disease genes (Table I).

The HLA locus on chromosome 6p is probably the most polymorphic region of the human genome. HA has been adopted for tissue matching of HLA class II genes (Bidwell *et al.*, 1993; Zimmerman *et al.*, 1993). It has been used for HLA-DR4 subtyping (Sorrentino *et al.*, 1992a). In combination with SSCP and RFLP methods, HA has been employed for HLA-DPB1 typing (Sorrentino *et al.*, 1992b), and for mapping of recombinants within the HLA class II region (Carrington *et al.*, 1992).

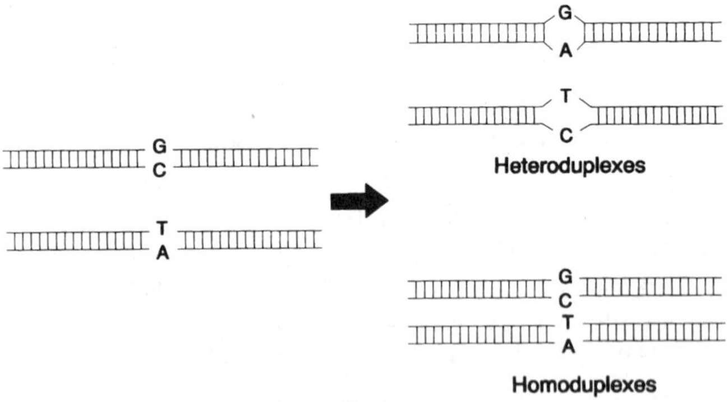

Figure 18.1. The principle of PCR heteroduplex analysis. Heteroduplexes are formed by the denaturation and annealing of two DNA molecules that differ in DNA sequence. Four species of molecules are generated, two heteroduplexes and two homoduplexes. Mismatched bases cause a bulge or bend in the DNA at the position of the mismatch and retard the heteroduplex molecules relative to the homoduplexes.

Table I
Genes Screened for Mutations by Heteroduplex Analysis

Gene	Literature
CFTR gene	White *et al.* (1990), Cuppens *et al.* (1992), Leoni *et al.* (1992), Claustres *et al.* (1993), Meitinger *et al.* (1993), Chevalier-Porst *et al.* (1993), Ravnik-Glavač *et al.* (1994), Dodson and Kant (1991)
Rhodopsin gene	Artlich *et al.* (1992), Inglehearn *et al.* (1992), Gayther *et al.* (1994)
Dystrophin gene	Saad *et al.* (1992), Prior *et al.* (1993b, 1994), Tsukamoto *et al.* (1994)
Neurofibromatosis type 1	Shen *et al.* (1993), Upadhyaya *et al.* (1992), Legius *et al.* (1994)
Neurofibromatosis type 2	Sainz *et al.* (1994), Bourn *et al.* (1994)
Adenomatous polyposis coli	Paul *et al.* (1993), Gayther *et al.* (1994), Mandl *et al.* (1994), Hamzehloei *et al.* (1994), Paffenholz *et al.* (1994)
Lactate dehydrogenase A (M)	Maekawa *et al.* (1994)
Human peripherin/RDS	Meins *et al.* (1993)
Tyrosinase	Fukai *et al.* (1994)
Procollagen type II	Winterpacht *et al.* (1994)
hMSH2	Mary *et al.* (1994)
α-Tocopherol	Ouahchi *et al.* (1995)
Pax-3	Tassabeji *et al.* (1992)
Microphthalmia	Tassabeji *et al.* (1994)
Phenylketonuria	Wood *et al.* (1993a)
Sickle-cell disease	Wood *et al.* (1993b)
Thromboembolic disease	Perry and Carrell (1992), Olds *et al.* (1993)
Glycogen storage disease type II	Hule *et al.* (1994)
Hemorrhagic telangiectasia type 1	McAllister *et al.* (1994)

1.1. HA and SSCP

It was shown that in many cases heteroduplexes are resolvable under the same conditions used for the SSCP technique. Therefore, there are numerous cases where the formation of heteroduplexes has been detected on gels used for SSCP screening. Examples where SSCP was used in combination with HA include the CFTR gene (White *et al.*, 1992; Claustres *et al.*, 1993; Ravnik-Glavač *et al.*, 1994; Cuppens *et al.*, 1992; Chevalier-Porst *et al.*, 1993; Leoni *et al.*, 1992; Meitinger et al, 1993), the adenomatous polyposis coli (APC) gene (Mandl *et al.*, 1993; Paul *et al.*, 1993), the neurofibromatosis 1 (NF1) gene (Upadhyaya *et al.*, 1992; Legius *et al.*, 1994), the neurofibromatosis type 2 (NF2) gene (Sainz *et al.*, 1994), and the lactate dehydrogenase A(M) gene (Maekawa *et al.*, 1994).

2. METHODS

Sequences that contain deletions/insertions or multiple mismatches often form heteroduplexes with their wild-type counterparts, phenomena that have been used as a simple way to detect these differences. Figure 18.1 shows the principle of heteroduplex analysis. Heteroduplex DNA may form during the late cycles in PCR amplification when two homologous DNA segments, or alleles, which have sequence differences are amplified. To enhance the formation of heteroduplex structures, wild-type and mutant (investigated) samples from PCR-amplified DNA are purposely crossmatched by mixing them in equal amounts of PCR-amplified DNA followed by a denaturation and reannealing. Four distinct species are generated by this reassortment: wild-type homoduplex, mutant homoduplex, and two different heteroduplexes. Ideally, all four species should be resolvable in electrophoretic analysis. However, in most cases only heteroduplex molecules are distinguishable from homoduplexes due to retardation in the gel.

2.1. Materials

1. 1× TBE electrophoresis buffer (90 mM/boric acid, 2.5 mM EDTA).
2. Acrylamide stock solution: 40% acrylamide (37.5:1 acrylamide:bisacrylamide). For 500 ml, dissolve 200 g electrophoresis-grade acrylamide and 5.35 g bisacrylamide in water to 500 ml.
3. Ammonium persulfate stock (10%): for 5 ml, use 0.5 g ammonium persulfate to 5 ml with water.
4. TEMED (*N,N,N′,N′*-tetramethylethylenediamine).
5. Gel loading solution: (15% Ficoll, 10 mM Tris pH 7.8, 1 mM EDTA, 0.05% bromophenol blue, 0.05% xylene cyanol). For 50 ml: 7.5 g Ficoll, 250 μL 2 M Tris pH 7.8, 100 μL 500 mM EDTA, 0.25 g xylene cyanol, 0.25 g bromophenol blue, and water to 50.0 ml. Sucrose or glycerol can be used instead of Ficoll.
6. Glass plates, spacers, and combs for pouring gels
7. Gel electrophoresis apparatus
8. Power supply with constant power capability

2.2. Gel Preparation

1. The glass plates should be clean and free of dust. Wash the plates with warm soapy water, rinse with deionized water and then with ethanol, and air-dry or wipe dry with a paper towel. If the gel is particularly thin (less than 1 mm), silanization of one plate with Repel silane will facilitate separation of the gel.

   Silanization: Rinse the shorter glass plate with 5 ml of Repel silane (2% dimethyldichlorosilane in 1,1,1-trichloroethane). Distribute it uniformly with a paper towel and leave to dry. Finally, rinse the glass with distilled water to remove the trace of hydrochloric acid formed in the treatment. Avoid skin contact with the Repel silane solution.
2. Clamp the sides and bottom of the plates to form a seal, as for a DNA sequencing gel. Use only two spacers, one for each side of the gel. Adjust the spacers so that both extend the full length of the gel from top to bottom. Grease is not needed.
3. Prepare the volume of gel forming solution appropriate for your plates. A 40 × 35 cm × 0.4 mm gel requires about 75 ml of gel forming solution, 450 ul of 10% APS solution, and 45 ml of TEMED.
4. Pour the gel. At room temperature, the solutions should polymerize in 10 to 30 min. (Caution: Unpolymerized acrylamide or other vinyl-based gel solutions act as a neurotoxin. Wear plastic gloves when handling these solutions.) Allow the gel to polymerize for 30–60 min before electrophoresis. Note: in most cases, if the gel fails to polymerize, the APS solution is often at fault. It is preferable to use APS solutions that are no more than 2 weeks old.
5. Mount the plates on the electrophoresis apparatus and prepare sufficient 1× TBE buffer to fill both the upper and lower buffer chambers. Use 100 ml 10× TBE and 900 ml deionized water to make 1 liter of 1× TBE buffer.

2.3. Sample Preparation and Electrophoresis

The products of the PCR should be evaluated for purity by electrophoresis in an agarose gel. It is important to optimize PCR conditions in order to minimize unwanted side products which may interfere with the identification of heteroduplex bands.

1. Mix 5 μl of the PCR product to be analyzed with an equal amount of wild-type PCR product and incubate the reaction mixture for 3 min at 95°C; slowly cool to room temperature (20–30 min). Mix 5–7 μl of each mixture with 3 μl of standard loading dye.
2. Attach plates to gel electrophoresis apparatus. Fill upper buffer chamber with 1× TBE buffer. Using a syringe with a bent needle, carefully rinse the wells with electrophoresis buffer. Carefully load each sample (2–3 μl) into a well using a drawn-out pipet tip. The optimal amount of DNA per lane is approximately 100 to 200 ng (usually 5–10% of total PCR volume). A positive control containing a known heteroduplex and a negative control containing only homoduplex DNA should be run on every gel. Appropriate DNA size markers can also be included on each gel.
3. Using a power supply, run the gel. To determine approximate run time it is useful to add to the loading buffer a marker dye. Usually two markers are added to the loading buffer. Xylene cyanol (light blue) comigrates with a 200-bp fragment while bromophenol blue (darker blue) at 70 bp on a 6% acrylamide gel.
4. Detach gel plates from electrophoresis apparatus. Stain gel attached to plate for 10 to 30 min in 0.5 μg/ml ethidium bromide. If necessary, soak in water for 10 to 30 min to destain. Document separation under UV and photograph. Alternatively, any DNA silver staining protocol can be used for detection (Blum *et al.*, 1987).

Considerable variation in the migration and resolution of the bands is observed under different electrophoretic conditions. Cold room gels are typically run for 4–12 hr, and room temperature gels are run for 12–30 hr, depending on the size of the fragment. The bromophenol blue dye should be run at least 30 cm into gels which are 40 cm long and 0.4 or 1 mm thick. For a 300-bp-long fragment, around 12 hr at 1000 V is needed for 30-cm migration; for a 900-bp fragment, this time is doubled. Generally, researchers have reported a decrease in sensitivity with increasing size of the fragment.

There are at least four factors affecting formation of heteroduplexes and their stability: the type of mismatch in the fragment, its position and neighboring sequence context, and the size of the fragment. The efficiency of detection is not the same for all possible types of single-base mismatches as the sequence surrounding the mismatch affects the mobility. Because of the combinations of these parameters it is difficult to plan electrophoretic conditions under which all heteroduplex molecules will separate from homoduplexes. Usually, large insertions or deletions (>3 bp) create stable heteroduplexes that are easily detected. Electrophoretic conditions involving heteroduplexes with single-base substitutions should be carefully selected and optimized. DNA fragment sizes from 100 to 500 bp with central mismatches are optimal for separation although in some cases good separation has been observed in PCR fragments up to 1 kb.

In comparison with single-strand conformers observed in the SSCP technique, double-stranded conformations seem less sensitive to environmental changes such as temperature, gel composition, ionic strength, and additives. The structure of the gel matrix can be varied by altering the concentration of acrylamide (% of T) and/or bisacrylamide in the gel (% of C). For optimization, 5% to 10% polyacrylamide gels in 1 × TBE buffer with variable cross-linking from 1% to 3% can be used. Low cross-linking (below 3% C) yields "long fiber gels" with increased pore size.

New gel matrices (Hydrolink and MDE from AT Biochem, Malvern, PA) have become available which enhance the ability to detect mutations in heteroduplex molecules. As for polyacrylamide gels, these gel matrices are poured between glass plates. Keen *et al.*, (1991) reported an increased sensitivity to single-base mismatches when Hydrolink D-5000 was substituted for acrylamide. An improved formulation of D-5000, MDE, was also used for the detection of mutations (Molinari *et al.*, 1993). Prior *et al.* (1993a) have demonstrated that a high percentage of known polymorphisms in the dystrophin gene can be detected by MDE.

For optimal sensitivity and reproducibility of results the temperature should be constant and controlled during electrophoresis. Temperature may be controlled with a device that precisely maintains gel temperature during electrophoresis by providing feedback control to the power supply. However, electrophoresis at constant power at either at room temperature or in the cold room is adequate. It has been found empirically that the presence of certain additives in the gel can create a mildly denaturing electrophoretic environment and increased the separation of heteroduplexes (White *et al.*, 1992; Ganguly *et al.*, 1993). The most commonly used additives in HA are urea, formamide, glycerol, and ethylene glycol. The choice of type of additive and its concentration in the gel is rather arbitrary and should be in most cases optimized for a given DNA fragment. Similarly, a mildly denaturing electrophoretic environment can be generated by elevated temperatures or with temperature gradients.

3. DISCUSSION

3.1. Structure and Stability of Heteroduplexes

Single-base mismatches occur where two non-Watson–Crick bases are in apposition. According to the definition of Bhattacharyya and Lilley (1989), multiple mismatches are referred to as a "bubble" and a mismatch with an additional base or bases on one side of the double helix as a "bulge." Mismatched molecules containing bulges, due to insertions or deletions, have rigidly defined structures and exhibit increased mobility in gels consistent with highly bent structures. The bent nature of this kind of heteroduplex has been confirmed by electron microscopic examinations of the ΔF508/wild-type heteroduplexes from cystic fibrosis carriers (Wang *et al.*, 1992). Little is known about the structure of the 12 possible single-base mismatches and about their effect on the conformation of double-stranded DNA. On the basis of NMR and X-ray data (Woodson and Crothers, 1988; Joshua-Tor *et al.*, 1992), it is believed that single-base mismatches are not very disruptive and that they are accommodated within the double helix with only local perturbation of the helical structure. In HA single mismatched bases are often not electrophoretically distinguishable from homoduplexes.

3.2. Sensitivity and Reliability of HA

Current methods for direct detection of mutations in disease genes are not 100% sensitive. Any mutation detection method used as a diagnostic tool must be sensitive and reliable. Known scanning methods such as DGGE, CCM, SSCP, and RNase protection are effective in detection of mutations on smaller fragments, from 100 to 400 bp. For HA only a limited amount of published data is available to estimate sensitivity. Generally, insertion or deletions can be detected on virtually any acrylamide gel. Since single-base changes are the most common type of mutation in the human genome, most of the variations of the heteroduplex technique have been directed to increasing the sensitivity of detection of single-base changes.

The sequence context of a base mismatch has an important effect on ease of detection by any physical, chemical, or enzymatic method. It was shown by Ganguly *et al.* (1993) that a heteroduplex containing a C:T mismatch was detected by differential migration of the heteroduplex when the cytosine was in the sense strand but not detected when the cytosine was in the antisense strand. It was also established that the mismatches are difficult to detect if they are within 50 bp of the end of the DNA fragment or if they are in a high-melting-temperature domain (Lerman and Silverstein, 1987).

3.3. Variations of HA

Several variations to increase the sensitivity and reliability of HA have been developed especially for diagnostic purposes. For example, the sensitivity of HA can be increased if a sample with a deletion mutant instead of the wild type is mixed with the investigated sample. This principle was used successfully with small 20% acrylamide gels in combination with silver staining (van den Akker *et al.*, 1992). In another approach (Guldberg and Güettler, 1993) the investigated sample was first mixed with a wild-type mutant and then with one or more control mutant samples. If the two mutant fragments were not identical, additional mismatched heteroduplexes were formed and revealed novel bands on the gel. Another variation involves the use of universal heteroduplex generators (UHG). A natural heteroduplex generator (ΔF508 in exon 10 of the CFTR gene) can be employed to detect other mutations in the same exon. Similarly, a UHG is constructed and mixed with the investigated samples. The major advantage of this approach is that a single UHG can detect known mutations within any given locus covered also by the UHG sequence (Bidwell *et al.*, 1993; Wood *et al.*, 1993a; Wood *et al.*, 1995). This method is useful only if the mutations are reasonably clustered in a relatively short fragment. Similarly, individual mutations may be detected by crossmatching PCR products with cloned naturally occurring pseudogenes containing small deletions (Zimmerman *et al.*, 1993).

Olivas and Maher (1994) have developed a method of oligonucleotide-directed DNA triple helix formation for the detection of alterations within homopurine sequences. This method might complement other assays and may be of value when screening many DNA samples for changes involving particular homopurine sites.

3.4. Conformation-Sensitive Gel Electrophoresis (CSGE)

Because bent mismatched structures exhibit large gel retardation, Ganguly *et al.* (1993) used a system of mildly denaturing solvents to enhance the tendency of single-base mismatches to produce conformational changes. A standard 6% polyacrylamide gel polymerized in 10% ethylene glycol/15% formamide/Tris-taurine buffer was used to increase the differential migration of DNA heteroduplexes and homoduplexes. Of the 12 possible single-base mismatches, they were able to resolve 60 of 68 single-base mismatches in some 59 different sequence contents. However, single-base mismatches within 50 bp of one end of a heteroduplex or in high-melting domains were not resolvable. A simplified CSGE protocol using bisacrolylpiperazine (BAP) as a cross-linker has recently been described (Williams *et al.*, 1995).

Recently another variation of HA was developed which relies on the ability of enzymes (bacteriophage resolvases) to recognize sequence mismatches within heteroduplexes. Bacteriophage resolvases T4 endonuclease VII and T7 endonuclease were used for detecting mutations in heteroduplexes. All four classes of possible single nucleotide mismatches were cleaved. One advantage of mismatch cleavage is that information about the location of the mutation is obtained (Dean, 1995).

3.5. HA and SSCP

Both HA and SSCP use the same PCR amplification products and are analyzed using the same or modified electrophoretic conditions. The sensitivity and convenience of these combined methods represents a powerful mutation detection system.

REFERENCES

Artlich, A., Horn, M., Lorenz, B., Bhattacharyya, S., and Gal, S. (1992). Recurrent 3-bp deletion at codon 255/256 of the rhodopsin gene in a German pedigree with autosomal dominant retinitis pigmentosa. *Am. J. Hum. Genet.* **50**:876–878.

Bhattacharyya, A., and Lilley, D. M. J. (1989). The contrasting structures of mismatched DNA sequences containing looped-out (bulges) and multiple mismatches (bubbles). *Nucleic Acids Res.* **17**:6821–6840.

Bidwell, J. L., Clay, T. M., Wood, N. A. P., Pursall, M. P., Martin, A. F., Bradley, B. A., and Hui, K. M. (1993). Rapid HLA-DR-Dw and DP matching by PCR fingerprinting and related DNA heteroduplex technologies, in: *Handbook of HLA Typing Techniques* (K. M. Hui and J. L. Bidwell, eds.), CRC Press, Boca Raton, pp. 99–116.

Blum, H., Beier, H., and Gross, H. J. (1987). Improved silver staining of plant proteins, RNA and DNA in polyacrylamide gels. *Electrophoresis* **8**:93–99.

Bourn, D., Carter, S. A., Mason, S., Evans, D. G. R., and Stachan, T. (1994). Germline mutations in the neurofibromatosis type 2 tumour suppressor gene. *Hum. Mol. Genet.* **3**:813–816.

Carrington, M., White, M. B., Dean, M., Mann, D., and Ward, F. E. (1992). The use of DNA heteroduplex patterns to map recombination within HLA class II region. *Hum. Immunol.* **33**:114–121.

Chevalier-Porst, F., Mathieu, M., and Bozon, D. (1993). Identification of three rare frameshift mutations in exon 13 of the cystic fibrosis gene: 1918delGC, 2118del4, and 2372del8. *Hum. Mol. Genet.* **2**:1071–1072.

Claustres, M., Laussel, M., Desgeorges, M., Giansily, M., Culard, J.-F., Razakatsara, G., and Demaille, J. (1993). Analysis of the 27 exons and flanking regions of the cystic fibrosis gene: 40 different mutations account for 91.2% of the mutant alleles in Southern France. *Hum. Mol. Genet.* **2**:1209–1213.

Cotton, R. G. H., Rodrigues, N. R., and Campbell, D. R. (1988). Reactivity of cytosine and thymine in single base pair mismatches with hydroxylamine and osmium tetroxide and its application to the study of mutations. *Proc. Natl. Acad. Sci. USA* **85**:4397–4401.

Cuppens, H., Loumi, O., Marynen, P., and Cassiman, J.-J. (1992). Identification of a new frameshift mutation and a duplication polymorphism in the CFTR gene in the Algerian population. *Hum. Mol. Genet.* **1**:283–284.

Dean, M. (1995). Resolving DNA mutations. *Nature Genet.* **9**:103–104.

Dodson, L. A., and Kant, J. A. (1991). Two temperature PCR and heteroduplex detection: Application to rapid cystic fibrosis screening. *Mol. Cell. Probes* **5**:21–25.

Friedl, W., Mandl, M., and Sengteller, M. (1993). Single-step screening method for the most common mutations in familial adenomatous polyposis. *Hum. Mol. Genet.* **2**:1481–1482.

Fukai, K., Holmes, S. A., Lucchese, N. J., Siu, V. M., Weleber, R. G., Schnur, R. E., and Spritz, R. A. (1994). Autosomal recessive ocular albinism associated with a functionally significant tyrosinase gene polymorphism. *Nature Genet.* **9**:92–95.

Ganguly, A., and Prockop, D. J. (1990). Detection of single-base mutations by reaction of DNA heteroduplexes with a water-soluble carbodiimide followed by primer extension: Application to products from the polymerase chain reaction. *Nucleic Acids Res.* **18**:3933–3939.

Ganguly, A., Rock, M. J., and Prockop, D. J. (1993). Conformation-sensitive gel electrophoresis for rapid detection of single-base differences in doublestranded PCR products and DNA fragments: Evidence for solvent-induced bends in DNA heteroduplexes. *Proc. Natl. Acad. Sci. USA* **90**:10325–10329.

Gayther, S. A., Wells, D., SenGupta, S. B., Champan, P., Neale, K., Tsioupra, K., and Delhanty, J. D. A. (1994). Regionally clustered APC mutations are associated with a severe phenotype and occur at a high frequency in new mutation cases of adenomatous polyposis coli. *Hum. Mol. Genet.* **3**:53–56.

Glavač, D., and Dean, M. (1993). Optimization of the single strand-conformation polymorphism (SSCP) technique for detection of point mutations. *Hum. Mutat.* **2**:404–414.

Guldberg, P., and Güttler, F. (1993). A simple method for identification of point mutations using denaturing gradient gel electrophoresis. *Nucleic Acids Res.* **9**:2261–2262.

Hamzehloei, T., West, S. P., Chapman, P. D., Burn, J., and Curtis, A. (1994). Four novel germ-line mutations in the APC gene detected by heteroduplex analysis. *Hum. Mol. Genet.* **3**:1023–1024.

Highsmith, W. E. (1993). Carrier screening for cystic fibrosis. *Clin. Chem.* **39**:706–707.

Hule, M. L., Chen, A. S., Brooks, S. S., Grix, A., and Hirschhorn, R. (1994). A de novo 13 nt deletion, a newly identified C647W missense mutation and a deletion of exon 18 in infantile onset glycogen storage disease type II (GSDII). *Hum. Mol. Genet.* **3**:1081–1087.

Inglehearn, C. F., Keen, T. J., Bashir, R., Jay, M., Fitzke, M., Bird, A. C., Crombie, A., and Bhattacharyya, S. (1992). A completed screen for mutations of the rhodopsin gene in a panel of patients with autosomal dominant retinitis pigmentosa. *Hum. Mol. Genet.* **1**:41–45.

Joshua-Tor, L., Frolow, F., Appella, E., Hope, H., Rabinovich, D., and Sussman, J. L. (1992). Three-dimensional structures of bulge-containing DNA fragments. *J. Mol. Biol.* **225**:397–431.

Keen, J., Lester, D., Inglehearn, C., Curtis, A., and Bhattacharyya, S. (1991). Rapid detection of single base mismatches as heteroduplexes on Hydrolink gels. *Trends Genet.* **7**:5.

Kerem, B. S., Rommens, J. M., Buchanan, J. A., Markiewicz, D., Cox, T. K., Chakravarti, A., Buchwald, M., and Tsui, L.-C. (1989). Identification of the cystic fibrosis gene: Genetic analysis. *Science* **245**:1073–1080.

Legius, E., Hail, B. K., Wallace, M. R., Collins, F. S., and Glover, T. W. (1994). Ten base pair duplication in exon 38 of the NFl gene. *Hum. Mol. Genet.* **3**:829–830.

Leoni, G. B., Rosatelli, M. C., Cossu, G., Pischedda, M. C., De Virgilliis, S., and Cao, A. (1992). A novel cystic fibrosis mutation: Deletion of seventeen nucleotides at the exon 10–intron 10 boundary of the CFTR gene, in a Sardinian patient. *Hum. Mol. Genet.* **1**:83–84.

Lerman, L. S., and Silverstein, V. (1987). Computational simulation of DNA melting and its application to denaturing gradient gel electrophoresis. *Methods Enzymol.* **155**:482–501.

McAllister, K. A., Grogg, K. M., Johnson, D. W., Gallione, C. J., Baldwin, M. A., Jackson, C. E., Helmbold, E. A., Markel, D. S., McKinnon, W. C., Murrell, J., McCormick, M. K., Pericak-Vance, M. A., Heutink, P., Oostra, B. A., Haitjema, T., Westerman, C. J. J., Porteous, M. E., Guttmacher, A. E., Letarte, M., and Marchuk, D. A. (1994). Endoglin, a TGF-beta binding protein of endothelial cells, is the gene for hereditary haemorrhagic telangiectasia type 1. *Nature Genet.* **8**:345–351.

Maekawa, M., Sudo, K., Kanno, T., Takayasu, S., Li, S. S.-L., Kitajima, M., and Matsuura, Y. (1994). A novel deletion mutation of lactate dehydrogenase A(M) gene in the fifth family with the enzyme deficiency. *Hum. Mol. Genet.* **3**:825–826.

Mandl, M., Paffenholz, R., Friedl, W., Caspari, R., Sengteller, M., and Propping, P. (1993). Frequency of common and novel inactivation APC mutations in 202 families with familial adenomatous polyposis. *Hum. Mol. Genet.* **3**:825–826.

Mandl, M., Kadmon, M., Sengteller, M., Caspari, R., Propping, P., and Fiedl, W. (1994). A somatic mutation in the adenomatous polyposis coli (APC) gene in peripheral blood cells—Implications for predictive diagnosis. *Hum. Mol. Genet.* **3**:1009–1011.

Mary, J.-L. M., Bishop, T., Kolodner, R., Lipford, J. R., Kane, M., Weber, W., Torhorst, J., Mueler, H., Spycher, M., and Scott, R. J. (1994). Mutational analysis of the hMSH2 gene reveals a three base pair deletion in a family predisposed to colorectal cancer development. *Hum. Mol. Genet.* **3**:2067–2069.

Mashal, R. D., Koontz, J., and Sklar, J. (1995). Detection of mutations by cleavage of DNA heteroduplexes with bacteriophage resolvases. *Nature Genet.* **9**:177–183.

Meins, M., Grüning, G., Blankenagel, A., Krastel, H., Reck, B., Fuchs, S., Schwinger, E., and Gal, A. (1993). Heterozygous 'null allele' mutation in the human peripherin/RDS gene. *Hum. Mol. Genet.* **2**:2181–2182.

Meitinger, T., Golla, A., Dörner, C., Deufel, A., Aulehla-Scholz, A., Boehm, I., Reinhardt, D., and Deufel, T.(1993). In frame deletion (deltaF311) within a short trinucleotide repeat of the first transmembrane region of the cystic fibrosis gene. *Hum. Mol. Genet.* **2**:2173–2174.

Molinari, R. J., Conners, M., and Shorr, R. G. L. (1993). Hydrolink gels for electrophoresis, in: *Advances in Electrophoresis*, Volume 6 (A. Chrambach, M. J. Dunn, and B. J. Radola, eds.), VCH Publishers, New York, pp. 44–60.

Myers, R. M., Larin, Z., and Maniatis, T. (1985). Detection of single base substitutions by ribonuclease cleavage at mismatches in RNA:DNA duplexes. *Science* **230**:1242–1249.

Myers, R.M., Maniatis, T., and Lerman, L. (1987). Detection and localization of single base changes by denaturing gradient gel electrophoresis. *Methods Enzymol.* **155**:501–527.

Nagamine, C. M., Chan, K., and Lau, Y.-F. C. (1989). A PCR artifact: Generation of heteroduplexes. *Am. J. Hum. Genet.* **45**:337–339.

Olds, R. J., Lane, D. A., Beresford, C. J., Abilgaard, U., Hughes, P. M., and Thein, S. L. (1993). A recurrent deletion in the antithrombin gene, AT106–109 (−6 bp), identified by DNA heteroduplex detection. *Genomics* **16**: 298–299.

Olivas, M. W., and Maher, L. J. (1994). Analysis of duplex DNA by triple helix formation: Application to detection of a p53 microdeletion. *Biotechniques* **16**:128–132.

Orita, M., Suzuki, Y., Sekiya, T., and Hayashi, K. (1989). Rapid and sensitive detection of point mutations and DNA polymorphisms using the polymerase chain reaction. *Genomics* **5**:874–879.

Ouahchi, K., Arita, M., Kayden, H., Hentati, F., Hamida, M. B., Sokol, R., Arai, H., Ionoue, K., Mandel, J.-L., and Koenig, M. (1995). Ataxia with isolated vitamin E deficiency is caused by mutations in the alpha-tocopherol transfer protein. *Nature Genet.* **9**:141–145.

Paffenholz, R., Mandl, M., Caspari, R., Sengteller, M., Propping, P., and Friedl, W. (1994). Eleven novel germline mutations in the adenomatous polyposis coli (APC) gene. *Hum. Mol. Genet.* **3**:1703–1704.

Paul, P., Letteboer, T., Gelbert, L., Groden, J., White, R., and Coppes, M. J. (1993). Identical APC exon 15 mutations result in a variable phenotype in familial adenomatous polyposis. *Hum. Mol. Genet.* **2**:925–931.

Perry, D. J., and Carrell, R. W. (1992). Hydrolink gels: A rapid and simple approach to the detection of DNA mutations in thromboembolic disease. *J. Clin. Pathol.* **45**:158–160.

Prior, T. W., Papp, A. C., Snyder, P. J., and Sedra, M. S. (1993a). Detection of an exon 53 polymorphism in the dystrophin gene. *Hum. Genet.* **92**:302–304.

Prior, T. W., Papp, A. C., Snyder, P. J., Burghes, A. H. M., Sedra, M. S., Western, L. M., Bartolo, C., and Mendell, J. R. (1993b). Exon 44 nonsense mutation in two Duchenne muscular dystrophy brothers detected by heteroduplex analysis. *Hum. Mutat.* **2**:192–195.

Prior, T. W., Bartolo, C., Papp, A. C., Snyder, P. J., Sedra, M. S., Burghes, A. H. M., and Mendell, J. R. (1994). Identification of a missense mutation, single base deletion and a polymorphism in the dystrophin exon 16. *Hum. Mol. Genet.* **3**:1173–1174.

Ravnik-Glavač, M., Glavač, D., and Dean, M. (1994). Sensitivity of single-strand conformation polymorphism (SSCP) and heteroduplex method (HA) for mutation detection in the cystic fibrosis gene. *Hum. Mol. Genet.* **3**:801–807.

Rommens, J. M., Kerem, B.-S., Greer, W., Chang, P., Tsui, L.-C., and Ray, P. (1990). Rapid nonradioactive detection of the major cystic fibrosis mutation. *Am. J. Hum. Genet.* **46**:395–396.

Saad, F. A., Vitiello, L., Merlini, L., Mostacciuolo, M. L., Olivero, S., and Danieli, G. A. (1992). A 3′ consensus splice mutation in the human dystrophin gene detected by a screening for intra-exonic deletions. *Hum. Mol. Genet.* **1**:345–346.

Sainz, J., Huynh, D. P., Figuerosa, K., Ragge, N. K., Baser, M. E., and Pulst, S. M. (1994). Mutations of the neurofibromatosis type 2 gene and lack of the gene product in vestibular schwannomas. *Hum. Mol. Genet.* **3**:885–891.

Sheffield, V. C., Cox, D. R., Lerman, L. S., and Myers, R. M. (1989). Attachment of a 40-base pair G+C-rich sequence (GC-clamp) to genomic DNA fragments by the polymerase chain reaction results in improved detection of single-base changes. *Proc. Natl. Acad. Sci. USA* **86**:232–236.

Shen, M. H., Harper, P. S., and Upadhyaya, M. (1992). Neurofibromatosis type 1 (NF1): The search for mutations by PCR-heteroduplex analysis on Hydrolink gels. *Hum. Mol. Genet.* **1**:735–740.

Sorrentino, R., Iannicola, C., Costanzi, S., Chersi, A., and Roberto, T. (1991). Detection of complex alleles by direct analysis of DNA heteroduplexes. *Immunogenetics* **33**:118–123.

Sorrentino, R., Cascino, I., and Tosi, R. (1992a). Subgrouping of DR4 alleles by DNA heteroduplex analysis. *Hum. Immunol.* **33**:18–23.

Sorrentino, R., Potolicchio, I., Ferrara, G. B., and Tosi, R. (1992b). A new approach to HLA-DPB1 typing combining DNA heteroduplex analysis with allele-specific amplification and enzyme restriction. *Immunogenetics* **36**:248–254.

Tassabehji, M., Read, A. P., Newton, V. E., Harris, R., Balling, R., Gruss, P., and Strachan, T. (1992). Waardenburg's syndrome patients have mutations in the human homologue. *Nature* **355**:635–636.

Tassabehji, M., Newton, V. E., and Read, A. P. (1994). Waardenburg's syndrome type 2 caused by mutations in the human microphthalmia (MITF) gene. *Nature Genet.* **8**:251–255.

Triggs-Raine, B. L., and Gravel, R. A. (1990). Diagnostic heteroduplexes: Simple detection of carriers of a 4-bp insertion mutation in Tay-Sachs disease. *Am. J. Hum. Genet.* **46**:183–184.

Tsukamoto, H., Inui, K., Matsuoka, T., Yanagihara, I., Fukushima, H., and Okada, S. (1994). One base deletion in the cysteine-rich domain of the dystrophin gene in Duchenne muscular dystrophy patients. *Hum. Mol. Genet.* **3**:995–996.

Upadhyaya, M., Shen, M., Cherryson, Farnham, J., Maynard, J., Huson, S. M., and Harper, P. S. (1992). Analysis of mutations at the neurofibromatosis 1 (NF1) locus. *Hum. Mol. Genet.* **1**:735–740.

van den Akker, E., Braun, J. E. F., Pals, G., Lafleur, M. V. M., and Retel, J. (1992). Single base mutations can be unequivocally and rapidly detected by analysis of DNA heteroduplexes, obtained with deletion-mutant instead of wild-type DNA. *Nucleic Acids Res.* **24**:6745–6746.

Wang, Y.-H., Barker, P., and Griffith, J. J. (1992). Visualization of diagnostic heteroduplex DNAs from cystic fibrosis deletion heterozygotes provides an estimate of the kinking of DNA by bulged bases. *J. Biol. Chem.* **267**:4911–4915.

White, M. B., Amos, J., Hsu, J. M., Gerrard, B., Finn, P., and Dean, M. (1990). A frameshift mutation in the cystic fibrosis gene. *Nature* **344**:665–667.

White, M. B., Carvalho, M., Derse, D., O'Brien, S. J., and Dean, M. (1992). Detecting single base substitutions as heteroduplex polymorphisms. *Genomics* **12**:301–306.

Williams, C. J., Rock, M., Considine, E., McCarron, S., Gow, P., Ladda, R., McLain, D., Michels, V. M., Murphy, W., Prockop, D. J., and Ganguly, A. (1995). Three new point mutations in type II procollagen (COL2A1) and identification of a fourth family with the COL2A1 Arg519→Cys base substitution using conformation sensitive gel electrophoresis. *Hum. Mol. Genet.* **4**:309–312.

Winterpacht, A., Schwarze, U., Mundlos, S., Menger, H., Spranger, J., and Zabei, B. (1994). Alternative splicing as the result of a type II procollagen gene (COL2A1) mutation in a patient with Kniest dysplasia. *Hum. Mol. Genet.* **3**:1891–1893.

Wood, N., Tyfield, L., and Bidwell, J. (1993a). Rapid classification of phenylketonuria genotypes by analysis of heteroduplexes generated by PCR-amplifiable synthetic DNA. *Hum. Mutat.* **2**:131–137.

Wood, N., Standen, G., Hows, J., Bradley, B., and Bidwell, J. (1993b). Diagnosis of sickle-cell disease with a universal heteroduplex generator. *Lancet* **342**:1519–1520.

Wood, N., Standen, G., Old, J., and Bidwell, J. (1995). Genotyping for haemoglobins S and C by DNA crossmatching with a universal heteroduplex generator. *Hum. Mutat.* **5**:166–172.

Woodson, S. A., and Crothers, D. M. (1988). Structural model for an oligonucleotide containing a bulged guanosine by NMR and energy minimization. *Biochemistry* **27**:3130–3141.

Youil, R., Kemper, B., and Cotton, R. G. H. (1995). Screening for mutations by enzyme mismatch cleavage using T4 endonuclease VII. *Proc. Natl. Acad. Sci. USA* **92**:87–91.

Zimmerman, P. A., Carrington, M. N., and Nutman, T. B. (1993). Exploiting structural differences among heteroduplex molecules to simplify genotyping the DQA1 and DQB1 alleles in human lymphocyte typing. *Nucleic Acids Res.* **21**:4541–4547.

Chapter 19

Mutation Analysis by Denaturing Gradient Gel Electrophoresis (DGGE)

Riccardo Fodde and Monique Losekoot

1. INTRODUCTION

The identification and characterization of single nucleotide variations in DNA still represents a common technical obstacle in the detection of DNA damage as well as in the genetic analysis of inherited disorders. The introduction of the polymerase chain reaction (PCR) (Saiki *et al.*, 1988) has greatly facilitated this technical problem: the *in vitro* production of large amounts of the DNA target makes the application of tedious molecular cloning steps obsolete, and allows the direct determination of the nucleotide sequence of the amplified fragment. However, when approaching the analysis of a large DNA segment, it is often desirable to first perform a screening of overlapping PCR-amplified fragments to establish where the putative nucleotide variant is located. To this aim, several protocols are available based on some physical features of DNA or on the activity of different chemical agents or enzymes capable of recognizing mismatches in double-stranded molecules. Of these protocols, denaturing gradient gel electrophoresis (DGGE) allows the identification of point mutations which alter the melting behavior of the DNA fragment to be analyzed (Fischer and Lerman, 1979, 1983; Myers *et al.*, 1985a).

Here, we will review some of the theoretical and practical aspects of DGGE and its applications to the analysis of mutations in genomic DNA.

1.1. Theory of DGGE

DGGE is based on the electrophoretic mobility of a double-stranded DNA molecule through linearly increasing concentrations of denaturing agents such as formamide and urea. As the DNA fragment proceeds through the gradient gel, it will reach a position where the concentration of the denaturing agent equals the melting temperature (T_m) of its lowest melting domain causing its denaturation (branching) and the consequent marked retardation of its electrophoretic mobility

Riccardo Fodde and Monique Losekoot • MGC-Department of Human Genetics, Leiden University, Leiden, The Netherlands.

Technologies for Detection of DNA Damage and Mutations, edited by Gerd P. Pfeifer, Plenum Press, New York, 1996.

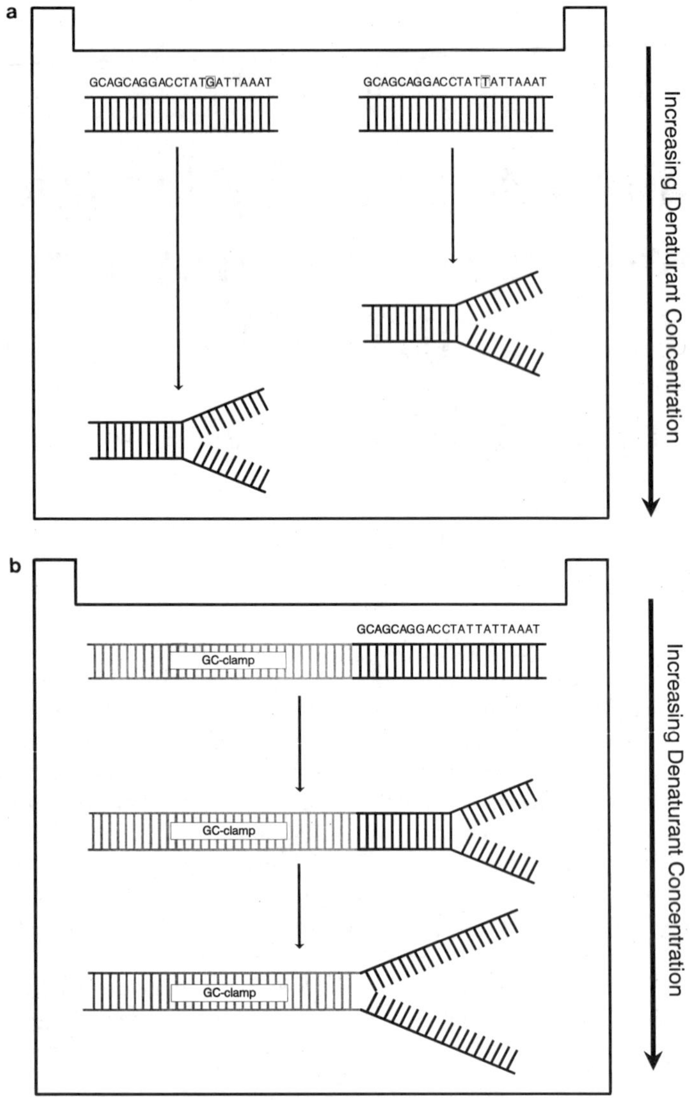

Figure 19.1. Theory of denaturing gradient gel electrophoresis. (a) Double-stranded DNA molecules differing by a single nucleotide in their lowest T_m domain will undergo denaturation (branching) at different positions along the denaturing gradient gel. (b) Introduction of a GC-rich domain, the so-called "GC-clamp," prevents complete denaturation of the fragment. (c) Enzymatic amplification by PCR of a genomic DNA sample heterozygous for a single base substitution results in the generation of two heteroduplexes and two homoduplexes which are resolved by DGGE.

(Fig. 19.1a). A *melting domain* is a region within the fragment in which all of the base pairs melt at approximately the same temperature. The *melting temperature* is here defined as the temperature at which each base pair of a DNA duplex is in perfect equilibrium between the denatured and helical state. Since stacking interactions between adjacent bases have a significant influence on the stability of the double helix, the T_m of any given DNA molecule is largely dependent on

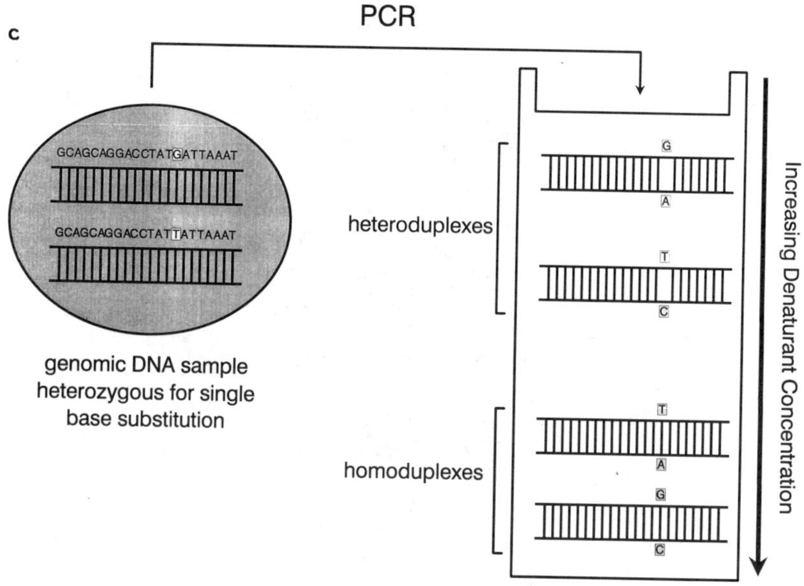

Figure 19.1. (*Continued*)

its nucleotide sequence. Therefore, when DNA fragments differing by a single nucleotide change in their lowest melting domain are electrophoresed through denaturing gradient gels, branching and consequent retardation of their mobility will occur at different positions along the gel allowing their separation (Fig. 19.1a) (Myers *et al.*, 1987).

1.2. GC-Clamps

In these experimental conditions, however, DGGE can only resolve a limited fraction of all of the possible point mutations. The majority of the DNA fragments of approximately 300–500 bp will in fact encompass more than one melting domain. Base variants located within the highest T_m domain will not be resolved by DGGE due to loss of sequence-dependent migration on complete strand dissociation. The problem can be overcome by introducing into the fragment to be analyzed a GC-rich domain (GC-*clamp*) whose high T_m will prevent complete dissociation of the two DNA strands (Myers *et al.*,1985b). Since a GC-clamp as short as 40 bp can efficiently serve as a high T_m domain for the DGGE analysis of most DNA fragments, one of the two PCR primers used to amplify the target DNA sequence can be designed with a 5′ GC-clamp tail so that it will be incorporated at one of the ends of the resulting PCR product (Sheffield *et al.*, 1989) (Fig. 19.1b). The introduction of the GC-clamp increases the percentage of mutations detectable by DGGE to close to 100% (Myers *et al.*, 1985b).

The PCR–DGGE combination is extremely powerful when applied to the detection of heterozygous nucleotide variants: continuous denaturation and reannealing of single strand molecules during PCR allows the formation of homoduplexes as well as hybrid heteroduplex molecules. The presence of a single mismatch within the latter greatly decreases their melting temperature allowing separation from the homoduplexes and an easier visual detection of the mutants. A total of four bands will be observed in a typical DGGE analysis of a DNA sample heterozygous for a single base alteration: two homoduplexes and two heteroduplexes (Figs. 19.1c and 19.2).

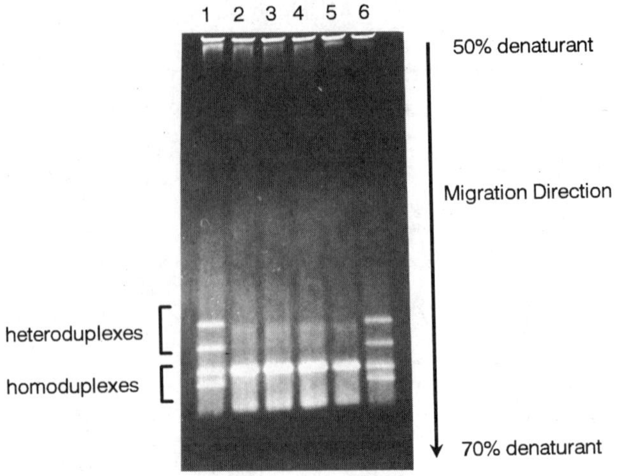

Figure 19.2. DGGE analysis of a β-globin mutation in a nuclear family affected by thalassemia. Samples 1 and 6 are heterozygous for the mutation and show the typical 4-band pattern.

1.3. Computer Simulation of Melting Behavior

An additional advantage of DGGE is represented by the possibility of simulating the melting behavior of any known DNA sequence by computer analysis. The programs MELT87 and SQHTX, as described by Lerman and Silverstein (1987), allow a preliminary examination of the properties of the DNA fragment to be analyzed: its melting map, the optimal gel running time, the expected effects of virtually any base change on the T_m map, and their consequences in terms of gradient displacement.

1.4. Research and Diagnostic Applications

Since its introduction, the PCR–DGGE protocol has been applied to the molecular analysis of a large number of genomic loci. In human molecular genetics, DGGE has proven particularly useful when approaching the analysis of inherited conditions caused by heterogeneous mutation spectra or by frequent *de novo* mutations. DGGE analysis of the β-globin gene in thalassemia patients has allowed the identification of a large number of different mutations in several populations (Cai and Kan, 1990; Losekoot *et al.*, 1990, 1991, 1992; Rosatelli *et al.*, 1992), and has offered a valid alternative to more conventional approaches based on allele-specific oligonucleotide (ASO) hybridization of PCR-amplified DNA (Saiki *et al.*, 1986).

DGGE has also proven to be very efficient in the detection of mutations in the genes coding for human clotting factor VIII (Kogan and Gitschier, 1990; Traystman *et al.*, 1990; Higuchi *et al.*, 1991a,b; Aly *et al.*, 1992) and IX (Attree *et al.*, 1989; Satoh *et al.*, 1993), responsible for hemophilia A and B, respectively. In these syndromes, the frequent occurrence of de novo mutations, the heterogeneous spectrum of molecular lesions, and the general paucity of intragenic or closely linked polymorphic markers have long hampered efficient diagnosis by linkage analysis and direct mutation detection (Hofker *et al.*, 1986; Antonarakis and Kazazian, 1988).

In cystic fibrosis (CF), where the most common mutation (ΔF508) accounts for about 70% of the CF cases worldwide, the remaining CF mutations are extremely rare and heterogeneous (more than 150 different mutations have been reported to date). Ferec *et al.* (1992) have characterized 98% of the CF mutations in 365 affected individuals by DGGE analysis, clearly indicating the potentials of this technique for such large-scale carrier detection and population screenings.

In cancer genetics, the identification of genetic alterations which lead through a multistep process to a malignant phenotype, has allowed the definition of important biological models such as the adenoma–carcinoma sequence in colorectal tumorigenesis (Fearon and Vogelstein, 1990). In these cases DGGE has been employed to rapidly detect germ-line and somatic mutations at "tumor-susceptibility" genes such as K-*ras*, *p53*, *hMSH2*, and APC, and to monitor their accumulation in tumor progression (Pellegata *et al.*, 1992; Fodde *et al.*, 1992; Olschwang *et al.*, 1993; Renault *et al.*, 1993; van den Broek *et al.*, 1993; Wijnen *et al.*, 1995).

In mutagenesis studies, DGGE has been successfully implemented to the establishment of the mutational spectra induced by different mutagens in vitro and in vivo (Thilly, 1985; Cariello and Thilly, 1986; Cariello *et al.*, 1990, 1991; Oller and Thilly, 1992; Keohavong and Thilly, 1992; Chen and Tilly, 1994).

2. METHODS

2.1. Setting Up DGGE

Once a decision has been made to analyze a given gene or any other DNA fragment by DGGE, some preliminary work is necessary to maximize the effectiveness of the strategy and to avoid predictable problems. First of all, the availability of the complete nucleotide sequence of the fragment(s) to be analyzed is desirable although not strictly essential (see Section 2.1.2).

2.1.1. Setting Up DGGE: Computer Predictions and Primer Design

As already mentioned, the MELT87 and SQHTX computer programs allow the prediction of the melting behavior of any DNA molecule (Lerman and Silverstein, 1987). A software package for the Macintosh® (MacMelt) is also commercially available from the Bio-Rad Laboratories. Such programs allow the identification of the different melting domains within the fragment and their specific T_m's. Based on this type of information, one can design PCR primers to encompass one or two melting domains within 100–500 bp. Three or more T_m domains within one PCR product should be avoided, since this usually results in a decreased sensitivity of mutation detection especially within the regions with the highest melting temperatures. Significant T_m differences between two melting domains within the same PCR fragment should also be avoided: denaturation and branching of the domain with the lowest T_m might cause retardation of the partially melted molecule to such a degree that it will not reach the position along the gel where the denaturant concentration is high enough to denature the second T_m domain. If unavoidable, these problems can be circumvented by (1) analyzing the same fragment on two different gradients (see below) or (2) by restriction enzyme digestion of the PCR product.

Computer simulation can also save both time and money: as shown in Fig. 19.3, the presence of the GC-clamp on either side of the DNA fragment might have profound effects on the T_m map and can affect the percentage of detectable changes (Myers *et al.*, 1985c). In most cases, the GC-clamp sequence described by Sheffield *et al.* (1989) should ensure prevention of complete denaturation. However, if unusually GC-rich sequences are being analyzed, longer "customized" clamps can be conceived and tested by computer simulation (Olschwang *et al.*, 1993).

2.1.2. Setting Up DGGE: Perpendicular Denaturing Gradient Gels

If the complete sequence of the fragment to be analyzed is not known (partial nucleotide sequences of the 5′ and 3′ ends are necessary to design the PCR primers), it is still possible to

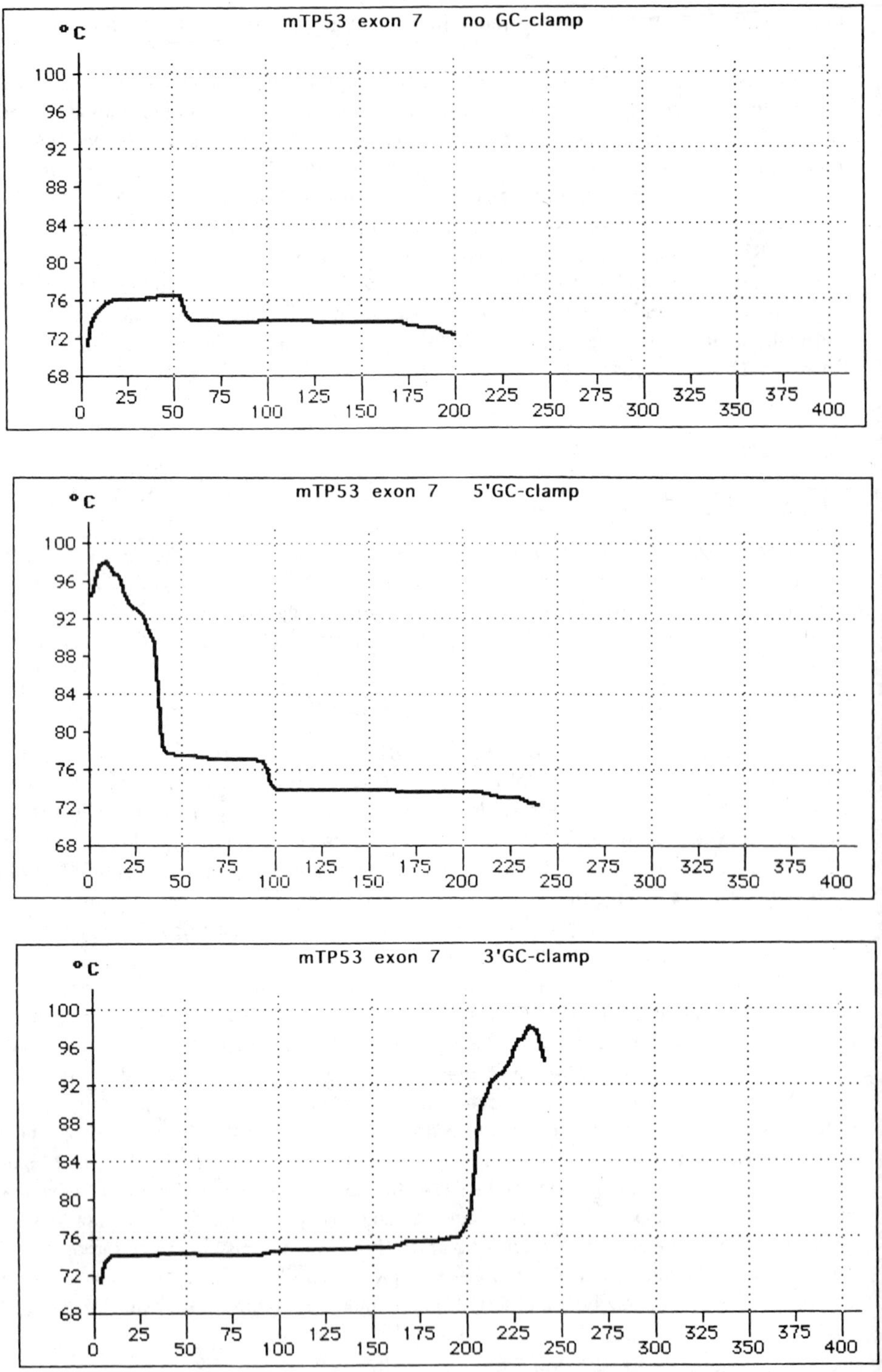
mTP53 exon 7 no GC-clamp
mTP53 exon 7 5'GC-clamp
mTP53 exon 7 3'GC-clamp
°C
100
96
92
88
84
80
76
72
68
0
25
50
75
100
125
150
175
200
225
250
275
300
325
350
375
400

a

b

c

experimentally determine its melting behavior by means of *perpendicular* DGGE. In this case the electrophoretic migration is perpendicular to the denaturant gradient and the sample is applied along the entire width of the gel. In this way each DNA molecule will travel along a path of constant denaturant concentration at a constant electrophoretic rate. The resulting curve will provide an indication of (1) the number of melting domains present and (2) the percentage of the denaturant agents at which denaturation of each domain occurs.

On the other hand, perpendicular gradients require the availability of the DNA fragment to be analyzed either in the form of cloned material or as PCR product (once primers have been designed and synthesized). Moreover, they do not provide "qualitative" information on the relative position of each domain within the fragment and on the effect of the presence of a GC-clamp on either side of the linear molecule.

A detailed protocol for the casting of perpendicular denaturing gradient gels has been reported by Myers *et al.* (1987) and will not be included here. However, the protocol for parallel gels may also be applied to the casting of perpendicular ones provided that the denaturing gradient is poured perpendicular to the direction of electrophoresis.

2.1.3. Setting Up DGGE: Parallel Denaturing Gradient Gels

Once the necessary computer (and/or experimental) simulations have been performed, primers can be designed to amplify the target sequence and the first denaturing gradient gel can be run. The choice of the denaturant range is made based on the T_m of the domain to be analyzed: the parallel gel should be initially chosen with a 25–30% difference in denaturant concentration centered around the T_m of the domain. The conversion factor between T_m and percentage denaturant for gels run at 60°C is given by the empirical formula: [% denaturant = $(3.2 \times T_m) - 182.4$]. For example, if the DNA fragment has a single domain with T_m = 75°C (57.6% denaturant), then it should be ideally analyzed on a gradient between 43 and 73% denaturant. Narrower denaturant gradients (10–15% difference) can also be employed with satisfactory results. If two distinct domains are present within the same fragment, two different gradients should be made for their optimal mutation analysis.

2.2. DGGE Protocol

From a methodological point of view, DGGE implies a conventional vertical electrophoresis where DNA molecules migrate through linearly increasing concentrations of denaturant agents (formamide and urea). Denaturant gradient gels are made in polyacrylamide and poured with conventional gradient mixers. In order to allow reproducibility of the electrophoretic runs, DGGE is generally performed at a constant temperature of 60°C. The latter was empirically chosen to exceed the T_m of an AT-rich DNA fragment in the absence of denaturant agents. Specific applications, however, may require temperatures other than 60°C. For extremely GC-rich sequences for example, temperatures up to 75°C can be employed (Harteveld *et al.*, 1996). To ensure uniform and constant maintenance of the chosen temperature during electrophoresis, the gel is placed in an aquarium tank and submerged with electrophoresis buffer kept at the desired temperature by a combination heater/stirrer thermostat.

←

Figure 19.3. Modification of the meltmap of a DNA fragment by attachment of a GC-clamp either at the 5′ or 3′ end. (a) Native meltmap of exon 7 of the mouse *p53* gene (204 bp). (b) Meltmap of the same fragment after attachment of a 40-bp GC-clamp (as reported in Sheffield *et al.*, 1989) at the 5′ end. (c) Meltmap of the same fragment after attachment of a 40-bp GC-clamp at the 3′ end.

2.2.1. Gel Apparatus

The gel assembly used in our laboratory is homemade, based on the original report by Myers *et al.* (1987). However, complete DGGE equipment can be purchased directly from different companies (D-Gene, Bio-Rad Laboratories, Hercules, CA, USA; IngenyPhorU, Ingeny B.V. Leiden, The Netherlands; CBS Scientific Company, Del Mar, CA, USA), or adapted from preexisting vertical electrophoresis equipment (Protean II, Bio-Rad Laboratories; SE 600 Series, Hoefer Scientific Instruments, San Francisco, CA, USA). A conventional gradient mixer is used to pour the denaturing gradient gel.

2.2.2. Reagents

The following list includes all of the chemicals and solutions necessary to pour and run denaturing gradient gels.

40% acrylamide stock solution (37.5:1 acrylamide:bisacrylamide): dissolve 100 g acrylamide and 2.7 g bisacrylamide in water to a final volume of 250 ml. Store in dark glass bottles at 4°C.

20× TAE electrophoresis buffer (0.8 M Tris base, 0.4 M sodium acetate, 0.02 M EDTA, pH 8.0): dissolve 96.912 g Tris base, 7.445 g Na_2EDTA, and 54.432 g sodium acetate in water to 1000 ml. Adjust pH to 8.0 with glacial acetic acid (approx. 36 ml).

80% denaturant stock solution (6%* acrylamide, 32% formamide, 5.6 M urea): For 500 ml: 170 g electrophoresis-grade urea, 75 ml of 40% acrylamide stock solution, 160 ml of 100% deionized† formamide, 25 ml of 20× TAE buffer, and water to 500 ml. Store in dark glass bottles at 4°C.

6% acrylamide stock solution (0% denaturant stock solution)*: For 500 ml: 75 ml of 40% acrylamide, 25 ml of 20× TAE buffer, and water to 500 ml. Store in dark glass bottles at 4°C.

10% ammonium persulfate stock solution: dissolve 10 g ammonium persulfate in 100 ml water. Store in small aliquots (1 ml) in Eppendorf tubes at −20°C.

TEMED (N,N,N',N'-tetramethylethylenediamine).

5× gel loading solution (0.25% bromophenol blue, 0.25% xylene cyanol, 20% Ficoll): dissolve 20 g Ficoll, 250 mg bromophenol blue, and 250 mg xylene cyanol in 100 ml water.

10 mg/ml ethidium bromide: dissolve 1 g ethidium bromide in 100 ml water. Store in dark glass bottles at 4°C.

2.2.3. Preparation of Denaturing Gradient Gels

Parallel denaturing gradient gels contain a linearly increasing concentration gradient of formamide and urea from top to bottom. The gels are used to analyze a number of samples loaded into wells at the top of the gel.

- Thoroughly clean the glass plates, spacers, comb, and gradient mixer with a strong detergent and rinse them with deionized water. Rinse the plates with ethanol and set them aside to dry.
- Lay the larger plate (non-eared) on the bench and arrange the spacers along the sides. Lay the smaller plate in position and clamp together with binder clips. Seal the sides and the bottom with gel-sealing tape.
- Put the comb into position and place the glass plates in the DGGE frame so that the eared

*Different acrylamide percentages may be employed depending on DNA fragment size.

†To deionize formamide: add 2 g of mixed bed resin (Baker) to 1000 ml formamide and stir for 30 min. at room temperature. Filter to remove resin and store at −70°C in dark glass bottles.

glass plate is against the rubber gasket, forming an upper electrophoresis chamber. Tighten the screws. Leave room for air to escape.

- Place the gradient maker on top of a magnetic stirrer, about 25 cm above the DGGE frame. Connect the exit tube to the top of the glass plates next to the comb. Close this tube as well as the one connecting the two chambers of the gradient mixer with stopcock. Prepare two solutions of equal volume (15 ml each; in our setup a 30-ml volume just fills the plates) which will give the desired denaturant concentration range. Add 16 μl TEMED and 160 μl APS to each solution. Mix well.
- Pour 15 ml of the solution with the higher concentration of denaturant in the chamber of the gradient maker that is directly connected to the plate (right chamber). Briefly open and close again the connection between the two chambers so as to allow the solution to fill the connecting tube. Make sure that no air bubbles block the passageway between the two chambers.
- Pour 15 ml of the solution with the lower percentage of denaturant in the other (left) chamber.
- While stirring the solution in the chamber with the higher concentration of denaturant, open the connection between the two chambers and the exit to the glass plates. Avoid air bubbles.
- The liquid passes by gravity through the plastic tubing into the cavity between the two glass plates. The gel should be poured in about 5 min.
- Allow the gel to polymerize for about 30–60 min.
- Remove the comb from the gel and the tape from the bottom of the glass plates.

2.2.4. Electrophoresis

- Place the frame with the gel into the bath containing the 1× TAE buffer heated to 60°C. Adjust the buffer level so that it rises just above the level of the wells. Avoid contact of the buffer with the upper electrophoresis chamber. Connect the combination thermostat so that buffer circulates from the aquarium into the upper buffer chamber (containing the cathode), while it overflows through a hole in the rear of the frame into the aquarium.
- Prerun the gel for about 30 min at 60 V (about 50 mA).
- Add gel loading solution to the samples; depending on the yield of the PCR reaction we usually load between one-tenth and one-half of the total PCR product. A small final volume (±10 μl) will result in sharper bands.
- With a syringe fitted with a needle, flush the wells with 1× TAE buffer. Load the samples and start the electrophoresis. For reasons of convenience, we usually perform our DGGE runs for about 16 hr at 60 V (50 mA); however, different times and running conditions can be applied.

2.2.5. Staining of the Gel

- Stop the electrophoresis and remove the frame from the aquarium tank. Remove the glass plates from the frame and gently lift the eared glass plate; use the other plate to support the gel during staining.
- Stain the gel for 20–30 min in 250 ml 1× TAE containing 0.5 μg/ml ethidium bromide with gentle shaking.
- In case a high background is observed, a destaining step of 15–20 min in 250 ml 1× TAE or water can be introduced.
- Examine the gel under UV (254 nm).

3. DISCUSSION

The direct detection of single base changes in DNA fragments relies on several protocols based on the differential behavior of DNA molecules differing by as little as a single base pair under specific experimental conditions. DGGE offers a valid technical approach for the identification of single base changes in genomic DNA, cloned material and in PCR-amplified fragments. Among the advantages of this technique are: (1) the high sensitivity of detection (~99%), (2) improved heterozygote detection (a pattern of two homo- and two heteroduplexes is observed), (3) the possibility to optimize the analysis by computer simulations, (4) a nonradioactive protocol, and (5) easy isolation of the mutant allele for its subsequent sequence determination. Disadvantages of the DGGE are: (1) laborious and time-consuming preliminary work prior to the actual analysis of the fragment (computer or experimental simulations), (2) costly synthesis of relatively long (~60 nt) PCR primers, and (3) the limited size of the largest DNA fragment (~500 bp) that can be efficiently analyzed by DGGE. While several alternatives to the use of GC-clamped primers have been successfully applied (Costes *et al.*, 1993; Top, 1992), DGGE analysis of a large genomic region undeniably requires substantial preliminary work. The introduction of the computer programs MELT87 and SQHTX, which allow simulation of the melting behavior of any DNA sequence, has considerably reduced the amount of preliminary experimental work needed.

3.1. Variations on the Theme

Since its introduction, the original DGGE protocol has been the object of many efforts to improve some of its features. In its basic and ideal form, GC-clamped DGGE will detect the great majority of base changes within an ~500-bp fragment encompassed within a single melting domain. However, when two or more T_m domains are present within the PCR fragment to be analyzed, the resolution of mutations located within the most thermostable melting domain of the native molecule may become a difficult task. In order to maximize the efficiency of the DGGE-based strategy, several investigators have explored the possibility to PCR relatively large (2–3 kb) DNA targets and to digest the PCR product into ~500-bp fragments prior to DGGE (Attree *et al.*, 1989). Satoh *et al.* (1993) implemented the same protocol by performing the PCR reactions with GC-clamped primers on both sides of the target DNA fragment. The average theoretical detectability of this method should be approximately 70–75%, based on the 100% detectability for the two GC-clamped fragments and 50% for the intermediate ones.

In genomic DGGE (gDGGE), genomic DNA is digested with a restriction enzyme, electrophoresed through a denaturing gradient gel, transferred to nylon filters, and hybridized to a unique DNA probe (Borresen *et al.*, 1988). Clear advantages of gDGGE over its parental protocol are that (1) it is not limited to any specific target sequence nor to its length (any available unique probe of any length can be used), (2) it does not require sequence information, and (3) covalent modifications in genomic DNA otherwise lost by enzymatic amplification, such as methylation, are detectable. On the other hand, because it relies on the presence of "natural clamps" (a melting domain with a T_m higher than that of the domain where the putative variant is located) within the restriction fragments to be analyzed, only a subset (20%–60%) of all of the possible single base variations will be detected by gDGGE (use of several restriction enzymes or combination of them might sensibly alleviate the latter problem). Moreover, since gDGGE is not PCR-based, heteroduplex formation is not feasible and one has to rely entirely on the resolution of two double-stranded DNA molecules differing by one base change. Nevertheless, gDGGE has been successfully applied for the identification of polymorphic sequence variations in human chromosome 21 (Burmeister *et al.*, 1991), and to screen for mutations in *Drosophila* (Doerig *et al.*, 1988;

Curtis *et al.*, 1989; Gray *et al.*, 1991). Abrams *et al.* (1990) have developed a modified protocol to generate heteroduplex molecules between a GC-clamped radiolabeled DNA probe and genomic DNA restriction fragments which are then analyzed by DGGE. A similar approach is based on the DGGE analysis of heteroduplexes obtained by hybridization of radiolabeled RNA probes to either genomic restriction fragments (Takahashi *et al.*, 1990, 1991) or GC-clamped PCR-amplified DNA (Theophilus *et al.*, 1989).

Two of the variations on the DGGE theme, e.g., two-dimensional fingerprint of complex genomes (2-D DNA Typing) and constant denaturing gradient gel electrophoresis (CDGE), are discussed elsewhere in this volume.

3.2. Comparison with Other Protocols

When compared with other mutation detection protocols such as RNase protection and hydroxylamine and osmium tetroxide (HOT) chemical cleavage, DGGE proved more reliable and sensitive (Theophilus *et al.*, 1989) combining a high sensitivity of mutation detection with a relatively less labor-intensive protocol. A direct comparison between SSCP and DGGE has shown that the latter is more sensitive since 100% of the variants tested could be detected versus 90% in the case of SSCP (Moyret *et al.*, 1994). However, it should be said that the decreased efficiency of SSCP could be improved by further optimalization of the protocols employed (Sheffield *et al.*, 1993).

In conclusion, DGGE has been employed for a wide spectrum of applications in both research and diagnostics. In molecular genetics, it has proven very useful in the analysis of inherited diseases caused by heterogeneous mutation spectra, or in the search for polymorphic sequences in linkage studies.

ACKNOWLEDGMENTS. R.F. is a fellow of the Royal Netherlands Academy of Arts and Sciences. The authors are grateful to J.Th. Wijnen, Ron Smits, and C. L. Harteveld for technical assistance and for their critical reading of the manuscript.

REFERENCES

Abrams, E. S., Murdaugh, S. E., and Lerman, L. S. (1990). Comprehensive detection of single base changes in human genomic DNA using denaturing gradient gel electrophoresis and a GC clamp. *Genomics* **7**:463–475.

Aly, A. M., Higuchi, M., Kasper, C. K., Kazazian, H. H., Jr., Antonarakis, S. E., and Hoyer, L. W. (1992). Haemophilia A due to mutations that create new N-glycosylation sites. *Proc. Natl. Acad. Sci. USA* **89**:4933–4937.

Antonarakis, S. E., and Kazazian, H. H., Jr. (1988). The molecular basis of haemophilia A in man. *Trends Genet.* **4**:233–237.

Attree, O., Vidaud, D., Vidaud, M., Amselem, S., Lavergne, J. M., and Goossens, M. (1989). Mutations in the catalytic domain of human coagulation factor IX: Rapid characterization by direct genomic sequencing of DNA fragments displaying an altered melting behavior. *Genomics* **4**:266–272.

Borresen, A. L., Hovig, E., and Brogger, A. (1988). Detection of base mutations in genomic DNA using denaturing gradient gel electrophoresis (DGGE) followed by transfer and hybridization with gene-specific probes. *Mutat. Res.* **202**:77–83.

Burmeister, M., diSibio, G., Cox, D. R., and Myers, R. M. (1991). Identification of polymorphisms by genomic denaturing gradient gel electrophoresis: Application to the proximal region of human chromosome 21. *Nucleic Acids Res.* **19**:1475–1481.

Cai, S. P., and Kan, Y. W. (1990). Identification of the multiple β-thalassemia mutations by denaturing gradient gel electrophoresis. *J. Clin. Invest.* **85**:550–553.

Cariello, N. F., and Thilly, W. G. (1986). Use of denaturing gradient gel electrophoresis to determine mutational spectra in human cells, in: *Mechanisms of DNA Damage and Repair: Implications for Carcinogenesis and Risk Assessment* (M. G. Simic, L. Grossmann, and A. C. Upton, eds.), Plenum Press, New York, pp. 439–452.

Cariello, N. F., Keohavong, P., Kat, A. G., and Thilly, W. G. (1990). Molecular analysis of complex human cell populations: Mutational spectra of MNNG and ICR-1191. *Mutat. Res.* **231**:165–176.

Cariello, N. F., Swenberg, J. A., De Bellis, A., and Skopek, T. R. (1991). Analysis of mutations using PCR and denaturing gradient gel electrophoresis. *Environ. Mol. Mutagen.* **18**:249–254.

Chen, J., and Thilly, W. G. (1994). Mutational spectrum of chromium (VI) in human cells. *Mutat. Res.* **323**:21–27.

Costes, B., Girodon, E., Ghanem, N., Chassignol, M., Thuong, N. T., Dupret, D., and Goossens, M. (1993). Psoralen-modified oligonucleotide primers improve detection of mutations by denaturing gradient gel electrophoresis and provide an alternative to GC-clamping. *Hum. Mol. Genet.* **2**:393–397.

Curtis, D., Clark, S. H., Chovnick, A., and Bender, W. (1989). Molecular analysis of recombination events in Drosophila. *Genetics* **122**:653–661.

Doerig, R. E., Suter, B., Gray, M., and Kubli, E. (1988). Identification of an amber nonsense mutation in the rosy516 gene by germline transformation of an amber suppressor tRNA gene. *EMBO J.* **7**:2579–2584.

Fearon, E. R., and Vogelstein, B. (1990). A genetic model for colorectal tumorigenesis. *Cell* **61**:759–767.

Ferec, C., Audrezet, M. P., Mercier, B., Guillermit, H., Moullier, P., Quere, I., and Verlingue, C. (1992). Detection of over 98% cystic fibrosis mutations in a Celtic population. *Nature Genet.* **1**:188–191.

Fischer, S. G., and Lerman, L. S. (1979). Length-independent separation of DNA restriction fragments in two-dimensional gel electrophoresis. *Cell* **16**:191–200.

Fischer, S. G., and Lerman, L. S. (1983). DNA fragments differing by a single base-pair substitution are separated in denaturing gradient gels: Correspondence with melting theory. *Proc. Natl. Acad. Sci. USA* **80**:1579–1583.

Fodde, R., van der Luijt, R., Wijnen, J. T., Tops, C., van der Klift, H., van Leeuwen-Cornelisse, J., Griffioen, G., Vasen, H., and Meera Khan, P. (1992). Eight novel inactivating germ line mutations identified by denaturing gradient gel electrophoresis. *Genomics* **13**:1162–1168.

Gray, M., Charpentier, A., Walsh, K., Wu, P., and Bender, W. (1991). Mapping point mutations in the *Drosophila rosy* locus using denaturing gradient gel blots. *Genetics* **127**:139–149.

Harteveld, C. L., Heister, J. G. A. M., Giordano, P. C., Losekoot, M., and Bernini, L. F. (1996). Rapid detection of point mutations and polymorphisms of the α-globin genes by DGGE and SSCA. *Hum. Mutat.* **7**:114–122.

Higuchi, M., Kazazian, H. H., Jr., Kasch, L., Warren, T. C., McGinnis, M. J., Phillips, J. A., III, Kasper, C., Janco, R., and Antonarakis, S. (1991a). Molecular characterization of severe haemophilia A suggests that about half the mutations are not within the coding regions and splice junctions of the factor VIII gene. *Proc. Natl. Acad. Sci. USA* **88**:7405–7409.

Higuchi, M., Antonarakis, S. E., Kasch, L., Oldenburg, J., Economou-Petersen, E., Olek, K., Arai, M., Inaba, H., and Kazazian, H. H., Jr. (1991b). Molecular characterization of mild-to-moderate haemophilia A: Detection of the mutation in 25 of 29 patients by denaturing gradient gel electrophoresis. *Proc. Natl. Acad. Sci. USA* **88**:8307–8311.

Hofker, M. H., Skraastad, M. I., Bergen, A. A. B., Wapenaar, M. C., Bakker, E., Millington-Ward, A., van Ommen, G. J. B., and Pearson, P. L. (1986). The X chromosome shows less genetic variation at restriction sites than the autosomes. *Am. J. Hum. Genet.* **39**:438–451.

Keohavong, P., and Thilly, W. G. (1992). Mutational spectrometry: A general approach for hot-spot point mutations in selectable genes. *Proc. Natl. Acad. Sci. USA* **89**:4623–4627.

Kogan, S., and Gitschier, J. (1990). Mutations and polymorphisms in the factor VIII gene discovered by denaturing gradient gel electrophoresis. *Proc. Natl. Acad. Sci. USA* **87**:2092–2096.

Lerman, L. S., and Silverstein, K. (1987). Computational simulation of DNA melting and its application to denaturing gradient gel electrophoresis. *Methods Enzymol.* **155**:482–501.

Losekoot, M., Fodde, R., Harteveld, C. L., van Heeren, H., Giordano, P. C., and Bernini, L. F. (1990). Denaturing gradient gel electrophoresis and direct sequencing of PCR amplified genomic DNA: A rapid and reliable diagnostic approach to β-thalassemia. *Br. J. Haematol.* **76**:269–274.

Losekoot, M., Fodde, R., Harteveld, C. L., van Heeren, H., Giordano, P. C., Went, L. N., and Bernini, L. F. (1991). Homozygous β^+-thalassaemia owing to a mutation in the cleavage-polyadenylation sequence of the human β-globin gene. *J. Med. Genet.* **28**:252–255.

Losekoot, M., van Heeren, H., Schipper, J. J., Giordano, P. C., Bernini, L. F., and Fodde, R. (1992). Rapid detection of a highly polymorphic fragment in the β-globin gene by denaturing gradient gel electrophoresis of PCR amplified material. *J. Med. Genet.* **29**:574–577.

Moyret, C., Theillet, C., Puig, P. L., Moles, J. P., Thomas, G., and Hamelin, R. (1994). Relative efficiency of denaturing gradient gel electrophoresis and single strand conformation polymorphism in the detection of mutations in exons 5 to 8 of the p53 gene. *Oncogene* **9**:1739–1743.

Myers, R. M., Lumelski, N., Lerman, L. S., and Maniatis, T. (1985a). Detection of single base substitutions in total genomic DNA. *Nature* **313**:495–498.

Myers, R.M., Fischer, S. G., Lerman, L. S., and Maniatis, T. (1985b). Nearly all single base substitutions in DNA fragments joined to a GC-clamp can be detected by denaturing gradient gel electrophoresis. *Nucleic Acids Res.* **13**:3131–3145.

Myers, R. M., Fischer, S. G., Maniatis, T., and Lerman, L. S. (1985c). Modification of the melting properties of duplex DNA by attachment of a GC-rich DNA sequence as determined by denaturing gradient gel electrophoresis. *Nucleic Acids Res.* **13**:3111–3130.

Myers, R. M., Maniatis, T., and Lerman, L. S. (1987). Detection and localization of single base changes by denaturing gradient gel electrophoresis. *Methods Enzymol.* **155**:501–527.

Oller, A. R., and Thilly, W. G. (1992). Mutational spectra in human B-cells. Spontaneous, oxygen and hydrogen peroxide-induced mutations at the hprt gene. *J. Mol. Biol.* **228**:813–826.

Olschwang, S., Laurent-Puig, P., Groden, J., White, R., and Thomas, G. (1993). Germ-line mutations in the first 14 exons of the adenomatous polyposis coli (APC) gene. *Am. J. Hum. Genet.* **52**:273–279.

Pellegata, N. S., Losekoot, M., Fodde, R., Pugliese, V., Saccomanno, S., Renault, B., Bernini, L. F., and Ranzani, G. N. (1992). Detection of k-ras mutations by denaturing gradient gel electrophoresis (DGGE): A study on pancreatic cancer. *Anticancer Res.* **12**:1731–1736.

Renault, B., van den Broek, M., Fodde, R., Wijnen, J., Pellegata, N. S., Amadori, D., Meera Khan, P., and Ranzani, G. N. (1993). Base transitions are the most frequent genetic changes at p53 in gastric cancer. *Cancer Res.* **53**:2614–2617.

Rosatelli, M. C., Dozy, A., Faa, V., Meloni, A., Sardu, R., Saba, L., Kan, Y. W., and Cao, A. (1992). Molecular characterization of ß-thalassemia in the Sardinian population. *Am. J. Hum. Genet.* **50**:422–426.

Saiki, R. K., Bugawan, T. L., Horn, G. T., Mullis, K. B., and Erlich, H. A. (1986). Analysis of enzymatically amplified β-globin and HLA-DQα DNA with allele-specific oligonucleotide probes. *Nature* **324**:163–166.

Saiki, R. K., Gelfand, D. H., Stoffel, S., Scharf, S. J., Higuchi, R., Horn, G. T., Mullis, K. B., and Erlich, H. A. (1988). Primer directed amplification of DNA with a thermostable DNA polymerase. *Science* **239**:487–491.

Satoh, C., Takahashi, N., Asakawa, J., Hiyama, K., and Kodaira, M. (1993). Variations among Japanese of the factor IX gene (F9) detected by PCR-denaturing gradient gel electrophoresis. *Am. J. Hum. Genet.* **52**:167–175.

Sheffield, V. C., Cox, D. R., Lerman, L. S., and Myers, R. M. (1989). Attachment of a 40 base-pair G+C-rich sequence (GC-clamp) to genomic DNA fragments by the polymerase chain reaction results in improved detection of single base changes. *Proc. Natl. Acad. Sci. USA* **86**:232–236.

Sheffield, V. C., Beck, J. S., Kwitek, A. E., Sandstrom, D. W., and Stone, E. M. (1993). The sensitivity of single-strand conformation polymorphism analysis for the detection of single base substitutions. *Genomics* **16**:325–332.

Takahashi, N., Hiyama, K., Kodaira, M., and Satoh, C. (1990). An improved method for the detection of genetic variations in DNA with denaturing gradient gel electrophoresis. *Mut. Res.* **234**:61–70.

Takahashi, N., Hiyama, K., Kodaira, M., and Satoh. C (1991). The length polymorphism in the 5′ flanking region of the human β-globin gene with denaturing gradient gel electrophoresis in a Japanese population. *Hum. Genet.* **87**:219–220.

Theophilus, B. D. M., Latham, T., Grabowski, G. A., and Smith, F. I. (1989). Comparison of RNase A, chemical cleavage, and GC-clamped denaturing gradient gel electrophoresis for the detection of mutations in exon 9 of the human acid β-glucosidase gene. *Nucl. Acids. Res.* **17**:7707–7722.

Thilly, W. G. (1985). Potential use of denaturing gradient gel electrophoresis in obtaining mutational spectra from human cells, in: *Carcinogenesis: The Role of Chemicals and Radiation in the Etiology of Cancer*, Volume 10 (E. Huberman and S. H. Barr, eds.), Raven Press, New York, pp. 511–528.

Top, B. (1992). A simple method to attach a universal 50-bp GC-clamp to PCR fragments used for mutation analysis by DGGE. *PCR Methods Appl.* **2**:83–85.

Traystman, M. D., Higuchi, M., Kasper, C. K., Antonarakis, S. E., and Kazazian, H. H., Jr. (1990). Use of denaturing gradient gel electrophoresis to detect point mutations in the factor VIII gene. *Genomics* **6**:293–301.

van den Broek, M. H., Renault, B., Fodde, R., Verspaget, H., Griffioen, G., and Meera Khan, P. (1993). Sites and types of p53 mutations in an unselected series of colorectal cancers in The Netherlands. *Anticancer Res.* **13**:587–592.

Wijnen, J., Vasen, H., Meera Khan, P., Menko, F. H., van der Klift, H., van Leeuwen, C., van den Broek, M., van Leeuwen-Cornelisse, I., Nagengast, F., Meijers-Heijboer, A., Lindhout, D., Griffioen, G., Cats, A., Kleibeuker, J., Varesco, L., Bertario, L., Bisgaard, M. L., Mohr J., and Fodde, R. (1995). Seven new mutations in *hMSH2*, an HNPCC gene identified by denaturing gradient gel electrophoresis (DGGE). *Am. J. Hum. Genet.* **56**:1060–1066.

Chapter 20

Constant Denaturant Gel Electrophoresis (CDGE) in Mutation Screening

Anne-Lise Børresen

1. INTRODUCTION

There is an increasing need for practical, efficient, and inexpensive ways to explore inherited mutations responsible for genetic diseases, acquired mutations involved in cancer development, and induced mutations in mutational spectrometry. Most of the current methods used for mutation screening are based on PCR, and this has solved the problem of target limitation. Simple and efficient methods such as allele-specific hybridization, allele-specific amplification, ligation, primer extension and artificial introduction of restriction sites, and variants of these methods have been developed for screening large numbers of samples for one particular mutation (for review see Cotton, 1993).

In many instances, however, screening for unknown mutations is required. Sequencing represents a reliable mutational screening technique, although PCR and direct sequencing is not very sensitive in detecting mutations present in only a fraction of the cells examined, which is often the case in tumor specimens. Sequencing is also very labor intensive unless expensive highly automated sequencers are applied. The development of simpler and less expensive screening techniques like single-strand conformation polymorphism (SSCP) and denaturing gradient gel electrophoresis (DGGE) has made it possible to screen larger series for mutations in several different genes. Nevertheless, improved screening techniques are still needed. Constant denaturant gel electrophoresis (CDGE), which is a modification of the conventional parallel DGGE (Hovig *et al.*, 1991), represents, in my opinion, such an improvement. DGGE was first described by Fischer and Lerman (1983) and is discussed in detail by Fodde and Losekoot (this volume). The separation principle of DGGE and CDGE is based on the melting behavior of the DNA double helix of a given fragment. This melting behavior is sequence dependent, and occurs at domains, rather than at single bases. The melting is detected as a reduction in the mobility of the DNA fragment as it moves throughout an acrylamide gel as a consequence of partial strand

Anne-Lise Børresen • Department of Genetics, Institute for Cancer Research, The Norwegian Radium Hospital, N-0310 Oslo, Norway.

Technologies for Detection of DNA Damage and Mutations, edited by Gerd P. Pfeifer, Plenum Press, New York, 1996.

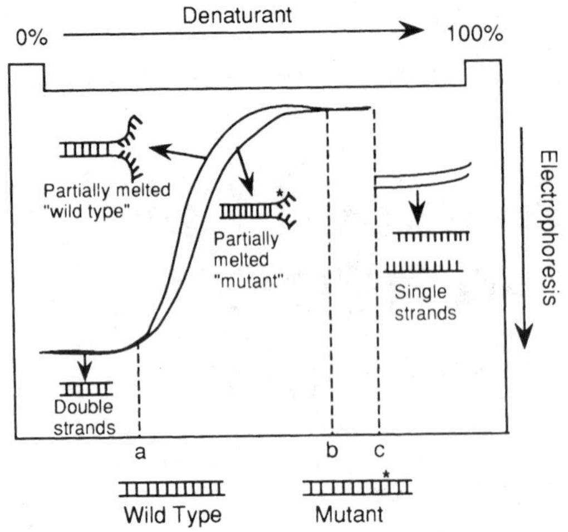

Figure 20.1. A schematic illustration of a perpendicular denaturing gradient gel in which the denaturing gradient is perpendicular to the electrophoresis direction. A mutant and wild-type fragment are loaded together into the long well along the top of the gel At low concentration of denaturant the DNA fragments remain double-stranded (to the left of point "a"), but as the concentration of denaturant increases, the DNA fragments begin to melt, creating branched shaped molecules. At higher concentrations of denaturant (to the right of "b"), both of the DNA fragments have come to a stage where they have the same configuration and migrate together. At even higher concentrations (to the right of "c") the DNA fragments are completely melted, creating two single strands.

separation. The separation principle is illustrated in Fig. 20.1 in a perpendicular DGGE. In a denaturing gradient polyacrylamide gel, DNA fragments migrate according to size until they reach the gradient where the molecule begins to denature (at point "a" in Fig. 20.1). When a domain of the DNA denatures, a branched shaped molecule which is anchored by a yet unmelted domain is created. At this point the migration rate is slowed down. Since the mutant and wild-type fragments have different sequences, their melting points are different and they will separate. As the fragments move they will continue to melt until they are almost completely denatured, only held together in the highest melting domain. At this point the migration again becomes a function of size (at point "b" in Fig. 20.1).

Perpendicular DGGE is not suitable for large-scale screening since only one sample can be analyzed on each gel. In parallel DGGE where the electrophoresis direction is parallel to the denaturing gradient and in CDGE where a constant percentage of denaturing agents is used throughout the gel, many samples can be loaded in different wells on top of the gel (Fig. 20.2a). Mutation analysis using parallel DGGEs are complicated by the difficulties in choosing the exact running time and running conditions to achieve the optimal separation. DGGE is based on the discontinuous phenomenon of strand dissociation. A molecule in a parallel gradient gel will migrate with a constant mobility until it reaches the position in the gel where the denaturant concentration is sufficient for the double-stranded molecules to undergo partial melting. Once double-stranded DNA becomes partially single-stranded, the molecules will start to migrate more slowly. However, the fragments do not necessarily stop completely, and the residual movement will cause the migration of the molecules into higher concentrations of denaturant, leading to new conformational changes and further alterations in the migration rate. Most fragments will have more than one melting domain, and in DGGE the migration of a fragment with multiple domains is difficult to predict because the net effect of several melted domains over time is discontinuous. Therefore, field strength, temperature, and time all need to be strictly controlled to achieve reproducible results. For example, by running the gel too long, a separation once achieved may decrease and even be lost, and mutations present in a homo- or hemizygote state, where heteroduplexes are not formed, may be missed. This is illustrated in Fig. 20.2b. These problems are eliminated by CDGE in which one single denaturing condition is used to melt a fragment. In contrary to the parallel DGGE the separation in CDGE is time dependent

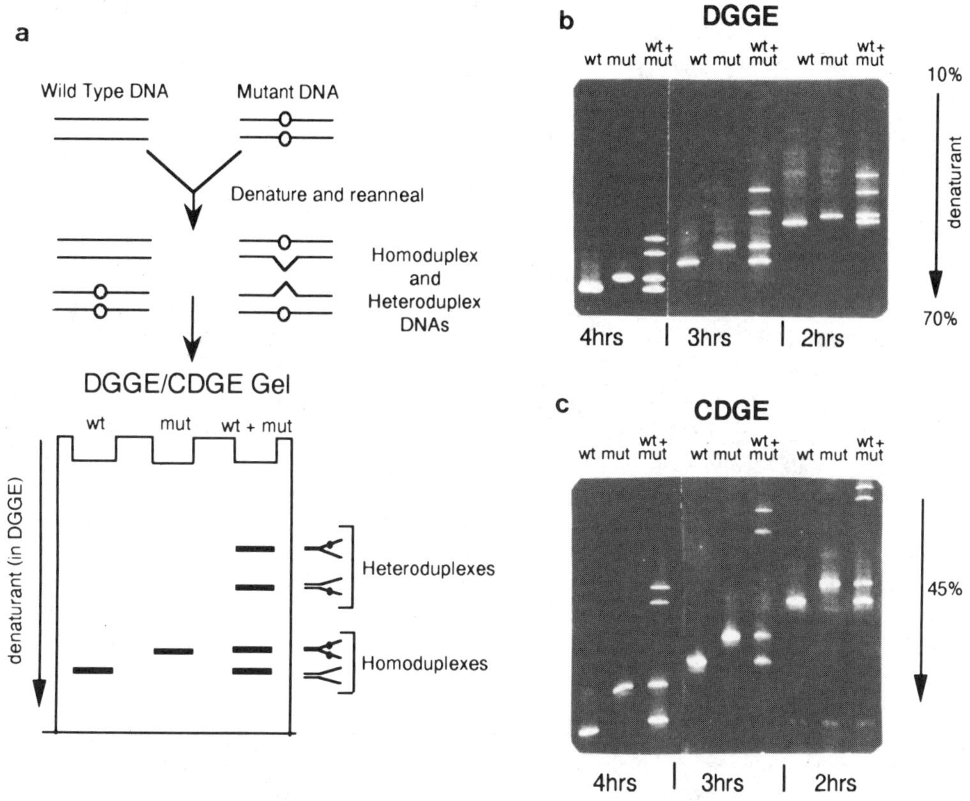

Figure 20.2. Wild-type and mutant DNA fragments from exon 8 of the *TP53* gene are denatured and reannealed to generate four possible fragments, two heteroduplexes and two homoduplexes. A schematic illustration of the separation principle is shown in (a) The difference between DGGE and CDGE is demonstrated in the time course experiment shown in (b) and (c), respectively. Note that the separation between the mutant and wild-type fragments decreases after 3 hr in the DGGE while the separation increases all the time in the CDGE. The time course had to be run on two constant and two denaturing gradient gels, respectively, since 2 hr of running prior to loading of the PCR products gave fuzzy bands.

only. In CDGE wild-type and mutant fragments will melt partially into a certain configuration immediately after entering the gel. This configuration will be kept throughout the run, and the fragments will migrate with a constant rate. Therefore, the longer the gel is run, the wider the separation between the mutant and wild type (illustrated in Fig. 20.2c).

The relative distance between the mutant and wild-type homoduplexes and the two heteroduplexes will be the same in different runs, making it possible to record and store the migration pattern for a given mutant in a computer. Hence, after building up a data base of CDGE migration patterns for different mutations in a specific gene, it may be possible to determine the sequence alteration without sequencing (see Section 4).

We and others (Børresen *et al.*, 1991, 1992; Hovig *et al.*, 1992; Husgafvel-Pursiainen *et al.*, 1992; Seruca *et al.*, 1992; Malkin *et al.*, 1992; Andersen *et al.*, 1993; Guldberg *et al.*, 1993; Heimdal *et al.*, 1993; Peng *et al.*, 1993; Ridanpää and Husgafvel-Pursiainen, 1993; Thorlacius *et al.*, 1993; Cornelis *et al.*, 1994; McIntyre *et al.*, 1994) have over the last few years utilized

CDGE for screening of inherited as well as acquired mutations in a number of different genes. The availability of the *D GENE* system from Bio-Rad makes the CDGE screening of large sets of samples relatively easy and reproducible after the PCR and running conditions for the PCR fragments of the gene in question have been established.

2. STRATEGY

Optimal mutation analysis by CDGE requires knowledge of the nucleotide sequence and melting profile of the DNA fragment to be analyzed. The theoretical melting profile of the fragment can be calculated using the MacMelt computer program (MedProbe AS, Norway) which is a simplified version of the Melt 87 (Lerman and Silverstein, 1987). The program is based on the thermodynamics of the transition of double-stranded to single-stranded DNA, on the statistical mechanical principles and algorithms described by Poland (1974), and the nearest-neighbor base-pair doublet parameters introduced by Gotoh and Tagashira (1981). The sequence to be analyzed for mutations including 20- to 30-bp flanking sequences (usually an exon with flanking intronic sequences) is entered into the computer, and the melting profile calculated. If cDNA is the source for the template, the theoretical profile of the total cDNA sequence (or parts of it) is calculated, and fragments with a sequence having a relatively flat melting profile differing only by 2–4°C, or with only increasing or decreasing melting along the fragment, are selected for further analyses (examples given in Fig. 20.3). Usually fragment sizes between 200 and 800 bp are analyzed, but it is the shape of the fragment's melting profile rather than the size which is critical. Primers are designed so that as much as possible of the target sequence resides in the lower melting domains. A GC-rich clamp (Sheffield *et al.*, 1989) is attached to one of the primers, usually to the primer in the highest melting domain end, to create a nonmelting sequence. Examples of a melting profile of a cDNA sequence, an exon with flanking intronic sequences, and the resulting designed fragment with a GC clamp are shown in Fig. 20.3a–c. The OLIGO primer analysis software program from National Biosciences (NBI) can be used to design the optimal PCR primers. One mutant primer for each domain, usually 35–40 bp long and overlapping one of the other primers, is also devised and used in conjunction with the opposite primer to generate a mutant exon fragment which serves as a positive control in the CDGE analyses. For a fragment with a profile as shown in Fig. 20.3c, three CDGEs, one for each melting domain, will be needed to detect mutations residing in all positions inside the primers. Alternatively, one or two of the lowest melting domains may be removed prior to the CDGE analysis by cutting the PCR product with a restriction enzyme with a unique restriction site in this area.

The optimal conditions for the CDGEs are determined using perpendicular DGGE (Fig. 20.3d). The PCR product from a control wild-type sample and the mutant control are mixed 1:1, heated for 3 min at 94°C and reannealed at 65°C for 1 hr to allow formation of heteroduplexes before loading into the long well along the top of the gel. The gel is run with the electrophoresis direction perpendicular to the denaturant gradient in the *D GENE* electrophoresis unit from Bio-Rad (catalogue No. 179-9000) for 2–4 hr, stained, and photographed. The running conditions for the further CDGE screening are determined from this perpendicular DGGE gel (Fig. 20.3d, top). In the example given in Fig. 20.3, three melting domains can be observed. The constructed mutant reside in the lowest melting domain, but affect both of the two lower domains (Fig. 20.3c and d, top). To screen the whole fragment for mutations, three different CDGEs should be run on each PCR product, one at 24%, another at 28%, and the last at 32% of denaturant, corresponding to the three steepest parts of the S-shaped melting profile. For the lower domains it can be seen that at 24 and 28% denaturant the control mutant clearly separates from the wild type (Fig. 20.3d, top). These concentrations of denaturant are optimal for CDGE analyses of this given fragment, and are to be used in the following screening (Fig. 20.3d bottom).

A flow chart diagram of the strategy is shown below:

Load the DNA sequence of interest (cDNA if that is the source for the analyses or an exon with 20- to 30-bp flanking intron sequences if genomic DNA is the source for the analyses) into the computer.

↓

Create a melting profile using MacMelt or Melt87 and select the part of the profile with flat or descending/ascending melting domains (maximum 3 domains) usually 200–800 bp long.

↓

Add the GC clamp sequence to the end of the selected fragment with the highest melting. Create a new melting profile, and design optimal PCR primers.

↓

Make a final melting profile. This profile should have no grooves, leave as much as possible of the sequence to be analyzed in the lower melting domains, and have as few low melting domains as possible. From this profile a judgment of how much of the fragment that will be analyzed by CDGE can be made.

↓

Construct one or more control mutants by PCR (preferably one in each melting domain) and amplify mutant and wild-type fragments.

↓

Run perpendicular DGGE on mutant and wild-type mixture and select optimal conditions for CDGE screening.

↓

Analyze the PCR products using CDGE with the optimal percentage of denaturant.

↓

Samples with aberrantly migrating bands can further be subjected to sequencing or to sequence-based CDGE analysis.

3. SENSITIVITY OF CDGE

The sensitivity of CDGE has been explored with different strategies. In a comparison study by Condie *et al.* (1993), SSCP, HOT, and CDGE were applied to screen DNA samples containing different point mutations in the *TP53* gene. Three different laboratories, each familiar with one technique, analyzed "blindly" the same 27 samples. The HOT technique detected 100% of the mutations while CDGE and SSCP detected 88 and 90%, respectively. However, the CDGE screening used were not optimized to screen for mutations in higher melting domains. When this was performed, 100% of the mutations were detected also by CDGE. In another study by Smith-Sørensen *et al.* (1993), seven different mutations, both transitions and transversions, were created within a 6-bp sequence residing in codon 247–249 of exon 7 of the *TP53* gene. All of these mutations were detected by CDGE at the homoduplex level.

In a large set of lung tumors both direct sequencing of PCR products and CDGE were performed to screen for *TP53* mutations (Ryberg *et al.*, 1994). Direct sequencing did not detect any additional mutations in the regions screened by CDGE. In fact, CDGE allowed the detection of several mutations that were missed by direct sequencing of PCR products. In these cases the mutations could be verified when direct sequencing was carried out on the aberrantly migrating

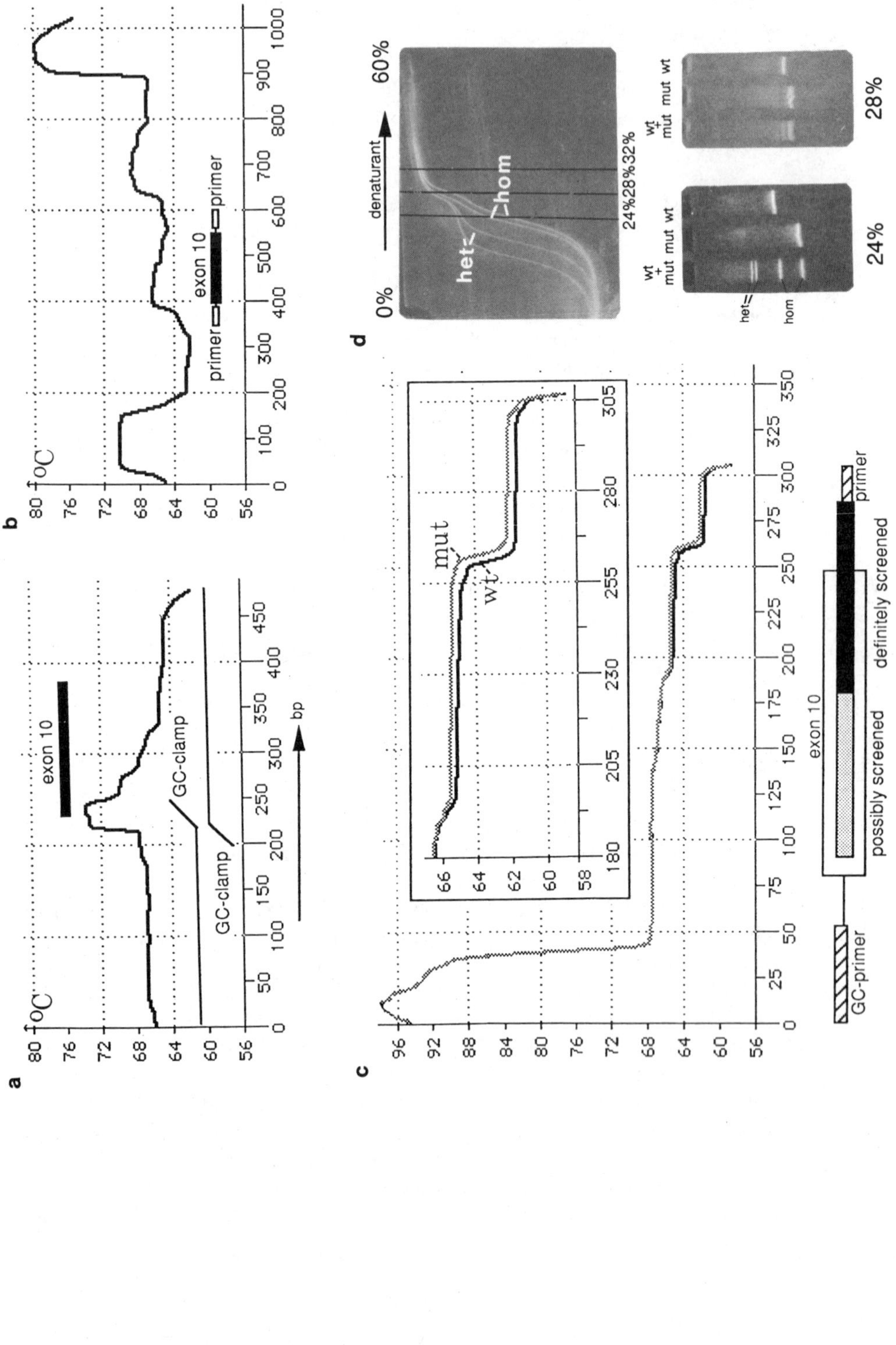
a
b
c
d
°C
exon 10
GC-clamp
GC-clamp
bp
primer
exon 10
primer
mut
wt
GC-primer
exon 10
possibly screened
definitely screened
primer
0%
denaturant
60%
het
hom
24%28%32%
wt + mut mut wt
24%
wt + mut mut wt
28%
het
hom

Figure 20.3. (a) Theoretical melting profile of a part of the cDNA sequence of the *hMSH2* gene (exon 8-11). Suggested fragments suitable for CDGE analyses using cDNA as template are indicated below the profile. (b) Melting profile of exon 10 of the *hMSH2* gene with flanking intronic sequences. (c) Melting profile of the final exon 10 fragment with a GC clamp included and after having chosen the optimal PCR primers. The location of the primers and an estimation of which part of the fragment will be screened for mutation by CDGE are indicated below. In the area "possibly screened" mutations causing deletions will be detected as well as transition mutations forming unstable heteroduplexes. The fragment has three melting domains, and to screen the whole fragment CDGE should be run at three different percentages of denaturants. The profile of a mutant (A→G) in position 264 is overlaid on the wild type (dotted curve). (D) Upper part: perpendicular DGGE of wild type + mutant (shown in C) shows separation in two of the three melting domains. Conditions for CDGE are determined from this gel. The percentage of denaturant corresponding to the steepest part of the S-shaped curve, and where mutant and wild type clearly separate (first domain for the mutant used here) are chosen for further CDGEs. Lower part: CDGE of exon 10 of the *hMSH2* gene fragment run at two different conditions, 24% of denaturant detecting mutations in the lower domain and 28% of denaturant detecting mutations in the middle melting domain.

bands cut out from the CDGE gels. These findings encouraged us to perform a titration experiment to study how large the fraction of mutant cells must be to be able to detect the mutation by CDGE. A cell line containing two mutant copies of the *TP53* gene (carrying a G→A mutation in codon 273 in exon 8) and no wild-type copies were mixed with normal lymphocytes in three parallel series with cell ratios from 1:1 down to 1:1000 (mutant : wild type). DNA was isolated and PCR and CDGE analyses as well as PCR and direct sequencing were performed. The results are shown in Fig 20.4. The homoduplex mutant band can clearly be seen down to 10% and the heteroduplexes are clearly detectable down to 1% in the CDGE analysis (Fig. 20.4a). PCR and direct sequencing did detect the mutation down to the 25% level although the mutant band was very weak. These results indicate that CDGE is superior to direct sequencing of PCR products in detecting low-frequency mutations.

To date, we have no indications that any mutation residing in the lower melting domains of a fragment has been missed by CDGE when run under optimal conditions. However, mutations residing in higher melting domains may be difficult to detect. An AT→TA and a GC→CG change in a higher melting domain close to the GC-clamp might be missed (Smith-Sørensen *et al.*, 1992). The estimates of how much of the PCR fragment the CDGE actually screens should thus be carried out conservatively. A control mutant in each melting domain should be applied to indicate the sensitivity of the separation.

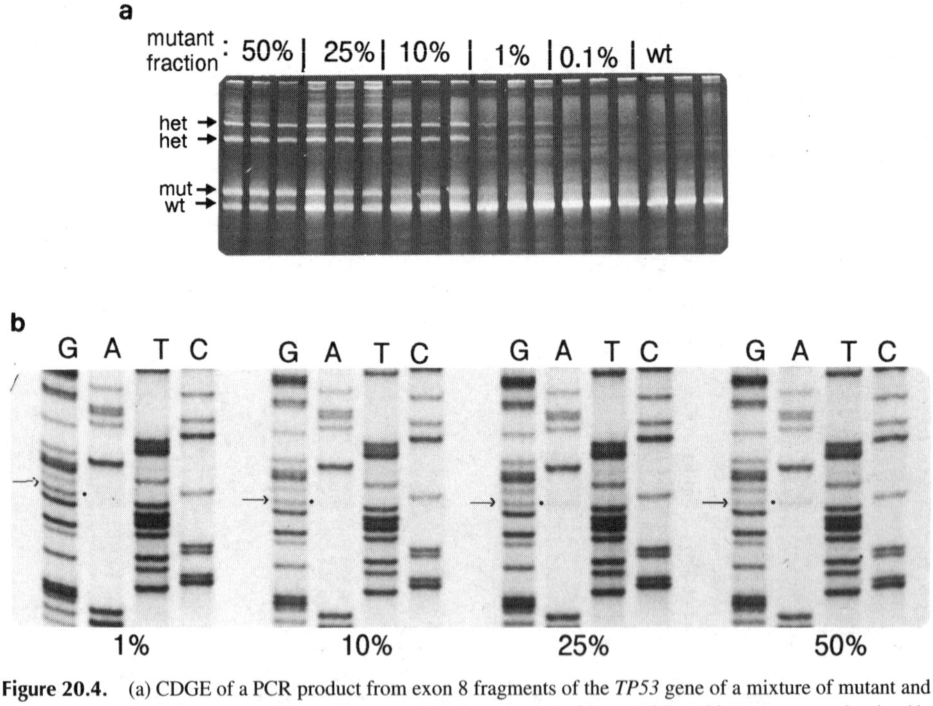

Figure 20.4. (a) CDGE of a PCR product from exon 8 fragments of the *TP53* gene of a mixture of mutant and wild-type DNA in different ratios. The gel is run at 47% denaturant for 2 hr at 56°C at 130 V constant and stained in SYBR green I. (b) Biotinylated PCR fragments of the same mixtures were subjected to direct sequencing using Dynabeads as solid support. Three parallel dilution series were analyzed with both techniques.

4. SEQUENCE-BASED CDGE WITHOUT SEQUENCING

If a number of different mutations occur within a certain gene and it is of importance to determine the nature of all of these mutations in large series, a data base of migration pattern can be compiled for all of the mutants. Since the separation principle of CDGE is time dependent and the migration rate of a certain fragment is constant, the relative distance between the two homoduplexes and the two heteroduplexes will be characteristic for that particular mutant. The pattern will be the same in different gels run under the same conditions. This may allow the comparison of a given migration pattern with a data base of known mutations, and hence allow prediction of the sequence of the mutation without performing sequence analyses. In Fig. 20.5 all of the different known mutations analyzed have their characteristic migration pattern. The sample with the unknown mutation (X) was predicted to have the G→A change at codon 273 since it had the same migration pattern as the sample in lane 15. Sequencing confirmed this result. CDGE may thus have the advantage to predict the sequence of the mutant without sequencing as compared to both SSCP where the mutant only creates one extra band with unpredictable migration, and parallel DGGE where much more care has to be taken to have the exact same running conditions. So far we have not detected two different mutations giving the same pattern although we may expect that this occasionally will occur. If two different mutations in the data base do give the same pattern, the PCR product under investigation can be mixed with the PCR product of either of them. No extra band will appear if identical PCR products are mixed, whereas additional heteroduplexes with different migration pattern will be created when mixing nonidentical PCR products.

5. METHODS

5.1. PCR

PCR is performed using 100–300 ng of template DNA in 50 mM Tris·HCl (pH 8.3) with 10 mM KCl, varying amounts of $MgCl_2$, 0.2 mM each dNTP, 25–50 pmole of each primer, and 1.25 units of Taq DNA polymerase (Amplitaq Cetus) in a 50-μl reaction. The mixture is

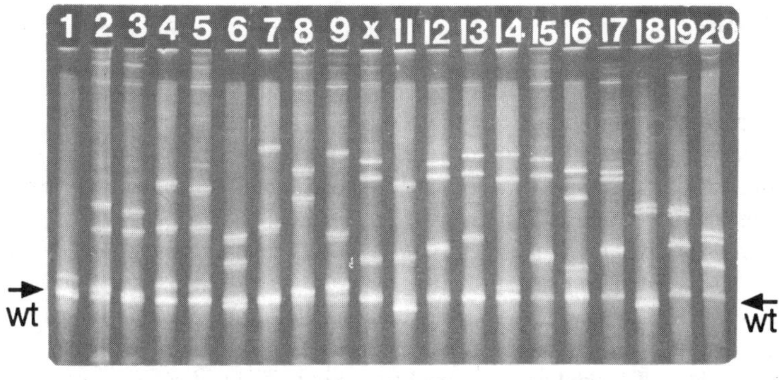

Figure 20.5. CDGE of PCR products of exon 8 of the *TP53* gene from 19 different tumor samples with different known mutations. The samples contained different amounts of the wild-type sequence in addition to the mutant. A sample with an unknown mutation is seen in lane 10 (X). The migration pattern in lane 10 (X) is identical to the migration pattern in lane 15 and the sequence alterations in sample X would hence be the same as the one in lane 15.

incubated in a Perkin–Elmer Cetus DNA Thermal Cycler for 25–35 cycles of denaturation 94°C for 1 min, annealing time and temperatures optimal for the particular fragment to be analyzed, followed by polymerization at 72°C for 1 min. The PCR reaction is initiated with a 3 min incubation at 94°C and ends with a 10-min incubation at 72°C, followed by 3 min at 94°C, and 60 min at 65°C to allow formation of heteroduplexes.

5.2. Denaturing Gel Electrophoresis

The CDGE and perpendicular DGGE analyses are performed using the *D GENE* system from Bio-Rad using 10 × 16-cm glass plates (catalogue No. 170-9000 through 170-9070). One perpendicular denaturing gradient gel of a mixed PCR product from normal and mutant control DNA is run for each fragment to optimize the CDGE conditions (see Section 2). The gel should contain 6–12% acrylamide (depending on fragment size) in 1× TAE buffer (40 mM Tris acetate, 1 mM EDTA pH 8.0) with bisacrylamide (in the ratio 37.5:1) as cross-linker, and varying concentrations of denaturant consisting of urea and formamide (100% denaturant corresponds to 7 M urea and 40% formamide). In the perpendicular DGGE usually a gradient from 0 to 70% of denaturant is applied. The gel is polymerized using ammonium persulfate (0.8 μg/ml gel solution) and *N,N,N′,N′*-tetramethylethylene diamine (1.6 μl/ml gel solution). Both the constant denaturant and the perpendicular denaturing gradient gels are run in 1× TAE buffer at 55–60°C depending on the melting behavior of the fragment. The temperature should be kept constant within ±0.2°C throughout the run. The electric field applied is usually 130 V constant for 2–3 hr. After electrophoresis the gels are stained in SYBR green I nucleic acid gel stain (Molecular Probes Inc.) and photographed using a UV transilluminator at 254 nm. In the CDGE analyses, 64 samples can be analyzed in each run. The constructed mutant control for the domain under analysis should be applied on each gel to ensure that optimal conditions have been applied. Samples with aberrantly migrating bands can be resubmitted to PCR followed by direct sequencing. If the mutation is present in a very small fraction of the cells examined, the sample should be reamplified by PCR, and a second CDGE gel should be run and the mutant band (avoiding staining) cut out from the gel for direct sequencing. If the abberrantly migrating bands are suspected to be PCR artifacts (seen from an unspecific PCR product), a perpendicular DGGE should be run. All true mutants will have the same type of S-shaped curve as the wild type whereas PCR artifacts usually have different melting profiles.

6. DISCUSSION

Denaturing gel electrophoresis allows the resolution of DNA fragments differing by as little as a single nucleotide substitution. Perpendicular DGGE has the highest sensitivity for detecting mutations in all of the lower melting domains in one gel, but is not suitable for large-scale screening since only one sample can be applied on each gel. The resolution of parallel DGGE depends on the choice of optimal parameters; the detection level depends heavily on sequence context and melting domains. The exact location of the mutation within the DNA fragment is uncertain because several domains of the fragment are being screened for mutations at once. CDGE seems to detect nearly all possible mutations within one melting domain, but has the disadvantage that only one domain can be screened at a time. Since CDGE is the easiest to perform and the same PCR product can be analyzed using different CDGE conditions, it seems to represent an easy way of screening a given sequence for mutations. However, all of the techniques based on denaturing gel electrophoresis have limitations. First, care must be taken to construct fragments with an optimal melting profile for mutation detection; the target sequence should reside in a lower melting domain. For some sequences with a high GC content this is impossible. GC-rich fragments are often found in the 5′ end of a gene, and are often present within genes coding for cyclin inhibitors. Such fragments should be screened using alternative strategies.

When screening large genes at both the cDNA and genomic level, the high number of fragments that have to be created represents another limitation. Usually the sequence rather than the size of the fragment is the limitation in CDGE and DGGE. Gene sequences with a flat melting domain usually do not exceed more than 600–800 bp. In most cases a GC-rich clamp has to be added on each fragment, and this makes the primers more expensive. Alternatively, psoralen derivative PCR primers called ChemiClamps can be used (Costes *et al.*, 1993). Psoralens are photoreagents that form covalent bonds with pyrimidine bases of nucleic acids. Psoralen–oligonucleotide conjugates allow cross-linking of DNA fragments at one end by photoinduction with a UV source. The efficiency of the cross-linking is, however, only 30–40%, and therefore more target DNA has to be used. Because ChemiClamps covalently link the two DNA strands at one end, they cannot be used if sequencing of a fragment isolated from the gel is required.

If two mutations are present in the same DNA fragment in a given sample, the migration pattern in CDGE as well as parallel DGGE will indicate only one mutation. Perpendicular DGGE, however, will clearly demonstrate the presence of two mutations in the same fragment since the heteroduplexes as well as the homoduplexes will cross each other as demonstrated in Fig. 20.6. Thus, submitting all samples demonstrating aberrantly migrating bands in the initial CDGE screen to a perpendicular DGGE will increase the knowledge of the nature of the mutant. In addition, the perpendicular DGGE will identify those of the aberrantly migrating bands caused by PCR artifacts.

Another limitation of the denaturing gel electrophoresis techniques is the problem with achieving clean PCR products when the long GC-rich primers are applied. This problem is particularly pronounced when amplifying fragments from crude templates like paraffin and formalin fixed and embedded tissues. Nested PCR using primers without GC clamps in the first PCR may help in such cases.

CDGE has wide applications in mutation screening in research as well as diagnostics. It has proven to be useful in determining the genotype/phenotype correlation in inherited diseases like the PKU deficiency (Guldberg *et al.*, 1993). We and others have used the CDGE in screening for mutations in tumor suppressor genes and oncogenes like *TP53*, *WT1*, *RB1*, *MTS1*, *WAF1*, *K-*, *H-*, *N-RAS*, *MSH2*, and others. The genotype of the tumor, with respect to these genes, may become useful molecular pathological markers. A tumor's genotype will determine its phenotype, and

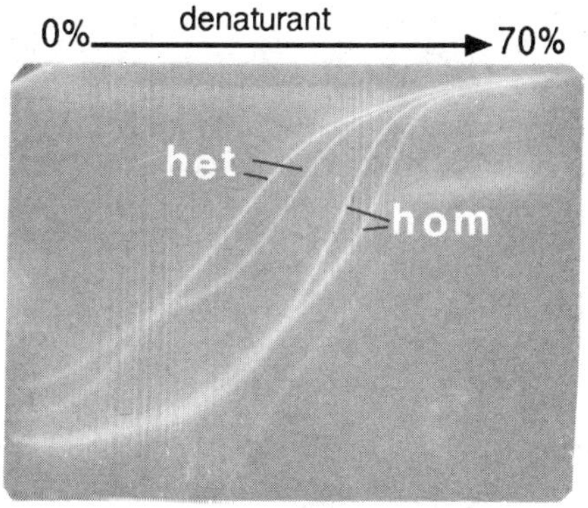

Figure 20.6. Perpendicular DGGE of a sample with two mutations in exon 13 of the *hMSH2* gene residing on the same DNA fragment. The crossing over of the heteroduplexes (het) and the homoduplexes (hom) indicates that the two mutations reside in a lower and a higher melting domain, respectively.

hence the behavior and resistance to different therapy. Mutation screening of both malignant and premalignant lesions may in the future become of clinical importance. It may further become important to detect mutations in tumor cells occurring in stool, urine, blood, and bone marrow as part of early detection and monitoring of tumor progression. Some mutations within a certain gene may be more critical than others, for example with regard to therapy. Hence, sequence-based diagnosis in cancer will not only have impact on the detection of inherited cancer cases by early diagnosis, but may also turn out to have a positive impact on patient survival.

CDGE has also been successfully applied in screening for sequence variation in bacterial and viral strains (Dworniczak *et al.*, 1992; Andersson *et al.*, 1993). Other applications are examinations of the fidelity of several DNA polymerases, and the analyses of the *in vitro* mutational spectra of several mutagens using constant capillary denaturing electrophoresis (CDCE) (Khrapko *et al.*, 1994).

ACKNOWLEDGMENTS. The excellent technical assistance of Sigrid Lystad, Eldri Due, Merete Hektoen, and Hilde Johnsen is highly acknowledged. Dr. Tone Ikdahl Andersen has contributed with critical discussion and evaluation of the manuscript. The work described has mainly been supported by The Norwegian Cancer Society.

REFERENCES

Andersen, T. I., Holm, R., Nesland, J. M., Heimdal, K. R., Ottestad, L., and Børresen, A.-L. (1993). Prognostic significance of *TP53* alterations in breast carcinoma. *Br. J. Cancer* **68**:540–548.

Andersson, B., Ying, J. H., Lewis, D. E., and Gibbs, R. A. (1993). Rapid characterization of HIV-1 sequence diversity using denaturing gradient gel electrophoresis and direct automated DNA sequence of PCR products. *PCR Methods Appl.* **2(4)**:293–300.

Børresen, A.-L., Hovig, E., Smith-S—rensen, B., Malkin, D., Lystad, S., Andersen, T. I., Nesland, J. M., Isselbacher, K. J., and Friend, S. H. (1991). Constant denaturant gel electrophoresis as a rapid screening technique for p53 mutations. *Proc. Natl. Acad. Sci. USA* **88**:8405–8409.

Børresen, A.-L., Andersen, T. I., Garber, J., Piraux, N. B., Thorlacius, S., Eyfjrd, J., Ottestad, L., Smith-Sørensen, B., Hovig, E., Malkin, D., and Friend, S. H. (1992). Screening for germ line *TP53* mutations in breast cancer patients. *Cancer Res.* **52**:3234–3236.

Condie, A., Eeeles, R., Børresen, A.-L., Coles, C., Cooper, C., and Prosser, J. (1993). Detection of point mutations in the p53 gene. Comparison of single-strand conformation polymorphism, constant denaturant gel electrophoresis, and hydroxylamine and osmium tetroxide techniques. *Hum. Mutat.* **2**:58–66.

Cornelis, R. S., van Vliet, M., Vos, C. B. J., Cleton-Jansen, A.-M., van de Vijver, M. J., Peterse, J. L., Meera Khan, P., Børresen, A.-L., Cornelisse, C. J., and Devilee, P. (1994). Evidence for a gene on 17p13.3, distal to TP53, as a target for allele loss in breast tumors without p53 mutations. *Cancer* **54**:4200–4206.

Costes, B., Girodon, E., Ghanem, N., Chassignol, M., Thuong, N.T., Dupret, D., and Goossens, M. (1993). Psoralen-modified oligonucleotide primers improve detection of mutations by denaturing gradient gel electrophoresis and provide an alternative to GC-clamping. *Hum. Mol. Genet.* **2(4)**:393–397.

Cotton, R. G. H. (1993). Current methods of mutation detection. *Mutat. Res.* **285**:125–144.

Dworniczak, B., Kalydjieva, L., Pankoke, S., Aulehla-Scholz, C., Allen, G., and Horst, J. (1992). Analysis of exon 7 of the human phenylalanine hydroxylase gene: A mutation hot spot? *Hum. Mutat.* **1**(2):138–146.

Fischer, S. G., and Lerman, L. S. (1983). DNA fragments differing by single base-pair substitutions are separated in denaturing gradient gels: Correspondence with melting theory. *Proc. Natl. Acad. Sci. USA* **80**:1579–1583.

Gotoh, O., and Tagashira, Y. (1981). Location of frequently opening regions of natural DNA's and their relation to functional loci. *Biopolymers* **20(5)**:1043–1058.

Guldberg, P., Henriksen, K. F., and Guttler, F. (1993). Molecular analysis of phenylketonuria in Denmark: 99% of the mutations detected by denaturing gradient gel electrophoresis. *Genomics* **17(1)**:141–146.

Heimdal, K. R., Lothe, R. A., Lystad, S., Holm, R., Foss, S. D., and Børresen, A.-L. (1993). No germline TP53 mutations detected in familial and bilateral testicular cancer. *Genes Chromosomes Cancer* **6(2)**:92–97.

Hovig, E., Smith-Sørensen, B., Brøgger, A., and Børresen, A.-L. (1991). Constant denaturant gel electrophoresis, a modification of denaturing gel electrophoresis, in mutation detection. *Mutat. Res.* **262**:63–71.

Hovig, E., Smith-Sørensen, B., Gebhardt, M. C., Ryberg, D., Lothe, R., and Børresen, A.-L. (1992). Alterations in retinoblastoma susceptibility gene exon 21 are not common in human tumours. *Genes Chromosomes Cancer* **5** (**2**):97–103.

Hovig, E., Smith-Sørensen, B., and Børresen, A.-L. (1994). Detection of mutation by denaturing gradient gel electrophoresis, in: *Current Protocols in Human Genetics* (N. Dracopoli, D. Moir, D. Smith, J. Haines, C. Morton, B. Korf, C. Seideman, and J. Seidman, eds.) Current Protocols, New York, Chapter 7.5.

Husgafvel-Pursiainen, K., Ridanpää, M., Hackman, P., Antilla, S., Karjalainen, A., Önfelt, A., Børresen, A.-L., and Vainio, H. (1992). Detection of ras gene mutations in human lung cancer: Comparison of two screening assays based on the polymerase chain reaction. *Environ. Health Perspect.* **98**:183–185.

Khrapko, K., Hanekamp, J. S., Thilly, W. G., Belenki, A., Foret, F., and Karger, B. L. (1994). Constant denaturant capillary electrophoresis (CDCE): A high resolution approach to mutational analysis. *Nucleic Acids Res.* **22**:364–372.

Lerman, L. S., and Silverstein, K. (1987). Computational simulation of DNA melting and its application to denaturing gradient gel electrophoresis. *Methods Enzymol.* **155**:482–501.

McIntyre, J. F., Smith-Sørensen, B., Friend, S. H., Kassell, J., Børresen, A.-L., Yu Xin Yan, Russo, C., Sato, J., Barbier, N., Miser, J., Malkin, D., and Gebhardt, M. C. (1994). Germline mutations of the p53 tumor suppressor gene in children with osteosarcoma. *J. Clin. Oncol.* **12**:925–930.

Malkin, D., Jolly, K. W., Barbier-Piraux, N., Look, T., Friend, S. H., Gebbhardt, M. C., Andersen, T. I., Børresen, A.-L., Li, F. P., Garber, J., and Strong, L. C. (1992). Germline mutations in the p53 tumor suppressor gene in children and young adults with second malignant neoplasms. *N. Eng. J. Med.* **20**:1309–1315.

Peng, H. Q., Malkin, D., Bailey, D., Gallie, B. L., Bulbul, M., Jewet, M., Buchanan, J., and Goss, P. E. (1993). Mutations of the p53 genes do not occur in testis cancer. *Cancer Res.* **53**:3574–3578.

Poland, D. (1974). Recursion relation generation of probability profiles for specific-sequence macromolecules with long-range correlations. *Biopolymers* **13**:1859–1871.

Ridanpää, M., and Husgafvel-Pursiainen, K. (1993). Denaturing gradient gel electrophoresis (DGGE) assay for K-ras and N-ras genes: Detection of K-ras point mutations in human lung tumour DNA. *Hum. Mol. Genet.* **2**(**6**):639–644.

Ryberg, D., Kure, E., Lystad, S., Skaug, V., Stangeland, L., Mercy, I., Børresen, A.-L., and Haugen, A. (1994). Mutations in lung tumors. Relationship to putative susceptibility markers for cancer. *Cancer Res.* **54**:1551–1555.

Seruca, R., David, L., Holm, R., Nesland, J. M., Fangan, B. M., Castedo, S., Sobrinho-Simoes, M., and Børresen, A.-L. (1992). P53 mutations in gastric carcinomas. *Br. J. Cancer* **65**(**5**):708–710.

Sheffield, V. C., Cox, D. R., Lerman, L. S., and Myers, R. M. (1989). Attachment of a 40-base-pair G + C-rich sequence (GC-clamp) to genomic DNA fragments by the polymerase chain reaction results in improved detection of single-base changes. *Proc. Natl. Acad. Sci. USA* **86**:232–236.

Smith-Sørensen, B., Hovig, E., Andersson, B., and Børresen, A.-L. (1992). Screening for base mutations in human HPRT cDNA using the polymerase chain reaction (PCR) in combination with constant denaturant gel electrophoresis (CDGE). *Mutat. Res.* **269**:41–53.

Smith-Sørensen, B., Gebhardt, M. C., Kloen, P., Aguilar, F., Friend, S. H., and Børresen, A.-L. (1993). Screening for TP53 mutations in osteosarcomas using the polymerase chain reaction (PCR) in combination with constant denaturant gel electrophoresis (CDGE). *Hum. Mutat.* **2**:274–285.

Thorlacius, S., Børresen, A.-L., and Eyfjörd, J. (1993). Somatic p53 mutations in human breast carcinomas in an Icelandic population, a prognostic factor. *Cancer Res.* **53**: 1637–1641.

Chapter 21

Single-Strand Conformation Polymorphism Analysis

Takao Sekiya

1. INTRODUCTION

Nucleotide sequences of genomic DNA in cells can be altered by a variety of factors. If the alterations do not influence the growth and development of cells, the nucleotide sequences present in germinal cells can be transferred to progenies. If the nucleotide sequence changes influence normal cellular functions, they result in diverse diseases, such as cancers and hereditary diseases in humans.

Alterations of nucleotide sequences in human DNA can be classified into two groups: (1) changes involving large DNA regions such as amplification, rearrangement, and deletion of genes and (2) small changes of DNA such as base substitutions, insertions, and deletions of short nucleotide sequences. Large DNA changes can be detected by conventional Southern blotting. On the other hand, to detect small DNA changes, Southern blotting is not helpful and methods using completely different principles are necessary. Several of these have been developed and in conjunction with polymerase chain reaction (PCR) they have detected DNA aberrations. The most straightforward method is PCR-based direct sequencing of the genomic DNA, for example by cycle sequencing (Murray, 1989). However, finding DNA abnormalities by direct sequencing is quite laborious, because the whole target region has to be screened. Therefore, the method should be used to identify DNA changes detected by other methods. To detect single nucleotide changes at a known position in the genome, hybridization of immobilized genomic DNA or PCR products with chemically synthesized oligonucleotide probes provides a simple method. These allele-specific oligonucleotides (ASO) hybridize to perfectly matched unique sequences, but their hybridization to a target molecule is destabilized by a single internal mismatch (Conner *et al.*, 1983). When a plasmid clone carrying a target region is available, a labeled RNA probe can be prepared and used for hybridization to PCR products from the genomic DNA. The RNA probe labeled at positions where it is mismatched is cleaved by ribonuclease (RNase) A (Myers *et al.*, 1985a; Winter *et al.*, 1985). When DNA fragments having a single-base substitution are separated by electrophoresis in a gel with an increasing gradient of denaturant, partly melted DNA fragments produced at a particular concentration of the denaturant depending on the nucleotide sequences have greatly reduced mobility (Myres *et al.*, 1985b). This denaturing gradient gel electrophoresis (DGGE) method detects about 50% of all possible single-base substitutions in

---

Takao Sekiya • Oncogene Division, National Cancer Research Institute, Tokyo 104, Japan.

Technologies for Detection of DNA Damage and Mutations, edited by Gerd P. Pfeifer, Plenum Press, New York, 1996.

DNA fragments of 50 to 1000 bp. Adding a GC-rich sequence of 40 bp (GC clamp) to one end of the DNA fragments by PCR increases the number of single-base changes that can be detected by this means (Sheffield *et al.*, 1989; Abrams *et al.*, 1990).

All of these methods are based on DNA/DNA or RNA/DNA hybridization. On the other hand, the principle of a method that we have developed is completely different as shown in Fig. 21.1. complementary single strands of a DNA fragment can be separated by electrophoresis in nondenaturing polyacrylamide gels (Hayward, 1972). This strand separation is used as a step in Maxam–Gilbert sequencing (Maxam and Gilbert, 1977) and is thought to be due to their unique tertiary structures depending on their nucleotide sequences. We found that single-stranded DNA fragments carrying the same nucleotide sequence but with a single-base substitution moved differently in nondenaturing polyacrylamide gels. Assuming that this mobility shift was due to an altered conformation of the single-stranded DNA by a single-base change, we named this feature of single-stranded DNA single-strand conformation polymorphism (SSCP) (Orita *et al.*, 1989a). Combined use of the PCR and SSCP analysis provides a simple and sensitive method for detecting single-base substitutions (Orita *et al.*, 1989b).

2. METHODS

Figure 21.2 shows the results of SSCP analysis of a region carrying exon 5 of the *p53* gene in DNAs from human cancer cell lines (Murakami *et al.*, 1991). The mobility of separated strands has shifted in the pancreatic cancer cell line PSN1, indicating the presence of a mutation. To identify the mutation, a single-stranded DNA fragment can be eluted from a dried gel corresponding to the shifted band after preparing an autoradiogram and amplified using the same primer (Suzuki *et al.*, 1991). Cycle sequencing of DNA fragment thus amplified revealed a single base substitution at codon 132. Furthermore, the absence of signals corresponding to the normal sequence in the SSCP analysis simultaneously reveals the loss of one of the p53 alleles in PSN1 cells. Abnormalities of mRNA can also be analyzed by SSCP, when the total RNA is converted to cDNA and amplified by PCR. In Fig. 21.2, mutated mRNA expression is indicated in PSN1 cells by mobility shift.

2.1. Instruments

1. Temperature cycler for PCR: We use a Thermal Cycler 480 (Perkin–Elmer Cetus) or Gene Amp PCR System 9600 (Perkin–Elmer Cetus). The latter is quite convenient, because it does not require the addition of mineral oil into the reaction tubes.
2. Apparatus for polyacrylamide gel electrophoresis: Any electrophoresis apparatus for sequencing can be used. We use an electrophoresis apparatus equipped with a water jacket, AE-6160-Genokencer SSCP (Atto, Tokyo, Japan). In this case, a circulating cooler system is necessary. When an apparatus without a cooling system is used, we cool both sides of gel surfaces with two fans, such as a cross-flow fan, FG-11027-AA (31 W) or FB-06017-AA (22 W) (Royal Electric Co., Tokyo, Japan).

2.2. Reagents

1. [γ-^{32}P]-ATP (160 mCi/ml, 7000 Ci/mmole, ICN) or [α-^{32}P)-dCTP (10 mCi/ml, 3000 Ci/mmole, Amersham)
2. Polynucleotide kinase (10 units/ml) (Boehringer-Mannheim Co.)
3. PCR kit: Gene Amp TM DNA Amplification Reagent Kit (Perkin–Elmer Cetus) with Taq DNA polymerase (5 units/ml)

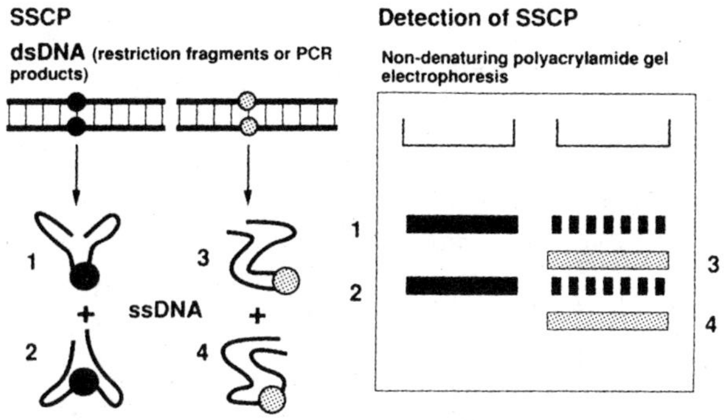

Figure 21.1. Principle of SSCP analysis.

4. 10 × PCR buffer: 500 mM KCl, 100 mM Tris-HCl (pH 8.3), 15 mM $MgCl_2$, and 0.1% (w/v) gelatin
5. 4 dNTP mixture: Four dNTP solutions in the kit are mixed and diluted to 1.25 mM each.
6. Mineral oil (Light) (Sigma). The oil is not necessary when Gene Amp PCR System 9600 is used.
7. 50% acrylamide: The ratio of acrylamide to *NN'*-methylene-bisacrylamide is 49 to 1 or 99 to 1.
8. 50% glycerol
9. 1.6% ammonium persulfate
10. *N,N,N',N'*-tetramethyl ethylene diamide (TEMED)
11. DNA sample: DNA is prepared by the proteinase K–phenol–chloroform extraction method (Blin and Stafford, 1976) and dissolved in TE buffer [10 mM Tris-HCl (pH 7.0), 1 mM EDTA]. The highly concentrated DNA solution is stored at 4°C. For PCR, a diluted solution (50 ng/μl) is prepared.
12. Primers: Oligonucleotides are prepared using a DNA synthesizer and a highly concentrated solution in TE buffer is stored at −20°C. For PCR, a diluted solution (10 mM) is used.

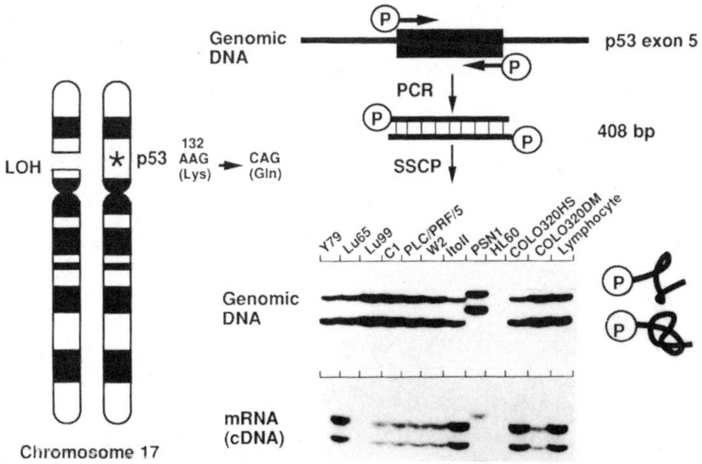

Figure 21.2. SSCP analysis of DNA fragments of 139 bp amplified by PCR and carrying the sequence of exon 7 of the *p53* gene. Genomic DNAs and cDNAs from human cancer cell lines were analyzed. Electrophoresis was performed at 25°C in a 6% polyacrylamide gel containing 10% glycerol at 30 W for 3–5 hr.

2.3. Buffers and Solutions

All reagents are dissolved in deionized water.

10× polynucleotide kinase (PNK) buffer

		Final concentration
1 M Tris-HCl (pH 8.3)	0.5 ml	500 mM
1 M $MgCl_2$	0.1 ml	100 mM
1 M dithiothreitol (DTT)	0.02 ml	20 mM
Deionized H_2O	0.38 ml	
Total	1 ml	

Aliquots of 200 μl are stored at −20°C.

Formamide-dye (F-dye) solution

		Final concentration
Formamide	95 ml	95%
0.5 M EDTA (pH 7.5)	2 ml	10 mM
Bromophenol blue	50 mg	0.05%
Xylene cyanol	50 mg	0.05%
Deionized H_2O	3 ml	
Total	100 ml	

Formamide (Merck) is deionized using Amberlite MB1. The solution is stored at 4°C.

10× TEB

		Final concentration
Trizma base	109 g	0.9 M
Boric acid	56 g	0.9 M
EDTA2Na	7.5 g	20 mM
Deionized H_2O	add to 1 liter	
Total	1 liter	

The solution is stored at room temperature.

2.4. Protocol for SSCP analysis

2.4.1. Labeling of PCR Products Using the 5′-^{32}P-Labeled Primer (Mashiyama *et al.*, 1990)

1. Preparation of the 5′-^{32}P-labeled primer (for 80 reactions) Polynucleotide kinase (PNK) mixture

		Final concentration
Deionized H_2O	1 μl	
10× PNK buffer	0.5 μl	1× PNK buffer
Primer 1	1 μl	2 μM
Primer 2	1 μl	2 μM
[γ-^{32}P]-ATP	1 μl	4.6 μM
Polynucleotide kinase	0.5 μl	1 unit/μl
Total	5 μl	

The reaction proceeds at 37°C for 30 min.

2. Labeled primer (LP) solution

		Final concentration in PCR mixture
PNK mixture	5 μl	25 pM primers
10× PCR buffer	40 μl	1× PCR buffer
4 dNTP	20 μl	62.5 μM
Deionized water	255 μl	
Total	320 μl	

The solution is stored at −20°C and can be used for several weeks.

3. PCR solution (for 10 reactions)

		Final concentration
LP solution	40 μl	
Taq DNA polymerase	0.5 μl	0.06 unit/μl
Total	40.5 μl	

Aliquots of 4 μl of the solution are transferred to 0.5-ml tubes and 1 μl of sample DNA solution (50 ng/μl) is added to each. Mineral oil (15 μl) is overlaid, if necessary. As a standard PCR condition, 30 cycles of reactions are performed using a thermal cycler; each cycle consists of denaturation at 94°C for 30 sec, reannealing at 55°C for 30 sec and 72°C for 60 min.

2.4.2. Labeling the PCR Products Using a Labeled Nucleotide Substrate

1. PCR mixture (for 20 reactions)

		Final concentration
10× PCR buffer	10 μl	1× PCR buffer
Primer 1	2 μl	200 pM
Primer 2	2 μl	200 pM
4 dNTP	2 μl	25 μM each
[α-^{32}P]dCTP	4 μl	1.3 μM
Taq DNA polymerase	0.5 μl	0.03 unit/μl
Deionized H_2O	60 μl	
Total	80.5 μl	

Aliquots of 4 μl of the mixture are transferred to 0.5-ml reaction tubes and 1 μl of DNA sample (50 ng/μl) is added to each. PCR is performed as described above.

2.5. SSCP Analysis

1. After PCR, 5 μl of the reaction mixture is added to 45 μl of F-dye solution. The diluted solution is heated at 80°C for 5 min to denature DNA fragments to single strands and 1 μl of the heated mixture is resolved by nondenaturing polyacrylamide gel electrophoresis.

2. Preparation of polyacrylamide gel for SSCP analysis
SSCP gel solution.

		Final concentration
10× TBE	1.5 ml	0.5× TBE
50% acrylamide solution	3 ml	5%
1.6% ammonium persulfate	1 ml	0.05%
(50% glycerol	3 or 6 ml	5 or 10%)
Deionized H_2O	21.5 or 18.5 ml	
Total	30 ml	

After adding of 30 μl of TEMED, the solution is poured into the space between two glass plates (20 × 40 cm) separated by strips of plastic (0.3 mm thick). The surfaces of the glass plates are cleaned using a detergent. After washing with water, the surfaces are wiped using tissue paper wetted with ethanol. The surface of one of the glass plates is siliconized. The flat side of a comb is placed at the top of the gel and the glass plates are placed horizontally for more than 1 hr. When polymerization of the acrylamide is completed, a set of glass plates containing polyacrylamide is fixed to an electrophoresis apparatus equipped with a cooling system. When an apparatus without a cooling system is used, an aluminum plate is attached to one side of the glass plates and both sides of the plates are cooled using two fans. The comb is removed and the other side of the comb with a serrated edge is placed into the top of the gel.

3. Electrophoresis proceeds in 0.5 × TBE at 30–40 W at room temperature (20°C) for several hours. We recommend using two polyacrylamide gels, one containing 5 or 10% glycerol and the other without glycerol. The time required for the electrophoresis differs depending on the temperature and the presence of glycerol. Therefore, it is necessary to determine the appropriate period for electrophoresis by a preliminary experiment. After electrophoresis, the gel is transferred to a sheet of Whatmann 3 MM paper, covered with Saran Wrap, then dried using a gel dryer. The dried gel is exposed to X-ray film. At the time of the exposure, the gel and the film are fixed at three positions using a stapler. When mobility shifts are small, a short exposure is helpful for their detection. On the other hand, if the radiochemical concentration in the gel is low, longer exposure is helpful to increase the intensity of the signals.

2.6. Identification of Mutations

2.6.1. Separation of Alleles

On the autoradiogram the dried gel is placed at the positions marked with a stapler during exposure. From the gel, small pieces (2 × 2 mm) at the positions corresponding to bands containing a mutated and normal DNA fragments are cut and put into 1-ml reaction tubes. After adding 20 μl of deionized water, the gel is melted by heating the mixture at 80°C for 15 min, then the mixture is centrifuged at 15,000 rpm for several minutes. One microliter of the supernatant containing the eluted single-stranded DNA fragment is amplified by PCR as follows.

2.6.2. Amplification of the Eluted Fragment

PCR reaction for the eluted DNA

		Final concentration
Eluted DNA fragment	1 μl	
10× PCR buffer	2 μl	1× PCR buffer
4 dNTP	2 μl	125 μM
Primer 1	2 μl	1 μM

Primer 2	2 μl	1 μM
Taq DNA polymerase	0.4 μl	0.1 unit/μl
Deionized H_2O	11 μl	
Total	20 μl	

Mineral oil (20 μl) is overlaid and PCR proceeds under the conditions described above. After the reaction, 20 μl of chloroform is added to the reaction tube and mineral oil extracted in the lower layer is removed.

2.6.3. Purification of Amplified DNA Fragments

Using a desalting and concentrating microcentrifuge system, Microcon 100 (Amicon), excess primer, substrates, salts, and others in the reaction mixture are removed by two cycles of 500 μl of deionized H_2O and centrifugation at 3000*g* for 10 min.

2.6.4. Cycle Sequencing

1. Sequencing is performed using about one-tenth of the PCR products amplified and purified from the eluted DNA fragments above and a commercially available kit such as the Cyclist Exo$^-$ *Pfu* DNA Sequencing Kit (Stratagene) and a double-strand DNA Cycle Sequencing System (GIBCO BRL).

When the Cyclist Exo$^-$ *Pfu* DNA Sequencing Kit is used, the protocol is as follows.

2. Reagents

- The same primers used for PCR-SSCP analysis (10μM)
- [γ-^{32}P]-ATP (160 mCi/ml, 7000 Ci/mmole, ICN)
- Polynucleotide kinase (10 units/μl, Boehringer-Mannheim)
- 10× PNK buffer
- Reagents provided in the kit

  10× sequencing buffer: 200 mM Tris-HCl (pH 8.8), 20 mM $MgSO_4$, 100 mM KCl, 100 μM $(NH_4)_2SO_4$, 1% Triton, 1 mg/ml bovine serum albumin, 20 μM dATP, 50 μM dCTP, 50 μM dGTP, and 50 μM dTTP

  Exo$^-$ *Pfu* DNA polymerase: 2.5 units/μl

  Stop dye mixture: 80% formamide, 1 mM EDTA, 50 mM Tris-HCl (pH 8.3), 0.1% (w/v) bromophenol blue, and 0.1% (w/v) xylene cyanol

  Dimethyl sulfoxide (DMSO)

  Dideoxy nucleotide solution: ddATP (1.5 mM), ddCTP (1.5 mM), ddGTP (1.5 mM), and ddTTP (1.5 mM)

3. Labeling of primers

Reaction mixture

		Final concentration
Primer (10 mM)	1.2 μl	3 μM
10× PNK buffer	0.4 μl	1× PNK buffer
Polynucleotide kinase	0.4 μl	1 unit/μl
[γ-^{32}P]-ATP	1 μl	5.6μM
Deionized H_2O	1 μl	
Total	4 μl	

The reaction is performed at 37°C for 30 min. Thereafter, 56 μl of deionized H_2O is added to the mixture and the diluted reaction mixture is heated at 55°C for 5 min, then placed on ice.

4. Cycle sequencing

Deionized H_2O is added to the DNA fragments (5–10 μl) recovered from the gel pieces, purified, and concentrated to a final volume of 20 μl.

Sequencing reaction mixture

		Final concentration
10× sequencing buffer	4 μl	1× buffer
DNA solution	5 μl	
Primer solution	6 μl	0.04 μM
Exo^- *Pfu* DNA polymerase	1 μl	0.08 unit/μl
DMSO	4 μl	
Total	20 μl	

Aliquots of 7 μl are transferred to four reaction tubes and 2 μl of one of ddTTP, ddCTP, ddGTP, and ddATP (430 μM) is added to one of the tubes. After mixing the solution, 10 μl of mineral oil is added to the tubes. The reaction is performed as described for PCR-SSCP. After the reaction, 5 μl of stop dye mixture is added to the mixture, which is then vigorously mixed and centrifuged.

2.6.5. Preparation of Polyacrylamide Gel for Sequencing

Mixture for sequencing

		Final concentration
Urea	13.5 g	7 M
Acrylamide solution	3.75 ml	6%
1.6% ammonium peroxide	1 ml	0.05%
10× TBE	3 ml	1× TBE
Deionized H_2O	add to 30 ml	
Total	30 ml	

By warming and mixing, all reagents are dissolved and the solution is filtered through a membrane (Acrodisk, 0.2 μm, Gelman Sciences) affixed to a syringe. After adding 30 μl of TEMED, the solution is poured into the space between two glass plates as described above and kept at room temperature for over 1 hr.

2.6.6. Polyacrylamide Gel Electrophoresis

Each sequencing reaction mixture containing stop dye mixture is heated at 90°C for 5 min, then immediately 1–2 μl is loaded onto a 6% polyacrylamide gel containing 7 M urea. When the apparatus is equipped with a water jacket, electrophoresis is performed in 1× TBE at 80 W, 55°C, for 1 hr. Although the period for electrophoresis is dependent on the chain length of the DNA samples, electrophoresis is usually stopped when bromophenol blue has just run out of the gel.

3. DISCUSSION

The most important advantage of PCR-SSCP analysis is that it can detect single-base substitutions at unknown positions of target DNA fragments. The efficiency of detection of single-base substitutions in fragments of less than 300–350 bp is estimated to be more than 95% when both strands are labeled and are separated in gel containing glycerol. Therefore, the upper and lower primers should be set to amplify a target region of the DNA fragments having a chain length less than 300–350 bp. The efficiency will decrease gradually in DNA fragments with increasing chain lengths above 300–350 bp. When the PCR gives heterogeneous products instead of a single product, the positions for the primer set should be moved slightly on the target sequences. Mobility shifts due to a single-base substitution can be obtained more efficiently by electrophoresis in a polyacrylamide gel with, than without glycerol. However, in some sequences the mobility shift due to one base change is much larger in the absence of glycerol. Therefore, we recommend using both types of gels for each analysis. Essentially the same effect is observed when glycerol is added at ratios of 5 and 10%. However, a single-base change in a sequence sometimes gives a larger mobility shift in gels containing 10% glycerol.

The temperature of gels during electrophoresis is also an important factor for detecting single-base substitutions. In some sequences, mobility shifts due to a single-base substitution are observed after electrophoresis at 4–10°C, but not at room temperature. At low temperature, a polyacrylamide gel without glycerol should be used, because of the extremely slow mobility of single-stranded DNA fragments in the presence of glycerol. Thus, the mobility of single-stranded DNA fragments in polyacrylamide gels is influenced by temperature and therefore an electrophoresis apparatus equipped with a cooling system is recommended.

In conclusion, the use of electrophoresis at room temperature in polyacrylamide gels with or without glycerol, and at low temperature in gels without glycerol, can provide an efficient analysis that minimizes false-negative results.

Another great advantage of PCR-SSCP analysis is that it separates and purifies mutated fragments with different mobilities in polyacrylamide gels (Hata *et al.*, 1990; Suzuki *et al.*, 1991). PCR-SSCP analysis can thus detect very faint signals due to mobility shifts caused by a mutation in a cancer specimen containing only a very low proportion of tumor cells. After autoradiography of the dried gel, a tiny piece corresponding to the position of the separated band is excised and the single-stranded DNA fragment is eluted. PCR then amplifies a large amount of the fragment. Nucleotide sequencing of the PCR products identifies the target gene and the mutation (Suzuki *et al.*, 1991).

When significant numbers of DNA samples are analyzed, labeled primers can be used for labeling the PCR products effectively and inexpensively. On the other hand, when a few DNA samples are analyzed or they are amplified with new primers, internal labeling of the PCR products with labeled nucleotide substrate is simple and convenient.

DNA fragments amplified by PCR often show more bands than those of the two strands separated by SSCP. One reason for this might be the presence of two or more equally stable, but different conformations of a single-stranded DNA. Similarly to identification of base substitution in DNA fragments showing a mobility shift, these isoconformers can be characterized by reamplification and sequencing the DNA fragments eluted from a gel piece corresponding to the signal on an autoradiogram.

The extensive application of PCR-SSCP analysis, especially in clinical laboratories, has been hampered by its requirement for radioactive nucleotides. To overcome this, nonisotopic detection of SCCP by silver or ethidium bromide staining of single-stranded DNA in polyacrylamide gel has been developed (Dockhorn-Dworniczak *et al.*, 1991; Mohabeer *et al.*, 1991; Yap and McGee, 1992). We found that fluorescence-labeled PCR products could be used for

SSCP analysis. The use of fluorescence-labeled primers for the PCR and analysis of SSCP with an automated DNA sequencer (F-SSCP) is as efficient as conventional PCR-SSCP (Makino *et al.*, 1992).

REFERENCES

Abrams, E. S., Murdaugh, S. E., and Lerman, L. S. (1990). Comprehensive detection of single base changes in human genomic DNA using denaturing gradient gelelectrophoresis and a GC clamp. *Genomics* **7**:463–475.

Blin, N., and Stafford, D. W. (1976). A general method for isolation of high-molecular-weight DNA from eukaryotes. *Nucleic Acids Res.* **3**:2303–2308.

Conner, B. J., Reyes, A. A., Morin, C., Itakura, K., Teplitz, R. L., and Wallace, R. B. (1983). Detection of sickle cell b^S-globin allele by hybridization with synthetic oligonucleotides. *Proc. Natl. Acad. Sci. USA* **80**:278–282.

Dockhorn-Dworniczak, B., Dworniczak, B., Brömmelkamp, L., Bülles, J., Horst, J., and Böcker, W. W. (1991). Non-isotopic detection of single strand conformation polymorphism (PCR-SSCP); a rapid and sensitive technique in diagnosis of phenylketonuria. *Nucleic Acids Res.* **19**:2500.

Hata, A., Robertson, M., Emi, M., and Lalouel, J.-M. (1990). Direct detection and automated sequencing of individual alleles after electrophoretic strand separation: Identification of a common nonsense mutation in exon 9 of the human lipoprotein lipase gene. *Nucleic Acids Res.* **18**:5407–5411.

Hayward, G. S. (1972). Gel electrophoretic separation of the complementary strands of bacteriophage DNA. *Virology* **49**:342–344.

Makino, R., Yazyu, H., Kishimoto, Y., Sekiya, T., and Hayashi, K. (1992). F-SSCP: A fluorescent polymerase chain reaction–single strand conformation polymorphism (PCR-SSCP) analysis. *PCR Methods Appl.* **2**:10–13.

Mashiyama, S., Sekiya, T., and Hayashi, K. (1990). Screening of multiple DNA samples for detection of sequence changes. *Technique* **2**:304–306.

Maxam, A. M., and Gilbert, W. (1977). A new method for sequencing DNA. *Proc. Natl. Acad. Sci. USA* **74**: 560–564.

Mohabeer, A. J., Hiti, A. L., and Martin, W. J. (1991). Nonradioactive single strand conformation polymorphism (SSCP) using the Pharmacia 'PhastSystem.' *Nucleic Acids Res.* **19**:3154.

Murakami, Y., Hayashi, K., and Sekiya, T. (1991). Detection of aberrations of the p53 alleles and the gene transcript in human tumor cell lines by single-strand conformation polymorphism analysis. *Cancer Res.* **51**:3356–3361.

Murray, V. (1989). Improved double-stranded DNA sequencing using the linear polymerase chain reaction. *Nucleic Acids Res.* **17**:8889.

Myers, R. M., Larin, Z., and Maniatis, T. (1985a). Detection of single base substitutions by ribonuclease cleavage at mismatches in RNA:DNA duplexes. *Science* **230**:1242–1246.

Myers, R. M., Lumelsky, N., Lerman, L. S., and Maniatis, T. (1985b). Detection of single base substitutions in total genomic DNA. *Nature* **313**:495–498.

Orita, M., Iwahana, H., Kanazawa, H., Hayashi, K., and Sekiya, T. (1989a). Detection of polymorphisms of human DNA by gel electrophoresis as single-strand conformation polymorphisms. *Proc. Natl. Acad. Sci. USA* **86**:2766–2770.

Orita, M., Suzuki, Y., Sekiya, T., and Hayashi, K. (1989b). Rapid and sensitive detection of point mutations and DNA polymorphisms using the polymerase chain reaction. *Genomics* **5**:874–879.

Sheffield, V. C., Cox, D. R., Lerman, D. R., and Myers, R. M. (1989). Attachment of a 40-base-pair G+C-rich sequence (GC-clamp) to genomic DNA fragments by the polymerase chain reaction results in improved detection of single-base changes. *Proc. Natl. Acad. Sci. USA* **86**:232–236.

Suzuki, Y., Sekiya, T., and Hayashi, K. (1991). Allele-specific polymerase chain reaction: A method for amplification and sequence determination of a single component among a mixture of sequence variants. *Anal. Biochem.* **192**:82–84.

Winter, E., Yamamoto, F., Almoguera, C., and Perucho, M. A. (1985). Methods to detect and characterize point mutations in transcribed genes: Amplification and overexpression of the mutant c-Ki-ras allele in human tumor cells. *Proc. Natl. Acad. Sci. USA* **82**:7575–7579.

Yap, E. P. H., and McGee, J. O. D. (1992). Nonisotopic SSCP and competitive PCR for DNA quantification: p53 in breast cancer cells. *Nucleic Acids Res.* **20**:145.

Chapter 22

Two-Dimensional Gene Scanning

Daizong Li, Nathalie van Orsouw, Chris Huang, and Jan Vijg

1. INTRODUCTION

1.1. Two-Dimensional Gene Scanning: Principles

The large size of many human genes, in combination with the frequent occurrence of many different mutations over the entire gene, severely limits direct diagnosis. Since gene diagnosis by sequencing on a large scale is not yet cost-effective, other methods such as the analysis of single-strand conformation polymorphisms (SSCP), heteroduplex analysis, and analysis by denaturing gradient gel electrophoresis (DGGE) are presently being employed (for a review, see Cotton, 1993; see also elsewhere in this volume). While these systems are very useful, they are difficult to adapt for large-scale analyses and most of them are restricted in their detection rates. Moreover, all currently available methods lack uniformity and standardization.

Based on DGGE, two-dimensional gene scanning (TDGS) was developed as part of a general approach to efficiently analyze multiple DNA fragments simultaneously for the presence of sequence variation (Fischer and Lerman, 1979a; Uitterlinden *et al.*, 1989; Uitterlinden and Vijg, 1994; Vijg, 1995a,b). TDGS involves the two-dimensional electrophoretic separation of DNA fragments, first on the basis of size in non-denaturant polyacrylamide (PAA) gels, and subsequently on the basis of sequence in PAA gels containing a gradient of denaturants, i.e., increased temperature or increased concentration of urea/formamide. The mutational target fragments are most easily obtained by PCR amplification from the relevant loci in genomic DNA.

Being based on DGGE, TDGS has a high accuracy of detecting point mutations. Indeed, evidence is now rapidly accumulating that DGGE is virtually 100% accurate in detecting such small DNA sequence variations, provided the target fragment is coupled to a GC-rich clamp fragment serving as the highest-melting domain (Myers *et al.*, 1985; Sheffield *et al.*, 1989; Abrams *et al.*, 1990; see also elsewhere in this volume). Heteroduplexing further increases the capability of the system to physically separate mutational variants (e.g., Uitterlinden and Vijg, 1994). In that format DGGE has proven to be superior to other mutation detection systems,

---

Daizong Li, Nathalie van Orsouw, Chris Huang, and Jan Vijg • Molecular Genetics Section, Gerontology Division, Department of Medicine, Beth Israel Hospital and Harvard Medical School, Boston, Massachusetts 02215.

Technologies for Detection of DNA Damage and Mutations, edited by Gerd P. Pfeifer, Plenum Press, New York, 1996.

including SSCP (Sheffield *et al.*, 1993; Moyret *et al.*, 1994), with an accuracy that appears to equal that of nucleotide sequencing (Guldberg *et al.*, 1993).

The advantages of two-dimensional instead of one-dimensional DGGE are: (1) the much higher resolution, allowing multiple fragments (between 100 and 600 bp) to be scanned for mutations in parallel; (2) the unequivocal identification of each target fragment on the basis of both size and melting temperature; and (3) the possibility to immediately distinguish between size changes and point mutations. Finally, the system lends itself well for complete automation, both the electrophoresis part and the 2-D spot pattern interpretation (Mullaart *et al.*, 1993).

1.2. Test Design: General Aspects

The potential of 2-D DGGE for comprehensive mutation analysis in single genes has been demonstrated by our earlier results with the cystic fibrosis transmembrane conductance regulator (CFTR) gene. Using primers mostly specified by others for use in 1-D DGGE (Costes *et al.*, 1993), we were able to detect 17 out of 17 previously identified mutations in a format consisting of 29 spots covering all 27 CFTR exons (Wu *et al.*, 1996). The CFTR 2-D format, however, was not based on extensive multiplexing and since the design of multiplex PCR reactions for large genes is not trivial we worked out a more generally applicable format, which is schematically depicted in Fig. 22.1. Instead of using total genomic DNA as direct template to obtain DGGE-optimized amplicons, two different PCR steps are carried out consecutively (Li and Vijg, 1996). First, all exon-containing genomic sequences are amplified by long-PCR. Then, DGGE-optimized exon fragments are coamplified with the long-PCR product(s) as template.

The purpose of the first preamplification step is to obtain overrepresentation of the exon-containing genomic sequences relative to all other genomic DNA. This greatly facilitates the selection of one set of PCR conditions using primers designed solely for optimal melting behavior of the fragments surrounded by them. Moreover, the increased amounts of target sequences prespecified by the long-PCR permit extensive multiplexing. Currently the most extensive multiplex reactions are the 9-fragment sets for the dystrophin gene designed by Chamberlain *et al.* (1988) and Beggs *et al.* (1990), while most multiplex sets are no more than 5 fragments (Edwards and Gibbs, 1994). The reason for this is that with each primer set added, the permissive reaction conditions allowing each fragment to reach its annealing temperature while evading spurious amplification products become increasingly less flexible. The two-step multiplex PCR has allowed us to generate as many as 25 DGGE-optimized RB1 exon fragments in a single reaction in one tube (see below and Li and Vijg, 1996). Subsequent separation of the fragment mixture on the basis of both size and base pair sequence permits the reliable detection of all kinds of mutations within large genes consisting of many exons.

Table I lists the different steps in the design of a TDGS test, based on our experience mainly with the RB1 tumor suppressor gene. First, the gene sequences are retrieved from a database (e.g., Genbank) and the target regions, i.e., exons, splice sites, regulatory regions, are defined. (TDGS designs can also be made for gene transcripts, based on the use of RT-PCR.) Then, primers are positioned to obtain all target regions as the smallest possible number of fragments that can still be amplified through long-PCR, i.e., up to at least 20 kb (TaKaRa LA PCR Kit, Product Insert). For the relatively small p53 gene, for example, this can only be one fragment, while the much larger RB1 gene required as many as six long-PCR amplicons (see below). Some general guidelines in choosing primer sequences for long-PCR have been described (Foord and Rose, 1994). In general, this aspect of the test design should be easy and multiplexing is not a problem; ample margin for adjustment of primer position is available.

Then, using the long-PCR fragments as template, primers for short-PCR are selected to yield fragments of between 100 and 600 bp. The main selection criterion here is the melting

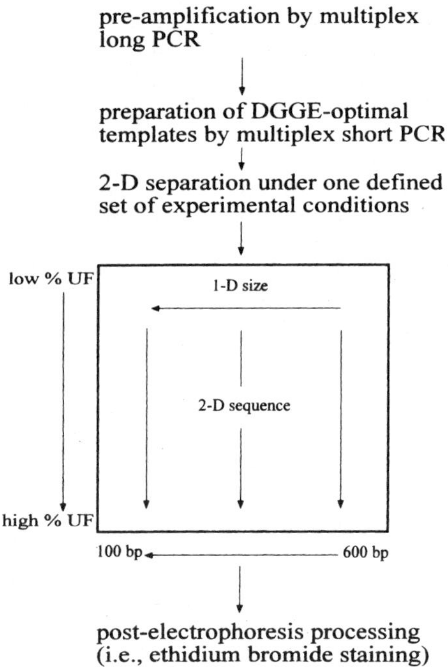

Figure 22.1. Schematic depiction of two-dimensional gene scanning (TDGS).

behavior of the fragments. In the ideal situation, each amplicon should comprise only one melting domain, which should be lower (less stable) than the GC-clamp attached to it. Attachment of a 30- to 40-bp GC-clamp is accomplished by making it part of one of the primers (Sheffield *et al.*, 1989). Optimal melting behavior is determined of each candidate target sequence by using a computer program, plotting T_m as a function of position along the molecule (e.g., MELT87; Lerman and Silverstein, 1987). An example of an amplicon with optimized melting behavior through GC-clamping is shown in Fig. 22.2. In general, a collection of primers is selected that allows an optimal distribution of fragments, in both size and DGGE dimension, over the 2-D gel. Due to the high resolution of 2-D gels (5- to 10-bp size differences are easily resolved) this is generally not too difficult. Indeed, with 50 fragments or less, spot distribution is hardly an issue and primers can simply be selected according to their melting behavior.

1.3. RB1 and p53 TDGS Test Designs

Figure 22.3 shows the collections of amplicons selected for the RB1 and p53 gene, together with the long-PCR fragments that served as templates. All primers are listed in Table II. Together the short-PCR fragments represent more than 90% and more than 70% of the RB1 and p53 coding regions, respectively. Exons 2–3 and 10–11 of p53 and exons 1 and 15–16 of RB1 were excluded from the current test designs (see below). Figure 22.4 shows the empirical spot distribution for the 24 exons of the RB1 gene covered by the amplicons shown in Figure 22.3, as compared to the predicted pattern. Although there are differences, our conclusion is that overall the melting program accurately predicts spot positions. The same is true for the seven amplicons representing the mutational hot spot region of the p53 gene; the theoretical pattern matches the empirical one

Table I
Design of a TDGS Test on Genomic DNA

1. Retrieve sequence from database
2. Position primers for long PCR to cover all desired regions (e.g., coding sequences, splice sites, regulatory regions) by the smallest possible number of amplicons; avoid complementarity among primers or between primers and genomic DNA
3. Position primers for short PCR according to the following criteria:
   a. Optimal coverage of the target sequences by amplicons of between 100 and 600 bp; overlapping fragments should be in different multiplex groups
   b. Amplicons should have optimal melting behavior and preferably consist of one lowest-melting domain in addition to the GC-clamp attached to one of the primers
   c. Optimal fragment distribution over the 2-D gel
   d. Similar PCR reaction kinetics, no complementarity with other primers or target sequences
4. Set up PCR conditions separately for each primer pair with one long-PCR product as template and develop multiplex coamplification conditions per long-PCR fragment; if necessary, reposition long-PCR primers
5. Develop multiplex coamplification conditions for long PCR
6. Develop short PCR multiplex coamplification conditions for all primer pairs by combining the long PCR groups and adjusting reaction components

very well (Fig. 22.5). The RB1 empirical pattern in Fig. 22.4 (right) was obtained from DNA extracted from the blood of a heritable retinoblastoma patient and indicates a mutation in exon 13. The p53 pattern was obtained from a control sample and contains no mutations (see also below).

A practical problem concerns the sometimes great differences in fragment melting behavior. Certain gene sequences can have very high melting temperatures. For many genes exon 1 has a high T_m, which was the reason to exclude it from the RB1 design. By contrast, RB1 exons 15/16 are AT-rich and melt very early. There is also an overall difference in melting temperature

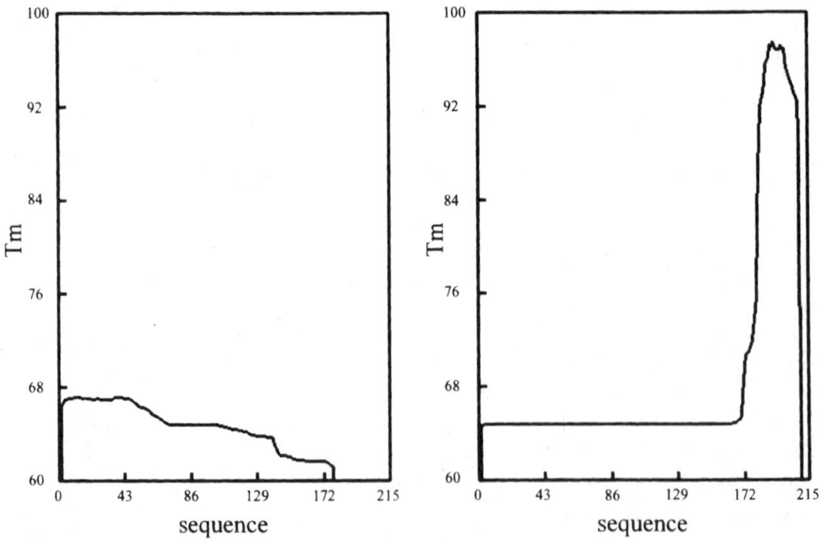

Figure 22.2. Melting curves for RB1 exon 12, with (right) and without (left) the GC-clamp.

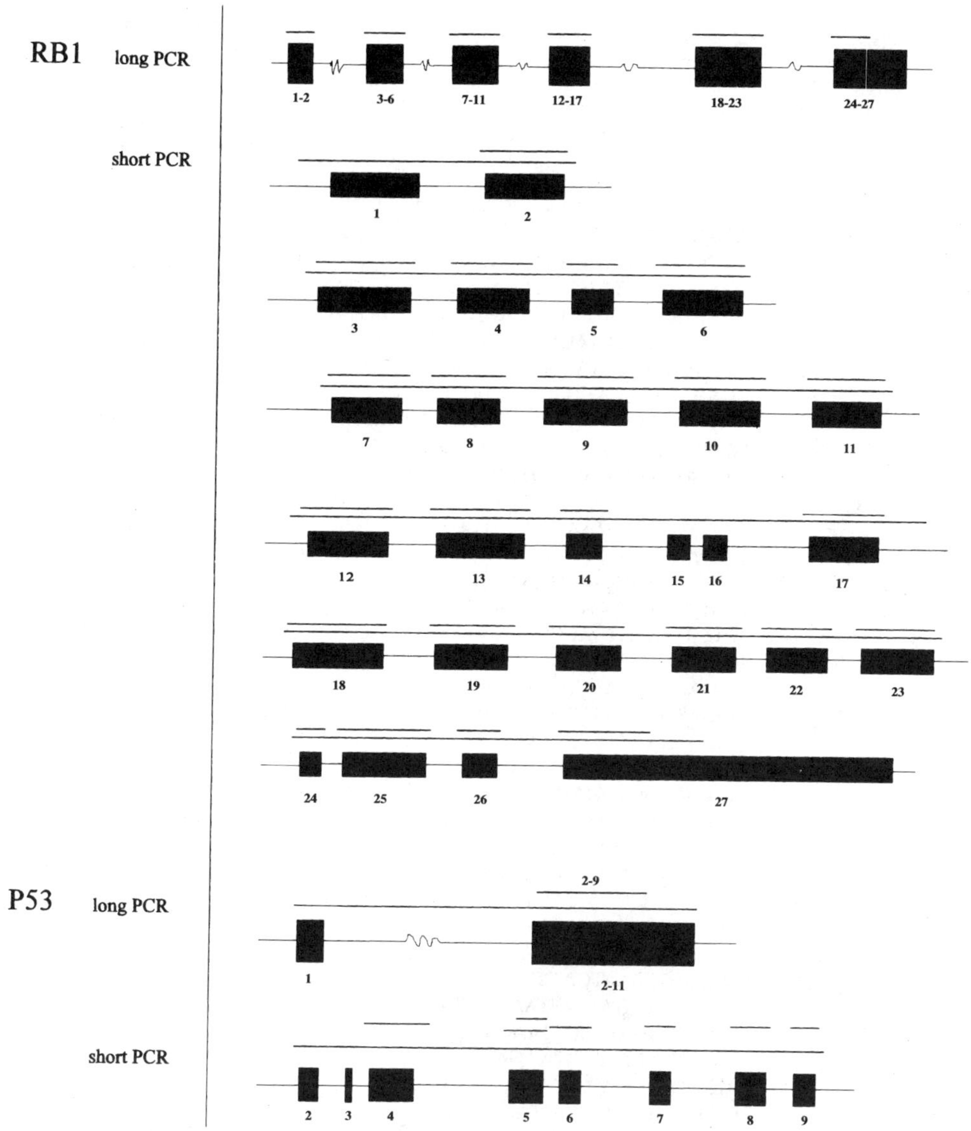

Figure 22.3. Maps of the tumor suppressor genes RB1 and p53 indicating the long- and short-PCR amplicons.

between the RB1 and p53 genes (66 and 76°C, corresponding to about 30 and 60% urea/formamide, respectively). Preliminary results indicate that under a different set of experimental conditions (i.e., 0–100% UF) it is possible to combine all exons of the two genes, including the relatively GC-rich ones, to form a single test. Under such conditions mutations can be detected, but the migrational differences between the mutational variants become very small. Although excluded from the RB1 design, a fragment representing exons 15/16 could be included in the PCR multiplex, which brought the total number of coamplified fragments (in the same reaction and

Table II
Primer Pairs for TDGS of RB1 and p53

	Exons	Primers 5′ → 3 ′	Size	T_m (% UF)
RB1				
Long PCR	1–2	TGTCAGGCCTGCCTGACAGACTTCTATTCAGCA ATGTTAGCAGAGGTAAATTTCCTCTGGGTAATGG	4.5 kb	
	3–6	GCAGTCATTTCCCAACACCTCCCCTCTGT AAGCCAAGCAGAGAATGAGGGAGGAGTACATTAC	9 kb	
	7–11	TCAGCAGTTTCTCCCTCCAAGTCAGAGAGGC GAGACCAGAAGGAGCAAGATCAGGTAGTAG	10 kb	
	12–17	ACCATTCCCCCTACTCTCCATGGTCCATG CTCACAGGAAAAATACACAGTATCCTGTTTGTGTGGC	12.4 kb	
	18–23	CCAGCCTTGCATTCTGGGGATGAAGC AGTCGTAAATAGATTTTCTTCACCCCGCCCC	14 kb	
	24–27	GCCTTTGCCCTCCCTAAATATGGGCAATGG CTGGGTTATCAGGACTCCCACTCTAGGGCC	7.3 kb	
Short PCR	2	[GC1]TTGATTTATAAGTATATGCCA CAAAACGTTTTAAGAAAATCC	229 bp	30
	3	[GC1]CCAGTGTGTGAATTATTTAA CCTTTTATGGCAGAGGCTTATA	239 bp	27
	4	[CG1]GAATTGAAATATCTATGATT ATCAGAGTGTAACCCTAATA	270 bp	24
	5	[GC1]TACTATGACTTCTAAATTACG GTGAAAAATAACATTCTGTG	157 bp	27
	6	TGGAAAACTTTCTTTCAGTG [GC1]GAATTTAGTCCAAAGGAATGC	237 bp	17
	7	[GC1]CCTGCGATTTTCTCTCATAC GCAACTGCTGAATGAGAAAG	257 bp	26
	8	GCGCCCGTCCCGCCCTTGTTCTTATCTAATTTACC [GC1]TTTTAAAGAAATCATGAAGTT	246 bp	27
	9	[GC1]AGTCAAGAGATTAGATTTTG ATCCTCCCTCCACAGTC	227 bp	20
	10	[GC1]GACATGTAAAGGATAATTGA GCAAATCAATCAAATATACC	212 bp	21
	11	AGTATGTGAATGACTTCACT [GC1]TATAATATAATTAAAAGTAGG	173 bp	24
	12	CTCCCTTCATTGCTTAACAC [GC1]TTTCTTTGCCAAGATATTAC	211 bp	24
	13	[GC1]GATTACACAGTATCCTCGAC GCAGTACCACGAATTACAATG	227 bp	34
	14	[GC1]GTGATTTTCTAAAATAGCAGG ACCGCGCCCGGCTGAAAT	179 bp	35
	17	[GC1]TTCTTTGTCTGATAATAAC CTCTCACTAACAATAATTTGTT	381 bp	26
	18	[GC1]TCAAATTATGCTTACTAATGTG GAATGGTCAACATAACATA	229 bp	31
	19	[GC1]CAACTTGAAATGAAGAC CGTCCCGCTGCTCTTGAAAATAATCATC	249 bp	31
	20	[GC1]AAAATGACTAATTTTTCTTATTCCC AGGAGAGAAGGTGAAGTGC	237 bp	44
	21	[GC2]CAATTCTGACTACTTTTACATC CGGGCTTACTATGGAAAATTAC	204 bp	28

(*continued*)

Table II
(Continued)

	Exons	Primers 5′ → 3 ′	Size	T_m (% UF)
	22	[GC3]CTTTTTACTGTTCTTCC CCAATCAAAGGATACTTTTG	196 bp	33
	23	[GC1]TCTAATGTAATGGGTCCACC CCCTACTTCCCTAAAGAGAAAAC	285 bp	35
	24	CGGAATGATGTATTTATGCTCA [GC1]TTCTTTTATACTTACAATGC	197 bp	22
	25	[GC1]ATGATTTAAAGTAAAGAATTCT CATCTCAGCTACTGGAAAAC	245 bp	38
	26	[GC1]TCCATTTATAAATACACATG ATTTCGTTTACACAAGGTG	161 bp	32
	27	[GC1]TACCCAGTACCATCAATGC TCCAGAGGTGTACACAGTG	191 bp	43
p53				
Long PCR	1–20	TCTTGGCGAGAAGCGCCTACG AGAACCTTGTGCCAGCCGTGG	20 kb	
	2–9	GCCAGACTGCCTTCCGGGTCACTGC CACTGCCCCCTGATGGCAAATGCC	3.2 kb	
Short PCR	4	[GC4]AGGACCTGGTCCTCTGACTG TCATGGAAGCCAGCCCCTCA	390 bp	62
	5a	GCCGTGTTCCAGTTGCTTTATC [GC4]GTCGTCTCTCCAGCCCCAGC	323 bp	51
	5b	[GC4]GGCCAAGACCTGCCCTGTGC GTCGTCTCTCCAGCCCCAGC	216 bp	70
	6	[GC4]CACTGATTGCTCTTAGGTC AGTTGCAAACCAGACCTCA	184 bp	52
	7	[GC5]ACAGGTCTCCCCAAGGCGCA CCAGTGTGCAGGGTGGCAAGTG	234 bp	58
	8	[GC4]GATTTCCTTACTGCCTCTTG CATAACTGCACCCTTGGTCT	271 bp	59
	9	[GC4]GTTATGCCTCAGATTCAC CCAAGACTTAGTACCTGAAG	176 bp	51

GC-clamps:
GC1: CGCCCGCCGCGCCCCGCGCCCGTCCCGCCC (30mer)
GC2: CGCCCCGCGCCGCCGCCCCGCCCCCGCCCGTCCCGCCC (38mer)
GC3: CGCCCCGCCGCGCCCCGCGCGCCCGGTCCCCGCGC (35mer)
GC4: CGCCCGCCGCGCCCCGCGCCCGTCCCGCCGCCCCCGCCCG (40mer)
GC5: CCCCGCTCCCCGCCCCCCTCCCCGCCCCGCCCCTCGCCGC (40mer)

reaction tube) to 25. No attempt was made to include the GC-rich RB1 exon 1 and promoter region in the multiplex PCR.

It is important to realize that mutations can only be detected by DGGE under optimal melting conditions, i.e., in two-domain fragments with the GC-clamp as the highest melting domain. Without the first long-PCR step the optimal melting criterion is often in conflict with other primer design criteria applied to PCR with total genomic DNA as template. Indeed, for the RB1 gene it was found to be impossible to select conditions suitable for both optimal separation in DGGE and optimal priming in PCR. The preamplification step represented by the long-PCR is apparently a sine qua non for the design of an optimal set of PCR primers in TDGS.

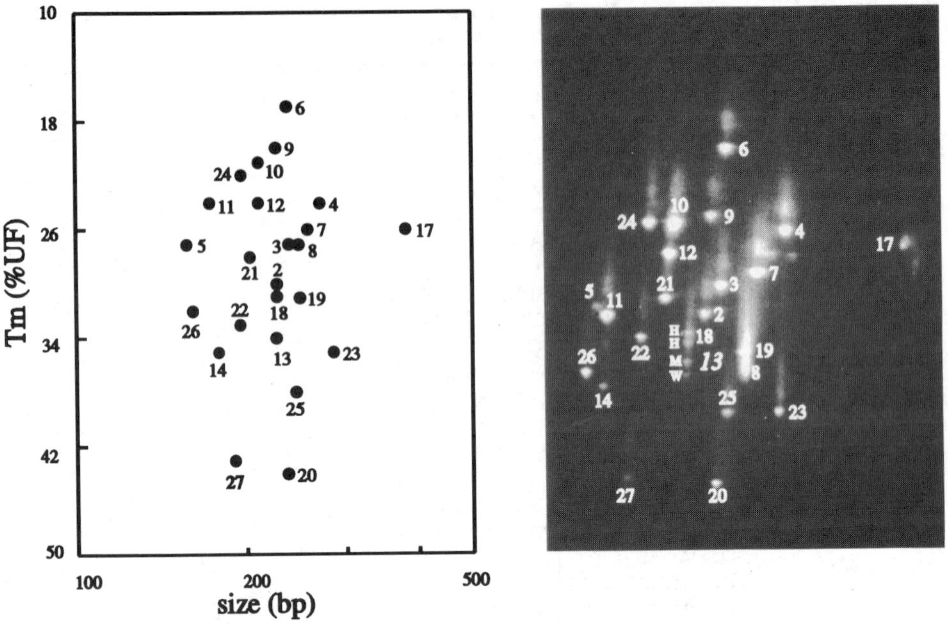

Figure 22.4. Theoretical (left) and empirical (right) TDGS patterns of the RB1 gene. The empirical pattern was obtained from genomic DNA of a heritable retinoblastoma patient and shows a mutation in exon 13. Note that exon 18 has comigrated with one of the heteroduplexes of exon 13.

The same is true for extensive multiplexing, which is not possible without the long PCR preamplification step. For RB1 TDGS all coding regions were amplified in a multiplex group of 6 long-PCR fragments. Then the 24 primer pairs (one of each coupled to a GC-clamp) were added to a part of the long-PCR product mixture to coamplify the 24 target fragments. The p53 hot spot region (exons 4–9) was amplified as one long-PCR fragment, followed by seven separate short-PCR reactions under identical conditions (Table II). (For this small number of fragments no attempt was made to multiplex, also in view of the overlap between some fragments.) After the second PCR, the fragments were allowed to undergo one complete round of denaturation/renaturation to facilitate the formation of heteroduplexes.

Subsequent to the PCR the mixture of fragments was subjected to 2-D electrophoresis. The availability of an automated instrument simplifies this process and makes the test less labor-intensive. The instrument used here allows one to run ten gels at a time without manual interference, i.e., cutting out lanes and loading these onto a second gel. However, it is our experience that using manual instruments a skilled technician can analyze up to 48 DNA samples a week. Experimental conditions (e.g., voltage, gradient, electrophoresis time) have to be specified in accordance with the selected short-PCR fragments. For RB1 and p53 specifications are provided in the Methods section, below. After 2-D electrophoresis the gels were released from between the glass plates and stained with ethidium bromide or other stains.

The resulting patterns were documented and evaluated (by eye) for the occurrence of mutations. Under the conditions applied, i.e., GC-clamping and heteroduplexing, heterozygous mutations result in four spots: the two homoduplex variants and the two heteroduplex variants. The empirical spot pattern in Fig. 22.4 indicates a mutation in exon 13. Details of mutations detected in the RB1 and p53 genes are shown in Fig. 22.6. The mutations selected demonstrate the

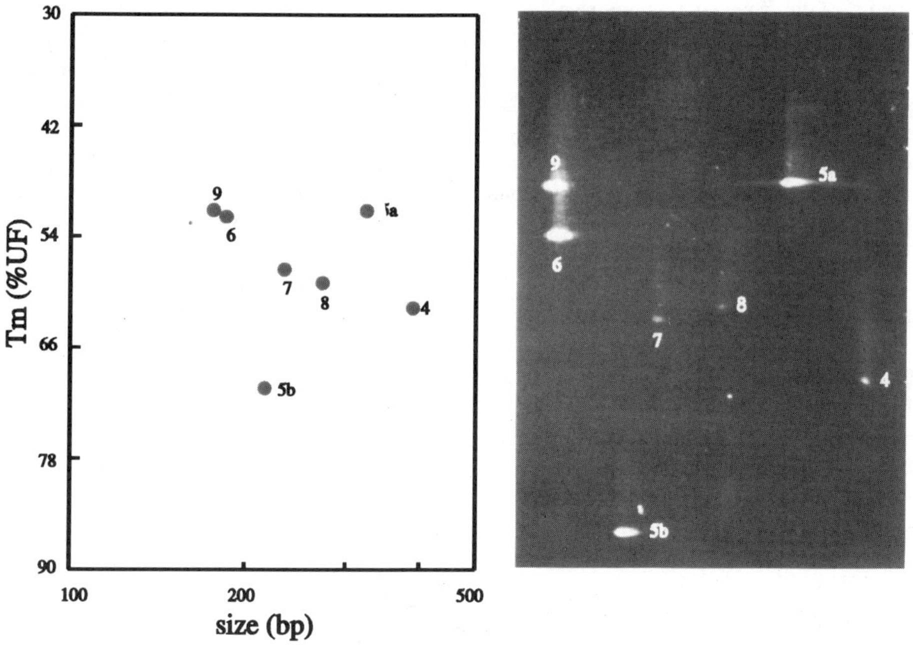

Figure 22.5. Theoretical (left) and empirical (right) TDGS patterns of the p53 gene. In this control sample no mutations or polymorphisms were detected.

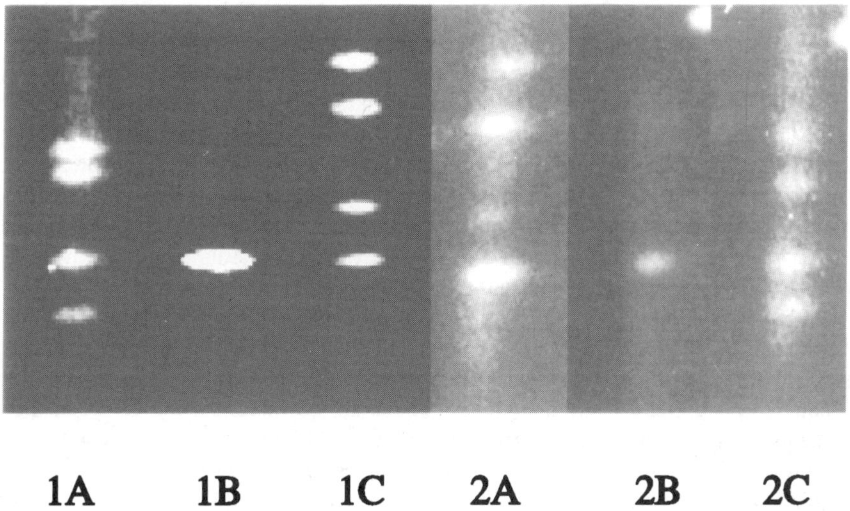

Figure 22.6. Details of wild-type (homozygous normal) and heterozygous mutant fragments obtained from the p53 (lanes 1A–C) and RB1 (lanes 2A–C) genes. The p53 and RB1 mutants were germ-line mutations in Li–Fraumeni and retinoblastoma patients, respectively (kindly provided by Dr. Frederick Li, Dana–Farber Cancer Institute, Boston). Each heterozygous mutation results in four spots: two homoduplex variants and two heteroduplex variants. Lane 1A: p53 exon 6, CGA→CGG mutation in codon 213; lane 1B: p53 exon 6, wild type; lane 1C: p53 exon 6, CGA → TGA mutation in codon 213; lane 2A: RB1 exon 18, C → T transition in codon 579; lane 2B: RB1 exon 18, wild type; lane 2C: RB1 exon 18, loss splice donor site in codon 604.

capacity of the system to distinguish between different mutations in the same exon on the basis of position alone. Indeed, the two p53 mutations were even in the same codon (Fig. 22.6).

Sometimes (reproducible) intensity differences were observed between two homoduplexes (see lanes 1A and 2A of Fig. 22.6) or between two heteroduplexes. Occasionally, mutations were revealed by only three spots, when either the two homoduplexes or the two heteroduplexes were not well separated (results not shown). Especially in tumors (e.g., retinoblastoma), homozygous or hemizygous mutations are regularly found. In such cases no heteroduplexes can be formed after PCR amplification and it becomes necessary to mix each sample before PCR with a control sample. Thus far a total of 35 previously identified point mutations and several polymorphisms have been detected in RB1 (van Orsouw *et al.*, 1996). Preliminary results obtained after screening about 100 heritable retinoblastoma patients for germ-line mutations in RB1, including exon 1, indicate a detection rate of more than 83% (C. Eng, manuscript in preparation). This compares favorably with the 20% detection rate found by Blanquet *et al.* (1995) in studying 232 patients. It should be noted that the absence of a mutation detectable by TDGS could be explained by the presence of a large deletion. Deletions of entire exons would go unnoticed in this system until it becomes possible to detect such events in a quantitative way as a 50% reduction of spot intensity. It goes without saying that also any mutation outside the defined target regions (i.e., introns, regulatory regions) will go undetected with the current format.

2. METHODS

2.1. Equipment

Instruments for DGGE have been described (Fischer and Lerman, 1979b; Myers *et al.*, 1987; Uitterlinden and Vijg, 1994). The procedures described below are based on the use of an automated 2-D electrophoresis instrument available from Ingeny B.V. [P.O. Box 685, 2300 AR Leiden, The Netherlands (telephone 31 71 214575)]. In this instrument, which can hold ten gels, the second dimension is run directly after the first, with no need for manual interference (Mullaart *et al.*, 1993). The instrument comes with a gel caster and gradient maker for pouring ten gradient gels simultaneously.

With some modifications the same procedures can be used in combination with the various DGGE instruments currently available. In these instruments it is necessary to first run the size separation, cut out the lane, and load this on the second-dimension gel for separation by DGGE. In this respect, we have obtained good results with the (manual) instruments of Ingeny B.V. (see above), who also provide a detailed manual for 2-D electrophoresis, and C.B.S. Scientific Co., Inc. [420 S. Cedros, Solana Beach, CA 92075 (telephone 619 755 0733)]. Equipment for 2-D electrophoresis can also be obtained from Hoefer Scientific Instruments; although we have never worked with those instruments, others apparently have had good experience with them (Sidman and Shaffer, 1994). Other instruments required for TDGS include a thermocycler. For our experiments we have used a GeneE thermocycler (Techne, Cambridge, U.K.) fitted with a Heated Lid, removing the need for an oil overlay on the samples. Good results were also obtained with the Perkin Elmer 480, albeit under slightly different reaction conditions.

Finally, a gel documentation system is required. For this purpose we have simply used a Polaroid camera in combination with a UV transilluminator to document the ethidium bromide-stained gel patterns. However, several alternative documentation formats are conceivable, varying from direct image analysis of the ethidium bromide-stained patterns to advanced fluorescence imaging (Molecular Dynamics, Technical Note #58). In the latter case, fluorophore-labeled primers or PCR-labeling protocols are required. As yet we have no complete information as to how such labels influence the separation. However, preliminary results obtained with end-labeled fluorescein primers suggest that the effects are marginal if at all present. This also seems to be the experience with the recently introduced Perkin Elmer ABI Prism system (Perkin Elmer ABI Prism Product Information).

2.2. Reagents

1. 20× TAE electrophoresis buffer: 0.8 M Tris, 0.4 M sodium acetate, and 0.02 M EDTA (adjust pH to 8.0 with acetic acid).
2. Denaturant stock solution (9A/0UF): 9% acrylamide. For 1 liter, 25 ml 20× TAE, 90 g acrylamide, 2.4 g bisacrylamide (37.5 : 1 acrylamide : bisacrylamide), and water to 1 liter. For higher or lower acrylamide concentrations, adjust amounts accordingly. Stored in a dark bottle it is stable for several weeks.
3. Denaturant stock solution (9A/100UF): 9% acrylamide, 100% urea/formamide (7 M urea/40% formamide). For 1 liter, 25 ml 20× TAE, 90 g acrylamide, 2.4 g bisacrylamide, 420 g urea, 400 ml deionized formamide, and water to 1 liter. For lower denaturant concentration, adjust amounts with 9A/0UF accordingly. Stored in a dark bottle at room temperature (do not store in a refrigerator) it should be stable for several weeks.

   To deionize formamide, gently mix 500 ml formamide with about 30 g mixed bed resin (Sigma). Store at −20°C or use immediately.
4. Ammonium persulfate stock (20%): For 50 ml, 10 g ammonium persulfate to 50 ml. Store at −20°C in small portions. After thawing, do not use again.
5. TEMED (*N,N,N′,N′*-tetramethylethylenediamine).
6. Gel Loading Solution: 0.25% xylene cyanol, 0.25% bromophenol blue, 15% Ficoll, 100 mM EDTA. Stable for months at 4°C.

2.3. PCR Primers

Primers (deprotected and desalted) were obtained from GIBCO BRL. For long-term storage, primers should be kept, for example, in a stock solution of 100 μM in ultrapure water, at −20°C. For short-term use, we kept them at −20°C as a solution of 12.5 μM in ultrapure water.

2.4. PCR Reactions and Heteroduplexing

We carried out our PCR reactions in thermowell tubes (Costar, Cambridge, MA) in a GeneE thermocycler (Techne) fitted with a Heated Lid, removing the need for an oil overlay on the samples. The PCR conditions specified below are for this instrument only and it may be necessary to adapt the protocol before using it in combination with other thermocyclers.

Long PCR. Long PCR reactions are carried out in a 50-μl volume with 100 ng genomic DNA as template and 0.25 μM of each primer, using the LA PCR Kit (TaKaRa). PCR reactions are performed according to the manufacturer's instructions. The conditions are as follows: one cycle of 94°C, 1 min, followed by 32 cycles of 98°C, 20 sec/68°C, 12 min with 15-sec incremental increases (5 min with no incremental increases for p53), and finally one cycle of 72°C, 5 min. The PCR products are stored at −20°C for further use.

Short PCR. Short PCR reactions are carried out, using the same GeneE thermocycler, in a 50-μl volume with 4 μl long-PCR product, 0.125–0.5 μM of each primer (0.125 μM for RB1 exons 3, 4, 10, 12, 13, 19, 24, and 27; 0.25 μM for RB1 exons 6, 7, 9, 11, 14, 25 and 26; 0.5 μM for RB1 exons 2, 5, 8, 17, 18, 20, 21, 22, 23 and each p53 individual exon), 0.25 mM dNTPs, 8 mM $MgCl_2$ (1.5 mM for p53), 5 units of *Taq* polymerase (GIBCO BRL) (3 units for p53), and 1%

DMSO (to increase specificity). The PCR conditions are as follows. For p53, one cycle of 94°C, 2 min, then 36 cycles of 94°C, 40 sec/55°C, 1 min/72°C, 50 sec with 1-sec incremental increase. For RB1, 5 cycles of 94°C, 45 sec/52°C, 40 sec/68°C, 2 min, then 5 cycles of 94°C, 45 sec/47°C, 40 sec/68°C, 2.5 min, then 32 cycles of 94°C, 50 sec/55°C, 10 sec/40°C, 40 sec/67°C, 2.5 min (with 3-sec incremental increases per cycle).

Heteroduplexing. After the short PCR, fragments were heteroduplexed by one complete round of denaturation/renaturation. For RB1: 67°C, 12 min/98°C, 12 min/52°C, 30 min/44°C, 30 min. For p53: 72°C, 10 min/98°C, 8 min/55°C, 30 min/40°C, 30 min.

2.5. Sample Preparation

After PCR and heteroduplexing part of the contents of the one tube (RB1) and part of the mixed contents of the six tubes (p53) are mixed with 1/10 vol of loading buffer. In our experience, there is usually enough sample for several runs. The electrophoresis conditions described below are for the RB1 and p53 gene separately.

2.6. Two-Dimensional Electrophoresis

2.6.1. Manual Instruments

In our laboratory, both the Ingeny and C.B.S. manual instruments are used; each provides satisfactory results. Gels are poured directly in the Ingeny instrument via the use of a U-shaped spacer, thereby avoiding the need to first pour an agarose plug (Vijg, 1995b). In the C.B.S. instrument the (time-consuming) use of an agarose plug is avoided employing silicone gaskets sealing the glass plate sandwich. In the Ingeny system a separate, double-wide version of the instrument is used for the second-dimension separation. The same C.B.S. instrument is used for the first and second dimension.

The mixtures of DNA fragments are first subjected to size separation using a 0.75-mm-thick 9% PAA gel at 50°C for 5–6 hr at 150 V in 0.5× TAE. The separation pattern is visualized by ethidium bromide staining for 10 min and UV transillumination of the gel, which lies on a glass plate to protect the DNA fragments from damage by the UV light. The 100- to 600-bp region in the middle part of the lane (so not including the edges) is quickly cut out and applied to a 1-mm-thick 10% PAA gel containing a 10–50% (RB1) or 30–90% (p53) urea/formamide (UF) gradient. For both systems gradients are poured using a simple gradient former (e.g., from GIBCO BRL). Electrophoresis is for 7.5 hr (RB1) or 11 hr (p53) at 60°C and 200 V. After electrophoresis the gels are stained with 0.5 μg/ml ethidium bromide for 15–20 min and destained in water for another 15 min. The patterns are documented under UV illumination using a Polaroid camera.

2.6.2. Automatic Instrument

Gels are poured, ten at a time, in the gel-casting device that comes with the automated 2-D electrophoresis instrument according to the manufacturer's instructions (Ingeny B.V.). This device is essentially a box in which the ten pairs of glass plates (cleaned with soap and ethanol), each one separated from the other by four small glass spacers glued to one of the plates, are placed. Each pair is separated from the other pairs by a foam sheet (otherwise they are difficult to separate later). The box is then closed by tightly pressing the lid onto the glass plates using a screw. With less than ten gels, dummy blocks are used.

After adding ammonium persulfate (about 1/500th volume) and TEMED (about 1/5000th) the appropriate volumes of the boundary solutions are poured into the chambers of the gradient maker (the solution of higher denaturant concentration should come out first). The exact volumes needed for this particular instrument are provided by the manufacturer.

The denaturing gradient gels are poured through a tube that comes from the gradient maker through a pump. The end of the tube is inserted near the top of the gel casting box between the side wall and the sides of the glass plates (there is little space left). After polymerization of the gradient gel the liquid on top (unpolymerized acrylamide) is discarded and the top-gel is poured. This top-gel is acrylamide of the same concentration, but without denaturants. After adding ammonium persulfate and TEMED it can be poured on top of the glass plates in the gel box. After insertion of the slot-formers (V-shaped to generate bands rather than blocks after the 1-D separation), the top-gel is allowed to polymerize. After polymerization the gels are removed from the gel-casting box and cleaned with a wet tissue. They are then placed in the instrument according to the manufacturer's instructions, that is, in two gel-holding cassettes with silicone-side sealings. The instrument containing buffer heated to 45°C is put in the 1-D mode with the power switched off.

After adding loading buffer, samples (up to 40 μl) are loaded in the V-shaped wells of the gels in the automated 2-D electrophoresis instrument. Gels of 19% acrylamide, 0.25× TAE are used with a gradient of 10–50% urea/formamide (RB1) or 30–90% urea/formamide (p53). The first dimension is run at 150 V for 5–6 hr at 50°C. The second dimension is run at 200 V for 7.5 hr (RB1) or 11 hr (p53) at 60°C. After electrophoresis the gels are stained with ethidium bromide and the patterns documented under UV illumination as described for the manual instruments.

3. DISCUSSION

Gene mutational scanning on a routine basis is severely hampered by the lack of methodology which is both highly accurate and cost-effective. The latter requires laboratory user-friendliness, low labor intensiveness and reagent costs, and a sufficiently high throughput. With the exception of nucleotide sequencing, which even in an automatic fashion is prohibitively expensive, DGGE appears to be the only method with an almost 100% accuracy in detecting point mutations in defined DNA fragments (e.g., Dianzani *et al.*, 1993). We have shown earlier that this method, in combination with a size separation in a 2-D format, can be applied to analyze the sequence integrity of multiple DNA fragments from the higher animal genome in parallel (for a review, see Vijg, 1995b). With the development of an automated 2-D electrophoresis system for running multiple gels in combination with rapid image analysis software to diagnostically analyze spot patterns, DGGE in two dimensions has become a method of scale to diagnostically analyze gene sequences (Mullaart *et al.*, 1993; Vijg, 1995a).

More recently, we have applied the 2-D principle to the complete diagnosis of the CFTR gene by mutation analysis (Wu *et al.*, 1996) using primers prespecified by others for 1-D DGGE (e.g., Costes *et al.*, 1993). In this chapter we have described the application of this methodology, termed two-dimensional gene scanning (TDGS), to mutationally scan the human RB1 and p53 tumor suppressor genes, thereby focusing on the generation of PCR gene fragments in a multiplex format with optimal melting behavior in the second DGGE dimension. Indeed, the difficulty of coamplifying many fragments in one PCR reaction could have become a major hindrance to the widespread application of TDGS in DNA diagnostics. Clearly, multiplexing prevents the number of pipetting steps and individual reactions to be carried out for a given TDGS test from becoming prohibitively high. It also saves on reagent costs and requires less patient material. Even more so than automation of the 2-D electrophoresis process, multiplexing is a necessary step to making TDGS laboratory user-friendly and thereby more suitable for clinical DNA diagnostic testing. The two-step PCR amplification strategy described here allows extensive multiplexing by first amplifying large parts of the target gene.

The use of TDGS as a practical alternative for sequence-based diagnosis can be illustrated by a cost comparison. While comprehensive mutation analysis of an approximately 10,000-bp coding region may cost between $500 and $1000, a TDGS test of the same region at the same accuracy is unlikely to cost more than $100 (including reagents, labor, and overhead). Once detected, mutations can easily be confirmed by sequencing the relevant fragment. For genes with recurrent mutations, positional prespecification of the mutations in a diagnostic database of an image analysis program would greatly simplify clinical interpretation, i.e., by the identification of each mutation on the basis of their *xy* coordinates in the gel. Indeed, as illustrated in Fig. 22.6, different mutations in the same fragment can clearly be distinguished, even when they involve the same codon. Although the possibility cannot be excluded that certain different mutations in a given exon will fortuitously comigrate, our experience with the CFTR, RB1, and p53 genes leads us to predict that most mutations can be identified on the basis of position alone. Extensive validation testing should reveal whether this prediction is correct.

Finally, on the basis of the high resolution of 2-D gels, it is possible to analyze more than one gene simultaneously. Such multigene tests could, for example, be composed for complex diseases like breast and colon cancer in which multiple genes are known to be involved. As is the case with protein 2-D gels, 2-D DNA typing gels can resolve hundreds of fragments (Vijg *et al.*, 1994). Assuming an average number of 20 fragments per gene, the current system should be able to scan several genes simultaneously for all kinds of mutations. Improvements in the resolution of acrylamide gels could increase this number even further.

ACKNOWLEDGMENTS. We thank Drs. Jeanne Wei and William Strauss for critically reading the text. This work was supported in part by a research grant from Toyobo Co., Ltd., Osaka, Japan.

REFERENCES

Abrams, E. S., Murdaugh, S. E., and Lerman, L. S. (1990). Comprehensive detection of single base changes in human genomic DNA using denaturing gradient gel electrophoresis and a GC clamp. *Genomics* **7**:463–475.

Beggs, A. H., Koenig, M., Boyce, F. M., and Kunkel, L. M. (1990). Detection of 98% DMD/BMD gene deletions by PCR. *Hum. Genet.* **86**:45–48.

Blanquet, V., Turleau, C., Gross-Morand, M.-S., Sénamaud-Beaufort, C., Doz, F., and Besmond, C. (1995). Spectrum of germline mutations in the RB1 gene: A study of 232 patients with hereditary and non hereditary retinoblastoma. *Hum. Mol. Genet.* **2**:975–979.

Chamberlain, J. S., Gibbs, R. A., Ranier, J. E., Nguyen, P. N., and Caskey, C. T. (1988). Deletion screening of the Duchenne muscular dystrophy locus via multiplex DNA amplification. *Nucleic Acids Res.* **16**:11141–11156.

Costes, B., Fanen, P., Goossens, M., and Ghanem, N. (1993). A rapid, efficient, and sensitive assay for simultaneous detection of multiple cystic fibrosis mutations. *Hum. Mutat.* **2**:185–191.

Cotton, R. G. H. (1993). Current methods of mutation detection. *Mutat. Res.* **285**:125–144.

Dianzani, I., Camaschella, C., Ponzone, A., and Cotton, R. G. H. (1993). Dilemmas and progress in mutation detection. *Trends Genet.* **9**:403–405.

Edwards, M. C., and Gibbs, R.A. (1994). Multiplex PCR: Advantages, development and applications. *PCR Methods Appl.* **3**:S65–S75.

Fischer, S. G., and Lerman, L. S. (1979a). Length-independent separation of DNA restriction fragments in two-dimensional gel electrophoresis *Cell* **16**:191–200.

Fischer, S. G., and Lerman, L.S. (1979b). Two-dimensional electrophoretic separation of restriction enzyme fragments of DNA. *Methods Enzymol.* **68**:183–191.

Foord, O. S., and Rose, E.A. (1994). Long-distance PCR. *PCR Methods Appl.* **3**:S149–S161.

Guldberg, P., Henriksen, K. F., and Guttler, F. (1993). Molecular analysis of phenylketonuria in Denmark: 99% of the mutations detected by denaturing gradient gel electrophoresis. *Genomics* **17**:141–146.

Lerman, L. S., and Silverstein, K. (1987). Computational simulation of DNA melting and its application to denaturing gradient gel electrophoresis. *Methods Enzymol.* **155**:482–501.

Li, D., and Vijg, J. (1996). Multiplex co-amplification of 24 retinoblastoma gene exons after pre-amplification by long-distance PCR. *Nucleic Acids Res.*, **24**:538–539.

Moyret, C., Theillet, C., Puig, P. L., Moles, J-P., Thomas, G., and Hamelin, R. (1994). Relative efficiency of denaturing gradient gel electrophoresis and single strand conformation polymorphism in the detection of mutations in exons 5 to 8 of the p53 gene. *Oncogene* **9**:1739–1743.

Mullaart, E., de Vos, G. J., te Meerman, G. J., Uitterlinden, A. G., and Vijg, J. (1993). Parallel genome analysis by two-dimensional DNA typing. *Nature* **365**:469–471.

Myers, R. M., Fischer, S. G., Lerman, L. S., and Maniatis, T. (1985). Nearly all single base substitutions in DNA fragments joined to a GC-clamp can be detected by denaturing gradient gel electrophoresis. *Nucleic Acids Res.* **13**:3131–3145.

Myers, R.M., Maniatis, T., and Lerman, L. S. (1987). Detection and localization of single base changes by denaturing gradient gel electrophoresis. *Methods Enzymol.* **155**:501–527.

Sheffield, V. C., Cox, D. R., Lerman, L. S., and Myers, R. M. (1989). Attachment of a 40-base-pair G+C-rich sequence (GC-clamp) to genomic DNA fragments by the polymerase chain reaction results in improved detection of single-base changes. *Proc. Natl. Acad. Sci. USA* **86**:232–236.

Sheffield, V. C., Beck, J. S., Kwitek, A. E., Sandstrom, D.W., and Stone, E.M. (1993). The sensitivity of single-strand conformation polymorphism analysis for the detection of single base substitutions. *Genomics* **16**: 325–332.

Sidman, C. L., and Shaffer, D.J. (1994). Large-scale genomic comparison using two-dimensional DNA gels. *Genomics* **23**:15–22.

Uitterlinden, A. G. and Vijg, J. (1994). *Two-Dimensional DNA Typing: A Parallel Approach to Genome Analysis.* Ellis Horwood, Chichester.

Uitterlinden, A. G., Slagboom, P. E., Knook, D. L., and Vijg, J. (1989). Two-dimensional DNA fingerprinting of human individuals. *Proc. Natl. Acad. Sci. USA* **86**:2742–2746.

van Orsouw, N., Li, D., van der Vlies, P., Scheffer, H., Eng, C., Buys, C. H. C. M., Li, F. P., and Vijg, J. (1996). Mutational Scanning of large genes by extensive PCR multiplexing and two-dimensional electrophoresis: Application to the RB1 gene. *Hum. Mol. Genet.*, in press.

Vijg, J. (1995a). Detecting individual genetic variation. *Bio/Technology* **13**:137–139.

Vijg, J. (1995b). Two-dimensional DNA typing: A cost-effective way of analyzing complex mixtures of DNA fragments for sequence variations. *Mol. Biotechnol.* **4**:275–295.

Vijg, J., Wu, Y., Uitterlinden, A. G., and Mullaart, E. (1994). Two-dimensional DNA electrophoresis in mutation detection. *Mutat. Res.* **308**:205–214.

Wu, Y., Hofstra, R., Scheffer, H., Uitterlinden, A. G., Mullaart, E., Buys, C. H. C. M., and Vijg, J. (1996). Comprehensive and accurate mutation scanning of the CFTR-gene by two-dimensional DNA electrophoresis. *Hum. Mutat.*, in press.

Chapter 23

Ligase Chain Reaction for the Detection of Specific DNA Sequences and Point Mutations

R. Bruce Wallace, Ching-I P. Lin, Antonio A. Reyes, Jimmie D. Lowery, and Luis Ugozzoli

1. INTRODUCTION

DNA diagnostics is a new field that utilizes the techniques of molecular biology to identify and characterize the specific genetic material associated with genetic, neoplastic, and infectious diseases. DNA diagnostics grew out of the discoveries made possible using molecular cloning and DNA sequencing techniques beginning in the mid-1970s. Despite the availability of exquisitely specific reagents such as restriction endonucleases, synthetic oligonucleotides, and cloned DNA and cDNA probes, as well as technologies such as electrophoretic methods to fractionate DNA and novel DNA hybridization methods (Southern and Northern blots, dot blot, etc.), DNA diagnostic procedures were not immediately adopted by clinical laboratories. This failure was due in part to the long turnaround times, the lack of sensitivity, and the lack of reliable nonradioactive detection procedures.

Major progress in the evolution of the DNA diagnostics field came with improvements in nonradioisotopic detection procedures and with the development of DNA amplification techniques; the latter making a major impact on the former. There are two kinds of amplification procedures: signal amplification and target amplification. Signal amplification involves the amplification of the probe signal during or after hybridization without altering the number of target molecules. Target amplification, on the other hand, increases the number of copies of a specific target DNA prior to a detection step. The use of amplification technologies has overcome the problems related to low sensitivity and specificity and has all but eliminated the use of radioisotopes from the clinical DNA diagnostic laboratories.

Perhaps the best studied of the target amplification methodologies is the polymerase chain reaction (PCR), a method that uses repeated cycles of *in vitro* DNA replication to produce an exponential amplification of a specific DNA segment from a sample containing a small amount of

---

R. Bruce Wallace, Ching-I P. Lin, Antonio A. Reyes, Jimmie D. Lowery, and Luis Ugozzoli • DNA Diagnostics Division, Bio-Rad Laboratories, Hercules, California 94547.

Technologies for Detection of DNA Damage and Mutations, edited by Gerd P. Pfeifer, Plenum Press, New York, 1996.

template or complex mixture of nucleic acids (Saiki *et al.*, 1988; Kleppe *et al.*, 1971). PCR has made a great contribution to both research and diagnostic laboratories and it remains the most popular and developed amplification technique.

2. THE LIGASE CHAIN REACTION

The ligation amplification reaction (LAR) (Wu and Wallace, 1989a), now called the ligase chain reaction (LCR) (Barany, 1991a), is a target amplification technique that is gaining popularity because it is especially suited to diagnostic applications. The reaction is useful for both the detection of specific DNA targets as well as the discrimination of targets that differ in sequence due to a point mutation, thus making this technique particularly useful for the diagnosis of human genetic diseases. Both LCR and PCR are cycling reactions, but unlike PCR, LCR does not involve the *de novo* synthesis of DNA. In PCR, two oligonucleotide primers hybridize to opposite strands of the target DNA and are extended by a DNA polymerase in the presence of deoxyribonucleoside triphosphates. The length of the amplification product is determined by the distance between the primers. In LCR, four oligonucleotides and a DNA ligase amplify only that portion of the template to which the oligonucleotides are complementary. The length of the amplification product is determined by the lengths of the oligonucleotides that are ligated together.

In this chapter we will discuss intrinsic aspects of LCR, some of its advantages and disadvantages, some variations of the reaction, different postamplification detection systems, and current and future applications. As will be described, specific reaction conditions for LCR are very target dependent, so rather than provide a detailed protocol for LCR, the principles will be discussed to give guidance to future practitioners.

2.1. Mechanism of LCR

LCR (LAR), described by Wu and Wallace (1989a), is based on template-dependent ligation (TDL) of adjacent oligodeoxyribonucleotides. The reaction is described schematically in Fig. 23.1. Two pairs of oligonucleotides, which are in molar excess over the target DNA sequence, are present in the reaction. The oligonucleotides W and X hybridize to the template immediately adjacent to each other and are complementary to the upper DNA strand. Likewise, the other two oligonucleotides Z and Y hybridize immediately adjacent to each other and are complementary to the lower DNA strand. Obviously, Z and Y are also complementary to the oligonucleotides W and X, respectively. The oligonucleotides W and Y are phosphorylated at their 5′ ends, the phosphate being either chemically incorporated into the oligonucleotide during its synthesis or introduced enzymatically with the enzyme T4 polynucleotide kinase. During the beginning of a typical cycle, the target DNA is denatured using elevated temperatures. In a second step, temperature is decreased to allow the oligonucleotides to hybridize to the target DNA. DNA ligase then covalently links W to X, and Y to Z. In the subsequent cycles, the products of the previous cycles will be used as template by the two pairs of oligonucleotides, thus amplifying the DNA target in an exponential manner.

As originally described, the LCR (LAR) procedure had several disadvantages related to the use of a mesophilic DNA ligase. Because heat denaturation inactivated the DNA ligase, the reaction required the addition of a new aliquot of enzyme at the beginning of each new cycle. This was a major drawback because it made for long reaction times, each addition resulted in the accumulation of protein and other components, and opening the tube at each cycle created the possibility of carryover contamination. Additional disadvantages included the requirements for a

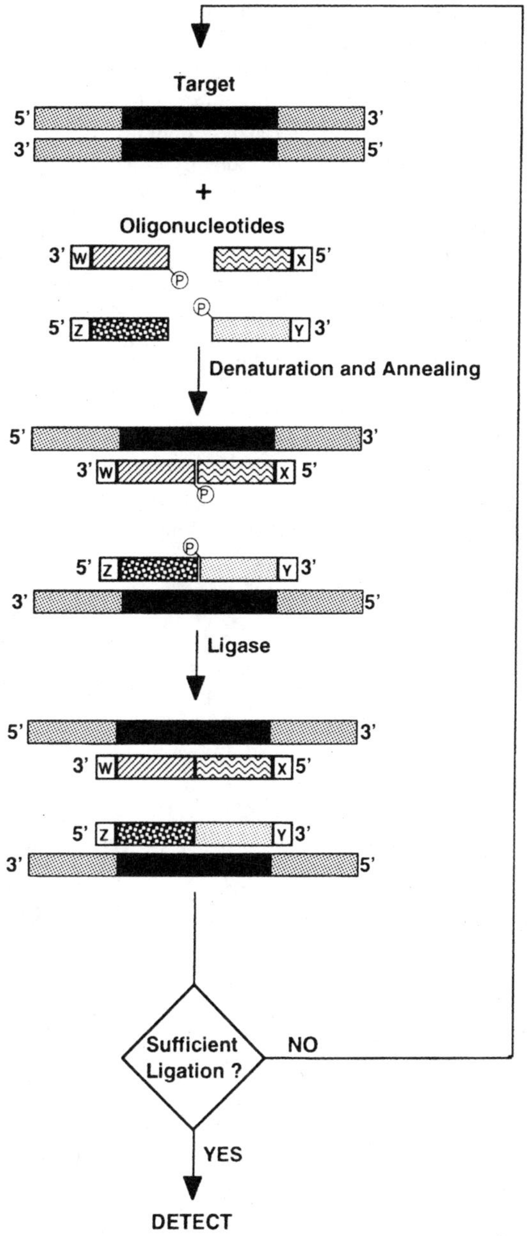

Figure 23.1. Schematic representation of the ligase chain reaction (LCR). The LCR involves the use of two pairs of oligonucleotides which are complementary to a particular DNA template. The first pair, W and X, hybridize to the template immediately adjacent to each other and are complementary to the upper DNA strand. The second pair, Z and Y, hybridize immediately adjacent to each other and are complementary to the lower DNA strand. Note that in this configuration, W is complementary to Z and X to Y. Oligonucleotides W and Y are phosphorylated at their 5' ends. The reaction is initiated by denaturing the target DNA using elevated temperatures. The temperature is then decreased to allow the two pairs of oligonucleotides to hybridize to the adjacent, complementary sequences on the target DNA. DNA ligase then joins the two pairs of oligonucleotides (X to W and Z to Y). In the subsequent cycles, the ligated molecules of previous cycles will be used as templates by the two pairs of oligonucleotides, thus amplifying the DNA target in an exponential manner. The reaction is continued until sufficient ligation is achieved to allow detection of the product of ligation.

low annealing temperature (30°C) and large number of cycles to achieve a detectable signal when human genomic DNA was used as a template.

The purification of thermostable DNA ligase (Taq ligase) (Takahashi *et al.*, 1984) and its subsequent cloning (Barany, 1991a; Barany and Gelfand, 1991) improved LCR. With the Taq ligase, enzyme is added once at the beginning of the LCR, eliminating the addition of new aliquots of enzyme after each successive cycle. In addition, the use of high annealing temperatures increased both the specificity of the reaction and the signal-to-noise ratio.

LCR is an allele-specific methodology. As discussed above, LCR requires complete complementarity between the four oligonucleotides and the template. The presence of a mismatch at any of the positions immediately flanking the ligation junction has an inhibitory effect on the ligation of two oligonucleotides hybridized on the DNA template (Wu and Wallace, 1989b; Landegren *et al.*, 1988). This property makes LCR a powerful tool to study human genetic diseases and genetic polymorphisms. To detect a point mutation, two LCRs can be set up. A first set of oligonucleotides specific to amplify allele 1 is used in one reaction , and a second set specific for allele 2 is included in the second reaction.

2.2. Template-Independent Ligation

LCR is a template-dependent reaction. However, because even low levels of template-independent joining of the oligonucleotides present in the reaction can lead to a ligation product indistinguishable from the template-dependent one, template-independent ligation (TIL) is one of the disadvantages of LCR. Two possible routes of TIL are shown in Fig. 23.2. DNA ligase can join duplex oligonucleotides in a blunt-end or cohesive-end ligation process (left side of Figure

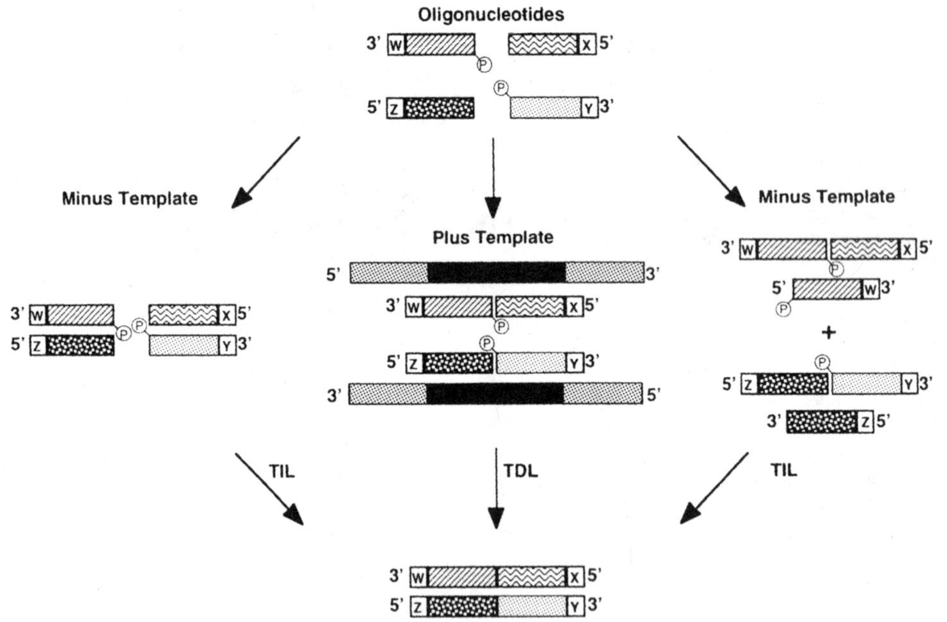

Figure 23.2. Schematic representation of template-independent ligation (TIL). Template-independent ligation is defined as the formation of a ligation product that is identical to that formed in the template-dependent process, but in the absence of template. Two pathways can result in TIL. On the left is depicted the pathway in which the oligonucleotides W, X, Y, and Z are annealed into two duplexes. These duplexes can then be joined by "blunt-end" ligation (or cohesive-end ligation depending on the configuration of the termini). On the right is depicted a pathway in which two oligonucleotides are ligated together using one of the oligonucleotides in the reaction as a template. This ligation will occur if there is substantial homology between the oligonucleotide pairs and a sequence within one of the oligonucleotides in the reaction. Both of these pathways result in a ligation product indistinguishable from that derived from the template-dependent process.

23.2). The product of this ligation event will be amplified in subsequent cycles. Alternatively, two oligonucleotide pairs can be ligated using one of the oligonucleotides as a template (right side of Fig. 23.2). This can happen if there is substantial complementarity between the ligation junction and a sequence within one of the oligonucleotides. Again, the product of this TIL pathway will be amplified in subsequent cycles. The latter mechanism has obvious implications to the choice of target sequence in designing an LCR.

TIL is a low-efficiency event. It is, however, influenced by several factors including ligase concentration, oligonucleotide concentration, and cycle number. Because of these influences, TIL has its major impact on the specificity of an LCR procedure. Figure 23.3 shows a graphical representation of the influence of several parameters on the specificity of LCR. Specificity is depicted as the ratio of TDL to TIL as a function of LCR parameters such as ligase concentration, cycle number, and oligonucleotide concentration. Since TIL is a low-efficiency process, it does not contribute significantly to the TDL signal measured at low parameter values. As parameter values are increased, however, TIL becomes increasingly more significant. For example, if one were to perform an LCR at very high cycle numbers (e.g., 50), the signal obtained from the template-containing sample might be nearly indistinguishable from the signal obtained from a minus-template control. The TDL product in a template-containing sample is produced at a high efficiency early in the reaction but as the cycle number increases the efficiency of its production decreases. Meanwhile, the minus-template control produces a low amount of TIL product in the early cycles which then is amplified in later cycles to levels similar to that of the TDL product in the template-containing sample.

2.3. Optimization of LCR

There is no universal set of conditions for LCR since conditions that efficiently amplify a particular DNA sequence will not necessarily achieve similar results when a different sequence is

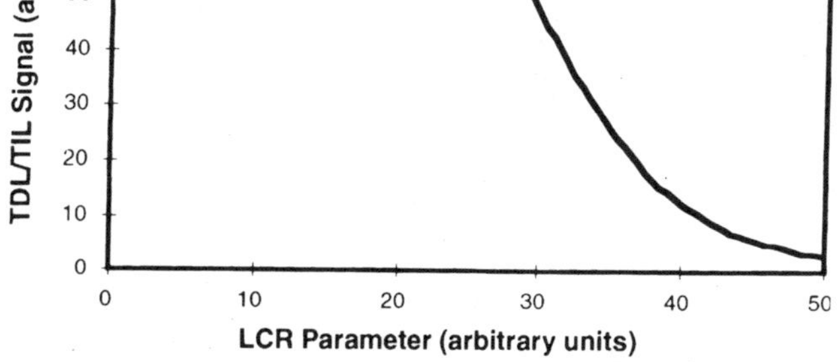

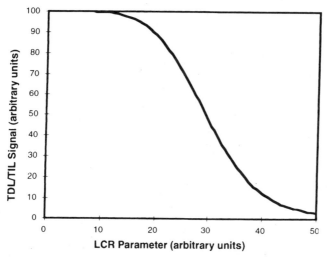

Figure 23.3. Schematic representation of the influence of LCR reaction parameters on TIL and the specificity of LCR. LCR is a template-dependent reaction. TIL only becomes a problem as the amount of amplification required becomes excessive. Therefore, TIL is increased by reaction parameters that increase the amount of amplification including cycle number, DNA ligase concentration, and oligonucleotide concentration. These parameters are depicted in this graph in arbitrary units. As any parameter or combination of parameters is increased, the ratio of TDL to TIL is reduced and the specificity of the LCR reaction decreased.

used as a target. However, when optimizing an LCR, several parameters must be considered in order to obtain specific and reproducible results. The goal in each case is the reduction of TIL. As described above, TIL is extremely sensitive to oligonucleotide concentration, enzyme concentration, and number of cycles. These parameters should be empirically determined to generate only TDL. Typically, the ligation reaction buffer contains 25–100 mM KCl, 10 mM $MgCl_2$, 1 mM nicotinamide adenine dinucleotide (NAD^+), and 20 mM Tris-HCl pH 7.6–8.3. The inclusion of high concentrations of salmon sperm DNA has been reported to decrease the background due to TIL (Barany, 1991a). Oligonucleotide annealing temperatures (e.g., T_m's determined using the Oligo Primer Analysis Software, National Biosciences, Inc., Plymouth, MN) should fall within the optimal temperature range (60–72°C) of the thermostable ligase (Takahashi *et al.*, 1984) although oligonucleotides with lower T_m's have been used successfully. Oligonucleotides should be free of self-complementary sequences that can result in the formation of secondary structure and should not serve as a template for other oligonucleotides since this would generate TIL (Fig. 23.2). The use of homogeneous oligonucleotide preparations ensures that oligonucleotide ends will be juxtaposed precisely at the ligation junction and the ligation products will have the predicted length.

2.4. Advantages and Disadvantages of LCR

LCR offers several advantages as a diagnostic amplification method. The reaction is easy to design and perform and the amplification products can be detected directly using different nonradioactive formats (see below). Moreover, even clinical samples that contain fragmented DNAs may be amplified successfully since the target region is relatively small.

As indicated above, one disadvantage of LCR is TIL. A large number of LCR cycles (more than 30) produces a background signal in the absence of DNA template. Thus, TIL limits the ultimate sensitivity of LCR for target DNA. LCR remains useful in cases where extremely high sensitivity is not required, such as the diagnosis of human genetic diseases. When the number of template molecules is very low (e.g., infectious diseases, minimal residual disease, and tumors where a small clone of mutated cells is present in a huge excess of normal cells), LCR in conjunction with a preamplification step can function as a detection technology. The preamplification step can be performed by PCR, strand displacement amplification, Q-β replicase, transcription-based amplification, or any other strategy that would increase the number of the starting molecules.

One important limitation to the use of LCR for mutation detection is the fact that the short target region to be amplified is defined by sequences immediately surrounding the mutation. Unlike PCR, where primers can be selected to hybridize to regions of the target gene that share little homology with related genes, LCR oligonucleotides must anneal to the target region precisely. Therefore, in LCR, sequences present elsewhere in the genome that are homologous to the target region (e.g., if the target gene is a member of a multigene family or if a pseudogene with related sequences exists) are templates for amplification as well. The presence of homologous sequences to the normal gene would be detected by LCR oligonucleotides specific for the normal allele even if the target gene itself is mutant type, producing a false positive result.

3. The Gap Ligase Chain Reaction

Gap ligase chain reaction (gLCR), also called repair chain reaction (RCR), is a variation of LCR. As shown in Fig. 23.4, gLCR utilizes two complementary oligonucleotide pairs that hybridize to adjacent positions in the denatured DNA template (Segev, 1992). However, unlike

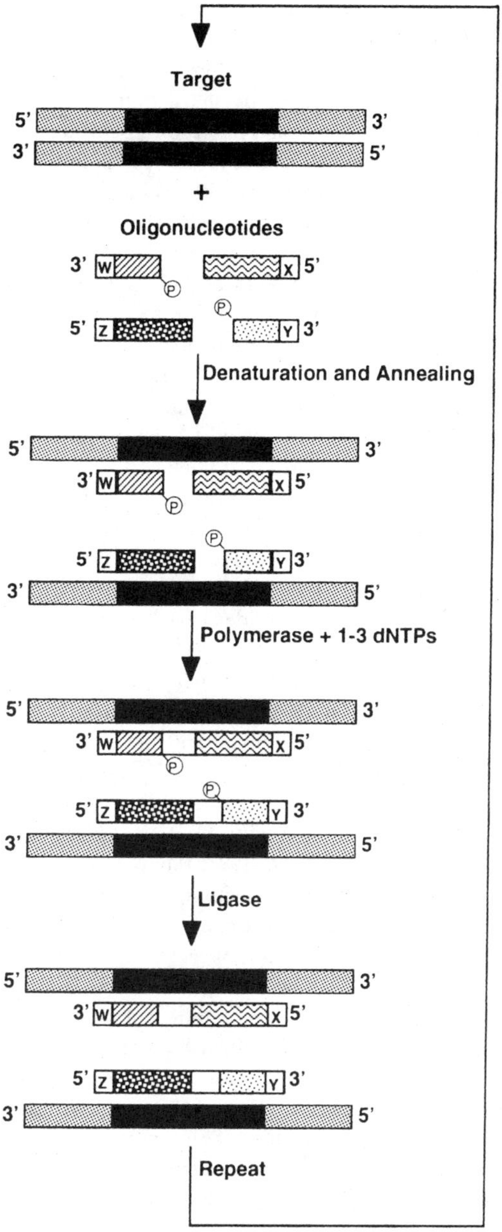

Figure 23.4. Schematic representation of the gap ligase chain reaction (gLCR). The gLCR involves the use of two pairs of oligonucleotides which are complementary to a particular DNA template. The first pair, W and X, are complementary to the upper DNA strand and hybridize to the template such that a gap exists between the two oligonucleotides. The second pair, Z and Y, hybridize to the lower DNA strand such that a gap exists between the two oligonucleotides. Note that in this configuration, W is complementary to Z and X to Y. When hybridized to W, Z forms a 3′ cohesive end. When hybridized to Y, X forms a 3′ cohesive end. Oligonucleotides W and Y are phosphorylated at their 5′ ends. The reaction is initiated by denaturing the target DNA using elevated temperatures. The temperature is then decreased to allow the two pairs of oligonucleotides to hybridize to the adjacent, complementary sequences on the target DNA. DNA polymerase then fills the gap in a template-dependent fashion. DNA ligase then joins the two pairs of oligonucleotides (X to W and Z to Y). In the subsequent cycles, the ligated molecules of previous cycles will be used as templates by the two pairs of oligonucleotides, thus amplifying the DNA target in an exponential manner.

LCR, the two oligonucleotides W and X when hybridized to the upper template strand, and the two oligonucleotides Y and Z when hybridized to the lower template strand, are separated by small gaps. The 5′ ends of the two "nonextending" oligonucleotides Y and W are phosphorylated. The outer 5′ and 3′ ends of the four oligonucleotides are normally synthesized either with an amino-terminal group to prevent unwanted ligation and extension reactions or with various haptens for product detection purposes. In one cycle of a gLCR procedure, the template is first denatured at elevated temperature, then the oligonucleotides are allowed to hybridize to the template at a lower temperature. The gap is filled with less than a total of four deoxynucleoside triphosphates (dNTPs) by the action of a thermostable DNA polymerase lacking 3′-to-5′ exonuclease activity and the resulting nicks on each strand are sealed by a thermostable DNA ligase. Repeated thermal cycling results in a gLCR amplification product, an oligonucleotide duplex approximately 40 to 60 bp in length with sequences dictated by the sequences of the gap region and the four oligonucleotides used in the amplification.

Most studies on gLCR have used oligonucleotides with similar T_m's although oligonucleotides with considerably different T_m's have also been used successfully (Grimberg *et al.*, 1993; Marshall *et al.*, 1994), and could result in higher amplification efficiency (P. Lin, unpublished data). The "extending" oligonucleotide X or Z (Fig. 23.4) is usually designed with a 3′-overhang of more than one nucleotide when annealed to its complementary nonextending oligonucleotide Y or W, respectively. This prevents the DNA polymerase from using oligonucleotides X and Z as templates for the extension of Y and W, thus preventing the formation of blunt-end duplexes that can participate in TIL. Likewise, the 3′-overhangs of the two oligonucleotide duplexes must be noncomplementary to prevent TIL.

As shown in Fig. 23.4, the gap between oligonucleotides W and X does not overlap the gap between Y and Z. However, overlapping gaps composed of AT or GC can also be used, particularly in some allele-specific applications. Gap sizes ranging from a few nucleotides to more than ten nucleotides have been reported. Less than four gap-filling dNTPs must be used so that uncontrolled extension by the polymerase does not occur.

In gLCR parameters such as cycle number, oligonucleotide and dNTP concentrations, cycling temperature and time, and buffer composition have to be optimized to achieve high efficiencies of both extension and ligation in the same solution. Although TIL events are probably less frequent in gLCR than in LCR, allele-specific gLCR also loses specificity and sensitivity after an optimal range of cycle numbers (Abravaya *et al.*, 1995). The same observation has been reported for allele-specific PCR (Ugozzoli and Wallace, 1991). Different thermostable polymerases (Mattila *et al.*, 1991; Lawyer *et al.*, 1989; Saiki *et al.*, 1988) and thermostable ligases (discussed in the LCR section) purified from different organisms or subsequently cloned are available. However, since enzymes from different sources may have very different properties and buffer requirements, optimization experiments need to be performed in order to achieve high-efficiency gLCR. The fidelity of the DNA polymerase is critical (Eckert and Kunkel, 1991; Ling *et al.*, 1991). In a gLCR procedure that uses an allele-specific extending oligonucleotide, the DNA polymerase must discriminate between a matched and mismatched base pair at or near the 3′ end of that oligonucleotide (Abravaya *et al.*, 1995).

gLCR shares several advantages with LCR as a clinical diagnostic tool, such as direct, nonradioactive detection of amplification products with proper design of oligonucleotides and the ability to use fragmented DNA as template. The relatively small size of gLCR oligonucleotides and gLCR ligation products can be an advantage in multiplex gLCR targeted at mutation hot spots (Jou *et al.*, 1995) and in subtyping different species of related genes (Dille *et al.*, 1993; Birkenmeyer and Armstrong, 1992; Segev, 1992).

Limitations in oligonucleotide selection discussed above for LCR also apply to gLCR. An additional constraint is that the gap sequence must be filled with less than four dNTPs. Compared to LCR, the gLCR cost per test could be higher since gLCR requires two additional components, namely, dNTPs and DNA polymerase. Template contamination and enzyme inhibition can exist when using clinical samples (Lee *et al.*, 1995); however, these problems are encountered in other target amplification-based DNA assays as well.

3.1. Amplification of Specific DNA Sequences Using gLCR

In the past few years, different gLCR amplification and detection methods have been developed for the detection of various viral, bacterial, and human genes. Detection levels of approximately 5 to 100 copies of target DNA molecules have been reported for gLCR using template DNAs from different clinical samples and using different detection methods (Schachter *et al.*, 1994; Birkenmeyer and Armstrong, 1992; Segev, 1992). For example, specific hepatitis B virus DNA (HBV) from immunogenically pretyped human serum samples (Grimberg *et al.*, 1993) and human papilloma virus (HPV) 16 DNA from tissue biopsy samples were specifically detected using gLCR followed by an immunocolorimetric microwell plate-based detection assay (Segev, 1992; Lin *et al.*, 1991). Other recent applications of gLCR to detect specific DNA targets are given in Table I.

3.2. Amplification of Specific Target RNA Sequences Using gLCR

Contrary to the success in recent years for developing direct gLCR methods for detecting specific DNAs, there has been only one report on using gLCR for the detection of RNA templates. One of the reaction design challenges lies in the fact that the reverse transcription (RT) reaction that generates cDNA from an RNA template typically uses four dNTPs for polymerization while gLCR uses three or less dNTPs for gap-filling. Marshall *et al.* (1994) have reported a modification of gLCR called asymmetric gap LCR (AGLCR) for the detection of hepatitis C virus (HCV) RNA that uses three dNTPs and four oligonucleotides. The gap region in

Table I
Recent Applications of LCR and Related Methodologies

Procedure	Target sequence	References
LCR	p53 gene	Nakazawa *et al.* (1994)
Multiplex LCR	β-globin gene	Wallace *et al.* (1994)
gLCR	*Chlamydia trachomatis*	Chernesky *et al.* (1994a,b), Schachter *et al.* (1994), Bassiri *et al.* (1995), Lee *et al.* (1995)
	HIV reverse transcriptase gene	Abravaya *et al.* (1995)
	Hepatitis B virus	Grimberg *et al.* (1993)
	Human papilloma virus 16	Lin *et al.* (1991), Segev (1992)
	Neisseria gonorrhoeae	Birkenmeyer and Armstrong (1992)
	Mycobacterium tuberculosis	Leckie *et al.* (1994)
Multiplex gLCR	Duchenne muscular dystrophy dystrophin gene	Jou *et al.* (1995)
AGLCR	Hepatitis C virus RNA	Marshall *et al.* (1994)
PCR+LCR	*Drosophila* optomotor-blind gene	Balles and Plfugfelder (1994)
	Bovine leukocyte adhesion deficiency gene	Batt *et al.* (1994)
	Lactococcus lactis ssp. *lactis*	Ward *et al.* (1994)
	Thanatophoric dysplasia gene	Tavormina *et al.* (1995)
	Cowpox virus	Pfeffer *et al.* (1994)
PCR+OLA	Finnish-type aspartylglucosylaminase gene	Delahunty *et al.* (1995)
PCR+ multiplex OLA	Cystic fibrosis transmembrane conductance regulator (CFTR) gene	Grossman *et al.* (1994), Eggerding *et al.* (1995)

this design contained two noncomplementary sequences with two different lengths for the upper and lower strands (2 and 11 nucleotides, respectively). In the first step, a specific limited length of cDNA was made from RNA templates in a reaction that included three dNTPs, an oligonucleotide complementary to a specific region of HCV RNA, and the Moloney murine lukemia virus reverse transcriptase. Subsequently, the reverse transcriptase was heat-denatured, the other three oligonucleotides, DNA polymerase, and DNA ligase added, and the reaction cycled as in gLCR. The net effect of the reverse transcription step was the filling in of the 11-nucleotide gap, which overlapped the 3′ end of the other extending oligonucleotide. The addition of the other three oligonucleotides to the cDNA produced in the first step resulted in the formation of a quaternary hybridization complex that could be amplified by gLCR. In this study, the AGLCR method was demonstrated to have a detection level of 20 copies of purified recombinant RNA templates. No reports of using RT coupled to LCR have been made. In such a coupled reaction the presence of the four dNTPs would not create a problem.

3.3. Detecting Mutations in DNA or RNA Using gLCR

In addition to gene amplification, gLCR has also been adapted for the detection of point mutations and deletion mutations in several genes. Depending on the particular point mutation to be tested, either allele-specific oligonucleotides or allele specific gap-filling can be used for single-base discrimination. In the former case, if the allele-specific base is located at the 3′ end (or penultimate 3′ end) of an extending oligonucleotide, then discrimination arises from the preferential extension by the polymerase of a perfectly matched over a mismatch-containing oligonucleotide (Abravaya *et al.*, 1995). In the latter case, discrimination is achieved by adding the allele specific dNTPs to the reaction. It should be noted that allele-specific gap-filling cannot be used to detect A-to-T (or vice versa) or G-to-C (or vice versa) mutations. Using allele-specific extending oligonucleotides designed with a single-base mismatch either at the ultimate 3′ or at the penultimate 3′ end, Abravaya *et al.* (1995) have reported the detection of single-base mutations in codon 215 of the cloned reverse transcriptase gene of HIV. The application of gLCR to the detection of deletion mutations was recently demonstrated by Jou *et al.* (1995) who used multiplex gLCR coupled with an immunochromatographic detection procedure to analyze nine mutations in the dystrophin gene. In this qualitative assay, the oligonucleotides were designed to detect only the normal alleles; absence of colorimetric signal was thus easily interpreted as a homozygous deletion. One limitation of this approach was that normals, carriers, and duplications all gave positive signals and could not be differentiated. Undoubtedly, further optimization of this and other gLCR designs will result in improved diagnostic assays for more infectious disease and genetic disease genes.

4. LIGATION OF TWO OLIGONUCLEOTIDES

The ligation of two oligonucleotides complementary to adjacent sequences on the same template strand can also be used to detect the products of other amplification techniques with a high specificity (Wu and Wallace, 1989b; Landegren *et al.*, 1988). This procedure has been called oligonucleotide ligation assay (OLA) (Nickerson *et al.*, 1990) or ligation detection reaction (LDR) (Barany, 1991b). The requirement for perfect base complementarity at the ligation junction makes this technique ideal for the discrimination of allelic variants. Since the target has been previously amplified (usually by PCR), one to five cycles (Nickerson *et al.*, 1990; Grossman *et al.*, 1994) of linear amplification are sufficient to generate signal detectable by nonradioactive methods.

5. RECENT APPLICATIONS OF LCR and RELATED TECHNOLOGIES

We have developed a colorimetric LCR assay for sickle-cell anemia (Wallace *et al.*, 1994; Reyes *et al.*, manuscript in preparation). A blood sample from each patient is tested in two reactions: one specific for the β-globin A allele and the other specific for the S allele. Each reaction also simultaneously amplifies the human growth hormone gene, allowing for the standardization of the β-globin signal to the amount of target DNA present in the sample. Amplification products are detected in a microwell plate format.

Other recent, representative applications of LCR and related methodologies are given in Table I. For a more comprehensive list, the reader is referred to Wiedmann *et al.* (1994).

6. POSTAMPLIFICATION DETECTION

LCR postamplification detection was reviewed in detail recently (Wiedmann *et al.*, 1994). In this section, only representative detection techniques will be discussed. Traditionally, LCR products were detected by using radio-labeled oligonucleotides in the reaction and analyzing the postamplification LCR mixture by slab gel electrophoresis. Amplification products were then detected by autoradiography. However, in order for LCR to be an effective diagnostic tool with widespread acceptance, the assay not only has to be reproducible, sensitive, and nonradioactive but also has to have a high throughput and be cost-effective.

At the conclusion of the LCR, the reaction mixture contains the following nucleic acids: amplification products, unreacted oligonucleotides, template, and carrier DNA (e.g., salmon sperm DNA). Analysis of LCR products usually involves two steps: *separation* of amplification products from the rest of the nucleic acids and other reaction components, and *detection* of the separated amplification products (Fig. 23.5). The basic detection strategy involves a "separation (or capture) tag" at a terminus of one oligonucleotide (e.g., at the 3′ end of W in Fig. 23.1) and a "detection tag" at a terminus of another oligonuleotide (at the 5′ end of X, Fig. 23.1); the two tags are combined in one molecule by ligation.

The two common separation methods are by size and by capture onto a solid support. Separation according to size, usually by electrophoresis, does not necessarily require a separation tag, since the longer amplification product migrates differently from the shorter unreacted oligonucleotides. However, in cases where multiplex amplification of different target sequences is desired, this approach has a limitation In multiplex LCR, the lengths of the different oligonucleotides have to fall within a narrow range so that amplification efficiencies are similar for all of the target regions at the specified annealing/denaturing temperatures. Although it is possible to label the different amplification products with different detection tags (e.g., fluorescent dyes), the number of usable dyes is limited. A solution to this problem was described by Grossman *et al.* (1994) for a PCR-OLA procedure, where the electrophoretic mobility of the OLA product was modified by attaching oligomeric nonoligonucleotide tails to one of the oligonucleotides. A wide range of oligomeric modifier length was tolerated by the ligation reaction with minimal effect on OLA reaction specificity and yield. Thus, it was possible to perform multiplex analysis of nine regions of the cystic fibrosis transmembrane regulator (CFTR) gene using a single fluorescent dye label. The different ligation products were resolved by gel electrophoresis and detected by fluorescence scanning. More recently, the same group has extended this approach to 31 regions of the CFTR gene using four fluorescent dyes (E. Winn-Deen, abstract, Advances in Genetic Screening and Diagnosis of Human Diseases Meeting, San Francisco, 1995). It should be noted that the post-PCR-OLA method used in these experiments involved a linear amplification of five ligation cycles using two ligation oligonucleotides. How the

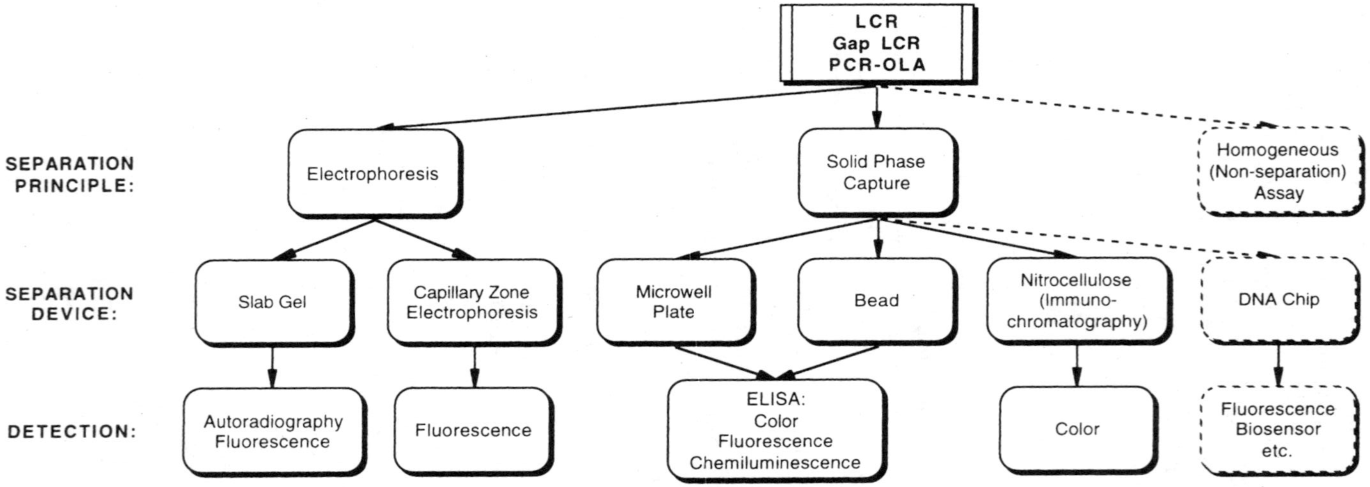

Figure 23.5. Flow chart for post amplification detection. Amplification products formed in LCR, gLCR or PCR-OLA assays are usually labeled with separation and detection tags. Using the separation tag, the ligation product is first separated from template, unreacted oligonucleotides and, in multiplex reactions, other ligation products. Next, a signal-generating process appropriate to the detection tag is performed. Dashes indicate methods that have not yet been described.

minimal effects of mobility modifiers on T_m and ligation kinetics observed in OLA will affect an exponential amplification reaction like LCR remains to be determined.

Capillary electrophoresis (CE) has been used in place of slab gel electrophoresis to analyze PCR fragments (Liu *et al.*, 1995; Schwartz and Ulfelder, 1992). CE coupled with laser-induced fluorescence detection has also been applied to PCR-OLA analysis (Grossman *et al.*, 1994). Advantages of CE over slab electrophoresis include ease of use, shorter run time, potential for high throughput by use of capillary arrays, and amenability to automation. However, despite these many good features, CE has not been used for routine detection to date.

Alternatively, separation can be achieved by designing the amplification product to contain a capture tag at one end and a detection tag at the other end. Both ligation products and unreacted oligonucleotides carrying the capture tag are captured onto a solid phase. Other reaction components are removed by a simple wash step. Capture tags like biotin, haptens for specific antibodies, and short DNA tails have been used for amplicon binding to avidin/streptavidin, cognate antibody, and complementary oligonucleotide, respectively. Usually, these capture reagents have been immobilized on microwell plates or beads. In the subsequent detection step, only amplification products are detected since only those molecules carry the detection tag.

The detection step in the solid phase often involves ELISA. Detection tags like biotin, haptens, and short DNA tails are reacted with enzyme-conjugated avidin/streptavidin, cognate antibody, and complementary oligonucleotide, respectively. A wide variety of chromogenic, luminescent and fluorescent substrates have been used.

A version of solid-phase capture using immunochromatography was recently reported (Jou *et al.*, 1995). In this method, the amplicon contains a hapten capture tag at one end and a biotin molecule at the other end. The postamplification reaction mixture is mixed with colored latex particles coated with antibiotin antibody, then allowed to chromatograph up a nitrocellulose strip that had been previously spotted with antihapten antibody. The presence of amplicon is shown by the appearance of color on the site on the strip where the antihapten antibody was spotted. The potential for multiplex detection using this method is limited only by the availability of hapten–cognate antibody pairs to use for the capture. For example, Jou *et al.* (1995) used this detection method to analyze nine exons of the Duchenne muscular dystrophy gene for deletion mutations by multiplex gLCR.

The detection step has been automated in some ligation-based assays. The Abbott *Chlamydia trachomatis* gLCR assay produces amplification products that are haptenated at both termini (Lee *et al.*, 1995). Capture of amplification products by beads coated with antibody against the first hapten, and fluorescent detection by enzyme-linked antibody against the second hapten, are done automatically in the LCx analyzer. The PCR-OLA assay described by Nickerson *et al.* (1990) uses a robotic workstation to perform the PCR amplification, oligonucleotide ligation, and microwell plate-based ELISA detection. The gel electrophoretic detection of ligation products in the CFTR assay described by Grossman *et al.* (1994) was performed automatically in a fluorescence-scanning DNA sequencer.

Factors affecting the choice between a solid phase capture or electrophoretic mode of separation/detection are ease of use, cost, and throughput. Microwell plate- or bead-based ELISAs are easy to perform and widely used in clinical laboratories while gel electrophoresis demands a higher level of technical skill and, in the case of automated fluorescence-based detection, requires expensive instrumentation. The disadvantage of an ELISA plate assay is that detection cannot be multiplexed, i.e., only one kind of amplification product can be detected in one well. Thus, a multiplex LCR mixture has to be divided into as many wells as the number of target loci, essentially reducing the amount of each amplification product available for detection, in addition to increasing the cost of the assay. This limitation is less important in electrophoretic detection, since the multiplex LCR mix can be loaded into a single gel well and multiple ligation products resolved by differences in mobility or fluorescent dye tags. The use of an oligonucleotide array immobilized on a chip (Chetverin and Kramer, 1994) to capture multiple amplification products, as well as the immunochromatographic strip method described above, are possible nonelectrophoretic alternatives to achieve multiplex detection of LCR products. Ultimately, the separation/detection method will have to be designed based on the number of patient samples (large population screen versus diagnosis) and the number of amplification targets per sample.

A big drawback of current LCR assay formats is the need to perform the amplification cycling steps and separation/detection steps in two different platforms (e.g., cycling reaction in a tube followed by detection in a 96-well plate or gel). In addition to being a potential source of carryover contamination in subsequent assays, this transfer step also increases the possibility of sample mix-up. A homogeneous, nonseparation assay where amplification and detection occur in a single tube will minimize these problems. For PCR, homogeneous detection schemes based on the increase in fluorescence as double-stranded amplicon accumulates (Higuchi *et al.*, 1993) or the unmasking of fluorescent signal due to the $5' \rightarrow 3'$ exonuclease activity of Taq DNA polymerase (Livak *et al.*, 1995) have been described. Although an analogous technique has yet to be described for the LCR, its development will result in tremendous improvement to current LCR assay formats.

7. CONCLUSION

In summary, LCR is a template amplification methodology that is both allele- and template-specific. One major advantage of LCR is the ease by which the amplified product can be directly detected and the potential for automating the entire process. In addition, because such a short template is required for the amplification, even rather degraded DNA samples are useful for LCR. One disadvantage of LCR is the fact that the sensitivity for template is limited by template independent ligation. Several ways around this problem have been found, including pre-amplification and gap LCR.

REFERENCES

Abravaya, K., Carrino, J. J., Muldoon, S., and Lee, H. H. (1995). Detection of point mutations with a modified ligase chain reaction (gap-LCR). *Nucleic Acids Res.* **23**:675–682.

Balles, J., and Plfugfelder, G. O. (1994). Facilitated isolation of rare recombinants by ligase chain reaction: Selection for intragenic crossover events in the Drosophila optomotor-blind gene. *Mol. Gen. Genet.* **245**:734–740.

Barany, F. (1991a). Genetic disease and DNA amplification using cloned thermostable ligase. *Proc. Natl. Acad. Sci. USA* **88**:189–193.

Barany, F. (1991b). The ligase chain reaction in a PCR world. *PCR Methods Appl.* **1**:5–16.

Barany, F., and Gelfand, D. H. (1991). Cloning, overexpression and nucleotide sequence of a thermostable DNA ligase-encoding gene. *Gene* **109**:1–11.

Bassiri, M., Hu, H. Y., Domeika, M. A., Burczak, J., Svensson, L.-O., Lee, H. H. and Mardh, P.-A. (1995). Detection of *Chlamydia trachomatis* in urine specimens from women by ligase chain reaction. *J. Clin. Microbiol.* **33**:898–900.

Batt, C. A., Wagner, P., Wiedmann, M., Luo, J., and Gilbert, R. (1994). Detection of bovine leukocyte adhesion deficiency by nonisotopic ligase chain reaction. *Anim. Genet.* **25**:95–98.

Birkenmeyer, L., and Armstrong, A. S. (1992). Preliminary evaluation of the ligase chain reaction for specific detection of *Neisseria gonorrhoeae*. *J. Clin. Microbiol.* **30**:3089–3094.

Chernesky, M. A., Jang, D., Lee, H., Burczak, J. D., Hu, H., Sellors, J., Tomazic-Allen, S. J., and Mahony, J. B.

(1994a). Diagnosis of *Chlamydia trachomatis* infections in men and women by testing first-void urine by ligase chain reaction. *J. Clin. Chem.* **32**:2682–2685.

Chernesky, M. A., Lee, H., Schachter, J., Burczak, J. D., Stamm, W. E., McCormack, W. M., and Quinn, T. C. (1994b). Diagnosis of *Chlamydia trachomatis* urethral infection in symptomatic and asymptomatic men by testing first-void urine in a ligase chain reaction assay. *J. Infect. Dis.* **170**:1308–1311.

Chetverin, A. B., and Kramer, F. R. (1994). Oligonucleotide arrays: New concepts and possibilities. *Bio/Technology* **12**:1093–1099.

Delahunty, C. M., Ankener, W., Brainerd, S., Nickerson, D. A., and Monomen, I. T. (1995). Finnish-type aspartylglucosaminuria detected by oligonucleotide ligation assay. *Clin. Chem.* **41**:59–61.

Dille, B. J., Butzen, C. C., and Birkenmeyer, L. G. (1993). Amplification of *Chlamydia trachomatis* DNA by ligase chain reaction. *J. Clin. Microbiol.* **31**:729–731.

Eckert, K. A., and Kunkel, T. A. (1991). DNA polymerase fidelity and the polymerse chain reaction. *PCR Methods Appl.* **1**:17–24.

Eggerding, F.fA., Iovannisci, D. M., Brinson, E., Grossman, P., and Winn-Deen, E. S. (1995). Fluorescence-based oligonucleotide ligation assay for analysis of cystic fibrosis transmembrane conductance regulator gene mutations. *Hum. Mutat.* **5**:153–165.

Grimberg, J., Lin, C. P., Sheridan, K., Tackney, C., and Waksal, J. (1993). Detection of serum *hepatitis B virus* DNA sequence by the repair chain reaction DNA amplification system. *Clin. Chem.* **39**:738.

Grossman, P. D., Bloch, W., Brinson, E., Chang, C. C., Eggerding, F. A., Fung, S., Iovannisci, D.A., Woo, S., and Winn-Deen, E. S. (1994). High-density multiplex detection of nucleic acid sequences: Oligonucleotide ligation assay and sequence-coded separation. *Nucleic Acids Res.* **22**:4527–4534.

Higuchi, R., Fockler, C., Dollinger, G., and Watson, R. (1993). Kinetic PCR analysis: Real-time monitoring of DNA amplification reactions. *Bio/Technology* **11**:1026–1030.

Jou, C., Rhoads, J., Bouma, S., Ching, S. F., Hoijer, J., Schroeder-Poliak, P., Zaun, P., Smith, S., Richards, S., Caskey, C. T., and Gordon, J. (1995). Deletion detection in the dystrophin gene by multiplex gap ligase chain reaction and immunochromatographic strip technology. *Hum. Mutat.* **5**:86-93.

Kleppe, K., Ohtsuka, E., Kleppe, R., Molineux, I., and Khorana, H. G. (1971). Studies on polynucleotides. XCVI. Repair replication of short synthetic DNAs as catalyzed by DNA polymerases. *J. Mol. Biol.* **56**:341–361.

Landegren, U., Kaiser, R., Sanders, J., and Hood, L. (1988). A ligase-mediated gene detection technique. *Science* **241**:1077–1080.

Lawyer, F. C., Stoffel, S., Saiki, R. K., Myambo, K., Drummond, R., and Gelfand, D. H. (1989). Isolation, characterization, and expression in *E. coli* of the DNA polymerase gene from the extreme thermophile, *Thermus aquaticus*. *J. Biol. Chem.* **264**:6427–6437.

Leckie, G., Cao, J., Davis, A., Facey, I., Lin, B.-C., and Lee, H. (1994). Ligase chain reaction (LCR) DNA amplification for direct detection of *Mycobacterium tuberculosis* (MTB) in clinical specimens, in: *Abstracts of the General Meeting of the American Society for Microbiology*, 189.

Lee, H. H., Chernesky, M. A., Schachter, J., Burczak, J. D., Andrews, W. W., Muldoon, S., Leckie, G., and Stamm, W. E. (1995). Diagnosis of *Chlamydia trachomatis* genitourinary infection in women by ligase chain reaction assay of urine. *Lancet* **345**:213–216.

Lin, C. P., Goldbard, S., Grills, G., Sheridan, K., Waksal, H. W., Tackney, C., and Segev. D. (1991). Amplification and nonradioactive detection of the human *papilloma virus* 16 DNA by the repair chain reaction, in *Abstracts of the Annual San Diego Conference on Nucleic Acids*.

Ling, L. L., Keohavong, P., Dias, C., and Thilly, W. G. (1991). Optimization of polymerase chain reaction with regard to fidelity: Modified T7, Taq and Vent DNA polymerases. *PCR Methods Appl.* **1**:63–69.

Liu, M. S., Zang, J., Evangelista, R. A., Rampal, S., and Chen, F.-T. A. (1995). Double-stranded DNA analysis by capillary electrophoresis with laser-induced fluorescence using ethidium bromide as an intercalator. *Bio-Techniques* **18**:316–323.

Livak, K. J., Marmaro, J., and Todd, J. A. (1995). Towards fully automated genome-wide polymorphism screening. *Nature Genet.* **9**:341–342.

Marshall, R. L., Laffler, T. G., Cerney, M. B., Sustachek, J. C., Kratochvil, J. D., and Morgan, R. L. (1994). Detection of HCV RNA by the asymmetric gap ligase chain reaction. *PCR Methods Appl.* **4**:80–84.

Mattila, P., Korpela, J., Tenkanen, T., and Pitkanen, K. (1991). Fidelity of DNA synthesis by the *Thermococcus litoralis* DNA polymerase—an extremely heat stable enzyme with proof reading activity. *Nucleic Acids Res.* **19**:4967–4973.

Nakazawa, H., English, D., Randell, P. L., Nakazawa, K., Martel, N., Armstrong, B. K., and Yamasaki, H. (1994).

UV and skin cancer: Specific p53 gene mutation in normal skin as a biologically relevant exposure measurement. *Proc. Natl. Acad. Sci. USA* **91**:360–364.

Nickerson, D. A., Kaiser, R., Lappin, S., Stewart, J., Hood, L., and Landegren, U. (1990). Automated DNA diagnostics using an ELISA-based oligonucleotide ligation assay. *Proc. Natl. Acad. Sci. USA* **87**:8923–8927.

Pfeffer, M., Meyer, H., and Wiedmann, M. (1994). A ligase chain reaction targeting two adjacent nucleotides allows the differentiation of cowpox virus from other *Orthopoxvirus* species. *J. Virol. Methods* **49**:353–360.

Saiki, R. K., Gelfand, D. H., Stoffel, S., Scharf, F. J., Higuchi, R., Horn, G. T., Mullis, K. B., and Ehrlich, H. A. (1988). Primer-directed enzymatic amplification of cDNA with a thermal stable DNA polymerase. *Science* **239**:487–491.

Schachter, J., Stamm, W. E., Quinn, T. C., Andrews, W. W., Burzak, J. D., and Lee, H. H. (1994). Ligase chain reaction to detect *Chlamydia trachomatis* infection of the cervix. *J. Clin. Microbiol.* **32**:2540–2543.

Schwartz, H. E., and Ulfelder, K. J. (1992). Capillary electrophoresis with laser-induced fluorescence detection of PCR fragments using thiazole orange. *Anal. Chem.* **64**:1737–1740.

Segev, D. (1992). Amplification of nucleic acid sequence by the repair chain reaction, in: *Nonradioactive Labeling and Detection of Biomolecules* (C. Kessler, ed.), Springer-Verlag, Berlin, pp. 212–218

Takahashi, M., Yamaguchi, E., and Uchida, T. (1984). Thermophilic DNA ligase—Purification and properties of the enzyme from *Thermus thermophilus* HB8. *J. Biol. Chem.* **259**:10041–10047.

Tavormina, P. L., Shiang, R., Thompson, L. M., Zhu, Y.-Z., Wilkin, D. J., Lachman, R. S., Wilcox, W. R., Rimoin, D. L., Cohn, D. H., and Wasmuth, J. J. (1995). Thanatophoric dysplasia (types I and II) caused by distinct mutations in fibroblast growth factor receptor 3. *Nature Genet.* **9**: 321–328.

Ugozzoli, L., and Wallace, R. B. (1991). Allele specific polymerase chain reaction. *Methods: A Companion to Methods in Enzymology* **2**:42–48.

Wallace, R. B., Ugozzoli, L., Lowery, J., and Reyes, A. A. (1994). The application of LCR to neonatal testing of variant hemoglobin genes. Proceedings, 10th National Neonatal Screening Symposium, Seattle, Washington, pp. 127–130.

Ward, L. J., Brown, J. C., and Davey, G. P. (1994). Application of the ligase chain reaction to the detection of nisinZ genes in *Lactococcus lactis ssp. lactis*. *FEMS Microbiol. Lett.* **117**:29–34.

Wiedmann, M., Wilson, W. J., Czajka, J., Luo, J., Barany, F., and Batt, C. A. (1994). Ligase chain reaction (LCR)—Overview and applications. *PCR Methods Appl.* **3**:S51–S64.

Wu, D. Y., and Wallace, R. B. (1989a). The ligation amplification reaction (LAR)—Amplification of specific DNA sequences using sequential rounds of template-dependent ligation. *Genomics* **4**:560–569.

Wu, D. Y., and Wallace, R. B. (1989b). Specificity of the nick-closing activity of bacteriophage T4 DNA ligase. *Gene* **76**:245–254.

Chapter 24

The Protein Truncation Test (PTT) for Rapid Detection of Translation-Terminating Mutations

Johan T. Den Dunnen, Pauline A. M. Roest, Rob B. Van Der Luijt, and Frans B. L. Hogervorst

1. INTRODUCTION

Most commonly used techniques to detect (point) mutations, e.g., SSCP, DGGE, chemical mismatch cleavage and heteroduplex analysis, are versatile techniques which have proven their usefulness (for references see other chapters in this book). Still, they all have their limitations. They use genomic DNA as starting material and reveal all sequence changes, including silent mutations and sequence variations in coamplified noncoding sequences. They can only be used to analyze small regions, 100–600 bp depending on the technique used. They do not pinpoint the site (except chemical mismatch cleavage) or the type of the mutation. Sequence changes identified in or close to, for example, splice sites or a promoter, require further analysis to verify their effect at a cellular level. Finally, several frequently mutated genes have been identified whose structure precludes an efficient analysis with the techniques mentioned; they are very large and split into many exonic fragments (sometimes over 70).

An increasing number of disease genes have been identified in which a major fraction of the patients contain mutations which cause premature translation termination (nonsense and frameshift mutations), including colon cancer, neurofibromatosis, Duchenne muscular dystrophy (DMD), breast cancer, and Emery Dreifuss muscular dystrophy (EMD). In such cases, an ideal method for mutation analysis would be a technique which selectively detects these chain-terminating mutations. Such a method, designated the protein truncation test (PTT), was developed in our laboratory, initially for the detection of point mutations in the dystrophin (DMD) gene (Roest *et al.*, 1993a). The techique is based on the *in vitro* translation of PCR products

Johan T. Den Dunnen, Pauline A. M. Roest, Rob B. Van Der Luijt, and Frans B. L. Hogervorst • MGC-Department of Human Genetics, Leiden University, 2333 AL Leiden, The Netherlands.

Technologies for Detection of DNA Damage and Mutations, edited by Gerd P. Pfeifer, Plenum Press, New York, 1996.

amplified from the target gene and the identification of shortened translation products. Recently, we and others have exploited this test as an efficient assay to detect the mutation involved in many different diseases, including adenomatous polyposis coli (Van Der Luit *et al.*, 1994; Powell *et al.*, 1993), neurofibromatosis [NF1 (Heim *et al.*, 1995) and NF2 (Hogervorst *et al.*, unpublished)], DMD (Roest *et al.*, 1993a; Gardner *et al.*, 1995), Hunter syndrome (Jonsson *et al.*, 1995), and breast cancer [BRCA1 (Hogervorst *et al.*, 1995)].

1.1. PTT Principle

PTT is based on the detection of truncated peptides after translation of the coding sequences amplified from a patient sample by PCR. To enable the translation of the PCR products, a tailed primer is used to introduce an RNA polymerase promoter, a translation initiation sequence, and an in-frame ATG triplet. A flowchart of the PTT procedure, outlining the essential steps, is presented in Fig. 24.1. First, the PCR template is isolated, usually from freshly drawn blood of a patient. Second, a reverse transcription reaction is performed to make a DNA copy of the messenger RNA. Third, PCR is used both to amplify the target sequence(s) and to introduce a tailed primer sequence. Fourth, the PCR fragments are electrophoresed to check their quality/quantity and to determine their size. Fifth, the products are transcribed using an RNA polymerase and the RNA transcripts are translated into peptides. Finally, the translation products are analyzed on SDS–PAGE gel to determine their length. Translation-terminating mutations will now appear as shortened peptides when compared to the fully translated control samples.

1.2. Templates

Normally, RNA is used as a starting template in the PTT assay. However, if the structure of the gene of interest is favorable (Section 3.1), i.e., contains large exons, DNA can also be used (Groden *et al.*, 1991; Miki *et al.*, 1994). However, the majority of genes do not contain such a favorable structure but rather are split into many small exons, leaving RNA as the only workable template.

Preferably, RNA should be isolated from a tissue where the gene of interest is expressed at significant levels. In many cases such a source is not easily available and RNA is isolated from either blood or cultured cells, e.g., fibroblasts obtained from skin biopsies. Depending on the level of expression in these samples, it is often necessary to perform a nested PCR reaction to amplify the target sequences to a visible level. This approach is very powerful and allows the amplification of nearly any gene from, for example, blood samples based on the minute levels of 'ectopic' ('illegitimate') transcription which have been observed for most genes in any tissue (Chelly *et al.*, 1989).

1.3. Reverse Transcription and PCR

Depending on the size of the coding region to be analyzed, the gene of interest is divided into segments which may be several kilobases in length and which partly overlap (Section 3.2). Although peptides can be translated efficiently from templates of 4 kb and larger, it is often difficult to coamplify such fragments in sufficient quantities unless specific precautions are taken (Cheng *et al.*, 1994). Furthermore, convenient lengths for the analysis of the translation products are determined mainly by the possibilities to detect small mobility shifts in SDS–PAGE, fragments of 1–2 kb giving optimal results.

Reverse transcription (RT) of the RNA is performed to synthesize cDNA, the template for the subsequent PCR. RT can be performed by using either random hexamers or gene-specific primers. Depending on the level of expression of the gene of interest, the PCR is performed

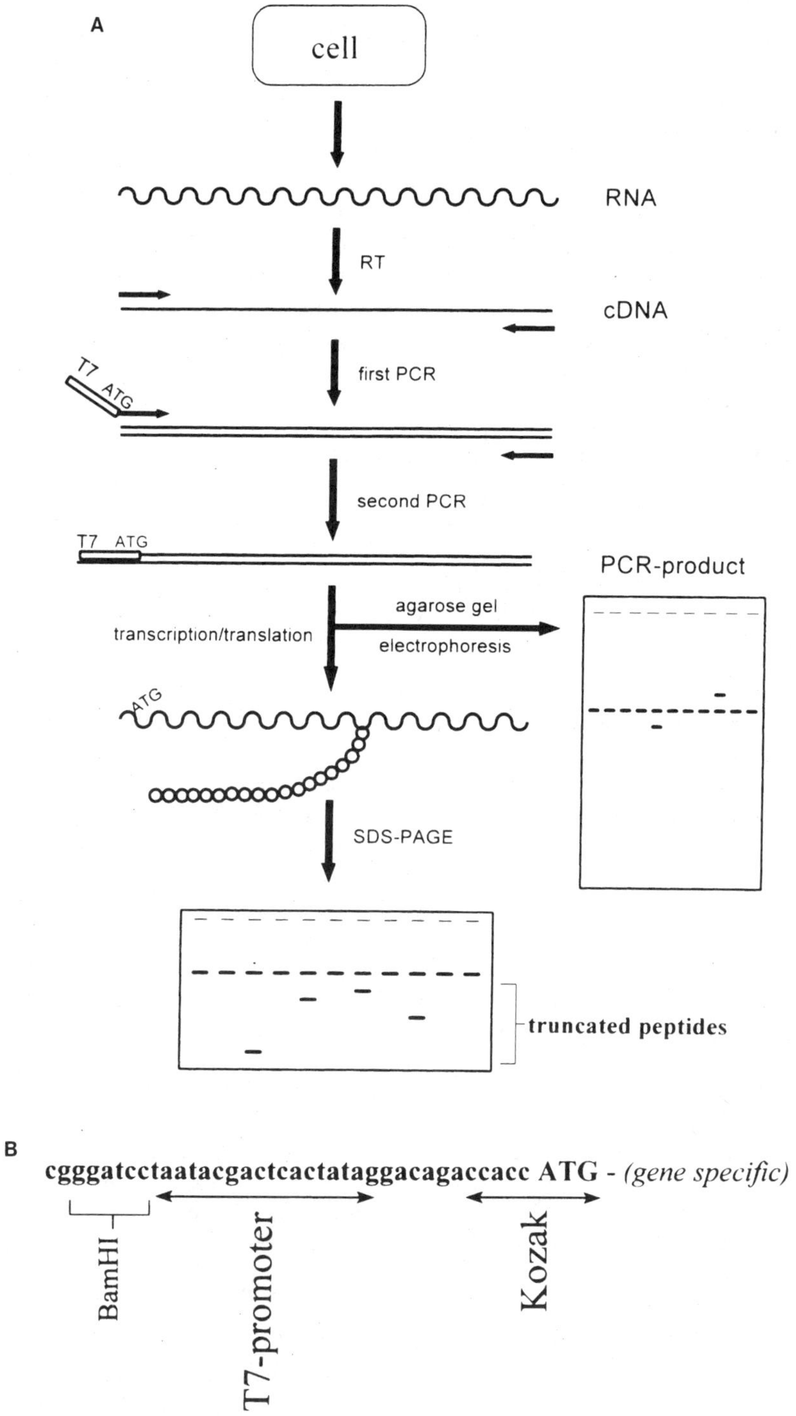

Figure 24.1. The PTT assay. (A) Schematic presentation of the individual steps in the PTT assay. RT is reverse transcription, T7 the T7 RNA polymerase promoter sequence, and ATG the Kozak translation initiation sequence. (B) Sequence of a tailed PTT primer and its elements.

directly with the tailed primer or the tailed primer is introduced in a second, nested PCR reaction. When DNA is used as a template, a nested PCR is not necessary and the tailed primer can be used directly.

To allow *in vitro* transcription and translation, the (nested) PCR is performed with a tailed gene-specific sense oligonucleotide, introducing an RNA promoter and in-frame translation initiation sequences at the 5′ end of the amplified fragment (Fig. 24.1B). Although other RNA polymerases can be used, T7-RNA polymerase is used most frequently. The tailed primers we have used (Roest *et al.*, 1993a; Van Der Luit *et al.*, 1994; Hogervorst *et al.*, 1995) are based on the sequence described by Sarkar and Sommer (Sarkar and Sommer, 1989), containing an 18-bp bacteriophage T7 promoter and an 8-bp eukaryotic translation initiation signal (or Kozak sequence). Our primers carry modifications at the spacer region, between the T7 promoter and translation initiation sequence, in the translation initiation signal, and, to facilitate cloning of the amplified fragments, we have added a restriction endonuclease site in the 5′ end (Fig. 24.1B, a *Bam*HI site). Since protein products are generated by run-off translation from the amplified segments, the reverse primers do not require the presence of a specific translation termination signal. Again, a restriction site may be added to facilitate cloning.

The construction of a set of overlapping segments together spanning the entire coding region of the gene of interest deserves careful attention. A primer pair consists of a (tailed) sense (or forward) primer and an antisense (or reverse) primer. Choice of both primers critically determines the ultimate sensistivity of the PTT analysis. To calculate the most effective sequence complementary to the target gene we use software like PRIMER (MIT, Cambridge, MA) and OSP (Hillier and Green, 1991). Next to these standard criteria to design primer pairs, several additional factors have to be considered:

1. To enable translation, the ATG-initiation codon in the tailed sense primer should be in frame with the coding sequence of the gene.
2. Flanking segments should contain a considerable overlap to ensure detection of mutations toward their ends, i.e., 150 to 200-bp overlaps (5–8 kDa after translation) for 1- to 2-kb fragments. In a specific set, early truncations are easily missed because the very short peptides produced run off the gel while truncations near the end might be missed because large peptides are not resolved near the top of the gel.
3. Primers of flanking segments should be located in different exons to decrease the danger of not amplifying, with both sets, one allele in which, for example, an exon is deleted or duplicated (Section 4.2.3). Primers which span bordering exons, i.e., genomically interrupted by introns, should not be used. Such primers double the risk of amplifying only one of the two alleles expressed when a mutation is present which influences the normal processing of the RNA. Consequently, the danger of missing essential mutations, e.g., at a splice site, increases considerably.
4. To enable the detection of a truncating mutation, the reverse primer should not be selected near the end of a region where a large open reading frame is present in one or both of the alternative reading frames; in such cases a frame shifting mutation would not cause premature translation termination.

1.4. Gel Analysis of PCR Products

After (RT-)PCR the fragments produced are analyzed on agarose gel. The quantity of the products made is used to determine the amount which should be added to the *in vitro* transcription/translation assay. Furthermore, their size will reveal if any of the products are altered in size, thereby revealing the presence of alternative splicing pathways and/or mutations which influence the processing of the primary RNA transcript. A decreased size usually indicates genomic deletions or mutations affecting splicing, while an increased size points to genomic duplications or, again, mutations affecting splicing.

1.5. Transcription and Translation

The PCR-amplified fragments can be efficiently analyzed using *in vitro* transcription/translation systems such as the TnT™ T7 Coupled Rabbit Reticulocyte Lysate System (Promega). The advantage of this system is that transcription and translation reactions are performed in a one-tube reaction. To enable the detection of the translation products, we have used incorporation of radiolabeled amino acids. The most commonly used labels are [^{3}H]leucine and [^{35}S]methionine. Depending on their frequency in the protein sequence, other amino acids may be preferred. Biotinylated amino acids can be used, giving the advantage of working with a nonradioactive system at the cost of an additional step, i.e., blotting of the gel and incubation of the blot.

Transcription and translation can also be performed in separate reactions, which, at a slight cost of speed, has some advantages. First, both reactions can be performed under optimal conditions, thereby increasing yields and decreasing the number of background products. Second, flexibility is increased, enabling the use of alternative RNA polymerases and other translation systems (e.g., wheat germ extracts).

1.6. Visualization of Truncated Peptides

Translation products are resolved by SDS–PAGE and visualized by autoradiography. The occurrence of a shortened polypeptide will show if the corresponding segment contains a premature translation termination sequence and thus a disease-causing mutation.

2. METHODS

PTT normally uses RNA as a template. In specific cases (Section 3.1), DNA can also be used. For DNA isolation we have used standard techniques (Maniatis *et al.*, 1989).

2.1. Collecting Cell Samples (for RNA Isolation)

In principle the RNA to be analyzed can be isolated from any source. Our experience involves two main sources, RNA isolated from either (freshly drawn) blood or cultured cells.

2.1.1. Preparation of Peripheral Blood Lymphocytes

- Collect two tubes of 10 ml of EDTA blood. Fill four tubes (white caps) with 5 ml Histopaque-1077 (Sigma). Slowly add 5 ml of blood on top of the Histopaque-1077 and spin at 2000 rpm for 20 min at room temperature in a swing out rotor.
- After centrifugation (do not use brakes), four layers should be visible; the top layer contains serum and thrombocytes, the second layer is white and contains lymphocytes, the third colorless layer is the Histopaque, and the fourth layer contains the erythrocytes. Discard most of the first layer and put the second layer in a fresh tube.
- Wash with 10 ml cold and sterile PBS. Spin at 1500 rpm for 10 min. A small white, sometimes red, pellet should be present.
- Remove the supernatant by carefully inverting the tube.

2.1.2. Preparation of Cultured Cells

- Take a tissue culture flask (75 cm^2) with cells grown to 80–90% confluency. Discard the culture medium and wash carefully with sterile and cold PBS.
- Add 5 ml PBS, put the flask on ice, and isolate the cells by scraping with a sterile wiper.
- Put the cells into a new tube, add 5 ml PBS to the original flask, and continue scraping. Collect all cells and centrifuge at 1500 rpm for 10 min.
- Remove the supernatant by carefully inverting the tube.

2.2. RNA Isolation Using RNAzolB

Continue directly after Sections 2.1.1 or 2.1.2 with the isolation of the RNA. For 5 ml blood or one 75 cm^2 flask we use 1–1.3 ml RNAzolB (Biotecx).

- Add RNAzolB to the freshly isolated cells. Lyse the cells by passing the lysate a few times through a pipet tip and transfer to a 1.5- to 2-ml Eppendorf tube.
- Add 0.1 ml chloroform per 1 ml of homogenate, shake vigorously for 15 sec, and put on ice for 5 min.
- Centrifuge the suspension at 12,000*g* for 15 min at 4°C. Two phases should be formed: an upper colorless, aqueous phase and a lower, blue phenol/chloroform phase.
- Carefully transfer the aqueous phase (0.5–0.7 ml) to a fresh tube and add an equal volume of isopropanol. Place 15 min on ice.
- Centrifuge 15 min at 12,000*g* and 4°C. The RNA precipitate should be visible as a pellet at the bottom of the tube.
- Remove the supernatant and wash the pellet with 200 μl 70% ethanol.
- Centrifuge 10 min at 12,000*g* and carefully remove all supernatant. Air dry the pellet (not for too long; otherwise the pellet will not resolve).
- Resolve the pellet in 50–100 μl TE and analyze an aliquot on gel (Section 2.3).
- Precipitate the RNA by adding 0.1 vol 3 M NaAc (pH 5.3) and 2.8 vol ethanol. Store at −20 or −70°C.

2.3. Analysis of RNA Samples

To check the quality and amount of the RNA isolated, it should be analyzed on gel using standard techniques (Maniatis *et al.*, 1989). Always use sterile pipets, tubes, and solutions, keep the RNA samples cold, wear gloves, and work with diethylpyrocarbonate (DEPC)-treated RNase-free solutions, i.e., add 0.1–0.05% v/v DEPC to a solution, leave O/N, and autoclave.

- Carefully clean an electrophoresis tank, the gel tray, and comb(s). Incubate for 1 hr with 1 M NaOH; wash extensively with sterile water.
- Pour a 1.5% agarose gel (SeaKam LE) using sterile TBE and add 0.2 μg/ml ethidium bromide.
- Add to an aliquot, i.e., 5–10% of the RNA sample, an equal amount of RNA loading mix. Incubate for 5 min at 65°C and load the gel.
- Run the gel and analyze the RNA on a UV transilluminator.

2.4. Reverse Transcription and PCR

2.4.1. Reverse Transcription (random primed, split in overlapping sets)

- Take 1–3 μg RNA (stored in ethanol) and centrifuge 10 min at 12,000 rpm and 4°C.
- Remove the ethanol using a sterile pipet and air dry the pellet (not for too long).

- Add directly (or make a premix and add) 2 μl random primer (0.5 μg/μl, Promega) and TE up to an end volume of 32 μl. Mix, incubate for 10 min at 65°C, and put directly on ice. NOTE: using specific primers (0.5 μl of 20 pmole/μl), we scale down the amount of RNA to 200–500 ng, anneal in 9.5 μl TE, and incubate 10 min at 65°C. Add for a final volume of 20 μl: 4 μl 5 × RT buffer (BRL), 2 μl 0.1 M DTT, 2.5 μl 10 mM dNTPs, 0.5 μl RNasin, and 200 U Superscript.
- Add directly (or make a premix and add) 12 μl 5× RT buffer (BRL), 6 μl 0.1 M DTT (BRL), 6 μl 10 mM dNTPs (Pharmacia), 1 μl 40 U/μl RNasin (Promega), and 2–3 μl 200 U/μl Superscript MMLV reverse transcriptase (BRL). Place (10 min at room temperature) 60 min at 42°C and put on ice.
- Optional: add 4 U RNase H (Promega) and incubate 20 min at 37°C. When necessary, store at −20°C.

2.4.2. First PCR

- Prepare a premix containing (per sample): 5 μl 10× PCR buffer, 2 μl (tailed) forward primer (20 pmole/μl), 2 μl reverse primer (20 pmole/μl), 0.3 μl Amplitaq (5 U/μl), and 32.7 μl H_2O.3.4.2b.
- Add 10 μl RT product, two drops of mineral oil and spin briefly.
- Perform PCR: once 3 min 93°C, 30 times 1 min 93°C/1 min 57°C/4 min 72°C, and once 7 min 72°C. NOTE: for difficult PCRs we saw improvement by adding 0.1–0.2 U Deep Vent. The total volume can be scaled down to 25 μl using only half of all the ingredients.

2.4.3. Second PCR (when necessary)

- Prepare a premix containing (per sample): 5 μl 10× PCR buffer, 5 μl 2 mM dNTPs, 2 μl tailed forward primer (20 pmole/μl), 2 μl reverse primer (20 pmole/μl), 0.3 μl Amplitaq, and 32.7 μl H_2O.
- Add oil and 3 μl first-round PCR product.
- Perform a PCR protocol as for round one (Section 2.4.2, step 3). NOTE: For the second PCR we normally use a cheaper polymerase, e.g., Supertaq (HT Biotechnology, U.K.).

2.5. Gel Analysis of the PCR Products

All (RT-)PCR products should be analyzed, e.g., using agarose gel electrophoresis following standard protocols (Maniatis *et al.*, 1989), to determine their amount and size. Product yields are used to determine the amount necessary for *in vitro* transcription/translation (Section 2.6, step 2). Abnormal sizes indicate the presence of genetic rearrangements (e.g., deletions, duplications) or mutations affecting RNA splicing.

2.6. *In Vitro* Transcription/Translation

The protocol is based on the TnT™ T7 Coupled Reticulocyte Lysate System (Promega), using [^{3}H]leucine as a radiolabeled amino acid. We were able to synthesize products up to 160 kDa (product of 3 kb), although the amount of background products increased considerably in such translations. Translation products of 50 ng of a 1.5-kb PCR product in a 12-μl reaction could be easily detected after overnight exposure. As a control for the *in vitro* transcription/translation reaction, the TnT™ Lysate System contains a luciferase-encoding control plasmid. When this template DNA is subjected to the *in vitro* transcription/translation, a 62-kDa protein should be synthesized.

- Remove the TnT reagents from −70°C. Put the TnT RNA polymerase directly on ice (not for too long!) and, after thawing, store all components on ice. Rapidly thaw the rabbit reticulocyte lysate by hand warming. NOTE: unused lysate should be refrozen as soon as possible.
- Pipet into a 1.5-ml microcentrifuge tube: 12.5 μl TnT rabbit reticulocyte lysate, 1 μl TnT reaction buffer, 0.5 μl TnT RNA polymerase, 0.5 μl amino acid mixture minus leucine, 2 μl [^{3}H]leucine (Amersham, 5 mCi/ml; 160 Ci/mmole), 0.5 μl RNasin Ribonuclease Inhibitor (Promega, 40 U/μl), 8 μl of PCR product (50–500 ng), and nuclease-free H_2O. NOTES: it is possible to further reduce the transcription/translation to a 12-μl reaction. ^{35}S-labeled amino acids can also be used.
- Incubate the reaction for 60 min at 30°C. NOTE: the temperature can be varied from 25 to 37°C without greatly affecting translation efficiency, but 30°C seems to be optimal.
- Store the reactions at −70°C (or −20°C) or add SDS–sample buffer for SDS–PAGE analysis (Section 2.7).

2.7. SDS–PAGE

For SDS–PAGE analysis of the samples we use the Miniprotean II Gel System (Bio-Rad). The system requires only a short electrophoresis time (1–1.5 hr) and uses small amounts of reagents to prepare and stain a gel. The gel described can be used to analyze peptides from 15 to 70 kDa (requiring 12–15% overlaps for segmented genes).

- Prepare a 12% gel mix (10 ml for two gels) and start polymerization by adding ammonium persulfate (100 μl 10% ammonium persulfate per 10 ml gel mix) and TEMED (10 μl per 10 ml gel mix).
- Pour the gel between the glass plates. The meniscus of the solution should be far enough below the top of the plate to leave enough space for the stacking gel to be added later, i.e., the length of the comb plus 0.5 cm. Carefully overlay the gel with water or isobutanol (previously saturated with water) to get a sharp meniscus. Leave at least 30 min for polymerization.
- After polymerization, wash the top of the gel several times with water. Remove as much water as possible.
- Pour a 3.75% stacking gel (4 ml for two gels) and place a cleaned comb into the solution. Avoid air bubbles underneath the comb teeth. The stacking gel is ready in 10 min.
- Remove the comb and wash the slots with running buffer. Place the gel in the electrophoresis tank and fill the chamber with running buffer.
- Add one volume of SDS-sample buffer (with or without reducing agents), mix, and boil 5 min. Centrifuge the sample for 30 sec in an Eppendorf centrifuge.
- Load the gel with the samples, controls, and protein markers. As molecular weight markers we use ^{14}C-labeled protein marker (Amersham) or prestained protein markers (Bio-Rad).
- Electrophoresis is performed at a constant current of 30 mA in the stacking gel and 40 mA in the separating gel. The run is stopped when the bromophenol blue dye reaches the bottom of the gel.

2.8. Coomassie Blue Staining and Autoradiography

When ^{14}C-labeled or prestained protein markers are used, the Coomassie blue staining (the first two steps below) can be omitted. If a mini-gel system is used, the times mentioned in the protocol can be reduced.

- Remove the gel and place it in a plastic box. Cover with staining solution and incubate 15–30 min under gentle agitation.
- Discard the staining solution. Rinse briefly with destaining solution, remove it, and repeat until the protein markers are clearly visible.
- Wash the gel for 30 min with water.
- Cover the gel with DMSO and wash for 30 min under gentle agitation. Repeat once. NOTE: because DMSO is a hazardous chemical, this and the next two steps should be performed in a fume hood.
- Treat the gel twice with DMSO/PPO for 60 min. NOTE: there are commercially available reagents which can be used to enhance the results for autoradiography (e.g., AMPLIFY from Amersham and ENHANCE from New England Nuclear).
- Wash the gel for 15–30 min with water to remove the DMSO/PPO. The gel becomes white.
- Put the gel on 3 MM paper and dry at 60–70°C for at least 1 hr (depending on the thickness of the gel and the percentage of acrylamide used).
- Expose with X-ray film (e.g., Kodak X-Omat AR). Usually, a clear signal is obtained after overnight exposure.

2.9. Solutions and Reagents

- PBS: 144 mM NaCl, 10 mM KH_2PO_4 (pH 7.8 with K_2HPO_4)
- TE buffer: 10 mM Tris-HCl, 0.1 mM EDTA, pH 8.0
- TBE buffer: 90 mM Tris, 90 mM boric acid, 1 mM EDTA, pH 8.3
- 2× RNA loading mix: 95% formamide, 20 mM EDTA, 0.05% bromophenol blue, 0.05% xylene cyanol
- 10× PCR buffer: 500 mM KCl, 100 mM Tris-HCl (pH 8.0), 20 mM $MgCl_2$, and 2.0 mg/ml BSA. Lately, we use a more universal PCR buffer [i.e., 16.6 mM $(NH_4)_2SO_4$, 67.0 mM Tris-HCl (pH 8.8), 6.7 mM $MgCl_2$, 10 mM β-mercaptoethanol, 6.8 μM EDTA] with 1.5 mM dNTPs, 40 pmole primers, 2 U Amplitaq, 10% DMSO, 0.2 mg/ml BSA, and optional 0.1–0.2 U Deep Vent DNA polymerase (BRL).
- AA/BA mix: 30% acrylamide and 0.8% *N′,N′*-bismethylene-acrylamide dissolved in an end volume of 500 ml, filtered, and stored at 4°C in the dark. NOTE: acrylamide is a neurotoxin. Always wear gloves when handling acrylamide, preparing solutions, or pouring gels (polyacrylamide is not toxic)
- Gel mix (12%, per 10 ml): 4 ml AA/BA mix, 3.3 ml distilled water, 2.5 ml 1.5 M Tris-HCl (pH 8.8), 100 μl 10% SDS, 100 μl 10% ammonium persulfate, and 4 μl TEMED (tetramethylethylenediamine).
- Stacking gel (3.75%, per 10 ml): 0.5 ml AA/BA mix, 3 ml distilled water, 0.5 ml 0.5 M Tris-HCl (pH 6.8), 40 μl 10% SDS 40 μl 10% ammonium persulfate, and 4 μl TEMED.
- SDS-sample buffer: 100 mM Tris-HCl (pH 6.8), 4% SDS, 0.1% (w/v) bromophenol blue, 20% glycerol, and, optional, 200 mM DTT or 8% (v/v) β-mercaptol (added immediately before use).
- Running buffer: 25 mM Tris base, 200 mM glycine, and 0.1% SDS.
- Staining solution: 50% methanol, 0.05% Coomassie brilliant blue R, 10% acetic acid, 40% H_2O.
- Destaining solution: 10% methanol, 10% acetic acid, and 80% H_2O.
- DMSO/PPO: 226 g PPO (2,5-diphenyloxazole) dissolved in 1000 ml DMSO.

3. EXAMPLES

3.1. Using DNA Templates: The APC and BRCA1 Genes

The PTT assay was developed to analyze mRNA templates. However, in exceptional cases, PTT can be used as an efficient tool to scan a large portion of a gene for truncating point mutations using genomic DNA as a template. Two prominent examples are the APC (Groden *et al.*, 1991; Nishisho *et al.*, 1991) and BRCA1 (Miki *et al.*, 1994) genes, involved in colon and breast cancer, respectively. In these cases, the genomic structure and the type of mutations involved favor the use of PTT. Both genes contain large exons carrying a substantial portion of the disease-causing mutations reported (Fig. 24.2), a major fraction of which cause premature translation termination. The APC gene has a large 7-kb 3′-terminal exon 16, encoding 77% of the coding region and 79% of the germ-line mutations reported, 97% of which cause premature translation termination (Nagase and Nakamura, 1993). Similarly, 84% of the mutations reported for BRCA1 are translation terminating (Shattuck-Eidens *et al.*, 1995), 50% of which were found in the 3.4-kb exon 11 containing 61% of the coding region of the gene (Shattuck-Eidens *et al.*, 1995).

3.2. Duchenne Muscular Dystrophy

For large multiexonic genes it is necessary to divide the entire coding region into several partly overlapping segments. In our hands, the most convenient size per segment lies in the range of 1 to 2 kb. Usually, such fragments both allow successful amplification of the reverse transcribed RNA and enable efficient *in vitro* transcription/translation and subsequent SDS–PAGE analysis of the peptide products. An example of such a segmentation, as used for the human dystrophin gene, is presented in Fig. 24.3.

The dystrophin gene is split into 79 exons, together spanning 2.3 Mb DNA (Roberts *et al.*, 1992; Den Dunnen *et al.*, 1989). For the PTT assay we divided the gene into ten overlapping segments (Fig. 24.3D) (Roest *et al.*, 1993a,b). Since the RNA template is isolated from peripheral blood lymphocytes, a tissue which normally does not express the dystrophin gene, a nested RT-PCR is necessary to amplify sufficient material for the subsequent transcription/translation assay. After RT, each segment is first amplified with primers A and B. For the nested PCR, the segment is split in two and amplified with primer sets CD and EF (Fig. 24.3D), where primers C and E are tailed with the T7/Kozak sequences.

Agarose gel electrophoresis of the PCR products is used to analyze the yield and quality of the PCR products (Fig. 24.3A), which vary in size from 1.0 to 1.4 kb. The crucial step of the PTT assay is the *in vitro* transcription/translation, SDS–PAGE, and autoradiography to visualize the encoded peptides (Fig. 24.3A). In general, the size of the PCR products nicely reflects the size of the translation products. However, for the DMD gene, the translation product of set 1EF (cDNA position 1.3–2.4 kb) migrates slower than expected (Fig. 24.3A, lane 2).

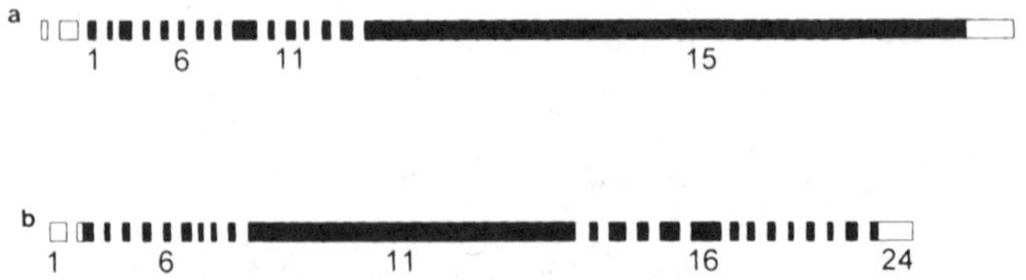

Figure 24.2. Two genes allowing screening by PTT using DNA templates. Gene structures are drawn schematically; coding regions are black and noncoding regions are blank.

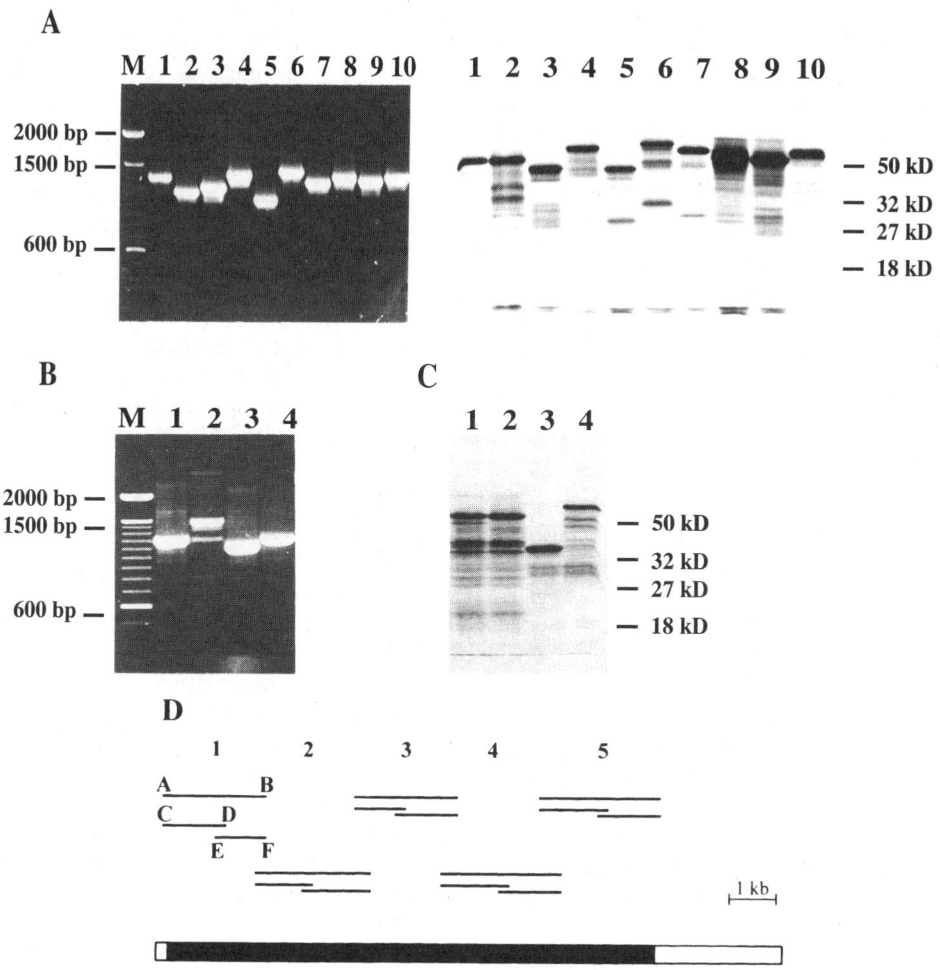

Figure 24.3. PTT of the human dystrophin gene. (A) Agarose gel analysis of the individually PCR-ed segments of the coding region of the dystrophin gene (left panel) and SDS–PAGE analysis of the encoded peptides (right panel). Loaded were segments 1CD, 1EF, 2CD, 2EF, 3CD, 3EF, 4CD, 4EF, 5CD, and 5EF, respectively. (B) Agarose gel analysis of dystrophin segment 2EF of DMD patients C8471 (lane 2) and BL214.2 (lane 3) and two controls (lanes 1 and 4). (C) SDS–PAGE analysis of dystrophin segments 2CD (lanes 1 and 2) and 2EF (lanes 3 and 4). Loaded were samples of a control (lanes 2 and 4) and patient BL153.1. (D) Segmentation of the dystrophin coding region in five overlapping segments (1 to 5). First-round PCR is performed with primers A and B, second-round PCR with pairs CD and EF. The bar (bottom) represents the dystrophin cDNA with coding sequences in black.

On agarose gel electrophoresis, the first mutations can be identified by the appearance of PCR products with an altered mobility. Two examples are shown in Fig. 24.3B. Segment 2EF should give a 1.2-kb fragment (lanes 1 and 4) but DMD patient C8471 gives a larger product of 1.45 kb (lane 2) while patient BL214.2 gives a smaller product of 1.1 kb. Both mutations turn out to be caused by mutations at the splice site. The first causes a shift to a cryptic intronic splice site while the second causes skipping of the entire exon. As frequently observed in mutations affecting splicing, low levels of secondary splice products can be seen in patient C8471 (lane 2).

For DMD, however, the majority of the mutations become visible as truncated peptides only after translation of the RT-PCR products. An example is presented in Fig. 24.3C. Agarose gel analysis of the PCR products of BL153.1 revealed no differences but translation of segment 2EF revealed a truncated band (lane 3) when compared to the normal product (lane 4). Translation of the flanking segments 2CD (Fig. 24.3C, lanes 1 and 2) and 3CD gave only the expected products. The example shown nicely demonstrates that most of the weak background translation products derive from secondary translation initiation and not from early translation termination; the larger peptides have disappeared while the smaller fragments remain unchanged.

3.3. Hunter Syndrome

Another example of the use of PTT for mutation detection is depicted in Fig. 24.4. Patients with Hunter syndrome contain mutations in the iduron-2-sulfatase (IDS) gene (Jonsson *et al.*, 1995; Wilson *et al.*, 1995). Analysis of this gene is possible using peripheral blood lymphocytes and a nested PCR (Hogervorst *et al.*, 1994). The IDS gene is an example of a gene where the entire coding region can be translated into one 60 kDa peptide (Fig. 24.4, fragment A).

Two mutations are shown: one residing early in the coding region, resulting in a short truncated peptide (lane 2), and a second near the end, giving a translation product which is only slightly smaller (lane 3). The position of these mutations is confirmed when only the 3′ coding region is analyzed (fragment B). No mutation is found in patient 76RD45 (lane 2); its translation product derives from 3′ of the nonsense mutation and is normal. However, in patient 78RD228 fragment B also gives a truncated peptide, confirming the analysis of fragment A.

One advantage of PTT is that the alteration detected pinpoints the actual site of the mutation. The mutation site can be estimated from the length of the truncated peptide. In patient 78RD228 the truncated peptide is 7 kDa shorter (25 versus 32 kDa), indicating that the site of the mutation should be around 0.2 kb upstream of the normal termination codon (7 kDa ≈ 0.18 kb), i.e., at cDNA position 1.6 kb. Similarly, the PTT assay indicates that the mutation in patient 2 lies around cDNA position 0.7 kb. Both estimates agree very well with the actual sites of the mutation, as determined by sequencing. Patient 78RD228 carries a 13-bp deletion at position 1617–1629 while patient 76RD45 has a GATTGGAG mutation at cDNA position 657.

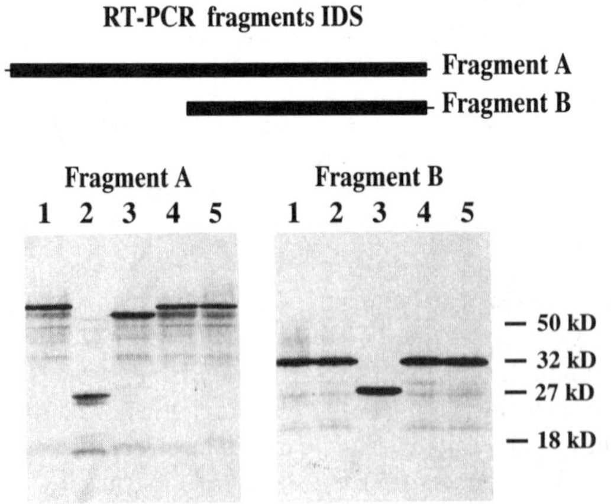

Figure 24.4. PTT of the IDS gene. Either the entire coding region of the IDS gene was amplified (fragment A) or the 3′ half (top). *In vitro* transcription/translation and SDS–PAGE (bottom panels) reveals the presence of truncating mutations in two samples: patient 76RD45 (lane 2) and 78RD228 (lane 3).

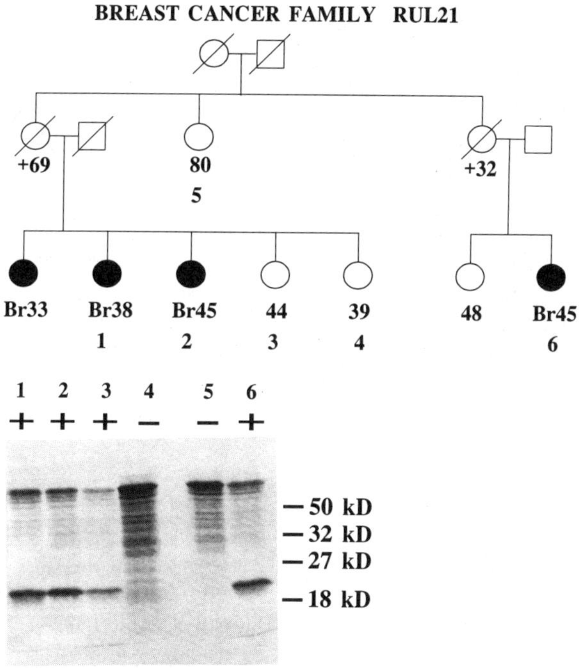

Figure 24.5. Detection of carriers using PTT. Family RUL21 (top), linked to the breast cancer 1 locus on chromosome 17 (BRCA1), was analyzed by PTT (bottom) using the central portion of exon 11 [segment B (Hogervorst *et al.*, 1995)]. A truncating mutation was identified and shown to segregate with the disease. Individual 3 carries the mutation and is at risk. Plus and minus signs indicate the presence or absence of the disease haplotype, respectively. Filled circles represent cases of breast cancer. Numbers indicate age, "Br" = age at which breast cancer was diagnosed, and "+" = age at death.

3.4. PTT and Direct Diagnosis

Another strength of the PTT assay, detecting clinically relevant mutations only, is the possibility to use it directly for diagnostic purposes. An example is shown in Fig. 24.5. In breast cancer family RUL21, the PTT assay revealed a truncated peptide in segment B (Hogervorst *et al.*, 1995) of the index patient. Screening of the other family members showed that the same mutation was carried by relatives 2, 3, and 6. These data were confirmed by the determined haplotypes (Fig. 24.5) and the appearance of breast cancer in individuals 2 and 6. The at risk haplotype of individual 3 could be unequivocally confirmed by the PTT assay (lane 3). Similarly, the risk of individual 4 could be decreased to the normal population risk.

4. DISCUSSION

4.1. Advantages of PTT

When in a specific disease a major fraction of the mutations involved are premature translation terminating, PTT has a number of clear advantages over other methods of mutation detection. One advantage of the method as presented here (not unique to PTT) is that it uses RNA templates and targets the gene coding region. This significantly simplifies mutation detection, especially for genes with a complex structure, such as the dystrophin gene (Roberts *et al.*, 1992; Den Dunnen *et al.*, 1989). The coding region of the dystrophin gene, measuring 2.3 Mb, is split into 79 exons. Mutation detection using PTT requires analysis of only 5–12 amplified fragments (Roest *et al.*, 1993b; Gardner *et al.*, 1995). In contrast, DNA-based methods would require the analysis of 79 fragments, one for each exon.

For diagnostic purposes, speed and reliability are important issues. Since PTT simplifies several important steps and since it is mainly based on PCR, mutation detection can be performed very rapidly; the assay from (RT-)PCR of the template to the visualization of truncated peptides can be performed within 2–3 days.

Advantages specific to PTT include the following:

4.1.1. Screening Large Regions in One Assay

Large coding segments (up to 2–4 kb) can be screened in a single assay. The use of other techniques, e.g., SSCP, DGGE, or heteroduplex analysis, only permits the analysis of (exonic) segments ranging in size from 100–600 bp. Genes containing small coding regions (below 4 kb) can be scanned in one experiment (Section 3.3).

4.1.2. No False Positives

All mutations introducing premature translation-termination identified to date have been proven to be disease causing. Only very few exceptions are known where the status of some of these is still unclear (Sobell *et al.*, 1995). Consequently, PTT can be expected to give no false positives and to detect clinically relevant mutations (i.e., disease causing) only. The result of a PTT analysis can thus be used directly for diagnostic purposes (Section 3.4). This is in contrast with other screening methods which detect all sequence changes present, including those which give no change in the protein sequence (coding triplet third base changes) and amino acid substitutions which are phenotypically silent. Although sequence analysis is always performed to analyze the nature of the PTT mutation detected, it is not strictly necessary. Furthermore, no PTT-detected alterations have been reported which were not confirmed by sequence analysis.

4.1.3. Pinpoints Mutation Site

Unlike other mutation detection methods, excluding chemical mismatch cleavage, PTT pinpoints the site of the mutation by the length of the truncated peptide produced (Section 3.3). This can be used to focus sequence analysis if further characterization of the mutation is desired.

4.2. Limitations and Potential Pitfalls

The most limiting factor for efficient use of PTT is the availability of RNA. Most diagnostic laboratories have usually stored DNA samples or tissue samples which do not allow isolation of high-quality RNA. When RNA-based methods are introduced, this means that former patients/families have to be contacted and resampled.

4.2.1. Missense Mutations

In its present form, PTT detects translation terminating mutations only. Although, in some respects, this is a major advantage of the technique, it also means that it does not detect all disease-causing mutations, e.g., it will not detect missense mutations (Section 4.4).

4.2.2. Expression of Both Alleles

When DNA samples are used, it is not necessary to verify if both alleles are amplified. Using RNA samples, however, this issue becomes critical. First, it is well documented that mutations which cause premature translation termination may produce unstable RNA molecules (Jones *et al.*, 1995; Baserga and Benz, 1988; Dietz *et al.*, 1993), resulting in no product or low yields of one, the mutated, allele. Second, if a mutation silences the promoter, no or less RNA is produced from the disease allele. Third, in a specific set only one allele might be amplified when a mutation alters the normal splicing pattern, e.g., preventing the excision of an intron thereby increasing the size of the product beyond amplifiable sizes. Allelic expression can be verified most effectively using polymorphisms which are present in the coding region (for 100% certainty it is necessary to check each amplified segment individually in this way).

In one aspect, the examples of DMD and Hunter syndrome are exceptional: they are X-linked diseases. Consequently, the analysis of patients is simplified since only one allele is present and PTT results are never obscured by those of the healthy allele. Thus, promoter mutations abolishing transcription are easily resolved by the absence of products. For autosomal diseases, it should be checked if both alleles were expressed and amplified, especially when no mutation has been identified. That these considerations are not only hypothetical was demonstrated recently for BRCA1, where a case was reported in which RNA analysis yielded transcripts from only one allele (Shattuck-Eidens *et al.*, 1995).

4.2.3. Genomic Rearrangements

Some mutations which introduce premature translation termination are missed by the PTT assay simply because they yield no PCR product. In cases where the primer binding site is mutated or resides in a deleted exon, no product of the allele will be amplified. Since PCR favors amplification of shorter products, duplications will be missed when the primer used lies in one of the duplicated exons. In fact, duplications will probably only be detected when the primers used span the entire duplicated region and when the duplication is relatively small, i.e., below 1 kb exonic sequences. Insertions will be missed when the incorporated sequences become too large to amplify. Finally, amplification across (and beyond) translocation and inversion breakpoints will also yield no products. The most convenient way to check for the potential presence of genomic rearrangements is an analysis of genomic Southern blots using cDNA probes.

4.2.4. N- and C-Terminal Mutations

The boundaries of the translated fragments are always the most critical region to miss mutations. This is especially true for the N- and C-terminal region of proteins since these cannot be double-checked using overlapping segments. Consequently, the detection of such mutations is critically dependent on the sensitivity of the SDS–PAGE analysis. For internal, overlapping segments, this issue becomes important only when the structure or sequence of a gene does not allow following the rules for primer design (Section 1.3).

4.2.5. Tissue-Specific Expression/Splicing

When RNA templates are used which were not derived from the disease-affected tissue, one must realize that mutations influencing tissue-specific expression only, including tissue-specific promoters and tissue specific splicing, cannot be detected. This problem increases further when, as in the DMD example, ectopic RNA transcripts are used to search for mutations. Therefore, before analysis starts, one should check whether the RNA template to be analyzed has the same structure as the transcript present in the affected tissue. For example, for DMD it has been reported that in RNA isolated from lymphocytes an additional exon "X" is present, spliced between exons 1 and 2, which cannot be found in muscle RNA (Roberts *et al.*, 1993).

4.2.6. Background Translation Products

Another problem which might obscure the detection of specific mutations relates to the background products which are present after translation. Although the origin of these products has not been investigated in great detail, they probably derive from secondary translation initiation near internal ATG codons (Section 3.2). On the other hand, the translation pattern of a specific set is fairly constant and the presence of a mutation results in the appearance of a new set of such side products thus increasing the chance of detecting truncated peptides.

4.2.7. Mobility of Translation Products

In an exceptional case, we ran into problems analyzing the translated peptides on SDS–PAGE. PTT of the exon 6–13 region of the cystic fibrosis gene, containing part of a transmembrane region, resulted in empty lanes. A strong signal near the slot of the gel indicated that the peptide(s) produced did not migrate. Addition of 5% β-mercaptoethanol to the sample, and omitting the boiling step, (partly) solved the problem and revealed the expected peptides. This example shows that specific peptides may require special treatments and/or running conditions to allow their visualization.

4.3. Troubleshooting

Our experience with the PTT assay has revealed several critical steps which easily cause technical problems. These include the following:

4.3.1. RNA Quality

In our hands, the quality of the RNA sample isolated was the most critical element in the PTT assay. We have tested several techniques and kits for the RNA isolation and obtained the best reproducible results with the procedure presented (Section 2.2). The quality of the sample for RNA isolation, i.e., its freshness and the storage buffer, has a significant influence on the results obtained. For blood samples we obtained best results with EDTA blood. Genomic DNA contaminations in the RNA samples often cause problems during PCR and should be minimized as much as possible.

4.3.2. Reverse Transcription

The RT reaction is very sensitive and one of the most frequent sources of failure for PTT. Fresh constituents and fresh batches of enzyme are the best safeguards against problems emerging later. We obtained our best results with random priming of the RT reaction (Roest *et al.*, 1993a; Hogervorst *et al.*, 1995) while others favor specific priming (Gardner *et al.*, 1995). Failure of the RT reaction is often caused by the use of RNA of bad quality.

4.3.3. Contaminations

Contamination of the sample and the ingredients of the RT-PCR reaction with trace amounts of cDNA and cloned genomic DNA are among those most difficult to prevent. Usually, the target gene is and has been a subject of study for many years in the laboratory performing the mutation detection. Consequently, contamination hazards are ubiquitous and it should be emphasized that a new "pre-PCR" laboratory should be established where the RNA sample and the primers can be stored and kept physically separated from the RT-PCR and subsequent steps of the procedure. A control reaction (no RNA or no enzyme added) should always be part of the PTT to exclude possible contaminations.

4.3.4. No Amplification of Specific Genes

RNA samples derived from peripheral blood lymphocytes can be used in most cases to amplify the gene of interest. We have experienced two exceptions to this rule. First, for reasons which are still unclear, in some DMD patients we were not able to amplify the dystrophin mRNA. Second, despite several attempts and the use of different primer sets, amplification of the OTC gene (Grompe *et al.*, 1989) failed.

4.3.5. Bad Translation

Occasionally, T7/Kozak-tailed PCR products give poor yields of transcription/translation products. This was mostly caused by poor quality of the tailed primer. Inefficient coupling steps during primer synthesis result in low quantities of full-size primers, especially of the rather long (>50 bp) tailed primers. Since primers are synthesized $3' \rightarrow 5'$, i.e., the gene-specific sequence first and T7-promoter last, such bad-quality primers usually give normal yields of (RT-)PCR products while *in vitro* transcription/translation fails. Reamplification of the tailed products with a T7-promoter primer usually solves the problem.

Specific sets may give many (strong) secondary translation products. To improve results there are several options. A first option is to redesign and shift the primers. Relocation can improve translation results considerably. Alternatively, primers can be used containing two or three in-frame Kozak sequences. A second option is to use alternative *in vitro* translation systems. Translation efficiency is influenced by the sequence of the PCR products and it has been reported that a wheat germ extract can be a good alternative when problems occur (Hope and Struhl, 1985). Finally, one could tag the tailed primer with a specific protein sequence (Ahn and Kunkel, 1995), enabling precipitation of only the correctly initiated peptides, using antibodies directed against the protein tag.

4.3.6. Spurious Truncated Peptides

Detection of a truncated fragment should always be verified by a second, completely independent assay starting with a new RT reaction. We have had several cases (only when a nested PCR was necessary) where a truncated fragment was observed only once. The cause of such artifacts probably resides in the low expression of the target RNA and the infidelity of the polymerases used for PCR; the erroneous incorporation of a specific nucleotide early during PCR (or in the RT) will be amplified in subsequent rounds. In the case of an X-linked disease like DMD, the presence of a full-size translation product next to a truncated product in patients is an obvious warning against such artifacts.

4.4. Future Developments

The PTT assay is still in its infancy and especially at the level of the last step, the protein analysis on SDS–PAGE, there are several possibilities to increase its sensitivity and to expand its detection range. Currently, only the length of the peptides is used as an analytical tool to detect mutations. Other methods of protein separation are available which probably extend the detection range to other types of mutations. Isoelectric focusing has been mentioned as an obvious candidate to expand the level of detection to that of amino acid substitutions and small insertions/deletions (Roest *et al.*, 1993a). Many thalassemia mutations were detected by mobility shifts observed on electrophoresis of the α- and β-globins. There are no obvious reasons why such mobility shifts would not be detected if these proteins were produced by *in vitro* translation. However, no publications are available where a successful combination of PTT and IEF has been reported. Alternatively, for genes containing small coding regions (Section 3.3), PTT should enable translation of the entire gene product and thereby facilitate direct functional assays of the protein, e.g., measurements of enzymatic activity.

An obvious application of the PTT assay resides in the identification of new disease genes. When only fragments of the sequence of a specific candidate disease gene are known, tailed primers can already be devised and used directly to scan large numbers of patient samples for the potential presence of truncating mutations. A first example where this approach has been used was the identification of the gene involved in Rubinstein–Taybi (Petrij *et al.*, 1995).

5. CONCLUSION

Undoubtedly, the number of genes in which premature translation termination constitutes a major fraction of the disease-causing mutations (e.g., most tumor suppressor genes) will rise further and the detection range of the technique will be expanded beyond that of truncations. Both developments will increase the value of the PTT assay as a valuable diagnostic tool for efficient mutation detection and thereby automatically stimulate laboratories to set up the system.

REFERENCES

Ahn, A. H., and Kunkel, L. M. (1995). Syntrophin binds to an alternatively spliced exon of dystrophin. *J. Cell Biol.* **128**:363–371.

Baserga, S. J., and Benz, E. J. (1988). Nonsense mutations in the human beta-globin gene affect mRNA metabolism. *Proc. Natl. Acad. Sci. USA* **85**:2056–2060.

Chelly, J., Concordet, J., Kaplan, J. C., and Kahn, A. (1989). Illegitimate transcription of any gene in any cell type. *Proc. Natl. Acad. Sci. USA* **86**:2617–2621.

Cheng, S., Fockler, C., Barnes, W. M., and Higuchi, R. (1994). Effective amplification of long targets from cloned inserts and human genomic DNA. *Proc. Natl. Acad. Sci. USA* **91**:5695–5699.

Den Dunnen, J. T., Grootscholten, P. M., Bakker, E., Blonden, L. A. J., Ginjaar, H. B., Wapenaar, M. C., Van Paassen, H. M. B., Van Broeckhoven, C., Pearson, P. L., and Van Ommen, G. J. B. (1989). Topography of the DMD gene: FIGE and cDNA analysis of 194 cases reveals 115 deletions and 13 duplications. *Am. J. Hum. Genet.* **45**:835–847.

Dietz, H. C., Valle, D., Francomano, C. A., Kendzior, R. J., Jr., Pyeritz, R. E., and Cutting, G. R. (1993). The skipping of constitutive exons in vivo induced by nonsense mutations. *Science* **259**:680–683.

Gardner, R.J., Bobrow, M., and Roberts, R. G. (1995). The identification of point mutations in Duchenne muscular dystrophy patients using reverse transcript PCR and the protein truncation test. *Am. J. Hum. Genet.* **57**:311–320.

Groden, J., Thliveris, A., Samowitz, W., Carlson, M., Gelbert, L., Albertsen, H., Joslyn, G., Stevens, J., Spirio, L., Robertson, M., Sargeant, L., Krapcho, K., Wolff, E., Burt, R., Hughes, J. P., Warrington, J., McPherson, J., Wasmuth, J., Le Paslier, D., Abderrahim, H., Cohen, D., Leppert, M., and White, R. (1991). Identification and characterization of the familial adenomatous polyposis gene. *Cell* **66**:587–600.

Grompe, M., Muzny, D. M., and Caskey, C. T. (1989). Scanning detection of mutations in human ornithine transcarbamoylase by chemical mismatch cleavage. *Proc. Natl. Acad. Sci. USA* **86**:5888–5892.

Heim, R. A., Kam-Morgan, L. N. W., Binnie, C. G., Corns, D. D., Cayouette, M. C., Farber, R. A., Aylsworth, A. S., Silverman, L. M., and Luce, M. C. (1995). Distribution of 13 truncating mutations in the neurofibromatosis 1 gene. *Hum. Mol. Genet.* **4**:975–981.

Hillier, L., and Green, P. (1991). OSP: A computer program for choosing PCR and DNA sequencing primers. *PCR Methods Appl.* **1**:124–128.

Hogervorst, F. B. L., Van Der Tuijn, A. C., Poorthuis, B., Kleyer, W., Bakker, E., Van Ommen, G. J. B., and Den Dunnen, J. T. (1994). Rapid identification of mutations in the IDS-gene of Hunter patients: Analysis of mRNA by the protein truncation test. *Am. J. Hum. Genet.* **55**:A223.

Hogervorst, F. B. L., Cornelis, R. S., Bout, M., Van Vliet, M., Oosterwijk, J. C., Olmer, R., Bakker, E., Klijn, J. G. M., Vasen, H. F. A., Meijers-Heijboer, H., Menko, F. H., Cornelisse, C. J., Den Dunnen, J. T., Devilee, P., and Van Ommen, G. J. B. (1995). Rapid detection of BRCA1 mutations by the protein truncation test. *Nature Genet.* **10**:208–212.

Hope, I. A., and Struhl, K. (1985). GCN4 protein synthesized in vitro, binds HIS3 regulatory sequences: Implications for general control of amino acid biosynthetic genes in yeast. *Cell* **43**:177–188.

Jones, C. T., McIntosh, I., Keston, M., Ferguson, A., and Brock, D. J. H. (1995). Three novel mutations in the cystic fibrosis gene detected by chemical mismatch cleavage: Analysis of variant splicing and a nonsense mutation. *Hum. Mol. Genet.* **1**:11–17.

Jonsson, J. J., Aronovich, E. L., Braun, S. E., and Whitley, C. B. (1995). Molecular diagnosis of mucopolysaccharidosis type II (Hunter syndrome) by automated sequencing and computer-assisted interpretation: Toward mutation mapping of the iduronate-2-sulfatase gene (IDS). *Am. J. Hum. Genet.* **56**:597–607.

Maniatis, T., Fritsch, E. F., and Sambrook, J., eds. (1989). *Molecular Cloning: A Laboratory Manual*, 2nd ed., Cold Spring Harbor Laboratory Press, Cold Spring Harbor, NY.

Miki, Y., Swensen, J., Shattuck-Eidens, D., Futreal, P. A., Harshman, K., Tavtigian, S., Liu, Q., Cochran, C., Bennett, L. M., Ding, W., Bell, R., Rosenthal, J., Hussey, C., Tran, T., McClure, M., Frye, C., Hattier, T., Phelps, R., Haugen-Strano, A., Katcher, H., Yakumo, K., Gholami, Z., Shaffer, D., Stone, S., Bayer, S., Wray, C., Bogden, R., Dayananth, P., Ward, J., Tonin, P., Narod, S., Bristow, P. K., Norris, F. H., Helvering, L., Morrison, P., Rosteck, P., Lai, M., Barrett, J. C., Lewis, C., Neuhausen, S. L., Cannon-Albright, L., Goldgar, R., Wiseman, R., Kamb, A., and Skolnick, M. H. (1994). A strong candidate for the breast and ovarian cancer susceptibility gene BRCA1. *Science* **266**:66–71.

Nagase, H., and Nakamura, Y. (1993). Mutations of the APC (adenomatous polyposis coli) gene. *Hum. Mutat.* **2**:425–434.

Nishisho, I., Nakamura, Y., Myoshi, Y., Miki, Y., Ando, H., Horii, A., Koyama, K., Utsonomiya, J., Baba, S., Hedge, P., Markham, A., Krush, A.J., Petersen, G., Hamilton, S. R., Nilbert, M. C., Levy, D. B., Bryan, T. M., Preisinger, A. C., Smith, K. J., Su, L. K., Kinzler, K. W., and Vogelstein, B. (1991). Mutations of chromosome 5q21 genes in FAP and colorectal cancer patients. *Science* **253**:665–669.

Petrij, F., Giles, R. H., Dauwerse, J. G., Saris, J. J., Hennekam, R. C. M., Masuno, M., Tommerup, N., Van Ommen, G. J. B., Goodman, R. H., Peters, D. J. M., and Breuning, M. H. (1995). Rubinstein–Taybi syndrome caused by mutations in the transcriptional co-activator CBP. *Nature* **376**:348–351.

Powell, S. M., Petersen, G. M., Krush, A. J., Booker, S., Jen, J., Giardello, F. M., Hamilton, S. R., Vogelstein, B., and Kinzler, K. W. (1993). Molecular diagnosis of familial adenomatous polyposis. *N. Engl. J. Med.* **329**:1982–1987.

Roberts, R. G., Coffey, A. J., Bobrow, M., and Bentley, D. R. (1992). Determination of the exon structure of the distal portion of the dystrophin gene by vectorette PCR. *Genomics* **13**:942–950.

Roberts, R. G., Bentley, D. R., and Bobrow, M. (1993). Infidelity in the structure of ectopic transcripts: a novel exon in lymphocyte dystrophin transcripts. *Hum. Mutat.* **2**:293–299.

Roest, P. A. M., Roberts, R. G., Sugino, S., Van Ommen, G. J. B., and Den Dunnen, J. T. (1993a). Protein truncation test (PTT) for rapid detection of translation-terminating mutations. *Hum. Mol. Genet.* **2**:1719–1721.

Roest, P. A. M., Roberts, R. G., Van Der Tuijn, A. C., Heikoop, J. C., Van Ommen, G. J. B., and Den Dunnen, J. T. (1993b). Protein truncation test (PTT) to rapidly screen the DMD-gene for translation-terminating mutations. *Neuromusc. Disord.* **3**:391–394.

Sarkar, G., and Sommer, S. S. (1989). Access to a messenger RNA sequence or its protein product is not limited by tissue or species specificity. *Science* **244**:331–334.

Shattuck-Eidens, D., McClure, M., Simard, J., Labrie, F., Narod, S., Couch, F., and Hoskins, K. (1995). A collaborative survey of 80 mutations in the BRCA1 breast and ovarian cancer susceptibility gene. *J. Am. Med. Assoc.* **273**:535–541.

Sobell, J. L., Lind, T. J., Sigurdson, D. C., Zald, D. H., Snitz, B. E., Grove, W. M., Heston, L. L., and Sommer, S. S. (1995). The D5 dopamine receptor gene in schizophrenia: Identification of a nonsense change and multiple missense changes but lack of association with disease. *Hum. Mol. Genet.* **4**:507–514.

Van Der Luit, R., Meera Kahn, P., Vasen, H., Van Leeuwen, C., Tops, C., Roest, P. A. M., Den Dunnen, J. T., and Fodde, R. (1994). Rapid detection of translation-terminating mutations at the adenomatous polyposis coli (APC) gene by direct protein truncation test. *Genomics* **20**:1–4.

Wilson, P. J., Morris, C. P., Anson, D. S., Occhiodoro, J., Clements, P. R., and Hopwood, J. J. (1995). Hunter syndrome: Isolation of an iduronate-2-sulfatase cDNA clone and analysis of patient DNA. *Proc. Natl. Acad. Sci. USA* **87**:8531–8535.

Chapter 25

Single Nucleotide Primer Extension for Analysis of Sequence Variants

Piroska E. Szabó, Gerd P. Pfeifer, Jeffrey R. Mann, and Judith Singer-Sam

1. INTRODUCTION

In the DNA single nucleotide primer extension (SNuPE) assay for detection of known mutations (Kuppuswamy *et al.*, 1991), the DNA segment containing the mutation is amplified by PCR and the product is isolated (Fig. 25.1). Then primer extension is done on aliquots of the PCR fragment with the SNuPE primer that ends just one nucleotide 5′ to the base difference. In this step only one radioactive nucleotide is used per aliquot, the one corresponding to the wild-type (Wt) sequence in one tube, and the one corresponding to the mutant sequence in the other. In one tube the radionucleotide corresponding to the Wt sequence can extend the primer on the Wt template but cannot extend the primer on the mutant fragment. In the other tube the opposite is the case. On a fragment amplified from an individual heterozygous for the given point mutation, each radionucleotide incorporates in the separate SNuPE reactions. The SNuPE primers are separated from unincorporated nucleotides by denaturing polyacrylamide gel electrophoresis and the results are visualized by autoradiography. The PCR amplification step before the primer extension makes this assay very sensitive. If the original DNA was a mix of mutant and Wt DNA, both nucleotides incorporate on a mixed template, and this enables one to screen pooled samples for mutation as performed by Krook *et al.* (1992). These authors also described a more economical variation of the method, called "multiplex SNuPE," that allows simultaneus examination of multiple loci in the same reaction and on the same gel. They amplified three different fragments in the same PCR reaction and did the SNuPE assay with three different SNuPE primers which differed in length by five nucleotides.

Piroska E. Szabó, Gerd P. Pfeifer, Jeffrey R. Mann, and Judith Singer-Sam • Department of Biology, Beckman Research Institute of the City of Hope, Duarte, California 91010.

Technologies for Detection of DNA Damage and Mutations, edited by Gerd P. Pfeifer, Plenum Press, New York, 1996.

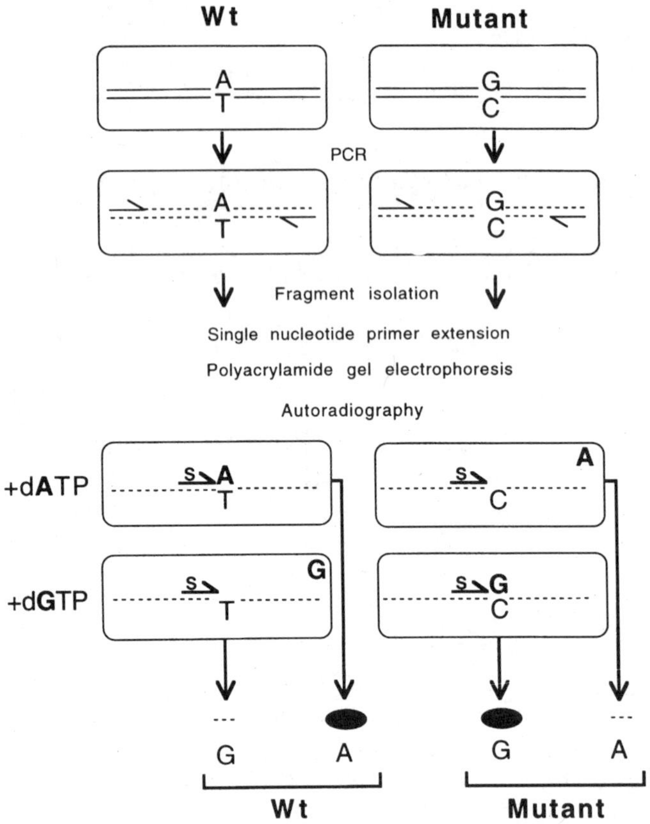

Figure 25.1. Outline of the DNA SNuPE technique, to distinguish wild-type (Wt) from mutant DNA. Genomic DNA is PCR amplified with primers spanning the known base difference. The products are isolated from agarose gels and equal aliquots are subjected to one primer extension step with the SNuPE primer (S) that abuts the base difference and with one or the other labeled nucleotide. Only the one nucleotide complementary to the template sequence is incorporated, visualized by autoradiography following denaturing acrylamide gel electrophoresis. (After Kuppuswamy *et al.*, 1991.) Rounded rectangle represents a tube.

2. THE QUANTITATIVE SNuPE ASSAY FOR MUTATION DETECTION

If the SNuPE reaction is done on a mix of PCR products containing Wt and mutant fragments, the ratio of the two nucleotides incorporated is proportional to the ratio of the two templates, because the SNuPE primer is equally complementary to both templates and will anneal to each of them according to their abundance in the mix. Here, we have used exon 7 of the human p53 gene and a codon 248 mutation as a model system to test the sensitivity of the SNuPE assay. DNA from the leukemia cell line CEM contains a mutation (CAG) at codon 248 on one allele (Cheng and Haas, 1990) and is Wt (CGG) on the other allele. DNA from HeLa cells is Wt at the same position. PCR products were made from both cell lines that contain all of exon 7 including codon 248 with primers 5′-GCACTGGCCTCATCTTGGGCCTG and 5′-CACAG-CAGGCCAGTGTGCAGGGT. A SNuPE primer was designed (5′-TGTGATGATGGTGA-GGATGGGCCTC, underlined below) that abuts the site of mutation and is complementary to the coding strand:

gcatgggcggcatgaaccggaggcccatcctcaccatcatcaca

cgtacccgccgtacttggcctccgggtaggagtggtagtagtgt

This primer was extended with Taq polymerase and a single ^{32}P-labeled nucleotide, either dCTP or dTTP (Fig. 25.2A). dTTP is specifically incorporated into the PCR product from CEM cells which contains the CAG sequence on one allele. dCTP is incorporated into both PCR products and twice as much is found in the PCR product from HeLa cells (homozygous for CGG). The assay is quantitative as shown after phosphorimaging (Fig. 25.2B). The background incorporation of dTTP into the CGG sequence is 0.5% as compared to CAG, which can be

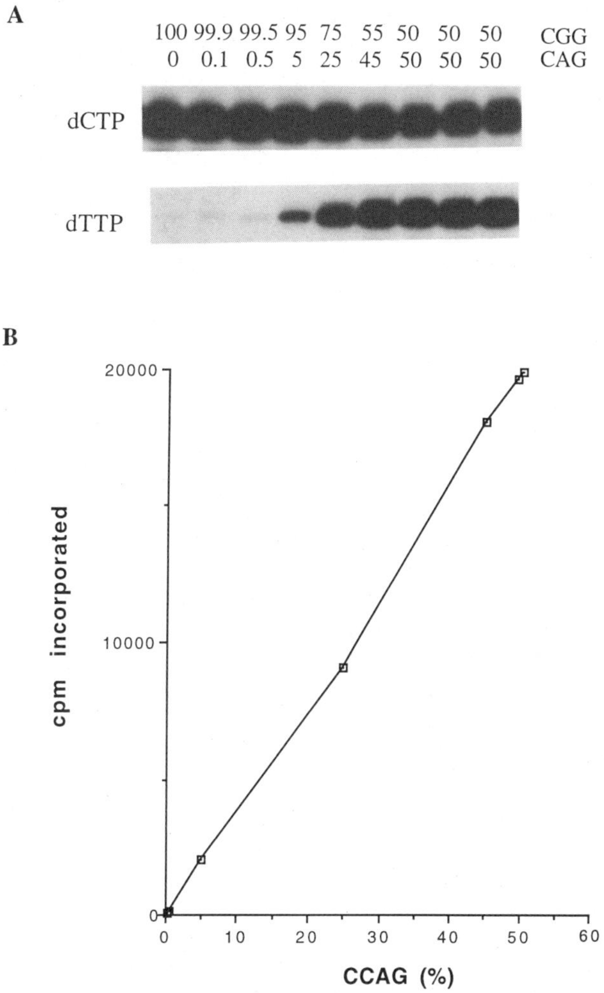

Figure 25.2. Sensitivity and quantitativeness of the DNA SNuPE assay. (A) PCR products from HeLa cells and CEM cells homozygous Wt (CGG/CGG) and heterozygous (CGG/CAG) mutant for codon 248, respectively, were mixed together in different ratios, representing allelic ratios in the reaction as indicated. SNuPE assay was done in the presence of [^{32}P]dCTP or [^{32}P]dTTP. The mutant allele can be easily detected even if it is represented only as 0.5% of the total mix. (B) The ratio of the two radionucleotides incorporated reflects the ratio of the two fragments present in the reaction.

subtracted. This shows that the SNuPE assay is highly specific and capable of detecting a mutant sequence even if it represents less than 1% of the total DNA present. The limits of sensitivity appear to vary with dNTP pairs used (Singer-Sam, 1995).

3. ASSAY CONDITIONS

This assay was performed essentially as described (Singer-Sam *et al.*, 1992).

3.1. PCR

The final concentrations of ingredients in 0.1-ml PCR reactions were: 10 mM Tris-HCl at pH 8.3, 50 mM KCl, 2 mM $MgCl_2$, 200 mM dNTPs, 0.5 mM of the lower and upper primers, and 2.5 U/reaction AmpliTaq DNA polymerase (Perkin Elmer Cetus). PCR conditions were 95°C–30 sec, 65°C–30 sec, 72°C–2 min for 35 cycles. PCR primers were designed with a computer program (Oligo 4.0, National Bioscience Inc., Plymouth, MN).

3.2. Isolation of Amplified DNA Products

These were subjected to agarose gel electrophoresis, and the single bands obtained were isolated with the Prepa-A-Gene DNA Purification Kit (Bio-Rad). DNA was suspended in water at approx. 2 ng/ml.

3.3. SNuPE

The reaction mix contained, in a total volume of 10 μl, 10 ng of the isolated DNA fragment, 1 mM of the SNuPE primer, 10 mM Tris-HCl at pH 8.3, 50 mM KCl, 2 mM $MgCl_2$, 0.75 U of AmpliTaq DNA polymerase, and 2 μCi of [^{32}P]dCTP or [^{32}P]dTTP. The radionucleotides (3000 Ci/mmole; 10 mCi/ml), were diluted tenfold in water, then 2-μl aliquots were added to each tube just prior to incubation. All other ingredients were added to the template as a master mix. Primers were kept in small aliquots to avoid repeated freeze–thawing. The SNuPE reaction consisted of one cycle of 95°C–30 sec, 42°C–30 sec, and 72°C–1 min, and was placed on ice to stop the reaction. Formamide loading dye (10 μl) was added, denatured at 98°C for 1 min, and chilled on ice. After electrophoresis on a 15% polyacrylamide/7 M urea gel for 1 hr at 15 mA, the gel was covered with Saran Wrap and the bands were visualized by autoradiography at room temperature for 20 min with Kodak X-OMAT™ AR film, and quantified with a PhosphorImager (Molecular Dynamics, Sunnyvale, CA).

4. RNA SNuPE

The RT-PCR SNuPE assay (Fig. 25.3) can be applied for detecting sequence polymorphisms in RNA in a quantitative fashion (Singer-Sam *et al.*, 1992). The sensitivity of the assay can be as good as recognizing one allele present in the RNA mix as little as 0.1% of the total (Singer-Sam *et al.*, 1992). First, total RNA is isolated from tissue samples, then reverse-transcription is done with the gene-specific lower primer, followed by PCR with the upper and lower primers that span the base difference and an intron, to make sure the isolated fragment is of RNA origin. Aliquots of the isolated fragment are subjected to the SNuPE assay. This method could be applied for diagnosis of exonic base difference not affecting transcription rate or RNA

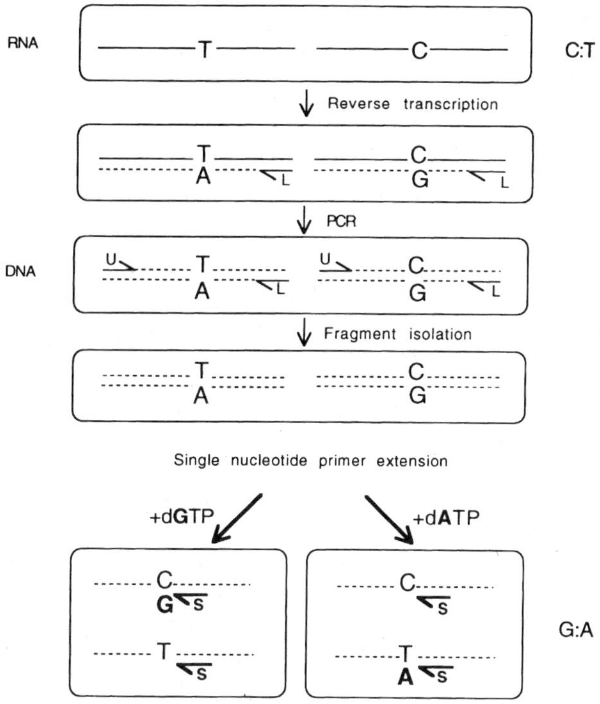

Figure 25.3. RT-PCR SNuPE (Singer-Sam *et al.*, 1992; Singer-Sam, 1994). The RNA segment harboring the base difference is amplified. The RNA is first reverse transcribed with the gene-specific lower primer followed by PCR with the lower and upper primers. The PCR fragment is isolated from an agarose gel and SNuPE is done with the SNuPE primer (S). RNA reconstitution experiments show that not only is the SNuPE assay quantitative but that both the PCR and the reverse transcription steps are proportional (Singer-Sam *et al.*, 1992; Szabó and Mann, 1995); the signal ratio equals the input RNA ratio (i.e., here G:A = C:T).

stability. This can be the method of choice when only very small amounts of tissue can be obtained, and the gene is known to be expressed at high levels in the given cell type, since the number of RNA molecules is much higher than that of DNA molecules in an expressing cell. Because of its high sensitivity, we successfully used the RT-PCR SNuPE assay for quantitation of allele-specific gene expression in single cells (Szabó and Mann, 1995b). For the quantitation of polymorphic alleles it is very important to have a standard curve. This can be obtained by mixing the two RNA species in different ratios and performing RT-PCR. Some points of the standard curve (0+100%, 50+50% and 100+0%) should be subjected to the SNuPE assay simultaneously with the experimentals to control for incorporation efficiencies of the radionucleotides in each experiment (Szabó and Mann, 1995a). The background incorporation levels (incorporation of the nucleotide on the opposite homogenous template) need to be subtracted from each sample. The ratio of the two alleles in the 50% samples must be used as a factor to adjust one allele to the other. In RNA titration experiments we found linearity of the assay over five magnitudes at a given amount of total RNA concentration (Szabó and Mann, 1995b). This method can also be adapted for the absolute measurement of changes in mRNA levels of a given transcript (Szabó and Mann, 1995b; Buzin *et al.*, 1994).

5. SOLID-PHASE MINISEQUENCING

Another elegant and powerful method has been developed for detection of known single base differences based on the same principle as the SNuPE assay, termed *primer-guided nucleotide incorporation assay* (Syvänen *et al.*, 1990) or *solid-phase minisequencing* (Syvänen *et al.*, 1992) (Fig. 25.4). The DNA fragment of interest is amplified with two oligonucleotides, one of which is biotinylated and captured on solid support (microtiter plate or coated microparticles) utilizing the biotin–streptavidin or biotin–avidin interactions. The nonbiotinylated strand is washed away after alkali denaturation. Single nucleotide primer extension is done with a primer abutting the base difference, the primer is eluted, and the radioactivity is measured by liquid scintillation counting. There is no need for fragment isolation or polyacrylamide gel electrophoresis. RNA can also be the starting material (Ikonen *et al.*, 1992). The method is also quantitative (Ikonen *et al.*, 1992; Syvänen *et al.*, 1992), and can be used for quantitation of absolute levels of RNA (Ikonen *et al.*, 1992). Two radionucleotides, emitting different radiation energy levels (e.g., ^{3}H and ^{32}P), can be added in the same tube and measured on two different channels simultaneously. Modification of the method to nonradioactive detection has been demonstrated using hapten-labeled nucleoside triphosphates that permit immunoenzymatic detection on microtiter plates (Harju *et al.*, 1993; Syvänen *et al.*, 1990). The method can be automated due to the microtiter-plate format. The principle of single nucleotide primer extension could also be used for detection of small deletion or insertion mutations if the base after the deletion is different from the first base of the deletion (Syvänen *et al.*, 1990).

The broad applicability of the solid-phase minisequencing method is reviewed by Syvänen (1994). The method has been successfully applied for screening individuals for dominant mutations before the manifestation of diseases or identifying carrier heterozygotes. It is routinely used in paternity and forensic cases. Due to its quantitative nature the frequency of certain alleles in the population can be determined.

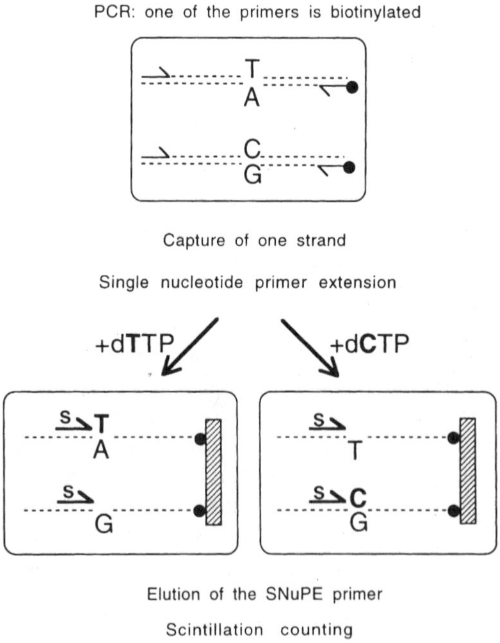

Figure 25.4. Primer-guided nucleotide incorporation method (Syvänen *et al.*, 1990) or solid-phase minisequencing (Syvänen *et al.*, 1992). One of the primers used in the PCR is biotinylated. The PCR product is bound to solid phase. The DNA fragment is denatured. With an oligonucleotide that abuts the base difference, primer extension is done on the captured strand with either of the radionucleotides. The oligonucleotide primer is eluted and the incorporated radiation is measured in a scintillation counter. The value ratios are corrected according to the standard curve.

The method of choice should rely on the scale of the planned experiments, available instrumentation, and whether there is need for visualization by autoradiography.

REFERENCES

Buzin, C. H., Mann, J. R., and Singer-Sam, J. (1994). Quantitative RT-PCR assays show *Xist* RNA levels are low in mouse female adult tissue, embryos and embryoid bodies. *Development* **120**:3529–3536.

Cheng, J., and Haas, M. (1990). Frequent mutations in the p53 tumor suppressor gene in human leukemia T-cell lines. *Mol. Cell. Biol.* **10**:5502–5509.

Harju, L., Weber, T., Alexandrova, L., Lukin, M., Ranki, M., and Jalanko, A. (1993). Colorimetric solid-phase minisequencing assay illustrated by detection of α1-antitrypsin Z mutation. *Clin. Chem.* **39**:2282–2287.

Ikonen, E., Manninen, T., Peltonen, L., and Syvänen, A.-C. (1992). Quantitative determination of rare mRNA species by PCR and solid-phase minisequencing. *PCR Methods Appl.* **1**:234–240.

Krook, A., Stratton, I. M., and O'Rahilly, S. (1992). Rapid and simultaneous detection of multiple mutations by pooled and multiplex single nucleotide primer extension: Application to the study of insulin-responsive glucose transporter and insulin receptor mutations in non-insulin-dependent diabetes. *Hum. Mol. Genet.* **1**:391–395.

Kuppuswamy, M. N., Hoffmann, J. W., Kasper, C. K., Spitzer, S. G., Groce, S. L., and Bajaj, S. P. (1991). Single nucleotide primer extension to detect genetic diseases: Experimental application to hemophilia B (factor IX) and cystic fibrosis genes. *Proc. Natl. Acad. Sci. USA* **88**:1143–1147.

Singer-Sam, J. (1994). Quantitation of specific transcripts by RT-PCR SNuPE assay. *PCR Methods Appl.* **3**: S48–S50.

Singer-Sam, J. (1995). Use of the SNuPE assay to quantitate allele-specific sequences differing by a single nucleotide, in: *PCR Primer: A Laboratory Manual* (C. W. Dieffenbach and G. S. Dveksler, eds.), Cold Spring Harbor Laboratory Press, Cold Spring Harbor, NY, pp. 339–343.

Singer-Sam, J., LeBon, J. M., Dai, A., and Riggs, A. D. (1992). A sensitive and quantitative assay for measurement of allele-specific transcripts differing by a single nucleotide. *PCR Methods Appl.* **1**:160–163.

Syvänen, A.-C. (1994). Detection of point mutations in human genes by the solid-phase minisequencing method. *Clin. Chim. Acta* **226**:225–236.

Syvänen, A.-C., Aalto-Setälä, K., Harju, L., Kontula, K., and Säderlund, H. (1990). A primer-guided nucleotide incorporation assay in the genotyping of apolipoprotein E. *Genomics* **8**:684–692.

Syvänen, A.-C., Ikonen, E., Manninen, T., Bengtström, M., Söderlund, H., Aula, P., and Peltonen, L. (1992). Convenient and quantitative determination of the frequency of a mutant allele using solid-phase minisequencing: Application to aspartylglucosaminuria in Finland. *Genomics* **12**:590–595.

Szabó, P. E., and Mann, J. R. (1995a). Biallelic expression of imprinted genes in the mouse germline: Implications for erasure, establishment, and mechanisms of genomic imprinting. *Genes Dev.* **9**:1857–1868.

Szabó, P. E., and Mann, J. R. (1995b). Allele-specific expression and total expression levels of imprinted genes during early mouse development: Implications for imprinting mechanisms. *Genes Dev.* **9**:3097–3108.

Chapter 26

Sequencing of PCR Products

Piroska E. Szabó, Jeffrey R. Mann, and Gerald Forrest

1. BASIC METHODS

Two basic methods are available for DNA sequencing, the Maxam–Gilbert and the Sanger method. The chemical method (Maxam and Gilbert, 1977) is based on base-specific chemical modification of the DNA and subsequent nicking of the sugar-phosphate bonds of the DNA at the modified bases. The enzymatic method (Sanger *et al.*, 1987) works by elongating a primer on a single-stranded DNA template by incorporating deoxynucleotides (dNTPs) with a polymerase enzyme and simultaneously terminating the chains by base-specific dideoxynucleotides (ddNTPs). Both methods use incomplete reactions (chemical or enzymatic, respectively) in order to generate fragment populations that give a sequence ladder after separation by denaturing electrophoresis. The sequence ladders are visualized by radioactive or nonradioactive means.

1.1. Maxam–Gilbert

Direct sequencing of PCR fragments (Fig. 26.1) can be done by the Maxam–Gilbert method (Tahara *et al.*, 1990; Stamm and Longo, 1990). This method could be the choice for very short fragments and when degenerate oligonucleotides, which cause mispriming in enzymatic sequencing, are used in the PCR reaction. The oligonucleotides used for PCR usually do not have a 5′ phosphate group. This makes it possible to end-label one of the oligos through phosphorylation with T4 polynucleotide kinase with ^{32}P γ-ATP before PCR and chemical treatment. The radioactive fragment has to be gel purified. This method did not become very popular, probably because of the inconvenience of this radioactive fragment isolation step and the toxic chemicals required in the base-modification reactions.

1.2. Sanger

The original Sanger method uses *Klenow* enzyme that incorporates a labeled nucleotide in the newly synthesized strand on a single-stranded DNA template (Fig. 26.1). The labeled nucleotide is provided at low concentration relative to the other three cold nucleotides (Sanger *et al.*, 1987). The chain termination is specified by four dideoxynucleotides given in four separate reactions. The ddNTP/dNTP ratio has to be high to result in early termination for obtaining

Piroska E. Szabó, Jeffrey R. Mann, and Gerald Forrest • Division of Biology, Beckman Research Institute of the City of Hope, Duarte, California 91010.

Technologies for Detection of DNA Damage and Mutations, edited by Gerd P. Pfeifer, Plenum Press, New York, 1996.

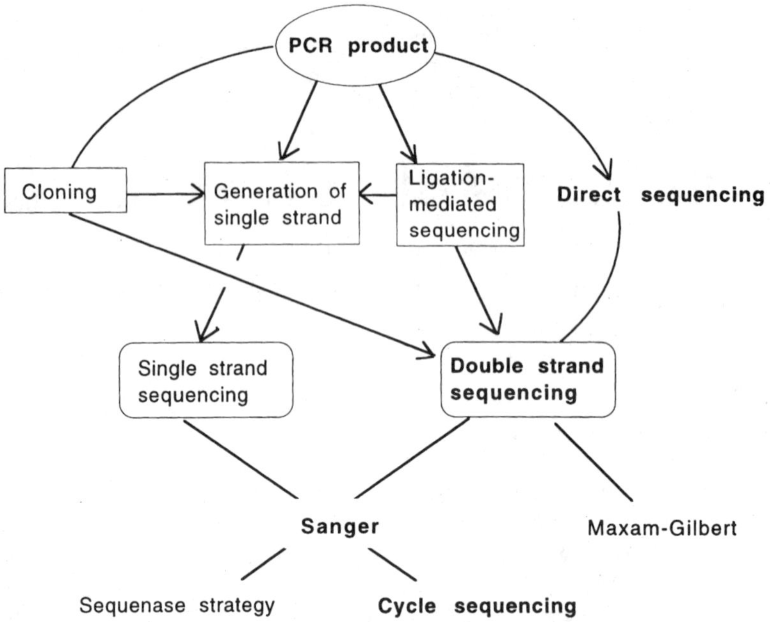

Figure 26.1. Alternative routes for sequencing PCR-generated fragments.

sequence information close to the primer, because the Klenow enzyme incorporates dNTPs with much higher efficiency than ddNTPs. Processivity of this enzyme is low, dissociating from the DNA after synthesis of ten nucleotides. Another enzyme, *reverse transcriptase*, used for enzymatic sequencing, has higher processivity (several hundred nucleotides before dissociation of the enzyme) but a slow synthesis rate.

One modification of the basic method was to use labeled sequencing primers instead of radioactive nucleotides (also called two-step protocol). DNA can be sequenced with this practice using different polymerases, like Klenow (Wrischnik *et al.*, 1987; Chen and Seeburg, 1985), AMV reverse transcriptase (Engelke *et al.*, 1988; Stoflet *et al.*, 1988; Sarkar and Sommer, 1988), Sequenase™ (Tabor and Richardson, 1989), and *Taq* polymerase (Innis *et al.*, 1988). Labeled primers require an additional, labeling step before sequencing, but also confer specificity to the sequencing reactions. In the case of sequencing several samples at the same time with the same primer, it is more economical to use labeled primers. When many different primers are required, the original one tube labeling/sequencing reactions are easier to perform in the presence of a labeled nucleotide.

2. STRATEGIES FOR SEQUENCING DOUBLE-STRANDED DNA

Discovery of new enzymes with different characteristics led to the development of new strategies based on the Sanger method. Two strategies, the Sequenase strategy and the cycle sequencing (Fig. 26.2), became quite popular. Both are used for sequencing of PCR fragments directly or after cloning into a plasmid vector (Fig. 26.1). These are available now in different kit formats: Sequenase™ 2.0 kit (US Biochemicals) and Autoread T7 Sequencing Kit (Pharmacia

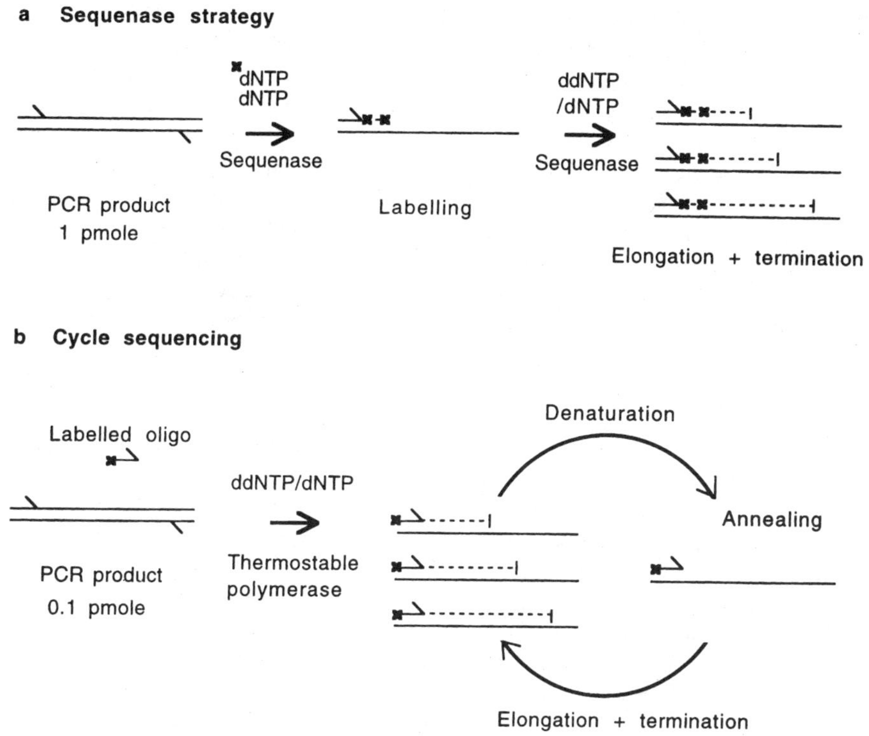

Figure 26.2. Two popular strategies based on the Sanger enzymatic DNA sequencing method: the Sequenase and the cycle sequencing strategies.

Biotech) or DTaq™ Cycle sequencing kit (US Biochemicals) and CircumVent™ Thermal Cycle Sequencing Kit (New England Biolabs), fmol™ DNA Sequencing System (Promega), respectively. Optimal conditions for directly sequencing double-stranded PCR products with the Sequenase kit, which was developed originally for plasmid sequencing, are described by Rao (1994a) and Casanova *et al.* (1990).

2.1. Sequenase Strategy

A new sequencing strategy has been developed for the Sequenase™ enzyme, (modified, exonuclease-free T7 polymerase). This polymerase has high processivity (incorporating thousands of nucleotides before dissociation) and a very rapid rate of nucleotide incorporation (Tabor and Richardson, 1987). In the Sequenase strategy, or three-step strategy (Tabor and Richardson, 1987), labeling and sequencing are done in successive steps (Fig. 26.2a). The double-stranded template is denatured, then an oligonucleotide is annealed to one end (step one). In the initial labeling step (step two), short, controlled primer extension is accomplished by polymerizing to an average chain length of 20–80 nucleotides in the presence of a radionucleotide. Most of the radioactive nucleotide is incorporated into this short strand, resulting in equal intensity bands on the sequencing ladders. This step requires suboptimal conditions: low temperature and low concentrations of all four dNTPs. This is followed by a processive extension (step three) in the presence of higher dNTP and ddNTP concentrations, where the four different termination/

extension mixes are added to four tubes, differing only in the terminating dideoxynucleotides. The extension of a strand continues until one ddNTP incorporates and terminates the strand. The range of the readable sequence on a gel can be influenced by changing the dNTP/ddNTP ratio (higher ratio results in longer fragments). Less sequence information is obtained than in the two-step protocol, because termination does not occur right from the end of the primer. When the sequences closer to the primers are of interest, Mn^{2+} ions also can be added to the reaction. The Mn^{2+} ions alter the relative reactivities of the deoxy- and dideoxynucleotides to Sequenase; this results in uniform terminations of DNA sequencing reactions (Tabor and Richardson, 1989). *Taq* polymerase can also be applied in the Sequenase strategy (Innis *et al.*, 1988). *Taq* polymerase was found to be superior when sequencing DNA with hairpin structures. This enzyme is active at high temperatures and at low salt concentration, conditions that resolve secondary structures.

2.1.1. Sequencing Linear Double-Stranded DNA

A problem associated with linear double-stranded DNA sequencing is that after denaturation the two complementary strands have a strong tendency to reanneal, competing with the primers and causing premature termination in the sequencing reaction. This problem is not present in sequencing double-stranded supercoiled plasmids, because the two strands do not easily reanneal after alkali denaturation (Chen and Seeburg, 1985). Single-stranded DNA, generated from M13 phage (Messing, 1983), will also eliminate the problem (Tabor and Richardson, 1987; Sanger *et al.*, 1980). Linear double-stranded PCR fragments can be cloned into plasmid vectors, or (after using special primers for the PCR) can be circularized before sequencing (Day and Walker, 1991; Gyllensten, 1989). To overcome the problem of reannealing of linear double-stranded DNA during nonthermal sequencing, numerous approaches have been developed (for review see also Bevan *et al.*, 1992). One approach lowers the possibility of reassociation of the two strands during sequencing, other strategies provide only one DNA strand in the sequencing reaction (see below). High-temperature sequencing can also resolve the problem (see Section 2.2).

2.1.1a. Keeping Strand Separation after Heat Denaturation. Stabilization of the single strands after denaturation is helped by fast cooling, e.g., placing the reaction quickly on dry ice (Casanova *et al.*, 1990). Alkali denaturation is followed by precipitation (Chen and Seeburg, 1985) or neutralization (Zimmermann *et al.*, 1990). Including DMSO (Winship, 1989), formamide (Zhang *et al.*, 1991), or the nonionic detergents Nonidet-P40 and Tween 20 in Sequenase reactions (Bachmann *et al.*, 1990), and the latter in *Taq* polymerase reactions (Innis *et al.*, 1988) gave good results. Addition of gene 32 protein (gp32, single strand binding protein, or SSBP) isolated from bacteriophage T4 facilitates sequencing with Klenow (Kaspar *et al.*, 1989) or Sequenase™ enzymes (Wanner *et al.*, 1992) by binding and stabilizing the single-stranded DNA (available from Pharmacia). Performing the annealing and the sequencing steps in agarose gels appears to inhibit template reannealing, and is possible with Sequenase™ (Khorana *et al.*, 1994; Kretz *et al.*, 1989). Competitive association of one of the strands with a complementary single strand, generated from a plasmid, can be used for routine sequencing of the same PCR fragment (Gal and Hohn, 1990).

2.1.1b. Generation of Single Strands from Double-Stranded DNA. Different techniques providing a single strand for the sequencing reaction and thus increasing sequencing quality are summarized in Fig. 26.3. A technique for removing one of the strands of the PCR fragment by *exonuclease* digestion was developed (Higuchi and Ochman, 1989) (Fig. 26.3a). The strand to be removed is specified by which oligonucleotide is phosphorylated. This technique requires a phosphorylated oligo and an additional enzymatic step. In another technique, only one of the strands is generated for sequencing by *asymmetric PCR* (Fig. 26.3b). This can be

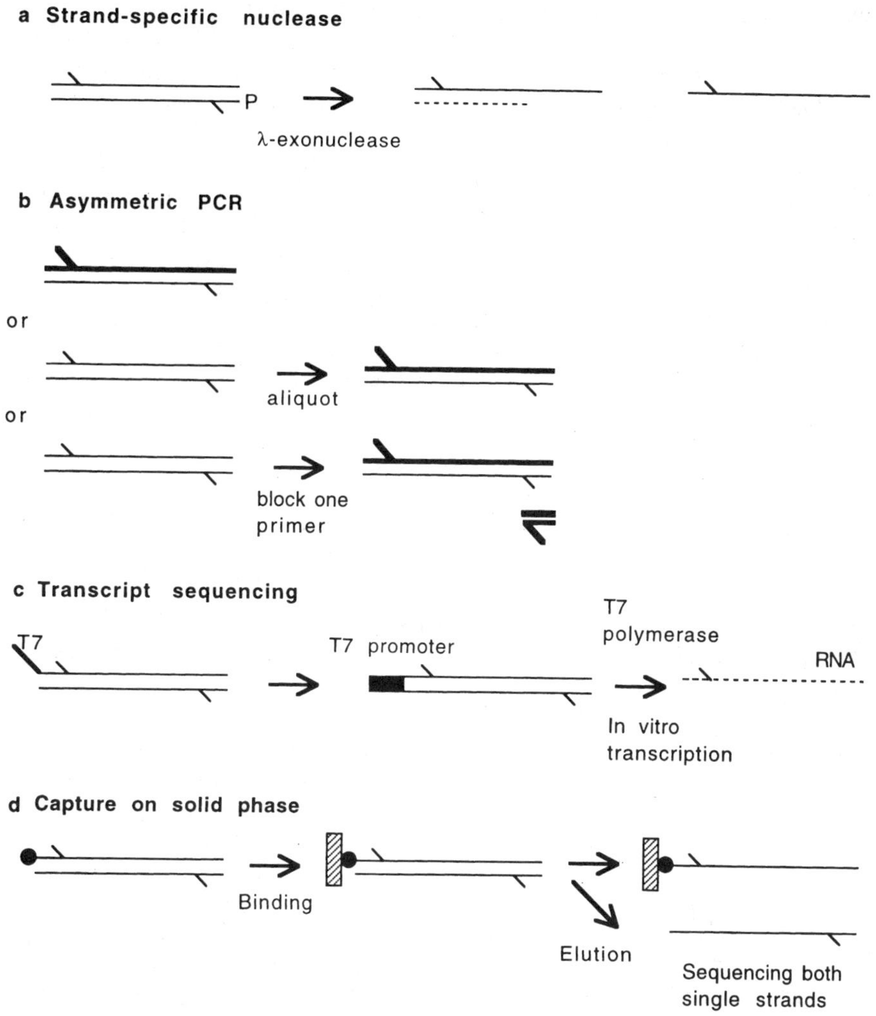

Figure 26.3. Different ways of generating single strands for DNA sequencing of PCR products with the Sequenase strategy. (a) Strand-specific exonuclease; (b) asymmetric PCR; (c) transcript sequencing; (d) capture on solid phase.

achieved by limiting one of the primers during the PCR (Gyllensten and Erlich, 1988; Innis *et al.*, 1988). One must consider that asymmetric PCR gives a much lower yield, because the template amount in consecutive cycles does not increase exponentially at limiting primer concentrations. Normal PCR is done in the first step, and an aliquot of this reaction is used in the asymmetric second step (Gibbs *et al.*, 1990). (In this article the normal PCR step was designed to amplify multiple fragments for screening deletion mutations, and the second asymmetric step provided single-stranded DNA for sequencing with the Sequenase strategy.) Alternatively, after an initial logarithmic amplification with two primers, one of the oligos is blocked by addition of the complementary oligonucleotide (Gyllensten, 1989). The asymmetric PCR can be optimized before direct sequencing in order to produce only a single band (Dowton and Austin, 1994). In the method of genomic amplification with transcript sequencing (GAWTS), PCR is done with an oligo carrying a *T7 phage promoter* sequence (Fig. 26.3c). A single-stranded RNA fragment is generated from the PCR fragments by *in vitro* transcription using T7 RNA polymerase (Stoflet *et al.*, 1988) and sequenced with reverse transcriptase. Similarly, RNA can be the starting material (RAWTS). In this case, reverse transcription is done before PCR (Sarkar and Sommer, 1988). One can separate the DNA strands on a *solid phase* by utilizing one biotinylated oligo in the PCR (Fig. 26.3d). Alternatively, the biotinylated fragment binds to a streptavidin column (Mitchell and Merill, 1989) or streptavidin-coated magnetic particles (Hultman *et al.*, 1989), the DNA fragment is denatured, and the nonbiotinylated strand is eluted. Both the strand that is bound on solid phase and the eluted strand can be sequenced separately (Hultman *et al.*, 1991). The biotinylated DNA strand can bind to streptavidin-coated microtiter plates, allowing performance of all reactions in a microtiter plate format together with a robotic workstation (Schofield *et al.*, 1993). "Oligo walking" was also demonstrated by these authors. Longer (2.4 kb) DNA fragments are immobilized on magnetic beads and sequenced with primers that bind about 300 bp apart from each other.

2.1.2. Difficult Templates

Compression of bands in the sequencing ladder of GC-rich DNA can be eliminated by substitution of dGTP by 7-deaza-dGTP or dITP (Tabor and Richardson, 1987). These analogues of dGTP are recognized by the various polymerases as dGTP, but they do not form very stable hydrogen bonds with dCTP, eliminating the secondary structure-caused premature chain termination. It is suggested to use 7-deaza-dGTP routinely when sequencing with Klenow (Mizusawa *et al.*, 1986), with Sequenase™ (Zimmermann *et al.*, 1990; Dierick *et al.*, 1993), or with *Taq* polymerase (Innis *et al.*, 1988), but with *Taq* polymerase dITP is not recommended (Innis *et al.*, 1988). For Sequenase™ enzyme, dITP was found to be even better (Tabor and Richardson, 1987). Analogues of dGTP could be included even in the initial PCR reactions prior to sequencing (Dierick *et al.* 1993) if PCR fragments with stubborn secondary structures are to be sequenced.

2.2. Cycle Sequencing

The cycle sequencing strategy (Murray, 1989) (Fig. 26.2b) combines the ideas of chain-termination sequencing and PCR (for review and protocol see Kretz *et al.*, 1994). Elongation/termination and denaturation steps are repeated several times. Enzyme addition is not required in the subsequent cycles, because thermostable polymerases are used. The DNA is denatured, a labeled oligo is annealed (Rao and Saunders, 1992; Carothers *et al.*, 1989; Murray, 1989) and extended by the polymerase, simultaneously terminating the chain base-specifically, when a ddNTP is incorporated. The newly synthesized, terminated strand is removed by denaturation, liberating the template molecule for more elongation/termination cycles. The dNTP concentration is lower in cycle sequencing than in normal PCR to achieve a sufficient termination rate by the ddNTPs. Instead of using a labeled primer, one radionucleotide can be included in the cycle sequencing reaction (Lee, 1991). The ddNTP/dNTP ratio in this case needs to be higher to balance the signal differences between short and long chains by increasing the number of short molecules relative to the longer chains, which have more radioactivity incorporated. One advantage of the cycle sequencing is that less template is sufficient compared to the Sequenase strategy for generating a strong signal. Less template purification is required, because less template is used and it is highly diluted in the cycle sequencing. The cycle sequencing protocol was demonstrated to work on a bacterial colony without time consuming purification steps (Kretz *et al.*, 1994). Thermal cycle reactions utilizing polymerases with high optimum temperature like *Taq* (Carothers *et al.*, 1989; Murray, 1989; Lee, 1991), *Tub* (Rao and Saunders, 1992), $\text{Vent}_{\text{Rexo-}}$ (Reeves *et al.*, 1995), or *Pfu* (Hedden *et al.*, 1992) eliminate most of the reannealing problems of the linear double-stranded fragment. If high-temperature sequencing does not resolve some stubborn secondary structures, then dGTP can be substituted with its analogue 7-deaza-dGTP in the sequencing reaction.

3. PRIMERS

Selection of primers for enzymatic DNA sequencing is of principal importance. There are several software programs available to help in primer selection. We routinely use the Oligo 4.0™ with good success. This is available commercially for both the PC and Macintosh (National Biosciences, Inc., Plymouth, MN). Advanced Gene Computing Technologies, Inc. (Irvine, CA) markets HYBsimulator™, a new oligonucleotide probe design software that allows investigators to design very specific probes and primers by directly comparing the sequence to the collection of gene sequences in GenBank and EMBL data bank with an algorithm (Mitsuhashi *et al.*, 1994). OSP (oligonucleotide selection program) is a free program used for selecting primers for DNA sequencing and PCR (Hillier and Green, 1991). The following rules are guidelines for optimization of primers:

1. The melting temperature (T_m) should be at least 45°C, but as high as possible to allow control of cross hybridization during the annealing step. According to De Bellis *et al.* (1992), sequencing primers with a T_m higher than 70°C give the best results. We find that oligos (designed also considering the other criteria below) having a T_m between 60 and 65°C give specific PCR products and are suitable also for sequencing. T_m is usually calculated by the software, but can be determined using the following formula (Sambrook *et al.*, 1989):

$$T_m\ ^\circ\mathrm{C} = 81.5 - 16.6(\log_{10}[\mathrm{Na}^+]) + 0.41(\%\mathrm{G+C}) - (600/N)^2$$

   where $[\mathrm{Na}^+]$ is the sodium ion concentration and N is the length of the oligo.
2. Primer length can be used to adjust annealing temperature and to increase specificity. Primers should be at least 18 nucleotides long with preferable lengths in the range of 20–25 nucleotides.
3. Hairpin structures should be eliminated. This problem is compounded if the 3′ end can anneal and be extended by the polymerase.
4. The possibility of each primer hybridizing with itself and with the second heterologous primer should be eliminated. Special attention should be given to the 3′ ends where three or four complementary bases could anneal and be extended in the presence of polymerase with the formation of primer-dimers.
5. A primer must not be chosen which has false priming sites on the template DNA.
6. The internal stability of the 5′ end of the primer should be higher than that of the 3′ end. The 5′ sequences are used to initiate specificity. If the 3′ end of the primer is less stable, there will be less chance of the 3′ end priming at nonspecific sites.
7. Strings of single bases should be limited to three or four with strings of G or C being less desirable.
8. The G–C content should be close to 50%.

4. TEMPLATE PURITY

The quality of the sequencing template is an important factor in the sequencing reaction. One needs to purify the template from nonspecific spurious PCR bands that can cause high background. In addition, one must remove dNTPs that change the ddNTP/dNTP ratio, and also remove the primers that can cause mispriming in the sequencing reaction. The best result can be achieved with gel-purified fragments which are devoid of all impurities, e.g., with Prep-a-Gen® matrix (Bio-Rad, Hercules, CA).

4.1. Nonspecific Bands

Sequencing is possible inside low-melting agarose gels (Khorana *et al.*, 1994; Kretz *et al.*, 1989), and with little effort one can get rid of spurious bands. The PCR should be optimized to obtain a single, specific band when using unpurified PCR products, or labeled sequencing primers should be used that are nested or located between the amplification primers (Gibbs *et al.*, 1990; Carothers *et al.*, 1989; Engelke *et al.*, 1988; Wrischnik *et al.*, 1987).

4.2. Contamination by Primers and Nucleotides

Primers from the PCR reactions carried over into the sequencing reaction cause mispriming that increases the background. If the sequencing reaction is primed by one labeled oligo, the other unlabeled primer(s) cannot give background activities with either sequencing strategy (Rao and Saunders, 1992; Innis *et al.*, 1988; Wrischnik *et al.*, 1987); however, they can tie up the polymerase and use up the dNTPs. Mispriming, caused by the PCR primers, can be eliminated by choosing the T_m of the sequencing primers 10°C higher than that of the PCR primers and performing high-temperature cycle sequencing with radionucleotides (Liu *et al.*, 1993). Simultaneous sequencing from the two ends of a PCR fragment with the two PCR primers, carried over from the amplification, is not a problem but an advantage in bidirectional cycle sequencing (Douglas *et al.*, 1993). The primers are unlabeled. Following electrophoresis and blotting, the membrane is probed with a labeled oligo, complementary to either primer used in the sequencing reaction. Rehybridization with the complementary oligo of the opposite primer provides sequence information from the other strand.

The nucleotides need to be removed before sequencing because the dNTP/ddNTP ratio would otherwise be altered. This is less critical in the cycle sequencing, because the PCR reaction components are more diluted in the sequencing reaction due to the low template requirement. However, reducing the concentration of dNTPs in the initial PCR is suggested (Rao and Saunders, 1992; Douglas *et al.*, 1993).

Primers and nucleotides can be removed by selective precipitation (Reeves *et al.*, 1995), spin column chromatography on Quiagen spin-20 or Quiaquick spin column (Rao, 1994a,b) or Microspin S-300 (Pharmacia) or fast enzymatic treatment with exonuclease I (exoI) combined with shrimp alkaline phosphatase (sAP) (Werle *et al.*, 1994). Oligonucleotides can also be removed by mung-bean nuclease digestion (Dowton and Austin, 1993). Centricon-30 spin columns (Amicon) remove primers, nucleotides, and buffer components (Gyllensten and Erlich, 1988).

5. VISUALIZATION

Visualization of the sequencing reaction is traditionally done by *radioactive* labeling, but recently nonradioactive methods are also available. For radioactive detection, incorporation of radioactive α-labeled ^{32}P, ^{35}S, or ^{33}P nucleotides or end-labeled primers are used. ^{32}P generates strong signals, ^{35}S results in weaker but sharper bands and better resolution, and its half-life is longer. ^{33}P generates strong signals and also sharp bands. *Taq* polymerase has a low labeling efficiency with ^{35}S. This can be compensated by increased cycle number (25 cycles) in the cycle sequencing (Lee, 1991). Alternatively, use of ^{35}S in the cycle sequencing is suggested with exonuclease-free *Pfu* polymerase. *Pfu* was shown to be nondiscriminative for thionucleotides (Hedden *et al.*, 1992).

Nonradioactive detection methods represent less hazardous options than radioactive methods. In fluorescent detection, a fluorescent label can be attached either to the primers (Smith *et al.*, 1986) or to the dideoxynucleotides (Schofield *et al.*, 1993; Prober *et al.*, 1987). During electrophoresis the bands pass a laser beam that generates fluorescent signals. These signals are detected, then analyzed and stored by a computer. Other nonradioactive alternatives, chemiluminescent and colorimetric sequencing are manual, not computerized, require hapten-labeled primers, but do not require special equipment like the fluorescent sequencing. Biotinylated primers are used in the sequencing reaction (Beck *et al.*, 1989), and the sequencing gel is blotted to a membrane. The detection is accomplished enzymatically, with streptavidin–alkaline phosphatase or streptavidin–horsradish peroxidase conjugates. These enzymes generate chemiluminescent products from specialized substrates. Chemiluminescent detection for cycle sequencing was described (Douglas *et al.*, 1993). The hybridization was done with the Millipore Plex Luminescent Detection Kit (e.g., Sequenase® Images™ nonisotopic DNA sequence detection kit, USB).

6. DIRECT SEQUENCING VERSUS SUBCLONING

Direct sequencing of PCR fragments has the advantage of rapidity. Cycle sequencing generates beautiful sequencing data with very small amounts of template without subcloning or other manipulations. The other advantage is that polymerase errors that are a feature of PCR are undetectable by directly sequencing a large population of fragments at the same time. When PCR DNA fragments are subcloned, several clones must be analyzed to determine the consensus sequence. Heterozygotes can also be identified with direct sequencing, because both sequences appear on the sequence ladder (Wrischnik *et al.*, 1987; Gibbs *et al.*, 1990) or in the automatic fluorescent sequencer output file (Gibbs *et al.*, 1990). The disadvantage of the direct chain termination sequencing is that some sequence is lost next to the primer, since the first chain termination occurs at a certain distance from the primer. This problem can be eliminated by cloning of the PCR fragment into a plasmid vector or M13 phage (Messing, 1983), and sequencing the double-stranded supercoiled plasmid or the single-stranded DNA from the M13 phage, using a universal primer (Sanger *et al.*, 1980) that binds outside of the sequence of interest. Direct sequencing of a heterogeneous population of fragments made with degenerate primers is not possible.

6.1. Cloning of PCR Products

Here we give an overview of different protocols dealing with cloning of PCR fragments (Fig. 26.4). Different cloning methods target either the blunt-ended fragments or the fragments with single base overhangs generated in the PCR. [The latter is due to the extendase activity (Clark, 1988) of *Taq* polymerase.] Others provide cohesive ends on the PCR fragment in different ways.

6.1.1. Blunt-End Ligation

Various protocols use blunt-end ligation (Fig. 26.4a). The efficiency of blunt-end ligation of *Taq* polymerase-generated fragments is low, not only because of the overall low efficiency of

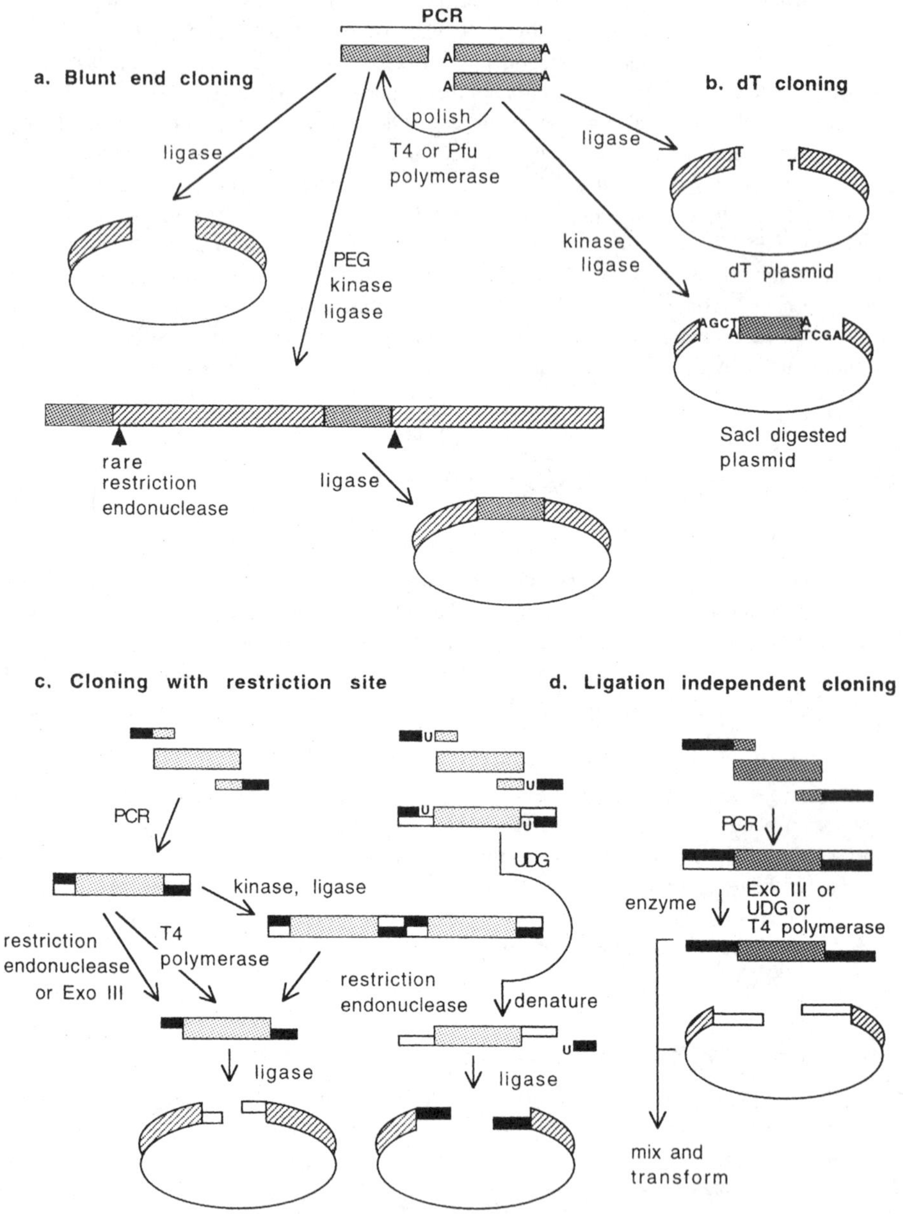

Figure 26.4. Strategies for cloning PCR fragments: (a) Blunt-end cloning; (b) dT cloning; (c) cloning with restriction sites; (d) ligation-independent cloning.

blunt-end ligation but also because the blunt-ended fragments are a minority among PCR products. The latter is due to the extendase activity of the *Taq* polymerase. *Taq* polymerase adds an extra 3′-terminal base (mostly A) to the PCR product, this way producing a non-blunt-ended fragment (Clark, 1988). The abundance of blunt-ended fragments can be increased by omitting the final extension step from the PCR (Haqqi, 1992), or by using Vent™ polymerase (NEB) in the PCR reaction instead of AmpliTaq™ (Perkin Elmer) polymerase (Lohff and Cease, 1991). Vent™ polymerase has 3′-to-5′ exonuclease activity and generates only blunt-end fragments. Polishing the ends of the PCR fragments with T4 or *Pfu* polymerase (Costa *et al.*, 1994; Costa and Weiner, 1994) after PCR increases the efficiency of blunt-end ligation. Klenow enzyme cannot be used for polishing because it also has extendase activity.

The blunt-end ligation can be facilitated by simultaneously cutting the plasmid with the blunt-end-producing restriction enzyme (e.g., SmaI) to reduce the background in the transformation (Liu and Schwartz, 1992) if the restriction site is not present on the fragment itself and the ligation does not re-create the site. The pCR-Script™ SK(+) cloning kit (Stratagene) is based on the same idea. Here the cloning plasmid has an *Srf*I site that has a very rare octanucleotide sequence which is highly unlikely to be present on any fragment to be cloned. A one tube ligation reaction was described (Chuang *et al.*, 1995) where digestion of vector DNA, polishing of PCR DNA ends, and blunt-end ligation were accomplished in a single step. Increasing the efficiency of blunt-end ligation by 1000-fold can be achieved by using 15% PEG under conditions of macromolecular crowding, which effectively produces vector-insert concatamers. Therefore, a subsequent cleavage step is also necessary to release vector-insert monomers using a rare cutting restriction enzyme, which does not cut the insert and cuts the vector only once (Upcroft and Healey, 1987). Turbo cloning (Boyd, 1993) also uses macromolecular crowding. Here single vector-insert units are circularized from concatamers with the Cre recombinase, which acts on pairs of lox sites within directly repeated vector molecules flanking insert DNA. This latter system is fast, and works for fragments with unknown sites, but requires special plasmids containing the lox site. Directional blunt-end ligation can be performed using the pCR Direct (Stratagene) plasmid (Costa *et al.*, 1994; Costa and Weiner, 1994; Weiner, 1993). One of the primers has to be phosphorylated in the PCR and similarly one end of the vector needs to be dephosphorylated. This is achieved by cutting the vector with *Srf*I, subsequent phosphatase treatment, and cleavage with *Sma*I that has a recognition site 15 bases 3′ from the *Srf*I site. Use of linkers or adaptors is also a possible way of cloning blunt-end fragments and avoids the need for restriction endonucleases (for ligation protocol see Lorens, 1991). Finally, PCR fragments can be dC-tailed and ligated into dG-tailed plasmid vector (Nicolas and Laliberté, 1991).

6.1.2. dT Cloning

dT cloning (Fig. 26.4b) utilizes the protruding dA at the 3′ end of the PCR fragments. The efficiency of the ligation is much higher than with blunt ends. Previous knowledge of restriction sites on the fragments is not required. Also there is no need for primer modifications and addition of extra sequences, or tricky sticky end generation methods. The only requirement for dT cloning is a vector that has a complementary dT overhang. Such a vector (pCR™II) is commercially available in a kit format (TA Cloning® Kit, Invitrogen, San Diego, CA). In our hands this kit performs extremely well. Among three picked white colonies we find two or three positives carrying the right insert. We routinely ligate eight fragments at a time, pick 24 white colonies, do miniprep, analyze the clones with restriction digestion, and sequence two positives with the cycle sequencing method. The dT vector can be generated by treating a blunt ended vector with *Taq* polymerase in the presence of dTTP (Marchuk *et al.*, 1991) or with terminal transferase in the presence of ddTTP (Holton and Graham, 1991). Cleavage of DNA with certain restriction enzymes can create this overhang, but these sites are not present in commercially available vectors. Synthetic *Xcm*I, *Asp*EI, or *Pfl*M1 recognition sites were engineered into the polylinker of the plasmid pBluescript (Testori *et al.*, 1994; Ichihara and Kurosawa, 1993; Mitchell *et al.*, 1992), or the *Xcm*I site was engineered into the M13 phage (Cha *et al.*, 1993), maintaining the reading frame of the β-galactosidase gene allowing blue-white selection.

The additional base at the end of the PCR fragment depends on the 3′-most base of the fragment, but mostly it is an A overhang (Clark, 1988). We successfully cloned PCR fragments into dT vector that ended in any of the four nucleotides before the extra A (Szabó and Mann, 1995).

Based on the dA overhang of PCR fragments, a single-step direct cloning has been developed that does not require specific dT overhang of the vector, but any restriction site with T as the last base of its sticky end. The PCR product must contain a 5′ phosphate. A single tube reaction including both cleavage and ligation can be performed using uncut vector DNA, provided the PCR product does not contain cutting site for the restriction enzyme used and ligation does not regenerate the cutting sites (Chuang *et al.*, 1995).

6.1.3. Cloning PCR Fragments with Restriction Sites

Utilizing restriction sites for ligation of PCR fragments (Fig. 26.4c) was achieved by amplifying the fragment with PCR primers harboring extra bases at their 5′ ends corresponding to a restriction site (Scharf *et al.*, 1986). Sticky ends for the ligation are generated by cutting the PCR product with the appropriate restriction endonuclease. This procedure requires the following considerations: (1) the restriction site used should not reside within the fragment and (2) the restriction site should be carefully selected by taking into account that some restriction endonucleases fail to cleave at sequences located near the extreme ends of a DNA fragment (Crouse and Amorese, 1986; Kaufman and Evans, 1990). In case of low-efficiency cutters, additional sequences should be designed 5′ from the restriction site to provide a spacer region, but this would result in extra cost for each oligo pair. Alternatively, for complete cleavage by the restriction endonuclease, the blunt-end PCR fragments can be phosphorylated and self-ligated, then the multimer fragments can be cleaved with the restriction enzyme (Lorens, 1991; Kaufman and Evans, 1990). The kinase and the ligase (and also the polishing) reactions can be performed in the same tube (Lorens, 1991). Defined cohesive termini can be created using the endogenous 3′ exonuclease activity of T4 DNA polymerase in the presence of selected nucleotides. This method allows directional cloning without restriction digestion, and with only short extensions of the PCR primers (Stoker, 1990). Use of controlled exonuclease III digestion for creating sticky ends (Kaluz *et al.*, 1992) has the same advantages. Use of oligos containing dU at specific positions allows generation of cohesive ends, mimicking restriction sites by nicking at dU by uracyl *N*-glycosylase (UDG) followed by heat- or alkali-denaturation (Smith *et al.*, 1993). These three procedures are particularly useful when cloning fragments with unknown sequences. A vector was developed for cloning and modifying of PCR fragments that has opposed class IIS restriction sites flanking blunt-cutting class I restriction site (Cease and Lohff, 1993). The fragment is blunt-end (or dT) cloned in this vector and then excised by the class IIS "reach-over" endonuclease, exactly at the site of the ligation, generating a 4-base 5′-overhang. The sequence of the overhang is determined by the primer sequence.

6.1.4. Ligation-Independent Cloning (LIC)

LIC of PCR fragments (Fig. 26.4d) requires a 12-base tail at the end of the PCR fragment with a complementary tail on the vector. These long overhangs anneal when mixed together, eliminating the need for a ligation step before transformation. One way to create such tails is to use primers with additional 5′ ends that lack one base, e.g., dCTP. The fragment is then treated with T4 DNA polymerase in the presence of dGTP (Aslanidis and de Jong, 1990) utilizing its 3′-5′ exonuclease activity. The vector has to be amplified similarly, but omitting the complementary base (dGTP) from the oligo for vector amplification, and then treating it with T4 DNA polymerase in the presence of dCTP. Incorporating specific sequences into the vector devoid of a particular nucleotide can eliminate the need for amplification of the vector itself (Kuijper *et al.*, 1992). Also, if the PCR primers have a 12-base overhang complementary to the restriction site used to linearize the vector, the vector amplification step can be omitted. Alternatively, both fragment and vector are subjected to controlled *Exo*III digestion that removes a stretch of 3′ nucleotides (Hsiao, 1993). One can eliminate the *Exo*III digestion of the fragment before LIC (Kaluz and Flint, 1994) by including a nonbase residue into the PCR primer that functions as a barrier for the polymerase. Another way of creating single-stranded sticky ends (Nisson *et al.*, 1991) is to PCR amplify the fragment and also the vector with oligos containing dUMPs instead of dTTPs, followed by digestion with UDG and denaturation.

6.1.5. *In Vivo* Cloning

In vivo cloning (Oliner *et al.*, 1993) of PCR products is an efficient method based on a high rate of homologous recombination between the PCR fragments containing engineered terminal sequences and cotransformed vector DNA utilizing the bacterial enzyme machinery.

6.2. Ligation-Mediated Sequencing

Two methods have been described that represent intermediates between direct sequencing and cloning of PCR fragments (Fig. 26.1). In *hybrid PCR sequencing* (Fig. 26.5a) (Berg and Olaisen, 1994) the PCR fragment is ligated to a blunt-end DNA fragment that contains a binding site for a universal M13 sequencing primer. Then a second PCR is performed with a phosphorylated M13 oligo and one of the nonphosphorylated fragment-specific primers. λ-exonuclease is used to remove the phosphorylated strand providing a single strand for the Sequenase™ reaction using the M13 primer. In the *Ligation-linked PCR* (LLR) (Reeves *et al.*, 1995) the PCR fragment is temporarily ligated into a dT plasmid vector, and a second, hybrid PCR is done with an oligo, specific to the plasmid molecule and another one that binds to the PCR fragment (Fig. 26.5b). This second PCR product is cycle sequenced with the universal sequencing primers present originally on the plasmid. This is particularly useful in cases where the PCR is done using arbitrary primers. These intermediate methods combine the advantages of direct sequencing and sequencing of cloned fragments. The template DNA comes from a pool of DNA molecules, making polymerase mistakes undetectable. Sequence information from the entire fragment is obtained using a universal primer that binds outside of the fragment and also gives high-quality results on any DNA fragment. Also there is no need for transformation of bacteria or plasmid purification steps, biotinylated oligos, or phosphorylation of one of the gene-specific PCR primers.

7. LARGE-SCALE SEQUENCING PROJECTS

7.1. Automatic Fluorescent Sequencing

The use of an automatic workstation for large-scale sequencing projects is worth considering. Several automatic sequencers are available that are based on fluorescent detection. The Applied Biosystem Automatic Workstation (Tracy and Mulcahy, 1991) is based on the use of

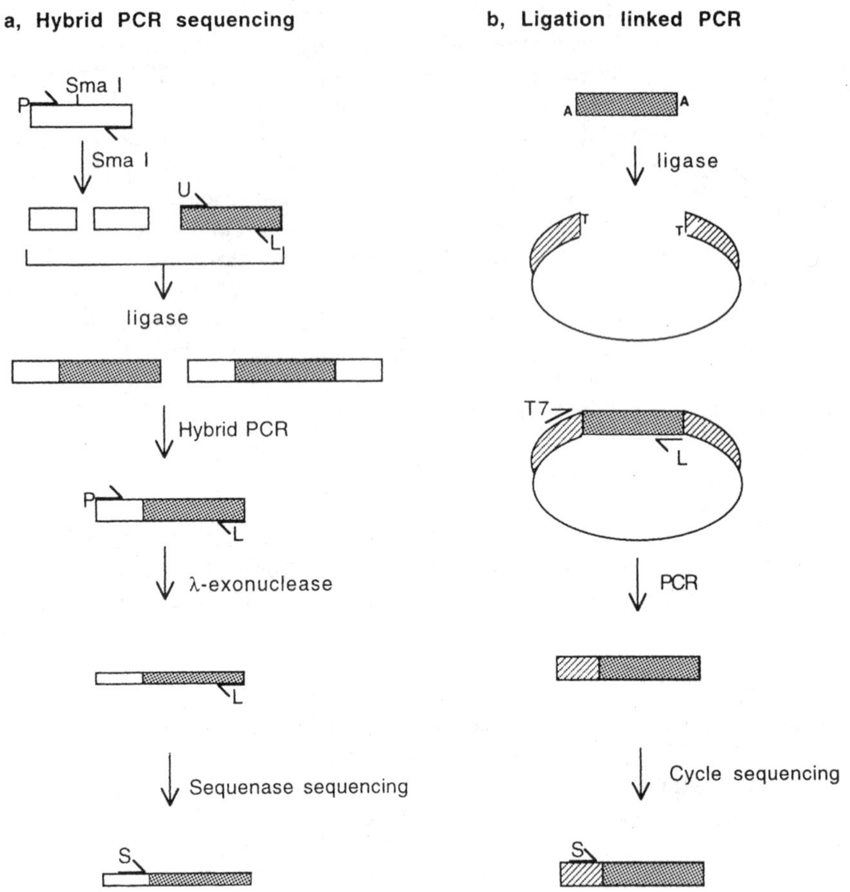

Figure 26.5. Ligation-mediated sequencing. (a) Hybrid PCR sequencing; (b) ligation-linked PCR (LLR).

fluorescent-labeled nucleotide analogues and their detection by laser scanning (Smith *et al.*, 1986). One advantage of the method is the use of nonradioactive detection. Also, because four different fluorophore molecules are available to tag the four reactions, it is possible to run four sequencing reactions in one lane instead of four, decreasing the number of gels to be handled. The third advantage is that the workstation is complete with a computer, attached to the laser reader to analyze and store the data with a computer (Smith *et al.*, 1986). The LI-COR automatic DNA sequencer (LI-COR Inc.) utilizes infrared fluorescence primer chemistry (see below) and two-dimensional imaging analysis, resulting in great sensitivity and resolution. The throughout is about 800 bp per reaction. The operating cost is low because no specific kits or glass plates are required (Middendorf *et al.*, 1992). The EMBL automatic workstation (Ansorge *et al.*, 1987) works using primer chemistry (see below) and is coupled to a computer. This automatic laser fluorescent DNA sequencer (ALF DNA Sequencer™) is available from Pharmacia-LKB.

Two types of chemistries are available for fluorescent sequencing. These include fluorescent primer chemistry and fluorescent terminator chemistry. Fluorescent primer chemistry uses labeled primers. The reactions must be carried out in four separate tubes which are then combined and analyzed in a single lane. This is possible when the primer is labeled with four different fluorophores, which are distinguished by the detection system. The reaction conditions are readily adaptable to previous radioactive sequencing reactions and give very good data. Primer chemistry can be somewhat limiting and inefficient for large-scale sequencing projects that require gene walking because of the need to label each primer with four different fluorochromes or run four sequencing lanes per sample. Fluorescent labeling of the primers results in extra cost; therefore, a labeled universal primer is more often used. This requires a universal primer binding site on the DNA (see Section 6.1). Fluorescent terminator chemistry uses four differently labeled terminators and offers several advantages over fluorescent primers. Each DNA template can be sequenced in a single tube and run in the same lane, reducing the work load fourfold. Only DNA fragments that have incorporated a terminator are labeled. Therefore, nonspecific termination caused by premature release of the polymerase will not be detected resulting in a lower background signal. Any good nonlabeled primer can be used, making large projects and especially gene walking fast and efficient.

7.1.1. Protocol for Automated Direct Sequencing of PCR Products

We have used the Applied Biosystems Model 373A DNA fluorescent sequencer (Perkin Elmer, Applied Biosystems Division, Foster City, CA) for direct sequencing of PCR fragments. It is important to have a clean template. PCR fragments are cleaned up before sequencing using Centri-Sep columns (Princeton Separations, Inc., Adelphia, NJ) or by agarose gel extraction using Gene-Clean (Bio-Rad).

Sequencing is normally performed using the ABI Prism™ Ready reaction Dye-deoxy terminator cycle sequencing kit (Perkin Elmer, Applied Biosystems Division). The standard conditions used with *Taq* polymerase are:

- 2 μl 5× buffer [100 mM $(NH_4)_2SO_4$, 400 mM Tris-HCl, pH 9.0, 10 mM $MgCl_2$]
- 1 ml of 20× dNTP mix (750 mM dITP, 150 μM each dATP, dCTP, dTTP)
- 1 μl of each (A, C, G, T) dideoxy terminator
- 4 units Amplitaq (Perkin Elmer)
- 5 μl (100–250 ng/μl) of PCR fragment (double stranded)
- 3.2 pmole primer
- dH_2O to 20 μl

Reaction components, except the fragment and the primer, are provided in the kit. They can be premixed for sequencing multiple samples. The cycle conditions are the following:

96°C, 30 sec denaturation
Anneal 50°C (or higher), 15 sec
Extend 60°C, 4 min
25 cycles

The sequencing reactions are purified on a Centri-Sep Spin Column (Princeton Separations), then denatured and electrophoresed.

Use one-half the DNA concentration for single-stranded fragments and reduce the primer concentration by 75%. Primer concentration can be calculated with the following equation (Applied Biosystems, Inc., 1991, Extension primers: Recommendations for sequence selection, synthesis and purification, User Bulletin 19):

$$\text{pmole}/\mu\text{l} = \frac{100 \times A_{260}/\text{ml}}{1.45n\text{A} + 0.75n\text{C} + 1.17n\text{G} + 0.92n\text{T}}$$

where n = number of dATP, dCTP, etc., respectively, and A_{260} = optical density at 260-nm wavelength.

7.1.2. Difficult Templates and Automatic Sequencing

The automatic cycle sequencing reaction can be modified for GC-rich templates, for templates with inverted repeats, and for templates with repetitive DNA. The denaturation temperature is raised to 98°C and the amount of *Taq* polymerase is raised from 4 units to 8–16 units. The buffer is changed to 10 mM KCl, 10 mM $(NH_4)_2SO_4$, 20 mM Tris-HCl, pH 8.8, 20 mM $MgSO_4$, and 0.1% Triton X-100. The number of cycles is increased to 30. The ABI Prism™ Ready reaction Dye-deoxy terminator cycle sequencing kit includes dITP instead of dGTP as a default condition to eliminate band compression caused by hairpin structures. One can try to substitute dGTP with the other analogue, 7-deaza-dGTP. Premature termination products of the sequencing reaction that occurs at GC-rich sequences can be eliminated before electrophoresis by extending the 3′ free hydroxyls with terminal transferase. This technique is normally used in the dye primer chemistry (Fawcett and Bartlett, 1990). Theoretically the premature termination products should not give a signal in the terminator-chemistry sequencing, but we have found that the terminal transferase treatment significantly increases the quality of the sequence for some GC-rich templates even when using dye terminators. A possible explanation for this could be that the unlabeled premature termination products, although not visualized, would indirectly influence the sequencing pattern. Due to their abundance at a certain fragment length, they could alter the running of the labeled, correct termination fragments of similar size. Sequenase enzyme combined with the terminator chemistry generates more uniform peaks in automatic sequencing than *Taq* polymerase. For this reason, in the case of AT-rich templates and templates with homopolymer areas one could get better sequence data with this enzyme.

7.2. Robotization

The need to obtain more and more sequence information from human and animal genomes led to development of robotic workstations (Harrison *et al.*, 1993), which are capable of generating large amounts of sequence data by a single person. Most of the reactions of the ABI fluorescent sequencing can be carried out with terminally biotinylated PCR products semi-automatically in a microtiter plate after immobilization (Schofield *et al.*, 1993). Robotization of even the gel handling steps of sequencing is possible with the capillary gel electrophoresis method (Luckey *et al.*, 1993), which is based on automatic fluorescent sequencing, when the traditional sequencing gels are replaced with gel-filled capillaries.

7.3. Multiplex Sequencing

One interesting concept for multiplying the sequencing data is multiplex sequencing (Church and Kieffer-Higgins, 1988). The fragments are cloned into different plasmid libraries, each library having its own identification tag. One clone from each library is pooled in one common sequencing reaction. Following blotting of the mutiplex samples on the same filter, it is hybridized subsequently with probes corresponding to each sequence tag. The filters can give sequence information even after 50 successive probings.

REFERENCES

Ansorge, W., Sproat, B., Stegemann, J., Schwager, C., and Zenke, M. (1987). Automated DNA sequencing: Ultrasensitive detection of fluorescent bands during electrophoresis. *Nucleic Acids Res.* **15**:4593–4602.

Aslanidis, C., and de Jong, P. J. (1990). Ligation-independent cloning of PCR products (LIC-PCR). *Nucleic Acids Res.* **18**:6069–6074.

Bachmann, B., Lüke, W., and Hunsmann, G. (1990). Improvement of PCR amplified sequencing with the aid of detergents. *Nucleic Acids Res.* **18**:1309.

Beck, S., O'Keeffe, T., Coull, J. M., and Köster, H. (1989). Chemiluminescent detection of DNA: Application for DNA sequencing and hybridization. *Nucleic Acids Res.* **17**:5115–5123.

Berg, E. S., and Olaisen, B. (1994). Hybrid PCR sequencing: Sequencing of PCR products using a universal primer. *BioTechniques* **17**:896–901.

Bevan, I. S., Rapley, R., and Walker, M. R. (1992). Sequencing of PCR-amplified DNA. *PCR Methods Appl.* **1**:222–228.

Boyd, A. C. (1993). Turbo cloning: A fast, efficient method for cloning PCR products and other blunt-ended fragments into plasmids. *Nucleic Acids Res.* **21**:817–821.

Carothers, A. M., Urlaub, G., Mucha, J., Grunberger, D., and Chasin, L. A. (1989). Point mutation analysis in a mammalian gene: Rapid preparation of total RNA, PCR amplification of cDNA, and Taq-sequencing by a novel method. *BioTechniques* **7**:494–499.

Casanova, J.-L., Pannetier, C., Jaulin, C., and Kourilsky, P. (1990). Optimal conditions for directly sequencing double-stranded PCR products with Sequenase. *Nucleic Acids Res.* **18**:4028.

Cease, K. B., and Lohff, C. J. (1993). A vector for facile PCR product cloning and modification generating any desired 4-base 5′ overhang: pRPM. *BioTechniques* **14**:250–255

Cha, J., Bishai, W., and Chandrasegaran, S. (1993). New vectors for direct cloning of PCR products. *Gene* **136**: 369–370.

Chen, E. Y., and Seeburg, P. H. (1985). Supercoil sequencing: A fast and simple method for sequencing plasmid DNA. *DNA* **4**:165–170.

Chuang, S.-E., Wang, K.-C., and Cheng, A.-L. (1995). Single-step direct cloning of PCR products. *Trends Genet.* **11**:7–8.

Church, G. M., and Kieffer-Higgins, S. (1988). Multiplex DNA sequencing. *Science* **240**:185–188.

Clark, J. M. (1988). Novel non-templated nucleotide addition reactions catalyzed by procaryotic and eucaryotic DNA polymerases. *Nucleic Acids Res.* **16**:9677–9686.

Costa, G. L., and Weiner, M. P. (1994). Protocols for cloning and analysis of blunt-ended PCR-generated DNA fragments. *PCR Methods Appl.* **3**:S95–S106.

Costa, G. L., Grafsky, A., and Weiner, M. P. (1994). Cloning and analysis of PCR-generated DNA fragments. *PCR Methods Appl.* **3**:338–345.

Crouse, J., and Amorese, D. (1986). Double digestion of multiple cloning site. *Focus* (Bethesda Research Laboratories) **8**:9.

Day, P. J. R., and Walker, M. R. (1991). Sequencing self-ligated PCR products using 3′ over-hangs generated by specific cleavage of dUTP by uracil-DNA glycosylase. *Nucleic Acids Res.* **19**:6959.

De Bellis, G., Manoni, M., Pergolizzi, R., Redolfi, M. E., and Luzzana, M. (1992). A more stringent choice of primers can improve the performance of fluorescent automated DNA sequencers. *BioTechniques* **13**:892–897.

Dierick, H., Stul, M., De Kelver, W., Marynen, P., and Cassiman, J.-J. (1993). Incorporation of dITP or 7-deaza dGTP during PCR improves sequencing of the product. *Nucleic Acids Res.* **21**:4427–4428.

Douglas, A. M., Georgalis, A. M., and Atchinson, B. A. (1993). Direct sequencing of double-stranded PCR products incorporating a chemiluminescent detection procedure. *BioTechniques* **14**:824–828.

Dowton, M., and Austin, A. D. (1993). Direct sequencing of double-stranded PCR products without intermediate fragment purification; digestion with mung bean nuclease. *Nucleic Acids Res.* **21**:3599–3600.

Dowton, M., and Austin, A. D. (1994). A simple method for finding optimal conditions for the direct sequencing of PCR products. *BioTechniques* **16**:816–817.

Engelke, D. R., Hoener, P. A., and Collins, F. S. (1988). Direct sequencing of enzymatically amplified human genomic DNA. *Proc. Natl. Acad. Sci. USA* **85**:544–548.

Fawcett, T. W., and Bartlett, S. G. (1990). An effective method for eliminating "artifact banding" when sequencing double-stranded DNA templates. *BioTechniques* **8**:46–48.

Gal, S., and Hohn, B. (1990). Direct sequencing of double-stranded DNA PCR products via removing the complementary strand with single-stranded DNA of an M13 clone. *Nucleic Acids Res.* **18**:1076.

Gibbs, R. A., Nguyen, P.-N., Edwards, A., Civitello, A. B., and Caskey, C. T. (1990). Multiplex DNA deletion detection and exon sequencing of the hypoxanthine phosphoribosyltransferase gene in Lesch–Nyhan families. *Genomics* **7**:235–244.

Gyllensten, U. B. (1989). PCR and DNA sequencing. *BioTechniques* **7**:700–708.

Gyllensten, U. B., and Erlich, H. A. (1988). Generation of single-stranded DNA by the polymerase chain reaction and its application to direct sequencing of the HLA-DQA locus. *Proc. Natl. Acad. Sci. USA* **85**:7652–7656.

Haqqi, T. M. (1992). Direct ligation of PCR products for cloning and sequencing. *Nucleic Acids Res.* **20**:6427.

Harrison, D., Baldwin, C., and Prockop, D. J. (1993). Use of an automated workstation to facilitate PCR amplification, loading agarose gels and sequencing of DNA templates. *BioTechniques* **14**:88–97.

Hedden, V., Simcox, M., Callen, W., Scott, B., Cline, J., Nielson, E., Mathur, E., and Kretz, K. (1992). Superior sequencing: Cyclist™ Exo-Pfu DNA Sequencing Kit. *Strat. Mol. Biol.* **5**:79.

Higuchi, R. G., and Ochman, H. (1989). Production of single-stranded DNA templates by exonuclease digestion following polymerase chain reaction. *Nucleic Acids Res.* **17**:5865.

Hillier, L., and Green, P. (1991). OSP: A computer program for choosing PCR and DNA sequencing primers. *PCR Methods Appl.* **1**:124–128.

Holton, T. A., and Graham, M. W. (1991). A simple and efficient method for direct cloning of PCR products using ddT-tailed vectors. *Nucleic Acids Res.* **19**:1156.

Hsiao, K. (1993). Exonuclease III induced ligase-free directional subcloning of PCR products. *Nucleic Acids Res.* **21**:5528–5529.

Hultman, T., Ståhl, S., Hornes, E., and Uhlén, M. (1989). Direct solid phase sequencing of genomic and plasmid DNA using magnetic beads as solid support. *Nucleic Acids Res.* **17**:4937–4946.

Hultman, T., Bergh, S., Moks, T., and Uhlén, M. (1991). Bidirectional solid-phase sequencing of in vitro-amplified plasmid DNA. *BioTechniques* **10**:84–93.

Ichihara, Y., and Kurosawa, Y. (1993). Construction of new T vectors for direct cloning of PCR products. *Gene* **130**:153–154.

Innis, M. A., Myambo, K. B., Gelfand, D. H., and Brow, M. A. D. (1988). DNA sequencing with Thermus aquaticus DNA polymerase and direct sequencing of polymerase chain reaction-amplified DNA. *Proc. Natl. Acad. Sci. USA* **85**:9436–9440.

Kaluz, S., and Flint, A. P. F. (1994). Ligation-independent cloning of PCR products with primers containing nonbase residues. *Nucleic Acids Res.* **22**:4845.

Kaluz, S., and Kölble, K., and Reid, K. B. M. (1992). Directional cloning of PCR products using exonuclease III. *Nucleic Acids Res.* **20**:4369–4370.

Kaspar, P., Zadrazil, S., and Fabry, M. (1989). An improved double-stranded DNA sequencing method using gene 32 protein. *Nucleic Acids Res.* **17**:3616.

Kaufman, D. L., and Evans, G. A. (1990). Restriction endonuclease cleavage at the termini of PCR products. *BioTechniques* **9**:304–306.

Khorana, S., Gagel, R. F., and Cote, G. J. (1994). Direct sequencing of PCR products in agarose gel slices. *Nucleic Acids Res.* **22**:3425–3426.

Kretz, K. A., Carson, G. S., and O'Brien, J. S. (1989). Direct sequencing from low-melt agarose with Sequenase. *Nucleic Acids Res.* **17**:5864.

Kretz, K., Callen, W., and Hedden, V. (1994). Cycle sequencing. *PCR Methods Appl.* **3**:S107–S112.

Kuijper, J. L., Wiren, K. M., Mathies, L. D., Gray, C. L., and Hagen, F. S. (1992). Functional cloning vectors for use in directional cDNA cloning using cohesive ends produced with T4 DNA polymerase. *Gene* **112**:147–155.

Lee, J.-S. (1991). Alternative dideoxy sequencing of double stranded DNA by cyclic reactions using Taq polymerase. *DNA Cell Biol.* **10**:67–73.

Liu, Y.-G., Mitsukawa, N., and Whittier, R. F. (1993). Rapid sequencing of unpurified PCR products by thermal asymmetric PCR cycle sequencing using unlabeled sequencing primers. *Nucleic Acids Res.* **21**:3333–3334.

Liu, Z., and Schwartz, L. M. (1992). An efficient method for blunt-end ligation of PCR products. *BioTechniques* **12**:28–30.

Lohff, C. J., and Cease, K. B. (1992). PCR using a thermostable polymerase with 3′ to 5′ exonuclease activity generates blunt products suitable for direct cloning. *Nucleic Acids Res.* **20**:144.

Lorens, J. B. (1991). Rapid and reliable cloning of PCR products. *PCR Methods Appl.* **1**:140–141.

Luckey, J. A., Drossmann, H., Kostichka, T., and Smith, L. M. (1993). High-speed DNA sequencing by capillary gel electrophoresis. *Methods Enzymol.* **218**:154–172.

Marchuk, D., Drumm, M., Saulino, A., and Collins, F. S. (1991). Construction of T-vectors, a rapid and general system for direct cloning of unmodified PCR products. *Nucleic Acids Res.* **19**:1154.

Maxam, A. M., and Gilbert, W. (1977). A new method for sequencing DNA. *Proc. Natl. Acad. Sci. USA* **74**: 560–564.

Messing, J. (1983). New M13 vectors for cloning. *Methods Enzymol.* **101**:20–78.

Middendorf, L. R., Bruce, J. C., Bruce, R.C., Eckles, R. D., Grone, D. L., Roemer, S. C., Sloniker, G. D., Steffens, D. L., Sutter, S. L., Brumbaugh, J. A., Patonay, G. (1992). Continuous, on-line DNA sequencing using a versatile infrared scanner/electrophoresis apparatus. *Electrophoresis* **13**:487–494.

Mitchell, D. B., Ruggli, N., and Tratschin, J.-D. (1992). An improved method for cloning PCR fragments. *PCR Methods Appl.* **2**:81–82.

Mitchell, L. G., and Merill, C. R. (1989). Affinity generation of single-stranded DNA for dideoxy sequencing following the polymerase chain reaction. *Anal. Biochem.* **178**:239–242.

Mitsuhashi, M., Cooper, A., Ogura, M., Shinagawa, T., Yano, K., and Hosokawa, T. (1994). Oligonucleotide probe design—A new approach. *Nature* **367**:759–761.

Mizusawa, S., Nishimura, S., and Seela, F. (1986). Improvement of the dideoxy chain termination method of DNA sequencing by use of deoxy-7-deazaguanosine triphosphate in place of dGTP. *Nucleic Acids Res.* **14**:1319–1324.

Murray, V. (1989). Improved double-stranded DNA sequencing using the linear polymerase chain reaction. *Nucleic Acids Res.* **17**:8889.

Nicolas, O., and Laliberté, J.-F. (1991). The use of PCR for cloning of large cDNA fragments of turnip mosaic potyvirus. *J. Virol. Methods* **32**:57–66.

Nisson, P. E., Rashtchian, A., and Watkins, P. C. (1991). Rapid and efficient cloning of Alu-PCR products using uracil DNA glycosylase. *PCR Methods Appl.* **1**:120–123.

Oliner, J. D., Kinzler, K. W., and Vogelstein, B. (1993). In vivo cloning of PCR products in E. coli. *Nucleic Acids Res.* **21**:5192–5197.

Prober, J. M., Trainor, G. L., Dam, R. J., Hobbs, F. W., Robertson, C. W., Zagursky, R. J., Cocuzza, A. J., Jensen, M. A., and Baumeister, K. (1987). A system for rapid DNA sequencing with fluorescent chain-terminating dideoxynucleotides. *Science* **238**:336–341.

Rao, V. B. (1994a). Direct sequencing of polymerase chain reaction-amplified DNA. *Anal. Biochem.* **216**:1–14.

Rao, V. B. (1994b). Strategies for direct sequencing of PCR-amplified DNA. *PCR Methods Appl.* **4**:S15–S23.

Rao, V. B., and Saunders, N. B. (1992). A rapid polymerase-chain-reaction-directed sequencing strategy using a thermostable DNA polymerase from Thermus flavus. *Gene* **113**:17–23.

Reeves, S. A., Rubio, M.-P., and Louis, D. N. (1995). General method for PCR amplification and direct sequencing of mRNA differential display products. *BioTechniques* **18**:18–20.

Sambrook, J., Fritsch, E. F., and Maniatis, T. (1989). Synthetic oligonucleotide probes, in: *Molecular Cloning: A Laboratory Manual*, 2nd ed., Cold Spring Harbor Laboratory Press, Cold Spring Harbor, NY.

Sanger, F., Coulson, A. R., Barrell, B. G., Smith, A. J. H., and Roe, B. A. (1980). Cloning in single-stranded bacteriophage as an aid to rapid DNA sequencing. *J. Mol. Biol.* **143**:161–178.

Sanger, F., Nicklen, S., and Coulson, A. R. (1987). DNA sequencing with chain-terminating inhibitors. *Proc. Natl. Acad. Sci. USA* **74**:5463–5467.

Sarkar, G., and Sommer, S. S. (1988). RNA amplification with transcript sequencing (RAWTS). *Nucleic Acids Res.* **16**:5197.

Scharf, S. J., Horn, G. T., and Erlich, H. A. (1986). Direct cloning and sequence analysis of enzymatically amplified genomic sequences. *Science* **233**:1076–1078.

Schofield, J. P., Jones, D. S. C., and Vaudin, M. (1993). Fluorescent and radioactive solid-phase dideoxy sequencing of polymerase chain reaction products in microtiter plates. *Methods Enzymol.* **218**:93–103.

Smith, C., Day, P. J. R., and Walker, M. R. (1993). Generation of cohesive ends on PCR products by UDG-mediated excision of dU, and application for cloning into restriction digest-linearized vectors. *PCR Methods Appl.* **2**:328–332.

Smith, L. M., Sanders, J. Z., Kaiser, R. J., Hughes, P., Dodd, C., Connell, C. R., Heiner, C., Kent, S. B. H., and Hood, L. E. (1986). Fluorescence detection in automated DNA sequence analysis. *Nature* **321**:674–679.

Stamm, S., and Longo, F. M. (1990). Direct sequencing of PCR products using the Maxam–Gilbert method. *GATA* **7**:142–143.

Stoflet, E. S., Koeberl, D. D., Sarkar, G., and Sommer, S. S. (1988). Genomic amplification with transcript sequencing. *Science* **239**:491–494.

Stoker, A. W. (1990). Cloning of PCR products after defined cohesive termini are created with T4 DNA polymerase. *Nucleic Acids Res.* **18**:4290.

Szabó, P. E., and Mann, J. R. (1995). Biallelic expression of imprinted genes in the mouse germline: Implications for erasure, establishment, and mechanisms of genomic imprinting. *Genes Dev.* **9**:1857–1868.

Tabor, S., and Richardson, C. C. (1987). DNA sequence analysis with a modified bacteriophage T7 DNA polymerase. *Proc. Natl. Acad. Sci. USA* **84**:4767–4771.

Tabor, S., and Richardson, C. C. (1989). Effect of manganese ions on the incorporation of dideoxynucleotides by bacteriophage T7 DNA polymerase and Escherichia coli DNA polymerase I. *Proc. Natl. Acad. Sci. USA* **86**:4076–4080.

Tahara, T., Kraus, J. P., and Rosenberg, L. (1990). Direct DNA sequencing of PCR amplified genomic DNA by the Maxam–Gilbert method. *BioTechniques* **8**:366–368.

Testori, A., Listowsky, I., and Sollitti, P. (1994). Direct cloning of unmodified PCR products by exploiting an engineered restriction site. *Gene* **143**:151–152.

Tracy, T. E., and Mulcahy, L. S. (1991). A simple method for direct automated sequencing of PCR fragments. *BioTechniques* **11**:68–75.

Upcroft, P., and Healey, A. (1987). Rapid and efficient method for cloning of blunt-ended DNA fragments. *Gene* **51**:69–75.

Wanner, R., Tilmans, I., and Mischke, D. (1992). Avoiding strand reassociation in direct sequencing of double-stranded PCR products with thermolabile polymerases. *PCR Methods Appl.* **1**:193–194.

Weiner, M. P. (1993). Directional cloning of blunt-ended PCR products. *BioTechniques* **15**:502–505.

Werle, E., Schneider, C., Renner, M., Vlker, M., and Fiehn, W. (1994). Covenient, single-step, one-tube purification of PCR products for direct sequencing. *Nucleic Acids Res.* **22**:4354–4355.

Winship, P. R. (1989). An improved method for directly sequencing PCR amplified material using dimethyl sulphoxide. *Nucleic Acids Res.* **17**:1266.

Wrischnik, L. A., Higuchi, R. G., Stoneking, M., Erlich, H. A., Arnheim, N., and Wilson, A. C. (1987). Length mutations in human mitochondrial DNA: Direct sequencing of enzymatically amplified DNA. *Nucleic Acids Res.* **15**:529–542.

Zhang, W., Hu, G., and Deisseroth, A. (1991). Improvement of PCR sequencing by formamide. *Nucleic Acids Res.* **19**:6649.

Zimmermann, J., Voss, H., Schwager, C., Stegemann, J., Erfle, H., Stucky, K., Kristensen, T., and Ansorge, W. (1990). A simplified protocol for fast plasmid DNA sequencing. *Nucleic Acids Res.* **18**:1067.

Part III

Mammalian Systems for Mutation Analysis

Chapter 27

Detection and Characterization of Mutations in Mammalian Cells with the pSP189 Shuttle Vector System

Michael M. Seidman

1. INTRODUCTION

1.1. The Development of Shuttle Vectors for Studying Mammalian Mutagenesis: Problems

In the early 1980s the attraction of recombinant DNA technology was beginning to be felt in fields that had previously been refractory to molecular analysis. One such field was mammalian cell mutagenesis. Cloning technology offered the opportunity to recover mutant genes for which there were effective selection protocols (such as HPRT). Sequence determination of the entire gene could then display the nature of the mutations. Despite the feasibility of such protocols, the time and effort required discouraged most investigators. An alternative approach, based on shuttle vectors, appeared more attractive. These were plasmids whose design was based on advances in two fields. The extensive analysis of DNA tumor viruses such as SV40 had defined the genetic information necessary for viral replication in monkey and human cells. The biology and molecular biology of bacterial plasmids was also well developed. Thus, constructs with SV40 virus replication functions (the T-antigen gene and an origin of replication) linked to components of bacterial plasmids (the plasmid origin and a drug resistance marker) had been shown to replicate in monkey cells, and could be recovered and introduced into bacteria (Peden *et al.*, 1980; Lusky and Botchan, 1981; these references also discuss the problem and resolution of the replication poison sequence found on pBR322). These experiments were primarily demonstrations of principle; there was no actual use of the shuttle technology. It seemed logical, however, to those interested in mammalian mutagenesis, to add a third component, a bacterial marker gene. The resultant vector, perhaps treated with a DNA damaging agent, would then be introduced into mammalian cells, allowed to replicate, recovered, and reintroduced into bacteria.

Michael M. Seidman • OncorPharm, Gaithersburg, Maryland 20877.

Technologies for Detection of DNA Damage and Mutations, edited by Gerd P. Pfeifer, Plenum Press, New York, 1996.

Bacterial colonies with mutations in the marker gene would be recognized by standard microbiological selection or screening procedures, and the nature of the mutations identified by direct sequence analysis. This logic was quite compelling, and a number of groups set out to develop the technology.

The first two to publish on these systems reported a disquieting observation which had not been considered in the original strategy. They found that during transfection and replication in mammalian cells a high frequency of plasmids with mutations in the marker gene appeared (Razzaque *et al.*, 1983; Calos *et al.*, 1983). The frequency (about 1% with the galK and lacI markers) was too high to permit clean experiments with DNA damaging agents. Analysis of the mutant plasmids indicated that the majority of the mutations were deletions, although insertions of cellular DNA, as well as point mutations, were found. These observations [confirmed by others such as Sarker *et al.* (1984)] appeared to discredit the entire approach.

1.2. Solutions

The two groups took somewhat different approaches to this problem. Calos and her collaborators found a cell line (Ad293) with a fortuitously low spontaneous background frequency (Lebkowski *et al.*, 1984). They were able to generate unambiguous mutational spectra with the original *lacI* vector in these cells, using several different mutagens (Lebkowski *et al.*, 1985). The value of this approach was that the wealth of data from the *lacI* system in *E. coli* could be compared with the same marker in human cells (Hsia *et al.*, 1989).

We, on the other hand, were interested in being able to do experiments in a variety of different cell types and lines. Our strategy required changing the design of the vector. The new construct, which arose from a collaboration with Kathleen Dixon and Michael Berman, carried a small gene, the *supF* suppressor tRNA, nested between two sequences necessary for replication and survival in bacteria (Seidman *et al.*, 1985). These were the plasmid origin of replication and the gene for ampicillin resistance. Deletions which would inactivate the marker would, in many instances, inactivate one of these necessary functions. These plasmids would simply drop out of the population at the bacterial screening stage. The new design proved to be successful. We found that the spontaneous background frequency dropped 50- to 100-fold in the same cells that had given such high levels in the initial experiments. In addition to the advantage of the low background, the small size of the marker gene made sequence analysis of mutants rapid. This vector, pZ189, has been used by many groups to characterize a broad range of mutagens and carcinogens. The key to the success of the vector was the small deletion target. This target size was reduced further with a derivative of pZ189, pS189, which had an even lower frequency of spontaneous mutagenesis (Seidman, 1989). Since then several other groups have developed workable shuttle systems. Some are based on SV40 (Menck *et al.*, 1987) while others have used Epstein–Barr virus genes and constructed vectors which could be maintained as stable episomes (Drinkwater and Klinedinst, 1986).

1.3. The Use of a Signature Sequence

Mutational spectra, regardless of marker gene, commonly show hot spots, where mutagenesis is much more common than other sites. This is true for the *supF* marker and underlies a problem in interpretation of data derived from the shuttle vector system. Transfection of adducted plasmid into the host cells brings many copies of the plasmid into a given cell. Mutagenesis occurs during the first few rounds of plasmid replication and the products of that mutagenesis are amplified during the remainder of the replication period (36–48 hr). Plasmids with an identical mutation could represent independent mutational events or siblings of an initial event. Typically, with UV as a mutagen, the harvest from a single transfected plate of human cells would contain many plasmids with the same mutation located at one of several hot spots. In order to eliminate the possibility of counting siblings, we routinely discarded all but one mutation from a given transfection. The resultant tallies clearly underrepresented the frequency of events at true hot spots.

In order to solve this problem, we developed the signature sequence. This was an oligonucleotide which carried a run of eight positions in which all possible bases were inserted during synthesis. The oligo was copied by Klenow polymerase so as to prepare a population of oligonucleotides which were highly heterogeneous. The oligo was then restriction digested so as to generate sticky ends and inserted downstream of the *supF* gene at the appropriate restriction site. The colonies from the transformation of the ligation mix were pooled and plasmid prepared from the population. This gave a plasmid prep that was clonal with respect to the *supF* gene and highly heterogeneous with respect to the "signature" oligonucleotide. The sequence of the signature can be read from the same run used for mutation identification (Parris and Seidman, 1992).

Results from our lab and several others that have used the signature system (in the plasmid prep pSP189) indicate that siblings (those plasmids with the same mutation and same signature sequence) are rare with strong mutagens and much more common with weak mutagens. However, an experiment with poor plasmid yields will also show appreciable sibling levels regardless of the mutagen. The use of the signature greatly increases the reliability of the counting of mutations and the assessment of hot spot intensity. Furthermore, it permits entire spectra to be done with the plasmid harvest from a single transfected plate of mammalian cells.

2. METHODS

2.1. Preparation of Signature Sequences

The pSP189 vector has a *Sac*I site located downstream of the *supF* gene and transcription terminator. The site is used for insertion of the signature oligonucleotide. A 38-mer oligonucleotide is synthesized by trityl off cyanoethyl phosphoramidite chemistry on an Applied Biosystems DNA synthesizer. A variable region is created 20 bases inside the oligo by supplying an equal mixture of all four precursors at each of eight synthesis steps. The use of the mixture is important. Simply directing the instrument to supply each precursor from separate bottles will produce oligos strongly biased toward the first base in. A 20-mer oligo is synthesized to act as a primer. After annealing, 1 μg of primer template complex is extended by Klenow polymerase in a buffer (10 mM Tris-HCl, pH 7.0, 10 mM $MgCl_2$, 1 mM DTT) compatible with the *Sac*I restriction enzyme, supplemented with all four dNTPs and a trace of ^{32}P dNTP. The reaction is allowed to proceed at room temperature for 15 min followed by heat inactivation at 75°C for 20–30 min. After cooling, the *Sac*I enzyme is added and incubated at 37°C for 1–2 hr. The large digestion product is purified away from the small end fragments by preparative electrophoresis on a 12.5% native acrylamide gel. The label can be followed as a monitor of the incorporation and cleavage reactions. The recipient pSP189 vector is digested with *Sac*I and then dephosphorylated. A ratio of 0.5 μg of oligo to 100–200 ng of vector in the ligation mixture provides a few hundredfold molar excess of oligo and should yield a population of vectors almost all of which contain one or more copies of a signature. Insertion of one copy in either orientation produces $2(4^8) = 131,072$ possible members in the plasmid population. The colonies from 10–12 dishes are pooled and plasmid isolated from the mixture. Although it may seem obvious in the context of this discussion, we find it useful to caution laboratories using the system for the first time to resist the urge to purify a single colony prior to plasmid purification.

2.2. Plasmid Modification

Most of our experiments have been done with plasmid modified *in vitro* with the mutagen of choice. This permits considerable control over the chemistry of the process and yields much stronger signal-to-background ratios than if modification follows transfection. It is important to be sensitive to the integrity of the DNA following modification. If the procedure introduces nicks or breaks in the DNA, these will have mutagenic consequences independent of any adducts (Seidman *et al.*, 1987). If the DNA is broken during the reaction, it may be necessary to repurify over CsCl to isolate the unbroken form I.

2.3. Cells and Transfection Method

A broad range of human cell lines has been used with this vector system. These have been derived from repair-proficient donors and several repair deficiency syndromes. Different cell types have been successfully used including fibroblasts, T and B lymphocytes, and epithelial cells. Most work has been with established cell lines, generally transformed or tumor derived. However, it is possible to work with activated primary T cells and so analysis of patient cells is feasible (Parris and Kraemer, 1992). The requirements for an appropriate cell host are two: the line must support SV40 T-antigen-driven replication and the cells must be transfectable. The replication restriction limits the application of the T-antigen-based vectors to cells permissive for SV40 viral replication. A derivative of the vector has been made in which the SV40 T-antigen gene has been replaced by the T antigen from polyoma virus and this has been used for studies in rodent cells (Zernik-Kobak *et al.*, 1990).

Different transfection methods have been used with the vector system. The choice reflects the vagaries of a particular cell line and the experience of the investigator. It is important to stay with one protocol when doing comparative studies. We have described some variability in the mutational spectra as a function of cell manipulation including the method of transfection (see below). Depending on the cell type we generally use either electroporation or a lipid-based transfection reagent. Our protocol for transfection of the xeroderma pigmentosum fibroblast line XP12Be is as follows:

Plate about 10^6 cells in 10 ml medium in a 100-mm dish the day before transfection. On the day of transfection, combine in polystyrene tubes (Falcon 2058) 20 μl of Transfectase (BRL) with 180 ml of the BRL transfection medium Optimem. In a separate tube, mix 5 μg of plasmid DNA (modified or unmodified) with Optimem to a total volume of 200 μl. Mix the two samples and incubate at room temperature for 10–15 min. Wash the cells 2× with 37°C Optimem. Add 1.6 μl of Optimem to the DNA-transfectase mixture and add to the cells. Place in the incubator for 90 min. The time in the incubator is a function of the specific cell line. For some lines we incubate for several hours. Remove the DNA lipid mixture from the cells and add growth medium. Incubate for 48–72 hr. The time of incubation is again dependent on the cells. For most cells we harvest after 48 hr.

2.4. Plasmid Harvest

Our best yields and cleanest preparations have been with an alkaline extraction protocol. We use the same protocol for extraction from both mammalian and bacterial cells:

Suspend the cells (from a 100-mm dish) in 100 μl of a 25 mM Tris-HCl pH 7.0, 10 mM EDTA, 50 mM glucose in a 1.5-ml microfuge tube. Add 200 μl of 0.2 mM NaOH, 1% SDS, mix by inversion several times, and immediately add 150 μl of 3 M K, 5 M acetate pH 5.5. After incubation on ice for 10–15 min, spin at 12,000 rpm in the cold. Remove the pellet with a toothpick and add 220 μl of 7.5 NH_4Oac. Mix and incubate on ice for 30 min. Spin in the cold and transfer the supernatant to a fresh tube. Add an equal volume of isopropanol, vortex, and stand 30 min. Then spin down the precipitate, wash with 70% EtOH, and dry. The sample is suspended in 50 μl of H_2O supplemented with the manufacturer's 10× Dpnl buffer, 5–10 units of Dpnl added, and incubated at 37°C for 30 min. This step eliminates input DNA (Peden *et al.*, 1980). RNase is then added and the incubation continued for another 30 min. One-half volume of 7.5 M NH_4Oac is added and the DNA precipitated by addition of 2 vol of EtOH. After spinning, the pellet is reprecipitated out of NH_4Oac and EtOH. The dried pellet is suspended in 10 μl of H_2O. This can be stored frozen or a portion used directly.

2.5. Electroporation of Competent Bacteria

We use the MBM7070 tester strain to screen progeny plasmids. Log-phase bacteria are rendered competent by a series of three cold washes in 10% glycerol, suspended in ice-cold 10% glycerol at an OD_{550} of 200–250. The cells should be aliquoted, frozen in a dry ice EtOH bath, and stored at −80°C. Electroporation with the Bio-Rad device can be done with 1 μl of sample and 20–25 μl of bacteria in the 0.2-cm cuvette with settings of 250 V, 25 μF, 200 ohms. We also have used the BRL electroporator on the medium setting. The cells are shaken for 15 min at 37°C and then plated. We reduced the expression time after the electroporation from the conventional 1 hr in an effort to discourage the generation of siblings during this step. We have not seen any reduction in yield with the shorter time before plating. The cells are plated on LB agar supplemented with ampicillin and X-gal and IPTG. The plates are incubated at 42°C overnight. The next day white and light blue colonies are selected for analysis. It is extremely important to be very careful to avoid contamination with plasmids that are not part of the experiment. It is easy to detect white colonies that contain plasmids from unrelated experiments, or even from other labs. We aliquot solutions and use plugged pipet tips, etc.

2.6. Thermal Cycle Sequence Analysis of Mutant Plasmids

Colonies are streaked out and used within a week. Those between 3 and 5 mm in diameter are suspended in 15 μl of lysis solution (10 mM Tris-HCl, pH 7.5, 1 mM EDTA, 100 μg/ml proteinase K). Heat for 15 min at 65°C, then at 80°C for 15 min. Centrifuge in a microfuge and transfer 9 μl of the supernatant to a new tube. Prepare an end-labeled sequencing primer by incubation, at 37°C for 60 min, of 15 ng of primer with [γ-^{32}P]-ATP (1 μl of 3000 Ci/mmole) with 10 units of T4 polynucleotide kinase in a final volume of 5 μl (this is for one reaction). Heat the reaction mixture at 65°C for 15 min. The template and primer components are then introduced into the protocols described by various manufacturers of thermal cycle sequencing kits. We have had good luck with the kit from Epicentre (Sequitherm cycle sequencing). Two microliters of the appropriate termination solution is placed in a well of a multiwell plate or in a thermal cycler tube. The following reagents are mixed: 2 μl of labeled primer, 2.5 μl of the 10× sequencing buffer, 2 μl of water, and 1 μl of Sequitherm polymerase; 7.5 μl of this mix is added to the 9 μl of lysate supernatant and 3-μl aliquots of this added to each of the wells or tubes with the termination solutions. The wells or tubes are covered with mineral oil and centrifuged briefly. The reaction vessels are placed in a thermal cycler preheated to 95°C and a standard three-step cycle run. Twenty to thirty cycles should be sufficient. It is possible to start the protocol in the morning and have a developed film by the end of the day. Using this approach a diligent worker can generate a complete mutational spectrum in less than 2 weeks.

2.7. Choice of Cell Line and Transfection Technology

We have analyzed mutational spectra from a number of cell lines and cells from different patient backgrounds (repair-proficient, xeroderma, etc.). It is tempting to interpret differences between two patterns in terms of the most obvious (and most newsworthy) differences between the cells. However, in our experience the differences may not be amenable to simple interpretation. For example, we initially described a UV hot spot in XP fibroblasts that was not seen in repair-proficient fibroblasts (Bredberg *et al.*, 1986). However, when we repeated the experiment in repair-proficient lymphoblasts, we saw the same hot spot (Seetharam *et al.*, 1991). Consequently, it is important to examine cells from several different sources from a particular syndrome and compare the results with data from several different control cells before drawing conclusions.

It is also important to be sensitive to the transfection technology. Experiments with DNA damage and cells are influenced by an unresolvable uncertainty. Treatment of cells with DNA damaging agents introduces adducts into DNA and also triggers a stress response. The processing of the damage occurs in the context of the stressed cells and the wealth of changes that that implies. While treatment of the shuttle vector with damaging agents *in vitro* might seem to avoid this problem, it should be noted that the transfection process is also stressful. We have shown that deliberate stressful treatments of the cells can modulate the hot spot patterns, and that different transfection protocols can also influence the patterns (Seetharam and Seidman, 1992). We think it likely that the biochemistry of mutagenesis is sensitive to the cellular environment, just as are many other biochemical processes. Consequently, it is necessary to standardize the transfection protocol for any experimental series.

3. APPLICATIONS OF SHUTTLE VECTOR TECHNOLOGY

The development of PCR-based approaches for determining mutations in chromosomal genes has greatly expanded the options available to the student of mutagenesis. Thus, it is important to understand the advantages and disadvantages of the shuttle vectors. For example, experiments with agents that are not direct modifiers, such as intercalators, have been difficult. In these cases it is necessary to treat the cells after vector transfection, and it has proven difficult to recover satisfactory signal-to-background ratios. Another concern is the integrity of the DNA after modification. If an agent introduces nicks or breaks, then these will provoke a variety of mutagenic events (Seidman *et al.*, 1987). Therefore, some caution must be exercised when interpretating data from experiments with agents such as X-radiation. The shuttle vectors as they are currently configured are probably not appropriate for studying the influence of transcription-linked strand-specific repair on mutagenesis. These experiments are better done with chromosomal targets and PCR-based analysis (Wei *et al.*, 1995).

On the other hand, the vectors do have some advantages that cannot be met by chromosomal markers. The assays are very rapid and so a family of related compounds can be studied without a massive effort. The chemistry of modification can be controlled and the technology is amenable to site-specific constructions. Furthermore, the developments in oligonucleotide synthesis and plasmid construction make it possible to consider an old problem that has received much discussion but little experimental attention. That is the issue of sequence context effects on mutational hot spots. The difficulties of manipulating sequence about a hot spot site while preserving marker gene activity have effectively precluded research on this problem. Recently we have begun to develop a family of variant tRNA genes with single base alterations in the vicinity of hot spots. The variants are active and it is possible to determine the influence of the changes on hot spot activity (Parris *et al.*, 1994; Levy and Seidman, unpublished). The shuttle vector system offers the speed of construction and experimental execution which would not be possible with chromosomal markers.

REFERENCES

Bredberg, A., Kraemer, K. H., and Seidman, M. M. (1986). Restricted ultraviolet mutational spectrum in a shuttle vector propagated in xeroderma pigmentosum cells. *Proc. Natl. Acad. Sci. USA* **83**:8273–8277.

Calos, M. P., Lebkowski, J. S., and Botchan, M. R. (1983). High mutation frequency in DNA transfected into mammalian cells. *Proc. Natl. Acad. Sci. USA* **80**:3015–3019.

Drinkwater, N., and Klinedinst, D. K. (1986). Chemically induced mutagenesis in a shuttle vector with a low background mutant frequency. *Proc. Natl. Acad. Sci. USA* **83**:3402–3406.

Hsia, C. H., Lebkowski, J. S., Leong, P. M., Calos, M., and Miller, J. H. (1989). Comparison of ultraviolet irradiation induced mutagenesis of the lacI gene in E. coli and in human cells. *J. Mol. Biol.* **205**:103–113.

Lebkowski, J. S., DuBridge, R. B., Antell, E. A., Greisen, K. S., and Calos, M. P. (1984). Transfected DNA is mutated in monkey mouse, and human cell lines. *Mol. Cell. Biol.* **4**:1951–1960.

Lebkowski, J. S., Clancy, S., Miller, J. H., and Calos, M. P. (1985). The lac I shuttle: Rapid analysis of mutational specificity of ultraviolet light in human cells. *Proc. Natl. Acad. Sci. USA* **82**:8606–8610.

Lusky, M., and Botchan, M. (1981). Inhibition of SV40 replication in simian cells by specific pBR322 DNA sequences. *Nature* **293**:79–81.

Menck, C. F. M., Sarasin, A., and James, M. R. (1987). SV40 based E. coli shuttle vectors infectious for monkey cells. *Gene* **53**:21–29.

Parris, C. N., and Kraemer, K. H. (1992). Ultraviolet mutagenesis in human lymphocytes: The effect of cellular transformation. *Exp. Cell Res.* **201**:462–469.

Parris, C. N., and Seidman, M. M. (1992). A signature sequence distinguishes sibling and independent mutations in a shuttle vector plasmid. *Gene* **117**:1–5.

Parris, C. N., Levy, D. D., Jessee, J., and Seidman, M. M. (1994). Proximal and distal effects of sequence context on ultraviolet mutational hotspots in a shuttle vector replicated in xeroderma cells. *J. Mol. Biol.* **236**:491–502.

Peden, K. W. C., Pipas, J. M., Pearson-White, S., and Nathans, D. (1980). Isolation of mutants of an animal virus in bacteria. *Science* **209**:1392–1396.

Razzaque, A. S., Mizusawa, H., and Seidman, M. M. (1983). Rearrangement and mutagenesis of a shuttle vector plasmid after passage in mammalian cells. *Proc. Natl. Acad. Sci. USA* **80**:3010–3014.

Sarker, S., Dasgupta, U. B., and Summers, W. C. (1984). Error prone mutagenesis detected in mammalian cells by a shuttle vector containing the supF gene of E. coli. *Mol. Cell. Biol.* **4**:2227–2230.

Seetharam, S., and Seidman, M. M. (1992). Modulation of ultraviolet mutational hotspots by cellular stress. *J. Mol. Biol.* **228**:1031–1036.

Seetharam, S., Kraemer, K. H., Waters, H. L., and Seidman, M. M. (1991). Ultraviolet mutational spectrum in a shuttle vector plasmid propagated in xeroderma pigmentosum lymphoblastoid cells and fibroblasts. *Mutat. Res.* **254**:97–105.

Seidman, M. M. (1989). The development of transient SV40 based shuttle vectors for mutagenesis studies: problems and solutions. *Mutat. Res.* **220**:55–60.

Seidman, M. M., Dixon, K., Razzaque, A., Zagursky, R. J., and Berman, M. L. (1985). A shuttle vector plasmid for studying carcinogen induced point mutations in mammalian cells. *Gene* **38**:233–237.

Seidman, M. M., Bredberg, A., Seetharam, S., and Kraemer, K. H. (1987). Multiple point mutations in a shuttle vector propagated in human cells. Evidence for an error prone polymerase activity. *Proc. Natl. Acad. Sci. USA* **84**:4944–4948.

Wei, D., Maher, V. M., and McCormick, J. J. (1995). Site specific rates of excision repair of benzo[a] pyrene diol epoxide adducts in the hypoxanthine phosphoribosyltransferase gene of human fibroblasts: Correlation with the mutation spectra. *Proc. Natl. Acad. Sci. USA* **92**:2204–2208.

Zernik-Kobak, M., Pirsel, M., Doniger, J., DiPaolo, J. A., Levine, A. S., and Dixon, K. (1990). Polyomavirus based shuttle vectors for studying mechanisms of mutagenesis in rodent cells. *Mutat. Res.* **242**:57–65.

Chapter 28

The *HPRT* Gene as a Model System for Mutation Analysis

Veronica M. Maher and J. Justin McCormick

1. INTRODUCTION

There is now very convincing evidence drawn from many different fields of study to support the hypothesis that carcinogenesis is a multistep process, and that mutations are causally involved in bringing about the changes required for a normal cell to become a malignant cell. (For review see McCormick and Maher, 1994). This understanding has served as our working hypothesis as we investigated the ability of physical or chemical carcinogenic agents to induce mutations in diploid human cells in culture and examined the effect of various DNA repair processes on such mutagenesis. We began by developing methods for culturing diploid human fibroblasts in culture and achieving high cloning efficiencies, using normal DNA repair-proficient fibroblasts, but also cells deficient in one or other DNA repair process, such as nucleotide excision repair derived from patients with genetic predispositions to develop cancer (Maher *et al.*, 1975, 1976). Once cloning efficiencies of 30 to 80% had been achieved, we developed methods to quantify the induction of mutations in the endogenous gene coding for hypoxanthine phosphoribosyltransferase (HPRT) (McCormick and Maher, 1981; Maher *et al.*, 1977, 1979) or diphtheria toxin resistance (Drinkwater *et al.*, 1982), and later for mutations induced in a bacterial gene, *supF*, when carcinogen-treated plasmids are allowed to replicate in human cells (Boldt *et al.*, 1991; Mah *et al.*, 1989, 1991; Yang *et al.*, 1987, 1988). Albertini and his colleagues (O'Neill *et al.*, 1990a,b; Albertini, 1985; Albertini *et al.*, 1982) developed techniques to allow similar studies of mutations in the *HPRT* gene of diploid human peripheral blood T lymphocytes.

The *HPRT* gene is an excellent model gene for studies on the mechanisms of mutagenesis induced by DNA damaging agents and the effect of DNA repair on these mutations and offers many advantages for mutational spectra analysis. Because it is carried on the X chromosome, there is only a single copy in cells from males and a single expressed copy in cells from females, which greatly facilitates detection of cells that have an inactivating mutation in the gene. The *HPRT* gene codes for a nonessential, salvage-pathway enzyme so cells that lack a functional HPRT protein are not at a selective disadvantage when growing in culture medium. However, loss of function of the *HPRT* gene by any kind of mutation—e.g., base substitution, frame shift,

Veronica M. Maher and J. Justin McCormick • Carcinogenesis Laboratory, Departments of Microbiology and of Biochemistry, The Cancer Center, Michigan State University, East Lansing, Michigan 48824.

Technologies for Detection of DNA Damage and Mutations, edited by Gerd P. Pfeifer, Plenum Press, New York, 1996.

deletion, insertion—distributed throughout the gene renders the cell resistant to 6-thioguanine (TG). The gene, composed of >45,000 bp, contains a coding region of only 654 bp distributed over nine exons. It has been the target gene for a very large number of mutagenesis studies in human cells and rodent cell lines. Once Yang *et al.* (1989) developed procedures to amplify *HPRT* cDNA and directly sequence the product to obtain the consensus sequence from as few as 100 human cells from a mutant clone, numerous investigators have used these techniques to determine the kinds and location (spectrum) of mutations induced by various agents in the *HPRT* gene. These data in the coding region, as well as data on mutations located in other areas of the gene have been collected and made available in data bases, such as that of Cariello *et al.* (1992).

Diploid human fibroblasts offer very important advantages to those interested in examining the effect of DNA repair on mutation frequency and/or spectra. First, repair-proficient fibroblasts are readily available from normal donors and fibroblasts deficient in specific pathways of DNA repair are also available for use in comparative studies. Second, diploid human fibroblasts are much more efficient in nucleotide excision repair than are rodent fibroblast cell lines (Tung *et al.*, 1996; Wang *et al.*, 1993; Chen *et al.*, 1992). Third, when diploid human cells reach confluence, they cease replicating (enter the G_0 state), but do not detach from the dish. Therefore, it is relatively easy to prepare large populations of cells in the resting state by density inhibition and mitogen deprivation (Watanabe *et al.*, 1985; Konze-Thomas *et al.*, 1982). When such cells are released from G_0 by being plated at lower density in medium containing serum, they proceed through G_1 and after 16 to 18 hr, enter S phase as a cohort. This elongated G_1 phase permits one to expose cells to DNA damaging agents at various times postrelease and thus manipulate the length of time available for DNA repair prior to the onset of S phase. One can also prepare large populations of tightly synchronized cells blocked at the G_1/S border by releasing cells from G_0, and plating them at lower density in medium containing serum and aphidicolin for 24 hr (Grossmann *et al.*, 1985). When aphidicolin is washed out, the whole population proceeds through S phase within ~6 hr and through mitosis after an additional 2 hr.

Using these protocols, we and our colleagues have treated cohorts of cells with carcinogenic agents at various stages of the cell cycle, determined the nature of the DNA damage induced (Patton *et al.*, 1986; Yang, *et al.*, 1980), the rate of repair of the damage from the genome overall (Wang *et al.* 1993; Yang *et al.*, 1982; Heflich *et al.*, 1980), or from the individual strands of the target *HPRT* gene (Tung *et al.*, 1996; Chen *et al.*, 1992), and most recently the location and rate of repair of damage at the level of individual nucleotides in the endogenous *HPRT* gene (Wei *et al.*, 1995). Using the same treated populations, we have determined the cell survival, the frequency of mutants, and the spectrum of mutations in the *HPRT* gene as a function of dose and time for repair. In this way we have determined that the frequency and the spectrum of mutations induced by ultraviolet radiation and by a series of structurally related polycyclic aromatic chemical carcinogens are directly related to the damage remaining unrepaired in the target gene at the time scheduled DNA replication occurs (Wang *et al.*, 1993; McGregor *et al.*, 1991; Chen *et al.*, 1990, 1991; Watanabe *et al.*, 1985; Konze-Thomas *et al.*, 1982).

2. METHODS

2.1. Cells and Cell Culture

A source of normal diploid human skin fibroblasts is foreskin material from newborns. Tissue samples immersed in serum-free medium supplemented with antibiotics are transferred into a small dish containing a few drops of medium and divided into small pieces with a sterile sharp iris scissors or scalpel. Fragments are transferred to a separate 35-mm-diameter glass culture dish containing a few drops of medium and minced finely with sterile scalpels. The pieces of tissue are made to attach to the dish by adding a drop of serum. After 24–36 hr the tissue will have attached and additional medium can be added. Foreskin cultures provide a very large initial population of cells which have great growth potential (>60 doublings) and, therefore, are exceptionally well suited for mutagenesis studies. However, skin biopsies from older persons, available from cell banks, are also a satisfactory source of fibroblasts from normal individuals and from patients with inherited defects in DNA repair, such as xeroderma pigmentosum patients.

The medium used routinely for cell culture is Eagle's minimal essential medium, supplemented with 10 or 15% fetal bovine serum, and 0.2 mM Asp, 0.2 mM Ser, and 1.0 mM sodium pyruvate, and penicillin (100 U/ml) and streptomycin (100 μg/ml). This same medium supplemented with 10% fetal bovine serum and 40 μM TG is used to select cells lacking HPRT activity. Extensive tests are carried out on samples from different lots of serum to identify one that will support high cloning efficiency and growth of cells to high density, while still allowing the designated concentrations of purine analogues to prevent nonresistant cells from undergoing even a single population doubling.

In subculturing cell stocks or cultures of carcinogen-treated cells during the expression period, care must be taken not to injure the cells by overtrypsinization or too vigorous pipetting. Cells are rinsed with phosphate-buffered saline (PBS), pH 7.2, and exposed to 0.25% trypsin at 37°C (1:250 Trypsin, Grand Island Biological Co., Grand Island, NY) for as short a period of time as necessary to release their attachment (<3 min). During the enzymatic digestion process the cells are viewed under an inverted microscope, and the trypsin is removed at the first sign of cell detachment. Digestion by the remaining film of trypsin is allowed to continue for a short time and then is stopped by covering the cells with culture medium containing serum. Gentle pipetting yields a suspension of single cells that form clones with high efficiency when plated at low density (i.e., 50–100 cells per 100-mm-diameter dish) and incubated in a CO_2 incubator that is continually monitored and the CO_2 level adjusted to maintain a pH of 7.2–7.4.

2.2. Exposing Cells to DNA Damaging Agents

Cells are exposed to ultraviolet radiation attached to the surface of culture dishes. The medium is removed, and the cells are rinsed with PBS and irradiated with the dish cover off. To avoid a shadowing effect caused by the edge of the dishes, the lamp is placed as far as possible above the dish and an incident dose of 0.1 or 0.2 J/m^2 per sec is used. The dish is rotated on a turntable during irradiation to equalize exposure. If cells are irradiated at high density rather than at low cloning densities and are then replated at cloning densities, prior to trypsinization the cells located around the circumference of the dish are removed using a sterile rubber policeman to avoid contribution of cells which may not have received the designated dose.

Cells are also exposed to chemical carcinogens while attached to the surface of culture dishes. Chemical carcinogens are stored in desiccators in a −20°C freezer in a locked weighing room. Quantities are weighed in still air using an electrobalance located within a vented glove box. The material is transferred to a sterile vial and protected from light by an aluminum foil cover. All operations with chemicals and with cells in culture are carried out under gold fluorescent lights (500–700 nm) to avoid possible chemical destruction or cell damage caused by white fluorescent light. Just prior to treatment of the cells, the chemical is dissolved in anhydrous dimethyl sulfoxide stored under nitrogen, anhydrous spectral-grade acetone, or ethanol (depending on the compound). The culture medium is replaced with culture medium lacking serum, and the test compound is introduced into the dishes by micropipette. (Final concentration of solvent is < 0.5%.) The control cells receive solvent only. At the end of the exposure period, the medium containing the carcinogen is removed, and the cells are rinsed and refed with culture medium containing 10 or 15% fetal bovine serum. Since proteins possess strong nucleophilic centers, the serum can be expected to inactivate remaining traces of electrophilic reactants. All operations with chemical carcinogens are carried out in a type 2, class B safety cabinet "hood" with an exhaust to the outside air.

2.3. Determining the Cytotoxic Effect of the Agent

To determine the cytotoxic effect of the treatment, the colony-forming ability (cloning efficiency) of the treated cells is compared with that of the untreated control. Two methods are employed: *in situ* and replating. For UV radiation, unless the cells are being irradiated at confluence, there is no density effect, and therefore, the *in situ* technique is more convenient and accurate to use. Cells are plated into a series of 100-mm-diameter dishes at cloning densities (i.e., densities calculated to yield ~50 colonies per dish) and allowed to attach for about 12 hr. Following irradiation, the cells are refed with freshly prepared culture medium and allowed to form colonies. The medium is renewed after 24 hr and after 7 days. When macroscopic colonies have formed (12–14 days), the colonies are stained with crystal violet and counted. The cloning efficiency of the treated cells divided by that of the untreated control cells multiplied by 100 gives the percentage of survival.

For chemicals, the replating technique is employed because cells to be used to assay survival must be treated at the same density as those to be assayed for mutations because the cell density can affect the extent of DNA damage that occurs. To determine the cytotoxic effect of exposure of a monolayer of exponentially growing cells or exposure of a confluent culture to radiation or chemicals, the same treatment procedures are used as described above. However, immediately following the treatment or after designated periods of time held in confluence, the cells are detached from the dishes with trypsin, pooled, and counted with an electronic counter (Coulter Corp., Hialeah, FL) and/or a hemacytometer. The cells are diluted appropriately, plated at cloning densities, and allowed to form colonies as above.

2.4. Determining the Frequency of Mutations Induced in the *HPRT* Gene

The general procedures for quantifying the induction of TG resistance in diploid human fibroblasts, including cells from cancer-prone individuals, were worked out in this laboratory over a period of several years and have been described in detail (e.g., see Chen *et al.*, 1990; Patton *et al.*, 1986; Wang *et al.*, 1986; Watanabe *et al.*, 1985; McCormick and Maher, 1981). Briefly, for each determination, sufficient target cells are plated into 150-mm-diameter dishes at the designated densities (usually 10^4 cells/cm^2) to ensure that after exposure to the DNA damaging agent, the number of surviving cells ("founder cells") will be large enough to contain at least 50 TG-resistant *clonable* cells, preferably more. The higher the cloning efficiency of the cells being assayed, and the more efficient the mutagen being tested, the fewer "founder cells" are needed. For example, if the cloning efficiency of the population is 50%, and the dose of a mutagenic agent will lower the cell survival to ~50% and induce 40 TG-resistant mutants per 10^6 viable cells assayed at the end of the expression period (i.e., 20 clonable mutants), then 5×10^6 target cells must be plated per assay in order to have 2.5×10^6 "founder cells." Also included for each dose and for the untreated control population are a set or sets of cells to be used to assay the cytotoxicity of the agent immediately after the exposure.

At the end of treatment, the control population, mock irradiated or exposed to solvent only, and the populations exposed to various doses of the agent are allowed to replicate three or four times in the original dishes. When the cells approach confluence, they are trypsinized, pooled, and subcultured at a lower density. The required number of cells (equal to or greater than the number of "founder cells" described above) are subcultured into 150-mm-diameter dishes at densities that allow the cells to continue exponential growth for an additional 4 or 5 days (total expression period of 8 days). (It is recommended to plate extra dishes of cells plated at the same density to be used to monitor the number of population doublings occurring during this expression period by electronic cell counting.) At the end of the expression period (six to eight population doublings), the cells for each determination are trypsinized, pooled, and sufficient cells (equal to or greater than the number of founder cells) are plated into culture medium at a density of 2500 cells/ml to be assayed for TG resistance. The rest are cryopreserved for future use. A 5-ml aliquot of cells is removed, diluted appropriately, plated into 100-mm-diameter dishes, and allowed to form colonies, with a refeeding after 7 days, to assay the cloning efficiency of the cells at the time of selection. (This information is used to correct the observed frequency of TG-resistant clones to what would be found at 100% efficiency, in order to make valid comparisons between cell lines with different cloning efficiencies or the same cell lines assayed in different experiments.) An appropriate aliquot of an 8 mM solution of TG is then added into the bottle containing the cells to be assayed for TG resistance to obtain a final concentration of 40 μM TG, and the cells are plated into 100-mm-diameter dishes using 10 ml per dish. After 7 days, the cells are refed with selection medium containing 40 μM TG. After 14 days, if there is no need to isolate individual clones for analysis of the nature and location of the mutation, the cells are fixed with methanol and stained with crystal violet. Colonies composed of more than 50 cells are counted and the frequency of observed mutants is corrected for the cloning efficiency at the time of selection. The induced mutant frequency is determined by subtracting the background frequency observed in the control population.

Reconstruction studies in this laboratory show that for human fibroblasts plated at 500 cells/cm^2, there is no decrease in mutant recovery caused by metabolic cooperation between mutant and nonmutant cells. Such studies also show that the cloning efficiency of TG-resistant cells plated in a population of 500 nonresistant cells/cm^2 in the presence of 40 μM TG is equivalent to the cloning efficiency of the total population plated at cloning density in a nonselective medium. Therefore, the frequency of TG-resistant cells is calculated from the number of observed mutant colonies corrected by the cloning efficiency of the cells plated at low density in a nonselective medium. For those experiments where the purpose is not to determine the frequency of mutants, but rather to obtain unequivocally independent mutants, the number of independent target populations is increased, and the cells in each treated dish are not pooled, but are selected independently of each other.

If representative colonies are to be isolated for further analysis, e.g., to be used for DNA sequencing to determine the location of the mutation in the *HPRT* gene, the dishes are scanned for clones by removing the top of the dish in a laminar flow hood in a darkened room and holding the dish above a focused beam of bright light. Clones that interrupt the beam are circled and subsequently verified by microscope examination. They are isolated and expanded by detaching the cells in a drop of trypsin and transferring them into a 35-mm-diameter dish containing selective medium. The remaining clones are then stained and counted as above to determine the frequency.

2.5. Achieving Synchronized Populations of Cells

Cells are inoculated at a density of 10^4 cells/cm^2 and grown to confluence. After they reach confluence, they are refed daily with medium containing serum for an additional 3 days and then not fed for 72 or 96 hr. At that time, such cells are designated "in G_0" and do not continue to divide (Maher *et al.*, 1979). The cells are released from confluence by trypsinization and plated in culture medium containing serum at a density of 10^4 cells/cm^2. They attach to the dish within a few minutes but require 4–5 hr to become completely flattened. Watanabe *et al.* (1985) showed that cells plated following this protocol begin DNA synthesis (S phase) after 16–17 hr, and that by 20 hr, 80% of the cells are in S phase. This method of synchronizing cells with its elongated G_1 phase permits one to treat cells with carcinogens in early S phase, just as they are about to replicate the *HPRT* gene (Grossmann *et al.*, 1985) so that they will have little or no time to repair the DNA damage prior to DNA replication or to treat them at various times in G_1 phase so as to vary the amount of time available for them to repair DNA damage before replication. To prepare populations synchronized at the G_1/S border, cells in the G_0 state are trypsinized and plated at a density of 10^4 cells/cm^2 into fresh culture medium containing 0.2 μg/ml aphidicolin for 24 hr.

2.6. Determining the Spectrum of HPRT Mutations Induced

Diploid human fibroblasts have a finite life span in culture. As a population, they can undergo greater than 60 population doublings, but when a single cell from the population, e.g., a TG-resistant mutant is isolated and cloned, the number of population doublings is significantly reduced. Therefore, Yang *et al.* (1989) developed and optimized methods to copy mRNA directly from a lysate of a small number of human cells (in fact, even from one cell), amplify the first and second strands of the cDNA of the gene of interest 10^{10}- to 10^{11}-fold using a two-stage polymerase chain reaction (PCR) strategy, and determine the consensus sequence by direct nucleotide sequencing of the PCR product without the need to isolate and purify mRNA or DNA. This method is used routinely in our laboratory, as well as many other groups, to determine the nature of the mutations induced in the *HPRT* gene and their location in the coding region.

For first-strand synthesis of cDNA directly from a clone of 100–200 cells, the cells are suspended in cold PBS pH 7.4 in a 0.5-ml Eppendorf tube and centrifuged for 10 min at 4°C. The PBS supernatant is removed using a Pasteur pipette (RNase-free) and the cell pellet is resuspended in 5 μl of first-strand cDNA cocktail, consisting of 50 mM Tris-HC1 prepared from a 1M stock, pH 8.55, 75 mM KC1, 3 mM $MgCl_2$, 10 mM dithiothreitol, 500 μM of each of the four dNTPs, 0.1 μg per μl BSA, 10 ng per μl oligo$(dT)_{12-18}$ primer, 1 unit per μl RNasin, 2.5 units per μl Moloney-MuLV reverse transcriptase, 2.5% NP-40. The cells are incubated at 37°C for 1 hr to allow the cell membranes to be lysed by NP-40 and the cytoplasmic mRNA to be copied into the first-strand cDNA.

The synthesized first-strand cDNA is used as a template for PCR amplification using two sets of amplification primers, one located just inside the other (set 1: −60 to −41 and 721 to 702; set 2: −36 to −17 and 701 to 682). An aliquot of cDNA is diluted in our optimized *Taq* buffer containing $MgCl_2$ (2.75 mM), KCl (60 mM), and Tris-HCl pH 8.8 (15 mM), the four dNTPs at 400 μM each, PCR primer set 1 at 0.15 μM, and 2.5 units of *Taq* enzyme in a total volume of 50 μl (Yang *et al.*, 1989). The PCR cycle consists of denaturation at 94°C for 1 min, annealing at 50°C for 1 min, and polymerization at 72°C for 2 min. Before the first cycle, the template is denatured at 94°C for 5 min and after the 30th cycle, the extension time is 7 min. At the end of 30 cycles, an aliquot of the amplified DNA is diluted 20-fold and 1 μl of the diluted DNA is reamplified with 20 cycles of PCR using primer set 2. The yield is 1 to 5 μg. For sequencing the final PCR product, we routinely use the three primers listed by Yang *et al.* (1989), but also a primer that is located at position 339 to 319. Several methods can be used, but that described by Yang *et al.* (1989) for sequencing the double stranded consensus product has proved very satisfactory.

Our current practice is to propagate the cells from each isolated TG-resistant colony for 7 to 8 population doublings (~50,000 cells), use 100–200 cells to prepare cDNA and PCR product as above, and distribute the rest into 0.5-ml Eppendorf tubes (5000 cells per tube). These tubes are stored at −80°C. If the subsequent sequencing of the PCR product indicates that an entire exon is missing, the DNA surrounding that missing exon is amplified from 5000 cells using an appropriate set of intron primers (Gibbs *et al.*, 1990) and the splice site region of the DNA product is sequenced directly to determine the nature of the mutation involved.

2.7. Determining the Effect of Repair on the Frequency and Spectrum of Mutations

The most common way to determine the effect of repair on the frequency and spectrum of mutations induced by an agent is to compare these parameters in cells that are proficient in repair of the DNA damage induced by that particular agent and in cells that are known to be deficient in such repair. We and our colleagues, including many of those whose research has been cited above, as well as Domoradzki *et al.* (1984), used this approach for mutation studies with human cells, as have many other groups, especially those using rodent cell lines. The underlying assumption is that the only significant difference between the target cell lines being compared is their ability to carry out a particular DNA repair process.

Our second approach eliminates that caveat. Cell survival, mutant frequencies, and the spectra of mutations in the target gene are determined and compared using only one repair-proficient cell line. However, in half of the target population, the capacity of cells to carry out repair of the DNA damage induced by a particular agent has been temporarily eliminated or blocked. For example, diploid human fibroblasts exhibit high levels of O^6-alkylguanine-DNA alkyltransferase (AGT) activity. The preexisting AGT can be completely inactivated (titrated away) by exposing the cells to 25 μM O^6-benzylguanine (O^6-BzG) for 2 hr (Lukash *et al.*, 1990). The two populations, one pretreated with O^6-BzG and the other not pretreated, are then exposed to *N*-methyl-*N*′-nitro-*N*-nitrosoguanidine for 20 to 60 min, with one-half of the target population still maintained in 25 μM O^6-BzG during treatment and for an additional 24 or 48 hr. Differences between the two populations in mutant frequency as well as in the kinds and location of the mutations in the *HPRT* gene, can be attributed to the failure of one set of cells to remove the methyl group from the O^6 position of guanine. The results of Lukash *et al.* (1990) support this conclusion.

A third approach is to use a synchronized population of repair-proficient cells and vary the length of time available for repair before the *HPRT* gene is replicated at the beginning of S phase. To extend the time beyond 12 or 13 hr, one can treat confluent cells in G_0 and after various periods of time, release them by plating them at lower density (10^4 cells/cm^2) so that they begin to cycle. [Examples using this method include McGregor *et al.* (1991), Chen *et al.* (1990), Watanabe *et al.* (1985), Konze-Thomas *et al.* (1982).]

A tremendous advantage of this method is that the delay of onset of S phase allows one to measure the rate of repair of damage from individual strands of the target gene and from the genome overall without having to control for apparent decrease in the frequency of DNA damage that would reflect scheduled DNA replication (e.g., by using incorporation of a density label into the DNA followed by density gradient centrifugation, etc.). This permits one to synchronize large populations of cells, expose one set in early S phase and the other sets in mid G_1, early G_1, and even in G_0, and then use some of the cells from each treatment to determine the frequency and spectrum of mutations and the rest for measuring DNA repair. For the latter determination, one can measure the frequency, kinds, and location of DNA damage in the target gene, either in the specific strands (Tung *et al.*, 1996; Chen *et al.*, 1992) or at the level of the individual nucleotides in the target gene (Wei *et al.*, 1995.) For these studies, part of the population treated in early G_1, or in G_0, can be harvested immediately and analyzed for such damage. The rest can be allowed various lengths of time for repair before being similarly harvested and analyzed.

2.8. Determining the Relationship between Site of DNA Damage and Spectra of Mutations

McGregor *et al.* (1991) showed that if human fibroblasts proficient in nucleotide excision repair are UV irradiated in early S phase, the majority of the mutations in the coding region of the *HPRT* gene arise from dipyrimidines located in the transcribed strand, but if the cells are irradiated in mid G_1 to permit them to have at least 6 hr for repair before S phase, the frequency of mutants is reduced to slightly less than half, and the majority of the mutations are from dipyrimidines located in the nontranscribed strand. No such decrease in mutant frequency or switch in strand bias for premutagenic lesions was seen when xeroderma pigmentosum (XP) cells virtually devoid of excision repair were similarly studied. The frequency of mutants induced in the latter population was significantly higher than that seen with the normally repairing cells.

Data of Tung *et al.* (1996) indicate that within the first 6 hr after UV irradiation of the same repair-proficient cells, 65% of the cyclobutane pyrimidine dimers (CPD) are removed from the transcribed strand of the *HPRT* gene and 45% of those in the nontranscribed strand (average, 55% removed). Their data also show that within 2 hr postirradiation, virtually all of the pyrimidine 6–4 pyrimidones (6–4's) have been removed from both strands of the *HPRT* gene. These data suggest that the majority of the mutations observed in the normal cells resulted from CPD and that the majority of the 6–4's had been eliminated by excision repair even before the cells that were irradiated in early S phase were able to recommence DNA replication. This would also be true for the cells irradiated in mid G_1 phase and allowed at least 6 hr to reach S phase.

Chen *et al.* (1990, 1991) used these same approaches with normal and XP cells exposed in early S or early G_1 to benzo[*a*]pyrene diol epoxide (BPDE), a reactive metabolite of benzo[*a*]pyrene, which binds covalently to the N^2 position of guanine. Allowing normal cells 12 or more hr for excision repair prior to onset of S phase reduced the mutant frequency threefold, and sequence analysis indicated preferential removal of premutagenic lesions located in the transcribed strand. No such decrease in mutant frequency or switch in strand bias was seen with the XP cells. Excision repair studies using the methods described above confirmed strand-specific repair of BPDE adducts in the target gene (Chen *et al.*, 1992).

Chen *et al.* (1990, 1991) also noted a *relative* increase in the frequency of mutations at certain sites in exon 3 of the *HPRT* gene in repair-proficient cells that were allowed 12 to 15 hr for repair prior to DNA replication. They hypothesized that this was the result of relatively rapid excision repair of adducts from the other sites in the gene and relatively slow repair at these particular sites. Recently, Wei *et al.* (1995) in this laboratory worked out conditions to detect the rate of repair of bulky adducts, such as those induced by BPDE, from the *HPRT* gene at the level of the *individual nucleotides* using normal human cells treated in early G_1 and allowed 0, 10, 20, or 30 hr for repair. The method involves using UvrABC endonuclease to cut the DNA at the site of the adducts, followed by use of ligation-mediated PCR. Using this method they showed that there was no difference in the original distribution of adducts in cells treated in early G_1 or S phase, but that the rate of excision of BPDE adducts varies significantly at individual sites in the gene, and that at the sites that showed a relative increase in mutation frequency with time for repair (Chen *et al.*, 1990), the rate is very slow. These results demonstrate the power of these methods and the advantage of measuring mutation induction and DNA repair in the same populations of carcinogen-treated cells.

ACKNOWLEDGMENTS. We thank our colleagues, especially those whose research is cited in this chapter, for their extremely valuable contributions to the research that has been carried out in the Carcinogenesis Laboratory. These investigations were supported by grants from the Department of Health and Human Services, National Cancer Institute and National Institute of Environmental Health Sciences and from the Department of Energy.

REFERENCES

Albertini, R. J. (1985). Somatic gene mutations in vivo as indicated by the 6-thioguanine resistant T-lymphocytes in human blood. *Mutat. Res.* **150**:411–422.

Albertini, R. J., Castle, K. S., and Borcherding, W. R. (1982). T-cell cloning to detect the mutant 6-thioguanine resistant lymphocytes present in human peripheral blood. *Proc. Natl. Acad. Sci. USA* **79**:6617–6621.

Boldt, J., Mah, M. C.-M., Wang, Y.-C., Smith, B. A., Beland, F. A., Maher, V. M., and McCormick, J. J. (1991). Kinds of mutations found when a shuttle vector containing adducts of 1,6-dinitropyrene replicates in human cells. *Carcinogenesis* **12**:119–126.

Cariello, N. F., Craft, T. R., Vrieling, H., van Zeeland, A. A., Adams, T., and Skopek, T. R. (1992). Human HPRT mutant database: Software for data entry and retrieval. *Environ. Mol. Mutagen.* **20**:81–83.

Chen, R.-H., Maher, V. M., and McCormick, J. J. (1990). Effect of excision repair by diploid human fibroblasts on the kinds and spectra of mutations induced by (±)-7β,8α-dihydroxy-9α,10α-epoxy-7,8,9,10-tetrahydrobenzo[a]pyrene in the coding region of *HPRT* gene. *Proc. Natl. Acad. Sci. USA* **87**:8680–8684.

Chen, R.-H., Maher, V. M., and McCormick, J. J. (1991). Lack of a cell cycle-dependent strand bias for mutations induced in the *HPRT* gene by (±)-7β,8α-dihydroxy-9α,10α-epoxy-7,8,9,10-tetrahydrobenzo[a]pyrene in excision repair-deficient human cells. *Cancer Res.* **51**:2587–2592.

Chen, R.-H., Maher, V. M., Brouwer, J., van de Putte, P., and McCormick, J. J. (1992). Preferential repair and strand-specific repair of benzo(a)pyrene diol epoxide adducts from the *HPRT* gene of diploid human fibroblasts. *Proc. Natl. Acad. Sci. USA* **89**:5413–5417.

Domoradzki, J., Pegg, A. E., Dolan, M. D., Maher, V. M., and McCormick, J. J. (1984). Correlation between O^6-methylguanine-DNA-methyltransferase activity and resistance of human cells to the cytotoxic and mutagenic effect of N-methyl-N′-nitro-N-nitrosoguanidine. *Carcinogenesis* **5**:1641–1647.

Drinkwater, N. R., Corner, R. C., McCormick, J. J., and Maher, V. M. (1982). An in situ assay for induced diphtheria toxin resistant mutants of diploid human fibroblasts. *Mutat. Res.* **106**:277–289.

Gibbs, R. A., Nguyen, P., Edwards, A., Civitello, A. B., and Caskey, C. T. (1990). Multiplex DNA deletion detection and exon sequencing at the hypoxanthine phosphoribosyltransferase gene in Lesch-Nyhan families. *Genomics* **7**:235–244.

Grossmann, A., Maher, V. M., and McCormick, J. J. (1985). The frequency of mutants in human fibroblasts UV-irradiated at various times during S-phase suggests that genes for thioguanine and diphtheria toxin resistance are replicated early. *Mutat. Res.* **152**:67–76.

Heflich, R. H., Hazard, R. M., Lommel, L., Scribner, J. D., Maher, V. M., and McCormick, J. J. (1980). A comparison of the DNA binding, cytotoxicity and repair synthesis induced in human fibroblasts by reactive derivatives of aromatic amide carcinogens. *Chem. Biol. Interact.* **29**:43–56.

Konze-Thomas, B., Hazard, R. M., Maher, V. M., and McCormick, J. J. (1982). Extent of excision repair before DNA synthesis determines the mutagenic but not the lethal effect of UV radiation. *Mutat. Res.* **94**:421–434.

Lukash, L., Boldt, J., Pegg, A. E., Dolan, M. E., Maher, V. M., and McCormick, J. J. (1990). Effect of O^6-alkylguanine-DNA alkyltransferase on the frequency and spectrum of mutations induced by *N*-methyl-*N*′-nitro-N-nitrosoguanidine in the *HPRT* gene of diploid human fibroblasts. *Mutat. Res.* **250**:397–409.

McCormick, J. J., and Maher, V. M. (1981). Measurement of colony-forming ability and mutagenesis in diploid human cells, in: *DNA Repair: A Laboratory Manual of Research Procedures*, Volume 1B (E. C. Friedberg and P. C. Hanawalt, eds.), Dekker, New York, pp. 501–521.

McCormick, J. J., and Maher, V. M. (1994). Analysis of the multi-step process of carcinogenesis using human fibroblasts. *Risk Anal.* **14**:257–263.

McGregor, W. G., Chen, R.-H., Lukash, L., Maher, V. M., and McCormick, J. J. (1991). Cell cycle-dependent strand bias for UV-induced mutations in the transcribed strand of excision repair-proficient human fibroblasts, but not in repair-deficient cells. *Mol. Cell. Biol.* **11**:1927–1934.

Mah, M. C.-M., Maher, V. M., Thomas, H., Reid, T. M., King, C. M., and McCormick, J. J. (1989). Mutations induced by aminofluorene–DNA adducts during replication in human cells. *Carcinogenesis* **10**:2321–2328.

Mah, M. C.-M., Boldt, J., Culp, S. J., Maher, V. M., and McCormick, J. J. (1991). Replication of acetylaminofluorene-adducted plasmids in human cells: Spectrum of base substitutions and evidence of excision repair. *Proc. Natl. Acad. Sci. USA* **88**:10193–10197.

Maher, V.M., Birch, N., Otto, J. R., and McCormick, J. J. (1975). Cytotoxicity of carcinogenic aromatic amides in normal and xeroderma pigmentosum fibroblasts with different DNA repair capabilities. *J. Natl. Cancer Inst.* **54**:1287–1294.

Maher, V. M., Ouellette, L. M., Curren, R. D., and McCormick, J. J. (1976). Frequency of ultraviolet light-induced mutations is higher in xeroderma pigmentosum variant cells than in normal human cells. *Nature* **261**: 593–595.

Maher, V. M., McCormick, J. J., Grover, P. L., and Sims, P. (1977). Effect of DNA repair on the cytotoxicity and mutagenicity of polycyclic hydrocarbon derivatives in normal and xeroderma pigmentosum human fibroblasts. *Mutat. Res.* **43**:117–138.

Maher, V. M., Dorney, D. J., Mendrala, A. L., Konze-Thomas, B., and McCormick, J. J. (1979). DNA excision-repair processes in human cells can eliminate the cytotoxic and mutagenic consequences of ultraviolet irradiation. *Mutat. Res.* **62**:311–323.

O'Neill, J. P., Sullivan, L. M., and Albertini, R. J. (1990a). In vitro induction, expression and selection of thioguanine-resistant mutants with human T-lymphocytes. *Mutat. Res.* **240**:135–142.

O'Neill, J. P., Hunter, T. C., Sullivan, L. M., Nicklas, J. A., and Albertini, R. J. (1990b). Southern-blot analyses of human T-lymphocyte mutants induced in vitro by γ-irradiation. *Mutat. Res.* **240**:143–149.

Patton, J. D., Maher, V. M., and McCormick, J. J. (1986). Cytotoxic and mutagenic effects of 1-nitropyrene and 1-nitrosopyrene in diploid human skin fibroblasts. *Carcinogenesis* **7**:89–93.

Tung, B., McGregor, W. G., Wang, Y.-C., Maher, V. M., and McCormick, J. J. (1996). Comparison of the rate of excision of major UV photoproducts in the strands of the human *HPRT* gene of normal and xeroderma pigmentosum variant cells. *Mutat. Res.* **362**:65–74.

Wang, Y., Parks, W. C., Wigle, J. C., Maher, V. M., and McCormick, J. J. (1986). Fibroblasts from patients with inherited predisposition to retinoblastoma exhibit normal sensitivity to the mutagenic effects of ionizing radiation. *Mutat. Res.* **175**:107–114.

Wang, Y.-C., Maher, V. M., Mitchell, D., and McCormick, J. J. (1993). Evidence from mutation spectra that the UV hypermutability of xeroderma pigmentosum variant cells reflects abnormal, error-prone replication on a template containing photoproducts. *Mol. Cell. Biol.* **13**:4276–4283.

Watanabe, M., Maher, V. M., and McCormick, J. J. (1985). Excision repair of UV- or benzo[a]pyrene diol epoxide-induced lesions in xeroderma pigmentosum variant cells is "error-free." *Mutat. Res.* **146**:285–294.

Wei, D., Maher, V. M., and McCormick, J. J. (1995). Site specific rates of excision repair of benzo(a)pyrene diol epoxide adducts in the hypoxanthine phosphoribosyltransferase gene of human fibroblasts—correlation with mutation spectra. *Proc. Natl. Acad. Sci. USA* **92**:2204–2208.

Yang, L. L., Maher, V. M., and McCormick, J. J. (1980). Error-free excision of the cytotoxic, mutagenic N2-deoxyguanosine DNA adduct formed in human fibroblasts by ($\pm$)-7β,8α-dihydroxy-9α,10α-epoxy-7,8,9,10-tetrahydrobenzo[a]pyrene. *Proc. Natl. Acad. Sci. USA* **77**:5933–5937.

Yang, J.-L., Maher, V. M., and McCormick, J. J. (1982). Relationship between excision repair and the cytotoxic and mutagenic effect of the "anti" 7,8-diol-9,10-epoxide of benzo[a]pyrene in human cells. *Mutat. Res.* **94**: 435–447.

Yang, J.-L., Maher, V. M., and McCormick, J. J. (1987). Kinds of mutations formed when a shuttle vector containing adducts of ($\pm$)-7β,8α-dihydroxy-9α,10α-epoxy-7,8,9,10-tetrahydrobenzo[a]pyrene replicates in human cells. *Proc. Natl. Acad. Sci. USA* **84**:3787–3791.

Yang, J.-L., Maher, V. M., and McCormick, J. J. (1988). Kinds and spectrum of mutations induced by 1-nitrosopyrene adducts during plasmid replication in human cells. *Mol. Cell. Biol.* **8**:3364–3372.

Yang, J.-L., Maher, V. M., and McCormick, J. J. (1989). Amplification and direct nucleotide sequencing of cDNA from the lysate of low numbers of diploid human cells. *Gene* **83**:347–354.

Chapter 29

Bacteriophage Lambda and Plasmid *lacZ* Transgenic Mice for Studying Mutations *in Vivo*

Jan Vijg and George R. Douglas

1. INTRODUCTION

The demonstration that genomic DNA fragments, including integrated vectors, can be efficiently rescued and cloned only into host-restriction negative bacterial strains, allowed the practical use of transgenic mice equipped with bacterial reporter genes for mutation analysis studies (Gossen and Vijg, 1988; Gossen *et al.*, 1989). This was later confirmed by Kohler *et al.* (1990). Since then numerous studies with *lacZ* and/or *lacI* transgenic mice have demonstrated the power of these model systems to detect spontaneous and induced mutations (Kohler *et al.*, 1991; Myhr, 1991; Suzuki *et al.*, 1993; Gossen and Vijg, 1993; Tinwell *et al.*, 1994; Douglas *et al.*, 1994, 1995a,b).

Several applications of transgenic mutation models in mutagenesis studies can be identified. First, their single most frequent use has thus far been in genetic toxicology testing. Indeed, there has been a long-standing gap in strategies for evaluating mutagenic hazards *in vivo* (Advisory Committee on Mutagenesis, 1988; Health Protection Branch Genotoxicity Committee, 1992). The problem is the current paucity of endogenous genes amenable to the quantitative analysis of somatic mutation in animal tissues. Accordingly, transgenic animal models harboring reporter genes that can be rescued from their integrated state and subjected to *in vitro* selection for the presence of mutations in host-restriction negative *E. coli* hosts provide surrogate target genes for the study of mutagenesis *in vivo*. Key issues here are the sensitivity and specificity with which these transgenic systems detect induced mutations by environmental mutagens and carcinogens.

A second field of application involves the study of mechanisms of mutagenesis *in vivo*. It has been argued that the use of cultured cells for studying mutational mechanisms might not always provide a realistic picture of the actual situation in various organs and tissues of a

Jan Vijg • Molecular Genetics Section, Gerontology Division, Department of Medicine, Beth Israel Hospital and Harvard Medical School, Boston, Massachusetts 02215. **George R. Douglas** • Mutagenesis Section, Environmental Health Directorate, Health Canada, Ottawa, Ontario K1A 0L2, Canada.

Technologies for Detection of DNA Damage and Mutations, edited by Gerd P. Pfeifer, Plenum Press, New York, 1996.

mammal (Lohman *et al.*, 1987). Transgenic mutation models, which allow molecular characterization of the mutations through sequencing and/or other methods of the rescued target genes, offer an ideal tool for such detailed mechanistic investigations (e.g., Gossen *et al.*, 1993a; Knöll *et al.*, 1994; Douglas *et al.*, 1994, 1995b). A key issue here is the question of whether integrated bacterial reporter genes, which are not expressed, offer a good mutational endpoint for mutagenesis compared to endogenous genes. While this is debatable, there are other potential pitfalls such as the influence of genomic regions which may act as mutational hot spots (Gossen *et al.*, 1991). These questions are especially relevant in view of the demonstrated specificities of DNA repair mechanisms (Bohr, 1991), which in part may be responsible for observed genomic heterogeneity in mutant frequencies.

Finally, a third application involves the testing of various hypotheses involving particular phenotypic endpoints to which somatic mutagenesis may be causally related. A major example here is the somatic mutation hypothesis of aging (for a review, see Vijg and Gossen, 1993), but mutations have also been implicated in other phenotypic endpoints (besides cancer), such as neurodegenerative disorders (Evans *et al.*, 1995) and overnutrition (Ames *et al.*, 1993).

In this chapter two transgenic mouse mutation models are described which, on the basis of a common positive selection system, allow the detection of mutations in different organs and tissues with great efficiency.

The bacteriophage lambda-based model is a CD-2-derived (BALB/c × DBA/2) transgenic mouse in which the *lacZ* gene is present as part of a recombinant lambda gt10 vector integrated head to tail in multiple copies on chromosome 3 (Blakey *et al.*, 1995). The plasmid model is a C57Bl/6-derived transgenic mouse with its *lacZ* gene integrated as part of the pUR288 plasmid (Rüther and Müller-Hill, 1983), also in a head-to-tail configuration. Figure 29.1 shows maps of the two vectors with some restriction sites.

The rescue procedures for the two models are compared in Fig. 29.2. For the lambda-based model this step is the simplest and involves the mixing of a sample of DNA extracted from any of the mouse organs or tissues with a commercially available packaging extract. The terminase enzyme in the extract excises each vector copy at the *cos* sites and encapsulates each viral genome. When this is subsequently plated on *lacZ*$^-$ *galE*$^-$ *E. coli* C hosts, only the mutants are able to form plaques on the bacterial lawn. The principle of this positive selection system (PSS), which can also be used for the plasmid model, is depicted in Fig. 29.3.

With the plasmid model the genomic DNA sample to be studied must be incubated with magnetic beads, previously coupled to a *lacZ–lacI* fusion protein, in the presence of *Hin*dIII restriction enzyme that cuts just outside the *lacZ* mutational target gene and generates multiple plasmid copies. These copies will subsequently bind to the operator sequence in front of the *lacZ* gene. On the basis of this high-affinity binding the plasmid sequences can be magnetically separated from the rest of the genomic DNA. The plasmids are then released from the beads (by IPTG and/or heat shock), ligated and, after ethanol precipitation, used to electrotransform *lacZ*$^-$ *galE*$^-$ *E. coli* C hosts. Similar to the bacteriophage lambda *lacZ* reporter genes, only mutant colonies will grow (Gossen *et al.*, 1993b).

From the above description it can be inferred that the procedure to rescue plasmids is somewhat more complicated than *in vitro* packaging. However, the single great advantage of the plasmid model is its sensitivity to deletion mutations (Gossen *et al.*, 1995). Indeed, one of the drawbacks of lambda-based models (this is the same for the *lacI* model discussed elsewhere in this volume) is their relative insensitivity to large deletion mutations. This can be readily understood when the size limitations of *in vitro* packaging are taken into consideration: only fragments of 42–52 kb are packaged with some efficiency. In addition, deletions encompassing the *cos* sites always go undetected. Finally, for still unknown reasons deletions of more than 50 bp, but smaller than 5 kb are also rarely found, either in *lacI* or *lacZ* in the lambda-based systems (Lee *et al.*, 1994; Douglas *et al.*, 1994). It is possible that the large stretch of prokaryotic DNA

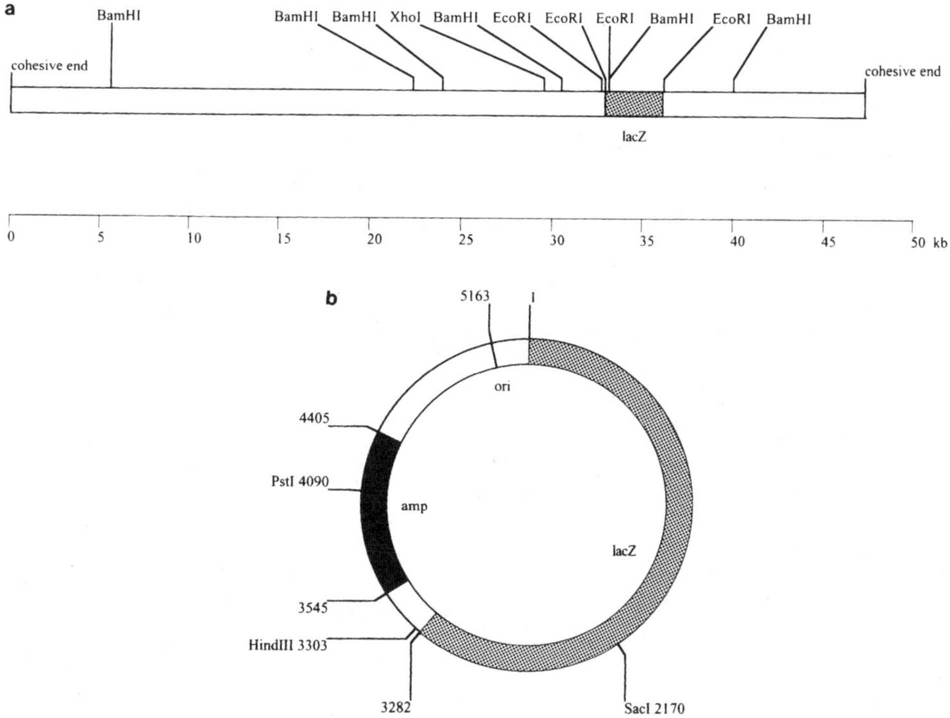

Figure 29.1. The λgt10LacZ (Gossen *et al.*, 1989) (a) and the pUR288 (Rüther and Müller-Hill, 1983) (b) shuttle vectors used for generating CD-2 (BALB/c × DBA/2)-derived and C57Bl/6 transgenic mice, respectively. The plasmid was linearized at the *Pst*I site. Integration of both vectors was head to tail as indicated by Southern hybridization experiments (not shown).

(millions of base pairs) represented by the phage cluster is not a good substrate for the induction of deletions.

Below we provide the methodology involved in the lambda and plasmid *lacZ* transgenic animal *in vivo* mutation assays. In the Discussion section some results illustrating the application of the systems will be presented.

2. METHODS

2.1. Bacteriophage Lambda System

2.1.1. Transgenic Animals

The transgenic mouse strain (40.6) used in this assay has been described in detail by Gossen *et al.* (1989). The transgene is based on a recombinant λgt10 vector containing the complete 3096-bp *E. coli lacZ* gene (Fig. 29.1). This phage vector was injected originally into the male pronucleus of a CD-2 (BALB c × DBA/2) F_1 mouse embryo, yielding a mouse with a 40-copy concatemer inserted at a single chromosomal locus in every cell (Gossen *et al.*, 1989). Subsequent breeding led to strain 40.6, which is disomic for the concatemer (n = 80; Gossen *et al.*, 1989), which is located on chromosome 3 (Blakey *et al.*, 1995). Offspring of strain 40.6 are of normal litter size and health.

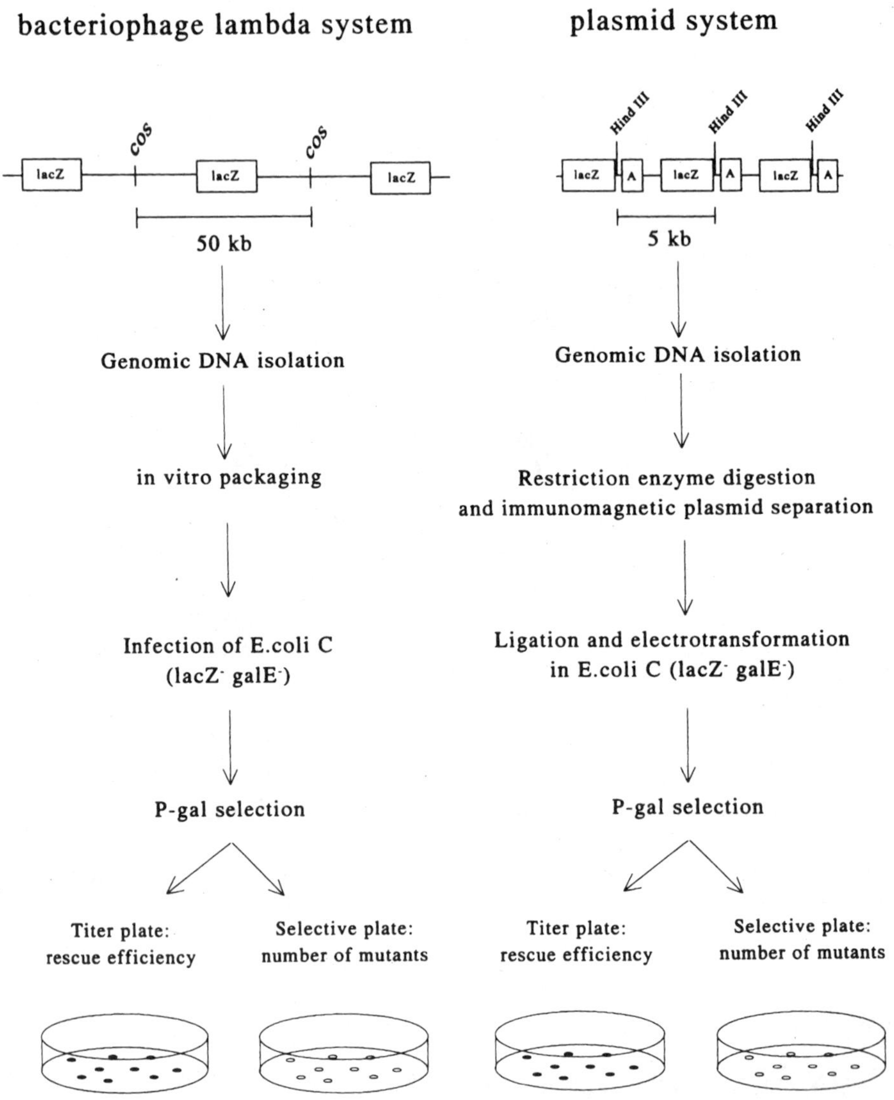

Figure 29.2. Schematic outlines of *lacZ* vector rescue and mutant frequency determinations with the bacteriophage lambda system and the plasmid system. On the titer plate 0.6% (for lambda system) or 0.1% (for plasmid system) of the recovered vectors are plated, the rest (99.9%) are plated on the selective plate containing the lactose analogue P-gal. Mutant frequencies are determined as the ratio of plaques (for lambda system) or colonies (for plasmid system) on the selective plates versus the number on the titer plates, times the dilution factor.

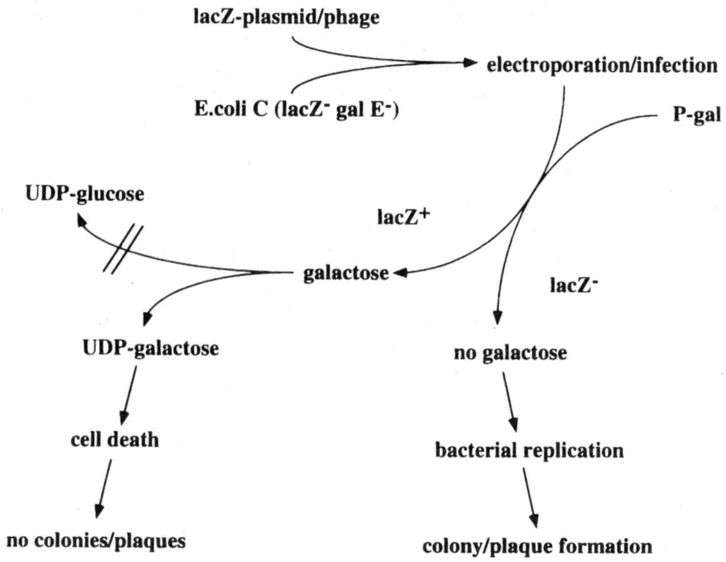

Figure 29.3. Positive selection system for *lacZ*⁻ plasmids or phages. *galE*⁻ *lacZ*⁻*E. coli* cells receiving the wild-type *lacZ* gene are unable to grow as they cannot convert the toxic UDP-galactose into harmless UDP-glucose. Only cells containing a *lacZ*⁻ plasmid or phage will be able to form colonies and plaques, respectively, on the selective plate. (Figure adapted from Dean and Myhr, 1994.)

2.1.2. *E. coli* Strain

The host cells used in this assay are *E. coli* C, $\Delta lacZ^-$, $galE^-$ $recA^-$, pAA119 (Gossen *et al.*, 1992; Mientjes *et al.*, 1994). They are kanamycin and ampicillin resistant and sensitive to galactose.

2.1.3. Reagents

1. PBS: Phosphate-buffered saline without Mg
2. Lysis buffer: 10 mM Tris, pH 7.6, 150 mM NaCl, 10 mM EDTA
3. Proteinase K: 100 mg proteinase K (GIBCO/BRL) in 5 ml H_2O
4. $TE_{7.6}$: 10 mM Tris, pH 7.6, 1 mM EDTA
5. TMST buffer: 50 mM Tris, pH 7.6, 3 mM $MgSO_4$, 250 mM sucrose, 0.2% Triton X-100
6. TEN buffer: 20 mM Tris, pH 7.4, 1 mM EDTA, 100 mM NaCl
7. Packaging extract: Gigapack II Gold (Stratagene)
8. SM buffer: 50 mM Tris, pH 7.6, 100 mM NaCl, 16 mM $MgSO_4$, 0.01% gelatin
9. Minimal agar: 1 part LB (GIBCO/BRL) to 3 parts 150 mM NaCl, 0.75% Bacto agar (Difco)
10. P-gal: Phenyl β-D-galactoside (Sigma)

2.1.4. Tissue Collection

Following sacrifice by cervical dislocation, remove all tissues, place in cryovials, and freeze in liquid nitrogen. Bone marrow and sperm from the epididymis and the vas deferens require special treatment (see below). Maintain samples at −80°C until used.

Collect *bone marrow* by flushing the femurs with 1 ml of PBS. Spin the cells down, discard the buffer, and freeze the pellet as above.

Remove the *epididymis*, cut into pieces, place in PBS containing 0.1% glucose, and store on ice. After all other tissues have been collected, warm the epididymis in a 37°C bath for 20–30 min. Vortex the sample, allow to settle for a few minutes, and then transfer the supernatant to a microtube. Centrifuge the cells for 2 min at 10,000 rpm and discard the supernatant. Meanwhile, add an additional 1 ml of PBS/glucose to the tissue and reincubate at 37° for 10–20 min. Repeat the collection procedure and freeze the combined pellet.

Place the *vas deferens* on a ground glass plate and expel the sperm by rolling the tissue with a small plastic roller. Suspend the tissue in PBS with a pipettor and transfer to a microtube. Pellet the cells by spinning at 12,000 rpm in a microcentrifuge for 1 min, discard the supernatant and freeze the pellet.

2.1.5. Isolation of Genomic DNA

Bone Marrow

1. Thaw the tissue on ice and resuspend in 5 ml of lysis buffer in a 15-ml centrifuge tube. Add 100 μl of 10% SDS (final concentration: 0.2%) and 100 μl of proteinase K stock solution (final concentration: 0.4 mg/ml) and digest the tissue overnight at 37–40°C and 40 rpm in a water bath.
2. The following morning add an additional 2 ml of lysis buffer along with 2 mg of proteinase K and continue incubation at 37–42°C for 2–3 hr.
3. Add 5 ml of phenol/chloroform and rotate the samples end over end at 20 rpm for 20 to 30 min. Centrifuge the samples at 1200*g* for 20 min.
4. Transfer about three-fourths of the aqueous fraction to a new tube while taking care to avoid transferring anything from the aqueous/organic interface.
5. Add 0.01 vol of 5 M NaCl to the sample (final concentration: 0.2 M) followed by 5 ml of chloroform:isoamyl alcohol (24:1).
6. Extract the samples as before. Transfer about two-thirds of the aqueous fraction to a new tube. This method recovers about half of the DNA available in the original sample.
7. Precipitate the DNA by adding 2 vol of ethanol and gently rocking and rolling the tube. Spool out the precipitate with a sealed glass Pasteur pipet, swish around in 70% ethanol to wash, allow to air-dry for a few minutes, then dissolve in 25 μl $TE_{7.6}$.

Spleen, Kidney

1. Thaw the tissues, then mince in a few milliliters of PBS with curved scissors.
2. Transfer the samples to centrifuge tubes, centrifuge, and process the pellets as for bone marrow.
3. After precipitation, dissolve the DNA in 50–100 μl $TE_{7.6}$.

Lung

1. Thaw the tissue, then mince in 2–3 ml of PBS. Place in a vacuum to remove air so that it will sink.
2. Centrifuge the tissue, wash once in PBS, centrifuge again, and process the pellet as for bone marrow.
3. Dissolve the DNA in 50 μl $TE_{7.6}$.

Liver

1. This procedure begins with the isolation of nuclei. Thaw the tissue and cut into small pieces. Homogenize in 5 ml cold TMST buffer using a conical glass homogenizer and Teflon pestle.

2. Centrifuge the resulting suspension at 800*g* for 6 min and discard the supernatant. Wash the nuclei twice in cold TMST, then process as for bone marrow.
3. Dissolve the DNA in 50–100 μl $TE_{7.6}$.

Testicular Germ Cells

1. Thaw the testis and remove the outer membrane. Strip germ cells from the seminiferous tubules using a small plastic roller to push and roll the tissue on a ground glass plate. If care is taken to keep the tubules largely intact, the creamy cell suspension which results from this procedure is highly enriched for germ cells.
2. Resuspend the preparation in 1 ml PBS and transfer to a microtube with a pipettor. Allow it to sit for 5–10 min. During this time, larger debris settles to the bottom of the tube.
3. Transfer the top 90% to a new tube and centrifuge at full speed for 10 sec. Discard the supernatant, resuspend the remaining pellet of cells in lysis buffer, and digest as for bone marrow.
4. Dissolve the DNA in 50 μl $TE_{7.6}$.

Spermatozoa

1. Thaw the frozen cell pellets quickly at room temperature and add 0.5 ml of 150 mM NaCl along with 5 μl of 10% SDS. Vortex the samples hard for 1–2 min to disrupt somatic cells. Centrifuge for 2 min at 10,000 rpm in a microcentrifuge. Discard the supernatant and resuspend the pellet in 1.0 ml of PBS by means of an additional 1–2 min of vortexing.
2. Transfer the samples to a 15 ml polypropylene centrifuge tube and spin at 800*g* for 10 min. Discard the supernatant and resuspend the cells in 1.0 ml of TEN buffer.
3. Add 100 μl of 10% SDS (final concentration: 1.0%), 40 μl of β-mercaptoethanol, and 40 μl of proteinase K stock solution (final concentration: 0.7 mg/ml). Digest the samples overnight at 30°C with gentle agitation.
4. The next morning, add 2.5 ml of TEN buffer along with an additional 40 μl of proteinase K stock solution and continue the incubation for 2–3 hr at 37°C.
5. Extract the samples with 4 ml of phenol/chloroform as for bone marrow. Add 0.02 vol of 5 M NaCl (final concentration: 0.2 M) and extract the samples with 4 ml of chloroform as for bone marrow.
6. Dissolve the precipitated DNA in 10–30 μl of $TE_{7.6}$.

2.1.6. Packaging Lambda Phage

Lambda phage are rescued from the genomic DNA preparation using Gigapack II Gold lambda packaging extract. The optimal amount of DNA depends on the tissue and how clean the DNA preparation is. Determine the DNA concentration using a spectrophotometer. A concentration between 1 and 5 μg/μl generally works well. For most tissues, more than 250,000 plaques can be obtained using 2 μl of DNA and one-half of a packaging extract. In the case of bone marrow and lung, one-quarter of an extract will usually suffice. The reaction is carried out under conditions recommended by the manufacturer except that it is allowed to proceed for 3 hr. When the time is up, add 500 μl of SM buffer and allow the samples to rotate end over end at 20 rpm for 1–2 hr before vortexing, centrifuging briefly, and testing.

2.1.7. Determination of Mutant Frequency

Cells

1. Set up an overnight culture using a colony from an LB plate to inoculate 10 ml of LB containing 0.2% maltose, 50 μg/ml ampicillin, and 20 μg/ml kanamycin in a 50-ml Erlenmeyer flask. Incubate the culture at 37°C and 250–300 rpm.

2. The next morning dilute the required volume of cells 1/100 in LB without antibiotics in an Erlenmeyer flask of a size appropriate to contain 8 ml for each sample to be tested. Grow the culture in a 37°C shaker bath as before.
3. After 3.5 hr spin the cells down (1000*g* for 10 min) and resuspend in 0.5 vol of cold LB + 10 mM $MgSO_4$. These cells may be kept on ice until used.

Plates

Pour plates with 8–10 ml of minimal agar as a bottom agar. Top agar is the same as the bottom with the addition of 5 mM $MgSO_4$. Keep this at 45°C until used.

Test for Mutants

1. Transfer 0.5 ml of packaged phage (a complete packaging reaction) to a 50-ml tube containing 2 ml of cells and allow to adsorb for 30 min at room temperature.
2. After adsorption, transfer a 15-μl sample to a second tube with 2 ml of cells to determine titer (see below).
3. Prepare an appropriate volume of a 100× stock solution (30%) of P-gal in dimethylformamide and add to the top agar to give a final concentration of 0.3%.
4. Add 30 ml of this top agar to the remaining cell/phage suspension, then divide among four plates (8 ml/plate).

Titer

1. Add 30 ml of top agar (without P-gal) to the tube containing cells and the 15-μl sample of cells/adsorbed phage. Divide this among four plates at 8 ml/plate.
2. Incubate all plates overnight at 37°C and score the next day.

Mutant frequency. Count the plaques on the titer plates and determine the number of PFU (plaque-forming units) in the original packaged phage preparation [number of plaques scored on four plates ÷ 15 μl × (2000 μl cells + volume of packaged phage preparation in μl)]. Typically, 2 μl of packaged genomic DNA (1–5 μg) yields 100,000 to 500,000 plaques. After scoring the mutant plaques on the P-gal selective plates, determine the mutant frequency by dividing total mutant plaques by the total number of PFU in the phage preparation.

2.1.8. Mutant Characterization: Complementation Assay for Locating *lacZ* Mutations

E. coli Complementation Strains

The β-galactosidase protein can be roughly divided into three complementation regions; α, β, ω. Mutations in these regions can be restored by the corresponding protein complementation. Three *E. coli* K-12 strains—DH5α (β + ω donor; GIBCO/BRL); W6480 (α + ω donor; Cook and Lederberg, 1962), and Hfr3000 × 90 (α + β donor; Newton *et al.*, 1965)—are used as specific region donors in this study. As mutants are selected in a $GalE^-$ *E. coli* C strain, which has a different DNA methylation system than *E. coli* K-12, it is necessary to start the complementation assay first in DH5α cells. A new phage stock is made from DH5α, and then used in both Hfr3000 × 90 and W4680 complementation assays.

Mutants which have some level of enzyme function will be light blue in all three strains. All three complementation plates should be compared for blue intensity. Mutations at some positions are not easily restored by protein complementation, and result in colorless plaques in all three strains.

Complementation Assay Procedures

1. Transfer 100 μl overnight donor strain (DH5α, Hfr3000 × 90, or W4680) cells into 10 ml LB with 0.2% maltose in a flask, and shake in 37°C water bath for 3 hr.
2. Mix 100 μl of donor cells with less than 1000 mutant phages (1/100 of a single phage plaque solution) in a 15-ml tube and then incubate at 37° for 15 min.
3. Transfer 8–10 ml of LB agar containing X-gal (80 μg/ml) into the tube and then pour onto an LB plate.
4. After overnight incubation at 37°C, plaque color is recorded. In the case of the DH5α assay, a few plaques are picked up to make a DH5α phage solution which will be used in the Hfr3000 × 90 and W4680 complementation assays. Add 5 μl of chloroform, vortex, and let stand for 2 hr at 4°C.

2.1.9. Mutant Characterization: DNA Sequencing

Lambda DNA Isolation

Lambda phage DNA can be prepared from either plate lysate or liquid culture. The following is a procedure for DNA isolation from liquid culture.

1. Mix 0.1 ml of a fresh *E. coli* C ($LacZ^-$) overnight culture with 10^6 PFU phage (⅕-½ of a single plaque) in a tube, and incubate for 15 min at 37°C to allow phage to absorb.
2. Transfer the cell/phage mixture into a 50-ml flask with 10 ml LB and shake vigorously until lysis occurs (10–12 hr, 37°C).
3. Add 20 μl of chloroform and continue shaking for 15 min.
4. Transfer all liquid into a 15-ml tube and centrifuge at 4000*g* for 10 min at 4°C.
5. Pour supernatant into a new tube and add 40 μl of mixture of RNase A (0.25 mg/ml) and DNase I (0.25 mg/ml). Incubate at 37°C for 15 min.
6. Add 4 ml of phage precipitate [33% polyethylene glycol (PEG-8000), 3.3 M NaCl], and place on ice for 30 min.
7. Centrifuge at 10,000*g* for 15 min.
8. Decant the supernatant. Resuspend the pellet in 500 μl of TE buffer by pipetting and transfer solution into a 1.5-ml tube. Centrifuge for 10 sec to remove insoluble particles and transfer the supernatant to a new tube.
9. DNA is extracted once with 1 vol of chloroform, once with 1 vol of phenol, and once with 1 vol of chloroform. DNA is precipitated by adding 30 μl of 3 M NaAc (pH 4.9) and 2 vol of 95% ethanol.

An alternative is using a Lambda DNA purification kit (Promega Corp., Madison, WI).

Amplification of lacZ Fragments

To amplify the *lacZ* region containing the putative mutation site (as determined by complementation analysis), either one of two PCR primer sets is used: P13/P8 (α-β) or P14/P2 (β-ω) (Table I).

Reagents

10× buffer: Mix 100 mM Tris-HCl (pH 8.3) with 500 mM KCl and autoclave.
dNTP mix: Mix equal volumes of dATP, dCTP, dGTP, and dTTP at final concentration of 2.5 mM each.
$MgCl_2$: 25 mM $MgCl_2$.

Table I
lacZ Sequencing Primers

Primer No.[a]	Complementation region	Sequence	Base position
P2	ω	5′-TGC TGT TGA CTG TAG CGG CTG-3′	2915–2895
P3	ω	5′-GGA TGC GGT GTA TCG CTC GCC-3′	2654–2634
P4	ω	5′-CGC CAA TCC ACA TCT GTG-3′	2332–2315
P5	ω	5′-CGC GTT CGG TTG CAC TAC GC-3′	2133–2114
P6	β	5′-TGG ATG CGG CGT GCG GTC-3′	1863–1846
P7	β	5′-CTT CAT CCA CGC GCG CGT AC-3′	1543–1524
P8	β	5′-CGC CGG TAG CCA GCG CGG-3′	1323–1306
P9	β	5′-AGC GTG CCG TCG GCG GTG T-3′	983–965
P10	α	5′-ATG CCG CTC ATC CGC CAC-3′	641–624
P11	α	5′-GTG TAG ATG GGC GCA TCG-3′	320–303
P1	ω	5′-GGC GAG CGA TAC ACC GCA TCC-3′	2634–2654
P13	α-β	5′-AAT GTG TGG AAT TGT GAG-3′	−46 to −28
P14	β-ω	5′-TAC ACG CTG TGC GAC CGC-3′	1213–1230

[a]Primers P2–P11 are used for sequencing the leading DNA strand. Primer P1 is used for sequencing the 3′-terminal 506 bp. Primers P13 and P14 are used, in combination with P8 and P2, respectively, to amplify the corresponding DNA regions containing the putative mutation site as determined by complementation analysis.

PCR Procedure

1. Mix the following components in a PCR tube:

H_2O	68.5 μl
10× buffer	10.0 μl
dNTP mix (200 μM)	8.0 μl
Primer mix (0.2 μM)	2.0 μl (P13P8 or P14P2)
Taq polymerase	2.5 U
$MgCl_2$ (2.5 mM)	10.0 μl
DNA <1 μg 1.0	μl
Total	100.0 μl

2. Add 50 μl mineral oil into the tube after mixing, and put the tube into a PCR machine. Run the following program:
   a. 95°C for 2 min, then
   b. 30 cycles of: 95°C for 40 sec
      50°C for 30 sec
      70°C for 1 min
   c. followed by cooling to 4°C
3. 5 μl of reaction mixture is run in 0.7% agarose gel to check the products.
4. The PCR products are purified using a purification kit (Wizard™ PCR Preps, Promega Corp) or by phenol–chloroform extraction.

DNA Sequencing Using an End-Labeled Primer

To sequence the *lacZ* region containing the putative mutation site, the appropriate primers are selected from the set listed in Table I.

Primer End-Labeling

1. Combine the following in a 0.5-ml microcentrifuge tube:

Primer (10 pmole)	2.5 μl
[γ-^{32}P]-ATP (10 pmole)	0.5 μl
T4 kinase 10× buffer	1.0 μl
T4 kinase (1–10 U/μl)	1.0 μl
H_2O	6.0 μl

2. Incubate at 37°C for 30 min and then inactivate the T4 kinase at 90°C for 2 min. Briefly spin in a microcentrifuge to collect any condensation. The end-labeled primers may be stored at −20°C for up to a month.

Extension/Termination Reactions

1. For each set of sequencing reactions, label four 0.5-ml thin-wall tubes (A, T, G, C). Add 2 μl of appropriate d/ddNTP mix to each tube. Cap the tubes and store on ice or at 4°C until needed.
2. For each set of four sequencing reactions, mix the following reagents in a microcentrifuge tube:

PCR products (4–40 fmole)	5.0 μl
Sequencing 5× buffer	5.0 μl
Labeled primer (1.5 pmole)	1.5 μl
Sterile H_2O	4.5 μl
Tag (5 U)	1.0 μl

3. Mix briefly by pipetting up and down. Add 4 μl of the mix to the inside wall of each tube containing d/ddNTP mix.
4. Add one drop (approx. 20 μl) of mineral oil or paraffin light oil to each tube and briefly spin in a microcentrifuge.
5. Preheat the PCR machine to 95°C and then place the tubes in the machine and start the cycling program:
   a. 95°C for 2 min, then
   b. 30 cycles: 95°C for 40 sec
      50–60°C for 30 sec*
      70°C for 1 min
   c. followed by cooling to 4°C.
6. Add 3 μl stop solution and centrifuge down. Heat the tube to 80°C for 2 min before loading.

   *Primer annealing conditions:

50°C 30 sec	P4
55°C 30 sec	P2, P3, P8, P11
60°C 30 sec	P1, P5, P6, P7, P9, P10

DNA Sequencing Using Direct Incorporation

Extension/Termination Reactions

1. For each set of sequencing reactions, label four 0.5-ml microcentrifuge tubes (A,T,G,C). Add 2 μl of the appropriate d/ddNTP mix to each tube. Cap the tubes and store on ice or at 4°C until needed.

2. For each set of four sequencing reactions, mix the following reagents in a tube:

PCR product (500 fmole of either P13P8 or P14P2 products)	5.0 μl
Primer (25 ng) 3.0 pmole	1.5 μl
[α-^{35}S]dATP (>1000 Ci/mmole, 10 μCi/μl)	0.5 μl
Sequencing buffer 5×	5.0 μl
H_2O	4.0 μl

3. Add 1.0 μl of sequencing-grade Taq DNA polymerase (5 U/μl) to the above mix. Mix briefly.
4. Add 4 μl of the mix to inside wall of each tube containing d/ddNTP mix.
5. Add one drop of mineral oil and briefly spin.
6. Place the reaction tubes in a thermal cycler which has been preheated to 95°C and start the cycling program.

2.2. Plasmid System

2.2.1. Transgenic Animals

The lacZ plasmid C57Bl/6-derived transgenic mouse strain used for *in vivo* mutation detection and characterization has been described by Gossen *et al.* (1995) and Dollé *et al.* (1996). The transgene is based on the pUR288 plasmid containing the complete 3096-bp *E. coli lacZ* gene (Fig. 29.1). Several transgenic lines were made by microinjection. Thus far, experiments were only performed with line 60, which harbors about 20 copies of the plasmid integrated in a head-to-tail fashion at two chromosomal sites: chromosome 3 and chromosome 4. Animals from line 60 were bred into homozygosity and maintained in the animal facilities of the Beth Israel Hospital.

2.2.2. *E. coli* Strain

The *E. coli* host strain used in this assay is *E. coli* C, Δ*lacZ*$^-$, *galE*$^-$. It is kanamycin resistant and sensitive to galactose.

2.2.3. Reagents

1. 5× binding buffer: 50 mM Tris-HCl (made from a 1 M stock solution, pH 7.5), 5 mM EDTA, 50 mM $MgCl_2$, 25% glycerol (v/v; 31.5 g/100 ml). Adjust pH to 6.8 with HCl and filter-sterilize the solution.
2. IPTG stock solution: 25 mg IPTG (isopropyl β-D-thiogalactopyranoside; Sigma) in ultrapure water. Filter-sterilize and store at −20°C.
3. IPTG-elution buffer: 10 mM Tris-HCl (made from a 1 M stock solution, pH 7.5), 1 mM EDTA, 125 mM NaCl. Filter-sterilize the solution.
4. ATP solution: 10 mM ATP (adenosine 5′-triphosphate; Sigma) in ultrapure water. Filter-sterilize and store at −80°C in 50 μl aliquots.
5. *Hin*dIII restriction enzyme (New England Biolabs; 20 U/μl).
6. Restriction endonuclease NEBuffer #2 (New England Biolabs).
7. Magnetic beads: Dynabeads M-450 sheep anti-mouse IgG (4 × 10^8 beads/ml; Dynal)
8. Anti-β-galactosidase monoclonal antibody (2 mg/ml; Promega).
9. LacI/LacZ fusion protein (available from Dynal, Norway, on request).
10. Phosphate-buffered saline (GIBCO BRL; without $MgCl_2$).
11. 0.1× T4 DNA ligase: T4 DNA ligase (GIBCO/BRL; 1 U/μl) is diluted 10× in 1× T4 DNA ligase buffer (GIBCO/BRL).
12. Glycogen: 20 μg glycogen (Boehringer)/μl ultrapure water.

13. Sodium acetate: 3 M sodium acetate, pH 4.9, in ultrapure water.
14. SOB medium: 20 g bacto-tryptone (Difco), 5 g bacto yeast extract (Difco), 0.5 g NaCl, 10 ml 250 mM KCl (add after everything else is dissolved) per liter distilled water. Adjust pH to 7.0 with 5 N NaOH and autoclave for 20 min (liquid cycle). Add $MgCl_2$ to a final concentration of 5 mM just before use ($MgCl_2$ should be autoclaved separately for 20 min).
15. LB medium: LB Broth Base (20 g/liter Lennox L Broth Base; GIBCO/BRL). Autoclave for 20 min (liquid cycle).
16. LB topagar: 6.125 g/liter LB Broth Base (GIBCO/BRL) and 6.125 g/liter Antibiotic Medium 2 (Difco).
17. Kanamycin: use as a 50 mg/ml solution (Sigma).
18. Ampicillin: make a stock solution of 25 mg/ml in ultrapure water (Sigma).
19. X-gal: 5-bromo-4-chloro-3-indolyl-β-D-galactoside; use as a 50 mg/ml solution (Promega).
20. P-gal: phenyl-β-D-galactoside; use as powder (Sigma).
21. Tetrazolium: 2,3,5-triphenyl-2*H*-tetrazolium chloride; use as powder (Aldrich).
22. $MgCl_2$: 1M $MgCl_2$ in ultrapure water.

Note: It is important that all reagents are absolutely clean and contain no traces of plasmids (the aim of the assay is to detect $lacZ^-$ plasmids at very low frequency).

2.2.4. Preparation of LacI–LacZ Magnetic Beads

One milliter of magnetic bead slurry is washed with 1 ml PBS, using a magnetic particle concentrator (Dynal) and resuspended in 850 μl PBS. Then 150 μl anti-β-galactosidase monoclonal antibody is added and the mixture is incubated for 1 hr at 37°C while rotating. After incubation, the beads are washed three times with 1 ml PBS, resuspended in 900 μl PBS plus 100 μl lacI–lacZ fusion protein. The mixture is incubated for 2 hr at 37°C while rotating. Subsequently, the beads are washed three times with 1 ml PBS and resuspended in 1 ml PBS. The lacI–lacZ beads can be stored at 4°C for at least 6 months.

2.2.5. Tissue Collection

Following sacrifice of the mice by CO_2 inhalation and decapitation, remove all tissues, place in 1.5-ml Eppendorf vials, and freeze on dry ice. Maintain samples at −80°C until used.

2.2.6. Extraction of Genomic DNA

Genomic DNA from mouse organs and tissues is isolated essentially as described for the bacteriophage lambda system. It is possible that, due to the much smaller size of the plasmid vector cluster in the mouse genome, less gentle and more rapid DNA extraction methods can be applied. Indeed, preliminary results obtained with the Qiagen Blood and Cell Culture DNA Maxi Kit indicated no significant differences in terms of rescue efficiency or mutant frequency.

2.2.7. Preparation of Electrocompetent Cells

1. Add 50 μl of the $lacZ^-$ $galE^-$ *E. coli* C strain glycerol stock and 5 μl kanamycin to 10 ml LB medium in a 50-ml Falcon tube. Grow the cells overnight in an Innova 4000 incubator shaker (New Brunswick Scientific) at 37°C at 250 rpm.

2. Add 1.5 ml of the overnight culture to each of two 1-liter Erlenmeyer flasks containing 500 ml LB medium. Grow the cells to an OD_{600} of 0.35–0.45. Distribute the cell suspensions over twenty 50-ml Falcon tubes and place on ice for 30 min.
3. Centrifuge for 15 min at 4000 rpm in a Beckman R3C3 centrifuge. Resuspend the pellets in 25 ml ultrapure water and combine two tubes resulting in a total of ten tubes. Centrifuge again for 15 min at 4000 rpm, resuspend in 25 ml ultrapure water, and repeat one more time (do not combine tubes).
4. Resuspend the pellets in 9 ml 10% glycerol and combine five tubes resulting in a total of two tubes only. Centrifuge for 20 min at 4000 rpm, resuspend each pellet in 1.5 ml 10% glycerol, and combine the contents of the two tubes. The OD600 should be 57, i.e., 10 μl cell suspension in 3 ml LB medium should have an OD_{600} of 0.19. If necessary, adjust the density of the suspension.
5. Distribute the suspension in portions of 250 μl in Eppendorf vials and freeze directly in a dry ice/ethanol bath. The electrocompetent cells can be kept at −80°C for several months.

2.2.8. Magnetic Bead Rescue of *LacZ* Plasmid from Mouse Genomic DNA

1. Pellet 60 μl of lacl-lacZ magnetic beads on the magnetic particle concentrator, discard the supernatant and resuspend in a pre-made mixture of 15 μl 5× binding buffer, 2 μl *Hind*III (40 U), 10–50 μg genomic DNA (in 58 μl); use a vortex. Incubate for 1 hr at 37°C while rotating.
2. After incubation, wash the beads once with 250 μl 1× binding buffer (vortex gently) and resuspend in 75 μl IPTG-elution buffer plus 5 μl IPTG stock solution (vortex gently). Add 20 μl NEBuffer #2, 1 μl *Hind*III, and 100 μl ultrapure water (vortex gently). Incubate for 30 min at 37°C while rotating.
3. Incubate at 65°C for 20 min and allow to cool to room temperature before adding 2 μl ATP solution (final concentration: 0.1 mM) and 1 μl 0.1× T4 DNA ligase (total amount: 0.1 U). Then, incubate for 1 hr at room temperature.
4. Spin all drops down, pellet the beads, and transfer the supernatant to a clean tube. Precipitate the DNA for 1 hr at −80°C, after adding 1.5 μl glycogen (30 μg), 0.1 vol sodium acetate (22 μl; vortex), and 2.5 vol 95% ethanol (560 μl; mix). Centrifuge for 30 min in an Eppendorf centrifuge at full speed, remove the ethanol, wash once with 250 μl 70% ethanol (vortex) and centrifuge 5 min in the same centrifuge at full speed. Remove all ethanol (use a pipet with a fine tip to remove the last traces) and allow the DNA pellet to dry for 10–15 min. Resuspend the DNA in 5 μl ultrapure water. The electrocompetent cells can now be added.

2.2.9. Electroporation, Plating, and Mutant Counting

1. Thaw electrocompetent cells on ice. Once thawed, directly add 60 μl cell suspension to the tubes containing the 5 μl DNA solutions which are placed on ice. For transfer of the samples to electroporation cuvettes, place one tube at a time on the magnetic particle concentrator (to prevent carryover into the cuvette of magnetic beads possibly still left with the DNA).
2. Transfer the cells with the DNA to a prechilled electroporation cuvette (0.1-cm electrode gap). Electroporate at 1.8 kV with 25 μF (Gene Pulser, Bio-Rad) and 200 Ω (Pulse Controller, Bio-Rad). Immediately add 935 μl ice-cold SOB medium. Then transfer into

a 15 ml tube with another 1 ml SOB medium. Incubate for 30 min at 37°C while shaking (225 rpm). The time constant of the electroporation event should be 4.4–4.8.

3. Add 2 μl of the transformed cells (1 : 1000) to 2 ml SOB medium and combine with 13 ml LB topagar, containing ampicillin (end concentration: 75 μg/ml), kanamycin (end concentration: 25 μg/ml), X-gal (end concentration: 75 μg/ml) and tetrazolium (end concentration: 75 ng/ml). Plate in 9-cm petri dish. This is the titer plate.
4. Add 13 ml LB topagar to the rest of the transformed cells. In addition to ampicillin, kanamycin, and tetrazolium, add P-gal in an end concentration of 0.3% (add P-gal directly as a powder to the topagar before adding the topagar to the transformed cells). X-gal is not necessary, but can be added also. Plate in 9-cm petri dish. This is the selective plate.
5. Both titer and selective plate(s) are grown overnight at 37°C, but not longer than 18 hr. The titer plate indicates the rescue efficiency. A typical yield is about 50,000 colonies (plasmid copies)/μg genomic DNA. Mutant frequencies are determined as the ratio of the number of colonies on the selective plates (visible as sharp dark-red points) versus the number of colonies on the titer plates (dark-red points with a much larger blue halo) times the dilution factor (1000 in this case). Mutant counting can best be done on a light table.

2.2.10. Mutant Characterization

For mutant characterization we thus far have experience only with restriction enzyme analysis to determine plasmid and target gene size changes. However, it is reasonable to assume that for the detection of point mutations comparable sequencing procedures as described for the bacteriophage lambda system can be applied.

1. Mutant colonies are picked from the P-gal plates and grown overnight in 3 ml LB medium with kanamycin and ampicillin in 15-ml polystyrene tubes (Fisher Scientific) in an Innova 4000 incubator shaker (New Brunswick Scientific) at 37°C at 250 rpm.
2. Pallet the cells by centrifugation for 10 min in a Beckman R3C3 centrifuge at 1000 rpm (about 1000*g*).
3. Remove the supernatant and prepare plasmid DNA (e.g., using the INSTAprep kit of 5 Prime → 3 Prime, Inc., Boulder, CO); for this purpose, resuspend the pellet in 100 μl TE buffer (provided with the kit) and continue according to procedure option II in the manual.
4. Of the 100 μl plasmid-containing solution obtained with the kit, 6 μl is digested with 20 U *Sac*I and 20 U *Pst*I in NEBuffer #1 (New England Biolabs) in a total volume of 20 μl, for at least 1 hr.
5. After adding 2 μl loading buffer, containing glycerol and bromophenol blue, the double digest is resolved on a 1% agarose (ultrapure, GIBCO/BRL) gel with ethidium bromide in the gel and buffer. On a UV transilluminator, size changes of 50 bp on average are detectable.

3. DISCUSSION

The development of the P-gal positive selection method for detection of *lacZ*$^-$ mutant plaques (Gossen *et al.*, 1992; Mientjes *et al.*, 1994), due to its much greater efficiency, has greatly facilitated the design and conduct of mutagenesis experiments using *lacZ* transgenic mice. The P-gal selection method for mutant plaques detects all mutants in which β-galactosidase activity is undetectable (clear plaques). However, because the P-gal selection is based on biochemical reaction, it is not unexpected that a subset of mutants with partial expression of β-galactosidase will be undetected (Table II). Nevertheless, the results obtained with the P-gal system closely parallel those obtained using visual identification with X-gal (Fig. 29.4).

With the plasmid system the results obtained are somewhat different. The *E. coli* strain used in this system does not contain the plasmid overexpressing the *galT* and *galK* gene. This could explain the observation of as much as 30% light-blue colonies among P-gal-resistant cells rescued from untreated animals (Dollé *et al.*, 1996). Although, the mutant status of these colonies has thus far not been confirmed by sequencing, we assume that they are color mutants due to incomplete inactivation of the *lacZ* gene. This is also suggested by our observation of a significantly higher percentage of such color mutants in ENU-treated animals. Since ENU may induce primarily point mutations, one would expect a higher chance of finding color mutants after treatment with this agent than without treatment. In fact, whereas about 35% of all spontaneous mutants appeared to be deletions of up to more than 3000 bp, in ENU-treated animals only about 5% deletions were found (Dollé *et al.*, 1996; Fig. 29.5). Color mutants were exclusively found among the nondeletion mutants, i.e., plasmid restriction digests indicating no change on an agarose gel.

An important characteristic of transgenic mouse models for mutation detection is their sensitivity to induced mutations. To some extent, this is determined by the background mutant frequency, that is, the mutant frequency observed in untreated animals. While the spontaneous mutant frequency in the lambda model appeared to be on the order of $2–4 \times 10^{-5}$, with the plasmid model values of $5–10 \times 10^{-5}$ are routinely found. It seems obvious to ascribe this to the higher sensitivity of this model to deletions, but other additional explanations should also be considered. For example, it is not inconceivable that the electroporation process itself induces damage to the vector which is processed in *E. coli* into mutations. Results obtained with purified plasmid prepared directly from bacteria, rather than derived from the mouse, and electroporated into the $galE^-$ host suggest that this is not the case. In these experiments the average mutant frequency observed was about 1×10^{-5}, which is less than the values found after replating wild-type blue plaques in the lambda-based transgenic mouse assay. These data suggest that in both the lambda phage and the plasmid system, *E. coli* contributes significantly to the mutation spectra obtained. With respect to the plasmid system some preliminary evidence indicates that most of the *E. coli* mutations originate during the culture period necessary to prepare the plasmid and not during or after electrotransformation (Dollé *et al.*, 1996).

Another possible explanation for higher mutant frequencies in the plasmid model is the chromosomal location of the vector, which appeared to be a significant factor in determining mutant frequencies of transgenes (Gossen *et al.*, 1991).

While only very few experiments have thus far demonstrated the usefulness of the plasmid model, ample experience with the bacteriophage lambda system has confirmed the applicability

Table II
Ability of P-gal Selection to Detect X-gal Mutants[a]

X-gal phenotype	No. detected originally with X-gal	No. selected subsequently with P-gal	Percent of X-gal mutants detected with P-gal
Clear	52	52	100
Light blue	62	52	84

[a]Mutants selected with X-gal were sequenced, and subsequently replated with P-gal.

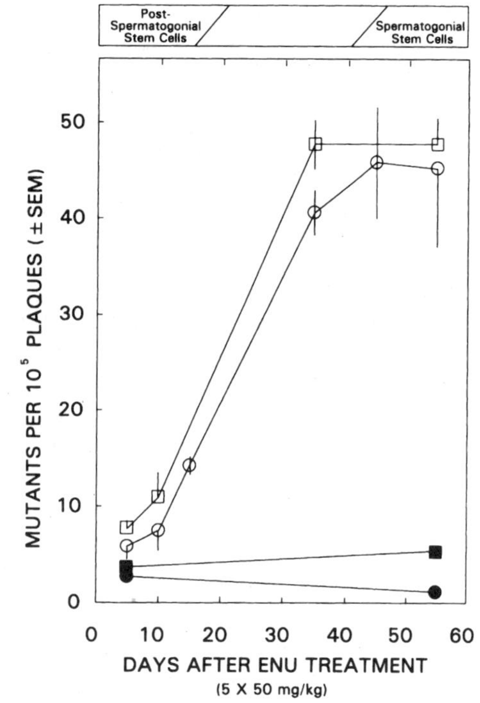

Figure 29.4. Comparison of the temporal response of testicular germ cells isolated from seminiferous tubules to ethylnitrosourea (ENU) in *lacZ* transgenic mice using the original visual (X-gal) and the new positive selection (P-gal) methods of mutant detection. Mice were treated with five daily i.p. doses of 50 mg/kg ENU and subsequently sampled at various times. The bar above the figure indicates the cell population thought to contribute to the mutant frequencies measured in the seminiferous germ cell population samples. X-gal data from Douglas et al. (1995b) with permission.

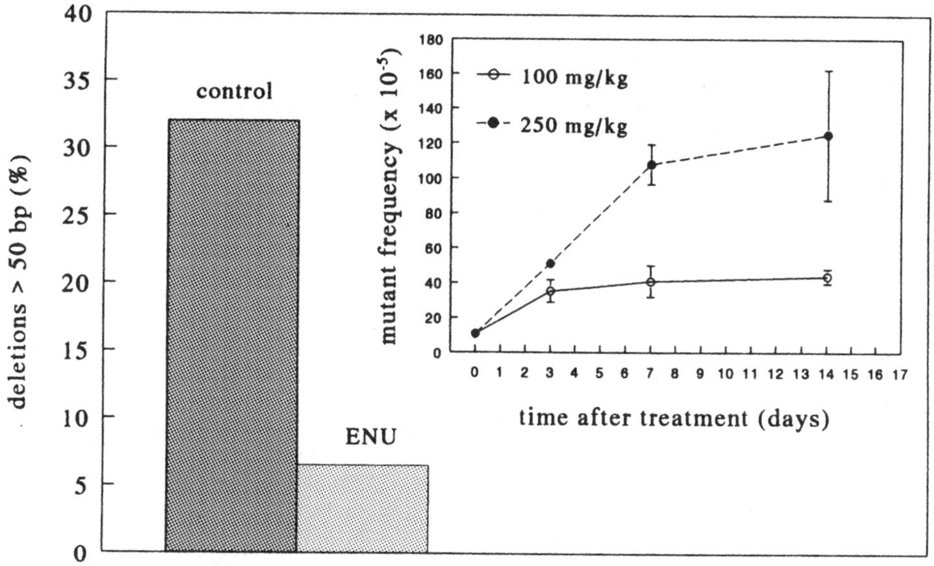

Figure 29.5. Percentage deletion mutations in pUR288 rescued from spleen genomic DNA of transgenic mice, untreated and 14 days after intraperitoneal injection of 250 mg/kg body wt ethylnitrosourea (ENU). The inset shows the induction of mutants by ENU with time.

of this latter model in genetic toxicology. Figure 29.4 demonstrates the utility of transgenic mouse methods for the study of gene mutations in germ cells. In this study, ENU is shown to induce both postspermatogonial and spermatogonial stem cell mutations (Douglas *et al.*, 1995b), as detected in spermatocytes and spermatids isolated from seminiferous tubules, and predicted from the extensive literature on this chemical (reviewed by Shibuya and Morimoto, 1993). With further validation, it is possible that such transgenic mouse assays may serve to complement the dominant lethal assay in the identification of germ cell hazards. It is anticipated, however, that tests involving offspring will still be required for the quantification of genetic risk to future generations.

The DNA sequencing of mutations has facilitated the validation of the *lacZ* transgenic mouse assays by comparison of mutation spectra with those of endogenous genes. Table III shows the spectra of spontaneous mutations in male germ cells isolated from seminiferous tubules, and ENU-induced mutation spectra from these cells 55 days posttreatment. The spontaneous spectrum is consistent with those from endogenous genes (with allowances for the fact that size-change mutations, e.g., deletions greater than about 50 bp are underrepresented in a phage-based mutation detection system), and consists of predominantly AT→GC transitions (Douglas *et al.*, 1994). The ENU-induced germ cell spectrum is quite different from the spontaneous spectrum in that it contains a preponderance of mutations in AT base pairs that is characteristic of the involvement of O^2– and O^4-thymine adducts in the etiology of mutation fixation (Douglas *et al.*, 1995b).

In conclusion, we have described two *lacZ* reporter gene transgenic mouse models for detecting and characterizing mutations *in vivo*. Both models employ the $galE^-$ positive selection system to rapidly score for mutations at a minimum of experimental time and effort. For the bacteriophage lambda-based model, there is an ever increasing list of publications describing studies on spontaneous and induced mutant frequencies and spectra in germ cells and somatic cells. In addition, the rescue assay associated with this model is very simple and also cost-

Table III
Summary of ENU-Induced $lacZ^-$ Mutations in Testicular Germ Cells of Transgenic Mice Obtained 55 Days After Treatment[a]

Mutation type		Spontaneous[b] *n* (CpG)[c]	%	ENU *n* (CpG)[c]	%
Transition	GC→AT	9(6)	37.5	10(3)	33.3
	AT→GC	1	4.2	5	16.1
	Subtotal	10	41.7	15	49.4
Transversion	CG→AT	5(5)	20.8	1	3.3
	TA→GC	1	4.2	2	6.5
	AT→TA	1	4.2	13	41.9
	GC→CG	3(2)	12.5	0	0
	Subtotal	10	41.7	16	51.7
Deletion		3	12.5	0	0
Insertion		1	4.2	0	0
Total		24	100	31	100

[a]Modified from Douglas *et al.* (1995b) with permission.
[b]Data for the number of spontaneous mutations are from Douglas *et al.* (1994).
[c]Number of mutations in CpG sites.

effective, in particular with the emergence of superior packaging extracts (Johnson, 1995). The plasmid model has the advantage of a higher sensitivity, in that it has the capability to detect large deletions and insertions including those that extend into the 3′ mouse flanking region (Gossen *et al.*, 1995). In addition, it has a very high rescue efficiency which cannot be surpassed by the lambda system. Indeed, at present up to 200,000 transformants/μg genomic DNA are obtained, with a theoretical maximum of 356,000/μg (transformation efficiency of 1×10^{10}/μg pure plasmid). Further validation of the plasmid model using a number of mutagens, including clastogens, is currently in progress.

ACKNOWLEDGMENTS. We thank Drs. Michael Boerrigter, Martijn Dollé, John Gingerich, Jan Glossen, Jianli Jiao, and Hans-Jörg Martus for their substantial contributions to the preparation of this chapter and to the development of methods for the two transgenic mutation models. Part of this work was supported by NIH grant 1PO1 AG10829-01.

REFERENCES

Advisory Committee on Mutagenesis, Department of National Health and Welfare and Department of Environment Canada. (1988). Guidelines for the use of mutagenicity tests in the toxicological evaluation of chemicals. *Environ. Mol. Mutagen.* **11**:261–304.

Ames, B. N., Shigenaga, M. K., and Hagen, T. M. (1993). Oxidants, antioxidants, and the degenerative diseases of aging. *Proc. Natl. Acad. Sci. USA* **90**:7915–7922.

Blakey, D. H., Douglas, G. R., Huang, K. C., and Winter, H. J. (1995). Cytogenetic mapping of λgt10 *lacZ* sequences in transgenic mouse strain 40.6 (Muta™Mouse). *Mutagenesis* **10**:145–148.

Bohr, V. A. (1991). Gene specific DNA repair. *Carcinogenesis* **12**:1983–1992.

Cook, A., and Lederberg, J. (1962). Recombination studies of lactose nonfermenting mutants of Escherichia coli K-12. *Genetics* **47**:1335–1353.

Dean, S. W., and Myhr, B. (1994). Measurement of gene mutation in vivo using MutaTMMouse and positive selection for lacZ⁻ phage. *Mutagenesis* **9**:183–185.

Dollé, M. E. T., Martus, H.-J., Gossen, J. A., Boerrigter, M. E. T. I., and Vijg, J. (1996). Evaluation of a plasmid-based transgenic mouse model for detecting *in vivo* mutations. *Mutagenesis* **11**:111–118.

Douglas, G. R., Gingerich, J. D., Gossen, J. A., and Bartlett, S. A. (1994). Sequence spectra of spontaneous *lacZ* mutations in transgenic mouse somatic and germline tissues. *Mutagenesis* **9**:451–458.

Douglas, G. R., Gingerich, J. D., and Soper, L. M. (1995a). Evidence for the nonmutagenicity of the carcinogen hydrazine sulfate in target tissues of *lacZ* transgenic mice. *Carcinogenesis* **16**:801–804.

Douglas, G. R., Jiao, J., Gingerich, J. D., Gossen, J. A., and Soper, L. M. (1995b). Temporal and molecular characteristics of mutations induced by ethylnitrosamine in germ cells isolated from seminiferous tubules and in spermatozoa of *lacZ* transgenic mice. *Proc. Natl. Acad. Sci. USA* **92**:7485–7489.

Evans, D. A. P., Burbach, J. P. H., and van Leeuwen, F. (1995). Somatic mutations in the brain: Relationship to aging? *Mutat. Res.* **338**:173–182.

Gossen, J. A., and Vijg, J. (1988). E. coli C: A convenient host strain for rescue of highly methylated DNA. *Nucleic Acids Res.* **16**:9343.

Gossen, J. A., and Vijg, J. (1993). Transgenic mice as model systems for studying gene mutations in vivo. *Trends Genet.* **9**:27–31.

Gossen, J. A., de Leeuw, W. J. F., Tan, C. H., Zwarthoff, E. C., Berends, F., Lohman, P. H. M., Knook, D. L., and Vijg, J. (1989). Efficient rescue of integrated shuttle vectors from transgenic mice: A model for studying mutations in vivo. *Proc. Natl. Acad. Sci. USA* **86**:7971–7975.

Gossen, J. A., de Leeuw, W. J. F., Verwest, A. M., and Vijg, J. (1991). High somatic mutation frequencies in a lacZ transgene integrated on the mouse X chromosome. *Mutat. Res.* **250**:423–429.

Gossen, J. A., Molijn, A. C., Douglas, G. R., and Vijg, J. (1992). Application of galactose-sensitive E. coli strains as selective hosts for LacZ plasmids. *Nucleic Acids Res.* **20**:3254.

Gossen, J. A., de Leeuw, W. J. F., Bakker, A. Q., and Vijg, J. (1993a). DNA sequence analysis of spontaneous mutations at the lacZ locus integrated on the mouse X chromosome. *Mutagenesis* **8**:243–247.

Gossen, J. A., de Leeuw, W. J. F., Molijn, A. C., and Vijg, J. (1993b). Plasmid rescue from transgenic mouse DNA using lacI repressor protein conjugated to magnetic beads. *BioTechniques* **14**:624–629.

Gossen, J. A., Martus, H.-J., Wei, J. Y., and Vijg, J. (1995). Spontaneous and X-ray induced deletions in a lacZ plasmid-based transgenic mouse model. *Mutat. Res.* **331**:89–97.

Health Protection Branch Geotoxicity Committee (1992). The assessment of mutagenicity: Health protection branch mutagenicity guidelines. *Environ. Mol. Mutagen.* **21**:15–37.

Johnson, M. G. (1995). Increased efficiency of rescue of a lambda shuttle vector from transgenic mouse DNA using MaxPlax™ packaging extract. *J. NIH Res.* **7**:84.

Knöll, A., Jacobson, D. P., Kretz, P. L., Lundberg, K. S., Short, J. M., and Sommer, S. S. (1994). Spontaneous mutations in lacI-containing λ lysogens derived from transgenic mice: The observed patterns differ in liver and spleen. *Mutat. Res.* **311**:57–67.

Kohler, S. W., Provost, G. S., Kretz, P. L., Dycaico, M. J., Sorge, J. A., and Short, J. M. (1990). Development of a short-term, in vivo mutagenesis assay: The effects of methylation on the recovery of a lambda phage shuttle vector from transgenic mice. *Nucleic Acids Res.* **18**:3007–3013.

Kohler, S. W., Provost, G. S., Fieck, A., Kretz, P. L., Bullock, W. O., Sorge, J. A., Putman, D. L., and Short, J. M. (1991). Spectra of spontaneous and mutagen-induced mutations in the lacI gene in transgenic mice. *Proc. Natl. Acad. Sci. USA* **88**:7958–7962.

Lee, T. A., DeSimone, C., Cerami, A., and Bucala, R. (1994). Comparative analysis of DNA mutations in lacI transgenic mice with age. *FASEB J.* **8**:545–550.

Lohman, P. H. M., Vijg, J., Uitterlinden, A. G., Slagboom, P., Gossen, J. A., and Berends, F. (1987). DNA methods for detecting and analyzing mutations in vivo. *Mutat Res.* **181**:227–234.

Mientjes, E. J., vanDelft, J. H. M., op'tHof, B. M., Gossen, J. A., Vijg, J., Lohman, P. H. M., and Baan, R. A. (1994). An improved selection method for *lacZ*$^-$ phages based on galactose sensitivity. *Transgenic Res.* **5**:67–68.

Myhr, B. C. (1991). Validation studies with Muta™Mouse: A transgenic mouse model for detecting mutations in vivo. *Environ. Mol. Mutagen.* **18**:308–315.

Newton, W. A., Beckwith, J. R., Zipser, D., and Brenner, S. (1965). Nonsense mutants and polarity in the Lac operon of *Escherichia coli*. *J. Mol. Biol.* **14**:290–295.

Rüther, U., and Müller-Hill, B. (1983). Easy identification of cDNA clones. *EMBO J.* **2**:1791–1794.

Shibuya, T., and Morimoto, K. (1993). A review of the genotoxicity of 1-ethyl-1-nitrosourea. *Mutat. Res.* **297**:3–38.

Suzuki, T., Hayashi, M., Sofuni, T., and Myhr, B. C. (1993). The concomitant detection of gene mutation and micronucleus induction by mitomycin C in vivo using lacZ transgenic mice. *Mutat. Res.* **285**:219–224.

Tinwell, H., Lefevre, P. A., and Ashby, J. (1994). Response of the Muta™Mouse lacZ/galE$^-$ transgenic mutation assay to DMN: Comparison with the corresponding Big Blue™ (lacI) responses. *Mutat. Res.* **307**:169–173.

Vijg, J., and Gossen, J. A. (1993). Somatic mutations and cellular aging. *Comp. Biochem. Physiol.* **104B**3:429–437.

Chapter 30

The Use of *lacI* Transgenic Mice in Genetic Toxicology

Johan G. de Boer, Heather L. Erfle, David Walsh, James Holcroft, and Barry W. Glickman

1. INTRODUCTION

Exposure to genotoxic agents can have a significant impact on human health. In each country, regulatory agencies have been created to establish safeguards to protect the public by minimizing these risks, usually by reducing or eliminating exposures. When this is not possible, estimates are required to quantitate risks so that a risk–benefit analysis is possible. To estimate risk, the field of genetic toxicology has embraced a number of assay systems to evaluate the genetic toxicity. Most well known is the Ames/Salmonella test, which uses a series of tester strains, with and without addition of metabolic activating S9 liver microsome extracts. This assay is generally considered to be the backbone of what has become known as "Tier One" testing and is used to screen all new chemicals and pharmaceuticals that may be introduced into the environment. The Ames assay is usually the first attempt to determine mutagenic potential (Maron and Ames, 1983). Cytogenetic toxicity is also evaluated using mammalian systems, such as the mouse lymphoma *in vitro* assay (MLA) (Oberly *et al.*, 1984; Hozier *et al.*, 1981), the micronucleus test (MN) (Wild, 1978), and chromosomal aberrations (CA) and sister chromatid exchange assay (SCE) (Latt, 1974). Damage to the germ line can be assessed in rodents by the specific locus test (Russell and Russell, 1992; Russell *et al.*, 1981). Carcinogenic potential can be determined in the National Toxicology Program assay (Chhabra *et al.*, 1990), a 2-year-long exposure of mice or rats to the potential carcinogen, a costly and time-consuming exercise that requires large numbers of animals.

One of the most relevant biomarkers of genotoxicity and potentially carcinogenesis is the occurrence of mutations in proto-oncogenes or tumor suppressor genes. Mutations are frequently found in these genes in cells in tumor tissues. The specificity of the changes found in oncogenes in tumors may provide a strong incentive for determining mutational specificity as well as genotoxic potency of compounds. Studies on mutational specificity have shown mutational spectra to

---

Johan G. de Boer, Heather L. Erfle, David Walsh, James Holcroft, and Barry W. Glickman • Center for Environmental Health, Department of Biology, University of Victoria, Victoria, British Columbia V8W 2Y2, Canada.

Technologies for Detection of DNA Damage and Mutations, edited by Gerd P. Pfeifer, Plenum Press, New York, 1996.

be unique for each chemical and physical agent (Glickman *et al.*, 1995). Consequently, the study of mutational specificity can provide insights into the mechanisms of mutation. The screening and sequence analysis of mutations in bacterial systems, e.g., *lacI* (Miller, 1972; Lavigueur and Bernstein, 1993; Zielenska *et al.*, 1993; Schaaper and Dunn, 1991; Gordon *et al.*, 1990, 1991; LeClerc *et al.*, 1988; Schaaper *et al.*, 1986), is relatively fast and a large data base has been accumulated. Mammalian systems for the study of mutational specificity include the *aprt* gene in Chinese hamster cell cultures (Drobetsky *et al.*, 1987; De Boer and Glickman, 1989; De Jong *et al.*, 1988; Mazur and Glickman, 1988; De Boer *et al.*, 1989; Meuth *et al.*, 1990), and the *hprt* gene for monitoring mutations in humans (Albertini *et al.*, 1989; Branda *et al.*, 1993; Cole *et al.*, 1990; Perera *et al.*, 1993).

The data indicate that many carcinogens are highly specific with regard to their target tissue. This may reflect dependence on tissue-specific metabolic activation, the organ-specific environment, or both. These factors make the results obtained from tissue culture difficult to interpret. Ideally, mutations should be determined *in vivo*, e.g., in a real animal. Such a system would require a mutational target that can be retrieved and assayed in a facile manner from any organ or tissue of the experimental animal. Transgenic technology has made such a system practical, and at least two systems are currently commercially available. One of these systems, Big Blue® (Kohler *et al.*, 1991a), utilizes the *lacI* gene, the other, MutaMouse™ (Gossen *et al.*, 1989), the *lacZ* gene. A clever system using a ϕX174 vector has also been developed (e.g., Burkhart and Malling, 1989).

In addition to determining the induced mutant frequencies after treatment, i.e., mutagenic potency, these systems also facilitate the analysis of the exact mutagenic event at the molecular level. These transgenic systems thus permit "mutational spectroscopy." This is extremely relevant as mutagens generally produce specific alterations at the DNA and mutational level that are thought to be indicative of exposure to the specific chemical. Moreover, a very modest increase in mutant frequency after treatment may not be significant at the statistical level, but when the induced changes show a high degree of specificity, a significant effect may be readily observed. This extends both the sensitivity and the value of the mutagenesis assay significantly.

2. SPECIES- AND TISSUE-SPECIFIC CARCINOGENESIS

Exposure to a selected chemical often results in tumors in only mice, or only in rats. The organ affected may also differ, sometimes depending on species, sex, or both (Gold *et al.*, 1989). The species, gender, and tissue specificity of induced cancer is of particular concern when extrapolating animal data to human exposure. It is extremely difficult to reconcile that rat data cannot be directly extrapolated to mouse and vice versa, let stand to humans.

As examples of tissue specificity, o-anisidine exposure results in bladder tumors in mice (NTP, 1978), streptozotocin causes kidney tumors (Liegibel *et al.*, 1992), while DMN causes liver tumors (International Agency for Research on Cancer, 1978). These differences may be the result of differences in metabolism in different animal species and tissues. Although many compounds are metabolized by mixed-function oxidases in the liver, chemical exposures often result in tumor development elsewhere. Benzene, for example, is initially metabolized in the liver to phenol, hydroquinone, catechol, and benzene-triol, after which some of its metabolites are further metabolized in the bone marrow by bone marrow-specific myeloperoxidases to semiquinones and hydroxy benzoquinone (Zhang *et al.*, 1993). It is in the bone marrow where the genotoxic effects are seen, likely because of associated free radical species. Mice metabolize benzene more effectively than do rats (Medinsky *et al.*, 1989), and there are differences in P450-mediated benzene metabolism between mice and rats that may explain some of the species differences

(Nakajima *et al.*, 1993). Mice are much more sensitive to 1,3-butadiene than are rats. We hypothesize that the species- and tissue-specific mutagenic response will parallel that observed for carcinogenicity. For example, such a relationship has been observed in the case of streptozotocin (Schmezer *et al.*, 1994) which is more mutagenic to the kidney (the carcinogenic target tissue) than the liver. A similar observation has been made for the bladder carcinogen *o*-anisidine which induces more mutations in the bladder than in the liver (Ashby *et al.*, 1994). The availability of transgenic animals for mutation research makes possible analysis of mutations in individual tissues. Now, as the same transgenic target is available in both mice and rats, interspecies differences can also be studied.

3. THE CHOICE OF TRANSGENIC SYSTEM

The transgenic animal systems have the unique advantage of permitting the study of real animals, i.e., the work can be conducted *in vivo*. In addition, the modern techniques permit the rapid recovery and assessment of mutation making these assays "short-term tests." Finally, they also permit the sequencing of the mutations so that mutational spectra can be produced where desired. We have chosen to work with the Big Blue® system (Kohler *et al.*, 1991a) in which the *E. coli lacI* is the mutational target. This makes these studies very much an extension of earlier work with mutational specificity in *E. coli* where the *lacI* gene was also the mutational target (e.g., Horsfall *et al.*, 1989; Zielenska *et al.*, 1993; Schaaper and Dunn, 1991; Gordon *et al.*, 1990, 1991; LeClerc *et al.*, 1988; Schaaper *et al.*, 1986). The *lacI* gene codes for the repressor of the Lac operon. Mutations in this protein permit the constitutive expression of β-galactosidase. Well over 25,000 LacI mutants have been genetically characterized after the treatment of the bacterial cells with a broad variety of chemical and physical agents. As a consequence, the *lacI* gene has been well characterized as a mutational target (Gordon *et al.*, 1988; Kleina and Miller, 1990).

Specifically, in Big Blue®, the *lacI* gene is present as part of a λ/LIZ construct. In addition to the target gene, the construct includes a bacterial origin of replication, the ampicillin resistance gene amp, and the first 675 bp of the *lacZ* gene. This fragment of the *lacZ* gene is essential to the selection systems as the encoded portion of the *lacZ* gene can complement (α-complementation) the bacterial host fragment (the ω-portion) to provide β-galactosidase enzyme activity which can then cleave the chromogenic compound 5-bromo-4-chloro-3-indolyl-β-D-galactopyranoside (X-gal) to produce an insoluble blue dye. The α-complementation assay is essential because it permits the identification of λ plaques carrying a mutated *lacI* gene. When the gene is mutated, the resulting λ plaques are blue. The *lacI* gene involves a 1254-bp fragment which can be amplified using the polymerase chain reaction (PCR) and directly sequenced. The convenient size of the *lacI* gene is an additional attraction. The *lacZ* gene in contrast is large and the MutaMouse™ incorporates a 3126-bp fragment of the *lacZ* gene which is too large to sequence conveniently. As a consequence, the *lacZ* mutational data base is currently limited.

4. THE *lacI* GENE AS A MUTATIONAL TARGET

The *lacI* gene has been used as a mutational target for almost 20 years (Miller, 1972; Lavigueur and Bernstein, 1993; Zielenska *et al.*, 1993; Schaaper and Dunn, 1991; Gordon *et al.*, 1990, 1991; LeClerc *et al.*, 1988; Schaaper *et al.*, 1986). Its extensive use in *E. coli* has provided a detailed understanding of its sensitivity to mutation. The gene codes for the LacI repressor protein which regulates the expression of the *Lac* operon. As a tetramer, the LacI protein binds to the *lacZ* operator sequence and represses the operon. The first 5′-terminal 200 bp of the *lacI* gene code for the DNA binding domain. This region, also called the NC^+ (negative complementation) region, is highly sensitive to base substitutions (Gordon and Glickman, 1988; Gordon *et al.*, 1988; Kleina and Miller, 1990). Consequently, numerous studies have been restricted to mutations in this part of the gene (Horsfall *et al.*, 1989; Schaaper and Dunn, 1987). Because of the costs of obtaining the *lacI* mutants in the mammalian system, however, we have generally chosen to examine mutations in the entire *lacI* coding and promoter sequence in the Big Blue® animals.

5. THE CONSTRUCTION OF THE *lacI* TRANSGENIC ANIMAL

The *lacI* system involved the design and construction of the shuttle vector/plasmid followed by its insertion/integration into the animal germ line. The construct of both the plasmid and the animal have been detailed elsewhere (Kohler *et al.*, 1991a), but are briefly restated here. The *lacI* gene, carried on a plasmid, was inserted into a bacteriophage λ shuttle vector. This resulted in the λ/LIZa vector (Kohler *et al.*, 1991a,b), which was then introduced into embryos of C57BL/6 mice by microinjection. Possible transgenic animals were then tested by Southern blotting of blood taken from the tail vein of offspring. A *lacI*-positive line (A1) from C57BL/6 was then crossed with an animal of the C3H line to produce the same genetic background as the National Toxicology Program bioassay test strain, B6C3F1. Approximately 40 copies of the λ shuttle vector are integrated at a single locus in a head-to-tail arrangement on chromosome 4 (Dycaico *et al.*, 1994). A parallel system has been developed using the Fisher F344 rat (Dycaico *et al.*, 1993, 1994). In addition to the potential study of species specificity, the transgenic rat also makes larger tissue samples available. In addition to the animals, a *lacI* transgenic rat cell line has been developed (Wyborski *et al.*, 1995).

6. DESCRIPTION OF THE λ/LIZ–LacI MUTANT PLAQUE RECOVERY AND ANALYSIS PROTOCOLS

The protocol for the analysis of mutations recovered from Big Blue® mice or rats involves several steps which have been standardized and are detailed in the Stratagene Big Blue® handbook that is included with the purchase of the system and is part of the Stratagene in-house training program (Rogers *et al.*, 1995). In short, after the Big Blue® animals are treated and sacrificed, the desired tissues are removed and stored in sealed packets at −80°C. Genomic DNA is then isolated from approximately 100-mg tissue samples using one of two methods. These involve either a gentle phenol extraction as described in the Stratagene Big Blue® manual (Rogers *et al.*, 1995) or the dialysis of the samples (Winegar *et al.*, unpublished). Extremely gentle methods are required as DNA of very high molecular weight is essential for efficient packaging into λ particles. Naturally, all handlings are done in the most gentle manner possible. For example, during pipetting small-bore pipette tips must be avoided as these would result in excessive shearing of the DNA. The purified mouse/rat DNA is then added to a λ packaging extract (Transpack®, Stratagene) which excises and packages individual λ genomes. The packaging can be carried out in smaller aliquots than recommended by the manufacturer. For example, three to five different samples can be packaged using a single tube producing a significant savings of materials, but each tube of packaging mix must be used in its entirety in a single set of reactions. The resulting bacteriophage particles can then be plated on SCS8, a modification-restriction deficient host bacterium which, in the presence of X-gal, will permit the identification of LacI mutants. Approximately 12,000 plaque-forming units (PFUs) are plated onto each 25 × 25-cm plate. Plating density is of critical importance. It is essential that the plaques do not overlap each other. Too high a plaque density impedes the accurate estimation of both wild-type and mutant plaque numbers which is essential for the reliable determination of mutant frequencies. In addition, if the plaques are too closely packed, their color becomes more difficult to determine. Plaque color is important and control plaques displaying a range of colors (CM-0 to CM-3) must be included in each experiment to ensure that plating conditions permit the detection of even the weakest of blue-tinted plaques. Blue plaques are most easily scored against either a yellow background (our laboratory) or through a red filter such as recommended by Stratagene. There does not appear to be a difference in efficiency between these two methods so it remains one of personal choice.

In addition to the above reasons, the plating of too high a phage concentration makes it difficult to distinguish pure and mosaic plaques. The evidence suggests that mosaic or sectored plaques are the result of mutation occurring during replication on the bacterial host. The logic is that at low multiplicities of infection, mutant phage will produce pure mutant clones whereas mutations that occur during the replication of phage in the bacterial hosts may often be mosaic as they will tend to occur in the presence of wild-type DNA. The size of the sector may thus be an indication of the round of replication in which the mutation occurred. The data are indeed consistent with this hypothesis as to the origin of mosaic plaques as under circumstances where the background mutation frequency is low, mosaic plaques are more prevalent. For example, they represent a significant portion of the blue plaques recovered in experiments involving germ line controls as these have the lowest mutant frequencies yet recorded (Kohler *et al.*, 1991b).

An alternative origin for mosaic plaques has been observed after mutagen treatment (Provost *et al.*, 1993). When the DNA is harvested too quickly following treatment, DNA can be packaged which contains nonprocessed DNA lesions. Under these circumstances, mosaic plaques can also be recovered as the mutations can arise through the processing of the lesions in the bacterial host. The use of SOS-deficient RecA bacteria minimizes the mutant yield, but they are nevertheless recovered at low frequency.

While mosaic plaques may at first appear to be a nuisance, they provide additional information about the course of the experiment, confirming low mutant frequencies, or the level of DNA damage and its repair in diverse tissues in the rodent. A spectrum of mutations has been obtained for LacI mutants occurring in the bacterial host (unpublished) so that their mosaic origin can be confirmed.

We must stress the importance of following the protocols established by Stratagene. These protocols were established only after extensive in-house and external testing and are essential in order to obtain results that are comparable to known standards. In addition, the protocols have been designed to minimize variation and maximize their statistical reliability. Recent inter- and intralaboratory comparisons have shown that errors introduced at various levels of the hierarchical sampling protocol can be minimized by adhering to certain minimum numbers of animals per data point (generally five) and plaque-forming units that are scored (generally 200,000 or so). More details can be found in a later section.

7. SELECTIVE SYSTEMS

The screening method originally developed by Stratagene involves the plating of packaged λ particles in the presence of the chromogenic compound, X-gal. As described above, mutants are identified as blue plaques among a background of clear plaques. Technical considerations dictate that perhaps fifteen to twenty 25 × 25-cm plates must be screened for each data point. While the screening of blue plaques against a white background is not as demanding as the reverse which is required for the plaque assay of MutaMouse™, direct screening assays in which only mutant phage are detected would greatly reduce the labor involved in screening mutants. A number of systems in which only mutant phage or bacteria lysogenic for mutant phage are detected have recently been described. The system developed for screening MutaMouse™-derived phage involves a *galE*-selective assay (Gossen *et al.*, 1992). Bacteria defective in the *galE* gene accumulate toxic intermediates in the galactose metabolic pathway. Since β-galactosidase converts lactose to glucose and galactose, this system can be used for *lacZ* mutant selection. For the Big Blue® mouse, similar selective systems are being developed (Lundberg *et al.*, 1993; Kretz *et al.*, 1992, 1993; Dycaico *et al.*, 1994). Expression of β-galactosidase results in cleavage of lactose into glucose and galactose. When supplied as the sole carbon source, the resulting glucose will permit the selective growth of mutant λ particles (Lundberg *et al.*, 1993). However, the initial results with this system (Knoll *et al.*, 1994) suggest that it is not straightforward. Mutant frequencies were found to increase and the mutational spectrum to change with the number of days of incubation of the bacterial selection plates (Knöll *et al.*, 1994, 1996). An alternative selection system has been developed placing the groE gene, which codes for an essential chaperone protein, under the control of the *lacI* repressor protein (Dycaico *et al.*, 1994). Derepression of the *lacZ* operator/*groE* construct through a mutation in the *lacI* gene allows λ to package and to form plaques. The addition of X-gal stains these plaques blue for confirmation. Initial results indicate again that the mutational spectra may depend on the system being used, and that selective systems allow for increased bacterial contributions to the recovered mutations (Kretz *et al.*, 1993).

While direct selection systems offer the advantage of ease of use, it must be remembered that the current plaque assay for Big Blue® has the advantage that mosaic plaques can indicate either the occurrence of mutations occurring in the bacterial host or the presence of DNA lesions in the packaged DNA. As might be expected, each system will have particular advantages and disadvantages and the researcher will have to make his or her own choice depending on their specific needs. Fortunately, the screening of blue plaques in the current system is not so arduous that it becomes a practical problem.

8. DNA SEQUENCING USING AUTOMATED SEQUENCERS

The production of mutational spectra requires the sequencing of substantial numbers of mutants. As this can be intensively labor intensive, the production of mutational spectra requires an efficient sequencing team, preferably with a division of responsibilities which might include plaque purification, PCR and DNA purification, the actual sequencing of the mutants, the analysis of the sequencing output, and, very importantly, the management of the mutational data. In our laboratory the sequencing is done using automated fluorescent DNA sequencers. Three of these are dedicated to the sequencing of *lacI* mutants. They are A.L.F. instruments manufactured by Pharmacia (Sweden).

The first step in the process is mutant plaque purification. All blue-tinged plaques are cored from the 25 × 25-cm plates with a Pasteur pipette and transferred to 100 μl SM buffer. Mutant plaques are then replating at low plaque density on SCS8 in the presence of X-gal so that a discrete blue plaque can be isolated for sequencing. An individual plaque is again cored and transferred to 100 μl SM buffer. The phage suspension can then be stored in sealed vials at 4°C until required for DNA sequencing.

The *lacI* DNA is directly amplified by PCR using a 5-μl aliquot of the phage suspension. Twenty picomoles of each of the two oligonucleotide primers specific for the 5′ and 3′ regions of the *lacI* gene are used. They are complementary to positions −53 to −37 (5′-CCCGACACCATCGAATG-3′)

and positions 1201 to 1185 (5′-ACAATTCCACACAACATAC-3′). Amplification is done in 30 cycles of 20 sec at 95°C, 30 sec at 59°C, and 60 sec at 72°C, after an initial denaturation step of 5 min at 95°C, with 10 units of *Taq* polymerase. The PCR buffer contains 15 mM Tris-HCl, pH 8.0, 2.75 mM $MgCl_2$, and 60 mM KCl. In our case the amplification is done in a PE-9600 thermal cycler. The amplification product is a 1254-bp fragment containing the 1080-bp *lacI* gene. The success of the PCR reaction is judged by examining 5 μl of the reaction on an agarose gel.

Effective DNA sequencing requires the partial purification of the PCR product. This can be efficiently done using Promega Wizard Prep columns attached to a vacuum manifold, according to the manufacturer's instructions. Twenty samples can be processed simultaneously. The processed DNA sample is then subjected to cycling sequencing with a primer labeled with fluorescine (Dalton Chemicals, Toronto, Canada). Two microliters of the purified DNA sample corresponding to approximately 50–100 ng is added to 1.5–2.0 μl primer (2.5 pmole/μl). The sequencing reactions are carried out with 5 units of *Taq* polymerase and 25 cycles of 10 sec at 94°C, 20 sec at 50°C, and 30 sec at 72°C in a PE-9600 thermal cycler. DNA sequencing is done on the Pharmacia A.L.F. automated sequencers as per the manufacturer's recommendations. The gel quality is critical to this process and extensive testing has shown that the most consistent results are obtained with Readymix Gel manufactured by Pharmacia.

The first round of DNA sequencing employs a primer that hybridizes at positions −41 to −20 in the *lacI* gene. This allows the reading of a minimum of at least the initial 300 bp at the 5′ end of the gene, and often over 600 bp. The 5′ end of the gene contains the DNA binding domain (± 200 bp), and approximately 70% of all mutants are found in the first 400 bp. A reverse primer, annealing at positions 1151–1128, is generally the second one used in the sequencing process as it usually provides the sequence of the second half of the gene. Other forward primers that work well and cover the remainder of the gene anneal to positions 249–271, 532–549, and 816–833. Additional reverse primers bind to positions 305–282, 573-550, and 848-825.

The initial analysis of the raw sequence data and base calling is performed following the completion of the sequence run, and the processed data are transferred by FTP to one of several Mac workstations. Sequence alignment of data from one or more runs from each mutant with the wild-type *lacI* sequence is done with Seqman v2.54 (DnaStar Inc., Madison WI). The inventory of mutant samples and the DNA sequence information is managed with a custom-built software package, developed in dBASE4 (De Boer, 1995). This program not only keeps track of the status of the mutant samples in each experiment, it also performs the analysis of the sequence data. The nature of each mutation and the associated amino acid alteration (if any) are tabulated. Mutational events for each treatment, animal, experiment, etc., are tabulated by class, and lists can be generated with the data broken down into the desired format. The data base now contains the sequences of over 5000 LacI mutants recovered from Big Blue® animals. The program can be obtained from Dr. J. de Boer on request.

For the first 1000 mutants, the entire gene was sequenced. Currently, we only sequence until a mutation is found. The remainder of the gene is sequenced only if there is a specific need. For example, if the first mutation identified were to be phenotypically silent, the rest of the gene would be sequenced. Similarly, in cases where the mutation has not been identified earlier, the mutation is first confirmed and the entire gene is sequenced. This is required to confirm that the mutation initially identified is responsible for the mutant phenotype. When no mutation is found, we sequence the remainder of the gene, starting from the 3′ end at primer position 1128–1151.

With a sequencing staff of three or four, we are generally able to complete the sequencing and analysis of approximately 100 mutant samples per week. This involves the running of up to two sequencing gels per DNA sequencer per day. Usually the sequence operation is handling between 300 and 500 mutants at some stage of the process.

9. LIMITATIONS OF THE SYSTEM

Despite its tremendous utility, the Big Blue® transgenic mutation assay has several specific limitations. One such limitation is related to the multiple copy integration of identical sequence units. The Big Blue® transgenic mouse has a single site of integration on chromosome 4 where approximately 40 tandemly repeated copies of the λ/*lacI* reside. The packaging of the bacteriophage λ chromosome requires between 38 and 51 kb of DNA between the λ *cos* ends (Dycaico *et al.*, 1994). The *lacI* gene, including the *Lac* promoter, is 1254 bp long. In order for mutants to be detected, the α-complement of the *lacZ* gene must be functional. In addition, a sufficient length of DNA is required for packaging. This limits recoverable "intragenic" deletions to a maximum of approximately 7500 bp (Dycaico *et al.*, 1994). However, "intergenic" deletions may have endpoints in the lacI sequence of different copies. Such events allow the recovery of deletions as large as 1.8 Mb (Dycaico *et al.*, 1994); however, since these deletions will appear as short intra-*lacI* deletions, or even as duplication events, they cannot readily be distinguished from "intergenic" deletions. A detailed discussion about the recovery of deletion mutants is given by Dycaico *et al.* (1994).

In mammalian cells methylation occurs at the 5-position of cytosine in 5′-CpG-3′ dinucleotide sequences. Methylated cytosines demonstrate increased susceptibility to spontaneous deamination resulting in their conversion to thymine (e.g., Holliday and Grigg, 1993). The resulting G:T mismatch, if uncorrected, can lead to G:C to A:T transitions on replication (Duncan and Miller, 1980). Indeed, such changes are the hallmark of spontaneous mutations in mammals. The high level of methylation of the *lacI* gene, however, also increases the background mutation frequency, thereby reducing the sensitivity of the assay when fold increases are considered. This is particularly the case when LacI mutants are not being sequenced, because when they are, the shift away from transitions at CpG sites is an extremely sensitive indicator of mutation induction.

An additional characteristic of the transgenic construct in this system is the fact that the *lacI* gene is not expressed in vivo. This is significant in that the repair of DNA by the nucleotide excision pathway has been shown to depend on transcription and to preferentially repair the transcribed strand (Sargentini and Smith, 1986; Vrieling *et al.*, 1991). The *lacI* gene is therefore not repaired in the same manner as an expressed endogenous gene and hence may produce spectra which do not resemble what would be recovered with a naturally occurring gene. This is particularly likely in terms of strand bias. On the other hand, the lack of repair may tend to increase the level of mutation and be an advantage in increasing the sensitivity of the system. Indeed, this is likely to be the case although under some circumstances the attempt at DNA repair may be required for mutation fixation.

Mutant clonality may also be a problem in tissues with a high mitotic index, i.e., where extensive cell proliferation occurs. Under such conditions, a mutant cell may clonally expand resulting in numerous identical mutants being recovered from the same tissue. Clonal expansion can thus cause an apparent increase in mutant frequency as well as skewing the mutational spectrum. In *in vitro* studies, this problem has traditionally been avoided by analyzing a single mutant from each culture. In the *in vivo* studies using the human *hprt* T-lymphocyte clonal assay, clonal relationships can be determined using the T-cell receptor gene rearrangements (De Boer *et al.*, 1993; Nicklas *et al.*, 1986, 1989). In the case of these *in vivo* transgenic animal studies, however, there is no obvious solution. Taking a single mutant per tissue would not appear to offer a practical option. We therefore stress the need to use several animals for each data point and then to estimate clonality by determining how many identical mutants are recovered from the same animal. The use of several animals not only improves the accuracy of the estimation of mutant frequencies, in combination with DNA sequencing, it permits clonal runs to be identified. Moreover, this kind of analysis permits the identification of mutational hot spots.

In practical terms we have already encountered several instances where the mutant frequency in one animal was severalfold higher than observed in other animals in the same study group. This was subsequently found to reflect clonal expansion. Indeed, in one case, all mutants recovered from lung, spleen, and bone marrow were identical, indicating an early mutational event during animal development (this laboratory, unpublished). This was accompanied by a mutant frequency of approximately 50-fold above background. Figure 30.1 illustrates the number of spontaneous base substitution mutants found at various positions in the first 400 bp of the *lacI* gene recovered from liver tissue of nine animals (unpublished results). Several hot spots are evident, e.g., in animals No. 3 and 4, sites where in several other animals no mutants were recovered.

As described earlier, the appearance of an excess of mosaic or sectored plaques can have significant ramifications. When this occurs following treatment, it likely reflects the packaging of DNA still containing DNA lesions. This may reflect either the tissue- or site-specific lack of DNA repair, or the processing of the DNA too quickly after treatment. The fraction of mosaic plaques may be as high as 45% depending on the mutant frequency (Piegorsch *et al.*, 1995) and was shown to be higher in ENU-treated animals compared to unexposed animals (Piegorsch *et al.*, 1995). In such cases, allowing for longer expression times should permit either more DNA repair, or the fixation of mutations *in vivo*. In either case, the consequence will be a smaller fraction of mosaic plaques.

10. DOSING CONSIDERATIONS

An important, and controversial, issue in genetic toxicology, recently addressed by Ashby and Liegibel (1992, 1993) and Mirsalis (1993), is the dosing regimen, i.e., a single high dose versus repeated low doses. Long-term doses, or even relatively short-term regimens (Mirsalis *et al.*, 1993a), can induce cell proliferation, and thereby may increase the fixation rate of DNA lesions during mitogenesis. Mitogenesis is thought to play an important role in mutation fixation and carcinogenesis. In addition, cell proliferation may result in an increase of the spontaneous mutation background, simply because of DNA synthesis activity. Realistic dosing regimes would probably mimic chronic exposures found in environmental situations. Current regimes call for five daily exposures followed by a 14-day expression period before sacrifice (Mirsalis, 1993).

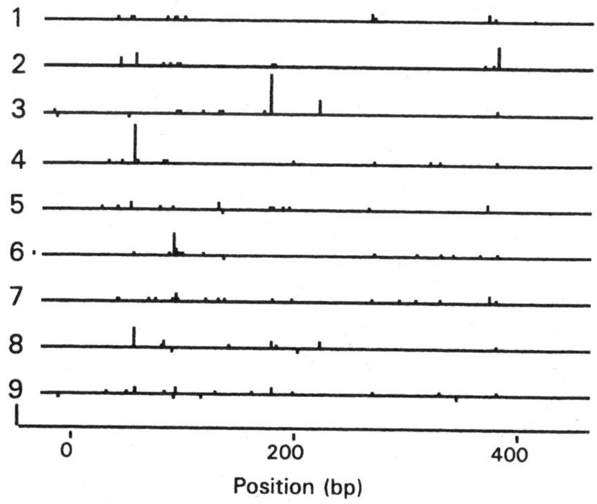

Figure 30.1. Distribution of base substitution mutations in the first 400 bp of the *lacI* gene recovered from nine animals. The length of the vertical bar on the horizontal base pair position axis indicates 5 mutants.

11. ASSAY STANDARDIZATION

The multistep procedure for the isolation of DNA, packaging, and screening for mutant plaques makes the control of technical variability critical. Variability may be introduced at each step. Rogers *et al.* (1995) have recently described an optimized protocol for mutant screening. An important aspect is the inclusion of known lacI mutants. The blueness of the plaques produced by these mutants [CM-0 = C(530)→T; CM-1 = C(179)→A; CM-2 = G(381)→G; CM-3 = C(977)→T)] increases progressively (Rogers *et al.*, 1995). CM-0 produces a very faint blue plaque under standard conditions. A red translucent filter, or a yellow background, enhances the detection of faint blue mutants such as CM-0 and CM-1. The consistent use of the recommended media components is critical to the uniform sensitivity of the plaque color assay.

Several reports have recently addressed the statistical aspects of the assay (Carr and Gorelick, 1995; Piegorsch *et al.*, 1994, 1995). Statistical significance naturally depends on the number of plaques screened, the number of animals used. A minimum of five animals per data point and a minimum of 100,000 to 200,000 PFUs, depending on mutant frequency, spread out over three packaging reactions per animal is considered optimal (Carr and Gorelick, 1994; Piegorsch *et al.*, 1995). There is little improvement in statistical power when a greater number of plaques are screened. However, this protocol may produce an insufficient number of mutants for sequence analysis, depending on the actual mutant frequency. A minimum of 80 to 100 mutants must be available per data point if mutational spectra are desired. The proper consideration of this factor is necessary in establishing the experimental design.

An important design consideration dictates the use of a "block design" in processing of the samples for the packaging reactions, in order to minimize interpackaging variability (Piegorsch *et al.*, 1995). This involves performing all packaging reactions of a "matched" set of control and treated animals at the same time. This reduces interexperiment variation. These conditions permit the detection of at least a doubling in the mutant frequency.

12. SENSITIVITY

The ability to detect mutations in the *lacI* gene still depends on a selectable phenotypic change. The ability to detect mutation is thus dictated by several factors including the degeneracy of the genetic code, the sensitivity of the protein to amino acid substitutions, and the availability of sequences that are targeted by the treatment. The severity of the effect of the amino acid alteration determines the intensity of the blue color of the plaque, and the ability to detect faint blue plaques will depend on various factors of the exact media composition. It should be amply clear that standardization of materials and methods is essential for intra- and interlaboratory comparisons.

In terms of the *lacI* gene as a mutational target, over 320 mutant positions have so far been identified (our data base). The number of detected mutational sites in the DNA binding region approaches 67%. Mutations have been recovered at a total of 122 sites out of 200 (61%) nucleotide positions and all three possible substitutions have been recovered at 37 of these sites (Glickman *et al.*, unpublished; Gordon *et al.*, 1988; Kleina and Miller, 1990). Thus, almost every nonwobble position in the sequence has been recovered! This confirms the observation that the DNA binding domain of the LacI protein is remarkably sensitive to amino acid substitutions (Kleina and Miller, 1990).

In Big Blue® mice, the sensitivity of the *lacI* transgene has been compared to that of the endogenous *hprt* gene following ENU treatment (Skopek *et al.*, 1995). The authors reported a greater fold increase in mutations after treatment in the *hprt* gene than in the *lacI* gene. This, however, largely reflects the lower spontaneous mutant frequency in the *hprt* gene. This can be at least partly explained by the difference in the number of CpG sites in the two genes and their methylation status. The density of 5′-CpG-3′ dinucleotide sequences per base pair in the coding sequence is 0.015 in the mouse *hprt* and 0.088 in the *lacI* gene, a 5.7-fold difference, reflecting the eukaryotic versus bacterial origins of the genes. Naturally, other yet unknown factors may play a role.

13. MUTATIONS IN GERM CELLS VERSUS SOMATIC CELLS

Unlike somatic mutation, mutations in germ cells have the potential to lead to inherited diseases. Lewis (1994) recently discussed the opportunities for transgenic animal systems in the study of germ cell mutation. The spontaneous mutant frequency in germ cells is significantly lower than that in somatic cells (Kohler *et al.*, 1991b). Dycaico *et al.* (1994) reported values between 0.6 and 1.1×10^{-5} for germ cells of *lacI* transgenic mice and rats. This compares to 4.1×10^{-5} in mouse liver, and 2.2×10^{-5} and 1.4×10^{-5} for the liver of two transgenic rat strains (Dycaico *et al.*, 1994). Mutation induction has been demonstrated in the Big Blue® mouse after treatment with ENU (Kohler *et al.*, 1991b; Provost and Short, 1994) and the nature of the recovered mutants is consistent with proposed mechanisms of ENU mutagenesis. A study is currently under way to address issues regarding aspects of mutation in germ cells in transgenic mice (Ashby, 1995).

14. PRELIMINARY DATA ON THE ANALYSIS OF MUTAGENS

To date, several chemical compounds as well as X-ray irradiation have been evaluated in the *lacI* and *lacZ* transgenic rodent systems. In addition, spontaneous spectra have been established in our laboratory for bladder, liver, spleen, bone marrow, kidney, stomach, skin, and lung. The hallmark of spontaneous mutation in mammalian tissues are G:C → A:T transitions which account for approximately 60 to 70% of all mutations recovered. As many as 80% of these are found at 5′-CpG-3′ sequences. As stated earlier, these mutations are thought to arise from the deamination of methylated cytosines which are found at 5′-CpG-3′ dinucleotide sequences. Cytosine, when methylated at the 5-ring position, is prone to spontaneous deamination, resulting in its conversion to thymine (Holliday and Grigg, 1993; Duncan and Miller, 1980). Replication of the resulting G:T mismatch can produce a G:C → A:T transition. Table I demonstrates the distribution of mutational classes found in the B6C3F1 and C57BL/6 strains. As can be seen, there is no significant difference in the spectra of these two strains of mice. Preliminary indications from the analogous system using Fisher 344 rats indicate that the background spectra do not differ significantly between the two rodent species.

The second greatest contribution to the spontaneous spectra are G:C → T:A transversions. These account for approximately 15–20% of all mutations. These may be the result of oxidative damage to guanine bases, producing 8-oxoguanine (Sakumi *et al.*, 1993; Michaels *et al.*, 1992), which may be a premutagenic lesion, possibly through the insertion of a adenine at an abasic site. Alternatively, these errors may reflect the action of a yet unknown repair pathway.

The analysis of the background spectra indicates that there are few differences between tissues. However, an excess of G:C → A:T transitions appears to be recovered in bladder and kidney tissue. This result is intriguing; bladder tissue is in a relatively low pH environment (Kadlubar *et al.*, 1991). Low pH values increase the spontaneous deamination rate of methylated cytosines (Holliday and Grigg, 1993), and may provide an explanation of the high fraction of these transitions.

Nearly 30 different chemicals have been evaluated in one of the *lacI* transgenic systems,

Table I
Comparison of Mutational Events between B6C3F1 and C57BL/6[a]

	B6C3F1				C57BL/6			
	Raw (264)		Corrected (205)		Raw (84)		Corrected (77)	
	%	%CpG	%	%CpG	%	%CpG	%	%CpG
Transitions	59.5		56.1		53.6		52.0	
G:C → A:T	53.8	81.7	49.3	74.3	50.0	73.8	48.1	75.7
A:T → G:C	5.7		6.8		3.6		3.9	
Transversions	25.0		26.3		33.3		33.8	
G:C → T:A	17.8	46.8	17.1	45.7	20.2	23.5	22.1	23.5
G:C → C:G	3.0	50.0	3.9	50.0	3.6	33.3	3.9	33.3
A:T → T:A	2.3		2.9		2.4		2.6	
A:T → C:G	1.9		2.4		7.1		5.2	
Others	15.5		17.6		13.1		14.3	

[a]Numbers after "Raw" and "Corrected" are the numbers of mutants in these groups, before and after correction for clonal expansion.

including the Big Blue® mouse, Big Blue® rat and rat cell lines, and this number is rapidly expanding. Table II shows many of the chemicals that have been tested in Big Blue® systems.

We will briefly describe the mutagenic specificity determined in the Big Blue® system for selected chemicals. 7,12-Dimethyl-benz[*a*]anthracene (DMBA), a polyaromatic hydrocarbon, and known to be a highly carcinogenic compound, induces mutations in skin tissue of mice and induces mammary carcinoma in rats and mice. DMBA requires metabolic activation to generate an active carcinogen, possibly involving a dihydrodiol epoxide, similar to benzo[*a*]pyrene (Dipple *et al.*, 1985). The mutations induced by this compound have been studied in skin tissue of Big Blue® (Gorelick *et al.*, 1993; Thompson *et al.*, 1992). The mutant frequency was increased sixfold over the spontaneous background 7 days after topical application of 100 μg (in acetone) onto shaved skin. DMBA was shown to induce primarily A:T → T:A and some G:C → T:A transversions. The mutations are very characteristically different from spontaneous mutations. A:T → T:A transversions have been seen in skin tumors from DMBA-treated mice (Quintanilla *et al.*, 1986; Nelson *et al.*, 1992). Transversions at A:T base pairs correlate with the frequent occurrence of DMBA adducts at adenine bases (Devanesan *et al.*, 1993; Vericat *et al.*, 1991). Mutations at G:C base pairs correlate with adduct formation at *N*2 and *N*7 atoms of guanine bases (Vericat *et al.*, 1991; Devanesan *et al.*, 1993). This compares well with the base substitution specificity found in *E. coli* for benzo[*a*]pyrene (Bernelot-Moens *et al.*, 1990), where -1 frame-shifts are found at G:C base pairs as well as G:C → T:A transversions at G:C base pairs that are 5′ preceded by a pyrimidine.

One mutagen commonly used in evaluating a mutagenesis system is ethylnitrosourea (ENU). It is a direct-acting alkylating agent in both eukaryotic and prokaryotic cells (Hu and Guttenplan, 1985; Rice *et al.*, 1987). Adducts at oxygen atoms of adenine and thymine comprise approximately 50% of the potentially mutagenic adducts (Singer *et al.*, 1978). Mutagenesis by ENU has been studied in the *lacI* gene in germ and spleen cells of Big Blue® (Provost *et al.*, 1992; Kohler *et al.*, 1991b; Skopek *et al.*, 1995; N. Gorelick, personal communication). The majority of substitutions in splenic T cells are A:T → T:A transversions. This change is very rarely seen in spontaneous mutations and is highly characteristic of exposure. Other mutations seen include transitions and transversions at G:C base pairs. In contrast to these findings, the majority of ENU-induced mutations in *E.coli* are changes at G:C base pairs.

Table II
Chemicals Evaluated in the *lacI* Big Blue® Transgenic System

Treatment	Species[a]	Tissue	References
2-Acetylaminofluorene	C57	Liver	Gunz *et al.* (1993), Shephard *et al.* (1994)
4-Acetylaminofluorene	B6	Liver	Provost *et al.* (1994)
Adozelesin	C57	Liver	Gunz *et al.* (1993)
Aflatoxin B1	Rat, C57	Liver	Dycaico *et al.* (1995)
Agaritine	C57	Forestomach, kidney, liver, lung	Shephard *et al.* (1995)
O-Anisidine	B6	Bladder	Ashby *et al.* (1991, 1994)
Benzene	B6	Lung, spleen	Mullin *et al.* (1995), Provost *et al.* (1994)
Benzo[*a*]pyrene	C57	Lung, spleen, liver	Kohler *et al.* (1991b), Shane *et al.* (1993)
1,3-Butadiene	B6	Bone marrow	Recio *et al.* (1993), Sisk *et al.* (1994a)
Carbon tetrachloride	B6	Liver	Mirsalis *et al.* (1993a)
CC-1065	C57	Liver	Gunz *et al.* (1993)
Cyclophosphamide	C57	Spleen	Kohler *et al.* (1991b)
Di(2-ethylhexyl)phthalate	C57	Liver	Gunz *et al.* (1993)
Dimethylbenzanthracene	C57	Skin	Gorelick *et al.* (1993), Thompson *et al.* (1992)
Dimethylnitrosamine	B6, C57	Liver	Shephard *et al.* (1994), Tinwell *et al.* (1994), Mirsalis *et al.* (1993b)
Ethylene oxide	B6	Lung, spleen	Sisk *et al.* (1994b)
Ethylnitrosourea	C57	Spleen, germ cells	Kohler *et al.* (1991b), Skopek *et al.* (1995a), Provost *et al.* (1992)
Fecapentaene-12	Rat cell line		Plummer (1995)
Gamma rays	B6, C57	Spleen	Winegar *et al.* (1994)
Heptachlor	C57	Liver	Gunz *et al.* (1993)
MeIQ	C57	Liver, bone marrow	Ushijima *et al.* (1994)
Methyl clofenapate	C57	Liver	Lefevre *et al.* (1994)
Methylmethanesulfonate	B6, C57	Liver, small intestine	Mirsalis *et al.* (1993b), Tao *et al.* (1993)
Methylnitrosourea	C57	Liver, lung, spleen	Kohler *et al.* (1991a)
	Rat cell line		Wyborski *et al.* (1995)
N-Nitrosomethylamine	C57	Liver	Shephard *et al.* (1994)
Phenobarbital	B6, C57	Liver	Gunz *et al.* (1993), Mirsalis *et al.* (1993a)
Radon		Lung	Jostes *et al.* (1995)
Streptozotocin	C57	Kidney, liver	Schmezer *et al.* (1994)
Tamoxifen	Rat, C57	Liver	Dycaico *et al.* (1995)
Urethane	C57	Lung	Shephard *et al.* (1994)

[a]B6, B6C3F1; C57, C57BL/6; rat, F344 transgenic; rat cell line, Big Blue® rat cell line.

In a recent study, Recio *et al.* (1993), and Sisk *et al.* (1994a) determined the sequence alterations in the *lacI* gene of male Big Blue® transgenic mice after inhalation exposure to 1,3-butadiene. This compound is an important environmental pollutant and suspected of causing cancer in humans. Bone marrow and spleen showed a 2- to 3.5-fold increase in mutant frequency, and sequence analysis indicated a 3- to 5-fold increase in the number of substitutions at A:T base pairs. Activated *ras* oncogenes have been found with G:C-to-C:G transversions at codon 13 (Goodrow *et al.*, 1995) in 9 of 11 liver tumors of mice, exposed to 1,3-butadiene, an event that has also been found with increased frequency in Big Blue® (Sisk *et al.*, 1994a).

Adozelesin and its analogue, CC-1065, compounds with potent antitumor activity, were evaluated in Big Blue® by Monroe and Mitchell (1993). This study demonstrates the utility of the *lacI* system for compounds with considerable minor groove sequence specificity. Seventy-three percent of all recovered mutants were found within a few base pairs from potential adduct sites.

15. FUTURE PROSPECTS

Several novel developments may make transgenic animal mutation systems more versatile. One of the more exciting prospects are animals with integrated activated oncogenes. These animals develop tumors rapidly on exposure to carcinogens, thereby reducing the duration of carcinogenesis testing. These constructs may represent major improvements in the arsenal of the bioassay user. An example of such an animal is the recently developed TG.AC mouse, which incorporates a v-Ha-*ras* gene (Spalding *et al.*, 1993; French *et al.*, 1995; Hansen and Tennant, 1995), and responds rapidly to nongenotoxic carcinogens (Tennant *et al.*, 1995). A second model system, the "knock-out" TSG-p53 mouse, is transgenic for the null allele of the *p53* tumor suppressor gene (Lavigueur *et al.*, 1989; J. M. Lee *et al.*, 1994), and is especially sensitive to genotoxic carcinogens (Tennant *et al.*, 1995). This strain has been crossed with the Big Blue® *lacI* mouse allowing the simultaneous analysis of both tumors and mutations from the same tissue. Other potential models include animals with metabolic deficiencies, and alterations in DNA repair pathways. The use of more than a single animal species may also be critical to the extrapolation of data to humans. The recently available Big Blue® rat is thus relevant in this sense. The first experimental data from this animal have appeared (Dycaico *et al.*, 1993, 1994, 1995). There are several parallel attempts currently being undertaken to produce Big Blue® fish. One of our colleagues has created a similar construct using a baculovirus, an insect-specific virus, that would permit the assessment of the mutagenic potential of pesticides in insects. We jokingly consider Big Blue® bunnies and primates, but are looking seriously into the construction of Big Blue® bivalves. Mussels are exceedingly attractive subjects for study as they are sessile and readily studied under laboratory and field environments.

We expect the development of new transgenic animal species to greatly further our understanding of genotoxicity and carcinogenicity of potentially hazardous compounds and to expand our knowledge of tumor development. In addition, *lacI* transgenic animals may also be useful in determining the role of mutations in cases other than genotoxic or nongenotoxic compounds. Lee *et al.* (1995) used Big Blue® mice to demonstrate mutagenicity associated with diabetes-induced teratogenicity and the effect of aging on spontaneous mutation (A. T. Lee *et al.*, 1994). These systems thus have the potential to permit the study of *in vivo* effects ranging from gender, age, organ, and tissue specificity to species specificity. While we may not yet have seen the final form transgenic testing animals will take, their importance to future genotoxic testing appears assured.

REFERENCES

Albertini, R. J., Gennett, I. N., Lambert, B., Thilly, W. G., and Vrieling, H. (1989). Mutation at the hprt locus. *Mutat. Res.* **216**:65–88.

Ashby, J. (1995). Transgenic germ cell mutation assays: A small collaborative study. *Environ. Mol. Mutagen.* **25**: 1–3.

Ashby, J., and Liegibel, U. (1992). Transgenic mouse mutation assays: Potential for confusion of genotoxic and non-genotoxic carcinogenesis: A proposed solution. *Environ. Mol. Mutagen.* **20**:145–147.

Ashby, J., and Liegibel, U. (1993). Dosing regimes for transgenic animal mutagenesis assays (response). *Environ. Mol. Mutagen.* **21**:120–121.

Ashby, J., Lefevre, P. A., Tinwell, H., Brunborg, G., Schmezer, P., and Pool-Zobel, B. (1991). The non-genotoxicity of rodents of the potent rodent bladder carcinogens o-anisidine and p-cresidine. *Mutat. Res.* **250**:115–133.

Ashby, J., Short, J. M., Jones, N. J., Lefevre, P. A., Provost, G. S., Rogers, B. J., Martin, E. A., Parry, J. M., Burnette, K., Glickman, B. W., *et al.* (1994). Mutagenesis of *o*-anisidine to the bladder of lacI-transgenic B6C3F1 mice: Absence of ^{14}C or ^{32}P bladder DNA adduction. *Carcinogenesis* **15**:2291–2296.

Bernelot-Moens, C., Glickman, B. W., and Gordon, A. J. E. (1990). Induction of specific frameshift and base substitution events by benzo[a]pyrene diol epoxide in exision-repair-deficient Escherichia coli. *Carcinogenesis* **11**:781–785.

Branda, R., O'Neill, J., Sullivan, L., and Albertini, R. (1993). Measurement of HPRT mutant frequencies in T-lymphocytes from healthy human populations. *Mutat. Res.* **285**:267–279.

Burkhart, J. G., and Malling, H. V. (1989). Mutagenesis of PhiX *am3 cs70* incorporated into the genome of mouse L-cells. *Mutat. Res.* **213**:125–134.

Carr, G. J., and Gorelick, N. J. (1994). Statistical tests of significance in transgenic mutation assays: Considerations on the experimental unit. *Environ. Mol. Mutagen.* **24**:276–282.

Carr, G. J., and Gorelick, N. J. (1995). Statistical design and analysis of mutation studies in transgenic mice. *Environ. Mol. Mutagen.* **25**:246–255.

Chhabra, R. S., Huff, J. E., Schwetz, B. S., and Selkirk, J. (1990). An overview of prechronic and chronic toxicity/carcinogenicity experimental study designs and criteria used by the National Toxicology Program. *Environ. Health Perspect.* **86**:313–321.

Cole, J., Green, M. H. L., Stephens, G., Waugh, A. P. W., Beare, D., Steingrimsdottir, H., and Bridges, B. A. (1990). HPRT somatic mutation data, in: *Mutation and the Environment. Part C: Somatic and Heritable Mutation, Adduction, and Epidemiology* (M. L. Mendelsohn and R. J. Albertini, eds.), Wiley–Liss, New York, p. 25.

De Boer, J. G. (1995). Software package for the management of sequencing projects using lacI transgenic animals. *Environ. Mol. Mutagen.* **25**:256–262.

De Boer, J. G., and Glickman, B. W. (1989). Sequence specificity of mutation induced by the anti-tumor drug cisplatin in the CHO aprt gene. *Carcinogenesis* **10**:1363–1367.

De Boer, J. G., Drobetsky, E. A., Grosovsky, A. J., Mazur, M., and Glickman, B. W. (1989). The Chinese hamster aprt gene as a mutational target. Its sequence and an analysis of direct and inverted repeats. *Mutat. Res. Lett.* **226**:239–244.

De Boer, J. G., Curry, J. D., and Glickman, B. W. (1993). A fast and simple method to determine the clonal relationship among human T-cell lymphocytes. *Mutat. Res.* **288**:173–180.

De Jong, P. J., Grosovsky, A. J., and Glickman, B. W. (1988). Spectrum of spontaneous mutation at the APRT locus of Chinese hamster ovary cells: An analysis at the DNA sequence level. *Proc. Natl. Acad. Sci. USA* **85**:3499–3503.

Devanesan, P. D., RamaKrishna, N. V. S., Padmavathi, N. S., Higginbotham, S., Rogan, E. G., Cavalieri, E. L., Marsch, G. A., Jankowiak, R., and Small, G. J. (1993). Identification and quantification of 7,12-dimethylbenz[a]anthracene–DNA adducts formed in mouse skin. *Chem. Res. Toxicol.* **6**:364–371.

Dipple, A., Moschel, R. C., and Pigott, M. A. (1985). Acid lability of the hydrocarbon-deoxyribonucleotide linkages in 7,12-dimethylbenz[a]anthracene-modified deoxyribonucleic acid. *Biochemistry* **24**:2291–2298.

Drobetsky, E. A., Grosovsky, A. J., and Glickman, B. W. (1987). The specificity of UV-induced mutations at an endogenous locus in mammalian cells. *Proc. Natl. Acad. Sci.* **84**:9103–9107.

Duncan, B. K., and Miller, J. H. (1980). Mutagenic deamination of cytosine residues in DNA. *Nature* **287**:560–561.

Dycaico, M. J., Ardourel, D. F., and Short, J. M. (1993). Spontaneous mutant frequencies in transgenic F344 rats. *Environ. Mol. Mutagen.* **21(Suppl. 22)**:18 (Abstract).

Dycaico, M. J., Provost, G. S., Kretz, P. L., Ransom, S. L., Moores, J. C., and Short, J. M. (1994). The use of shuttle vectors for mutation analysis in transgenic mice and rats. *Mutat. Res.* **307**:461–478.

Dycaico, M. J., Rogers, B. J., and Provost, G. S. (1995). The species-specific difference of mutation sensitivity of transgenic lambda/lacI rats. *Environ. Mol. Mutagen.* **25(Suppl. 25)**:13.

French, J. E., Libbus, B. L., Hansen, L., Spalding, J., Tice, R. R., Mahler, J., and Tennant, R. W. (1994). Cytogenetic analysis of malignant skin tumors induced in chemically treated TG.AC transgenic mice. *Mol. Carcinogen.* **11**:215–226.

Glickman, B. W., De Boer, J. G., and Kusser, W. C. (1995). Molecular mechanisms of mutagenesis and mutational spectra, in: *Environmental Mutagenesis* (D. H. Phillips and S. Venitt, eds.), Bios Scientific Publishers, Oxford, pp. 33–59.

Gold, L. S., Bernstein, L., Magaw, R., and Slone, T. H. (1989). Interspecies extrapolation in carcinogenesis: Prediction between rats and mice. *Environ. Health Perspect.* **81**:211–219.

Goodrow, T., Reynolds, S., Maronpot, R., and Anderson, M. (1995). Activation of K-ras by codon 13 mutations in C57BL/6 × C3H F1 mouse tumors induced by exposure to 1,3-butadiene. *Cancer Res.* **50**:4818–4823.

Gordon, A. J. E., and Glickman, B. W. (1988). Protein domain structure influences observed distribution of mutation. *Mutat. Res.* **208**:105–108.

Gordon, A. J. E., Burns, P. A., Fix, D. F., Yatagai, F., Allen, F. L., Horsfall, M. J., Halliday, J. A., Gray, J., Bernelot-Moens, C., and Glickman, B. W. (1988). Missense mutation in the lacI gene of Escherichia coli. Inferences on the structure of the repressor protein. *J. Mol. Biol.* **200**:239–251.

Gordon, A. J., Burns, P. A., and Glickman, B. W. (1990). N-methyl-N′-nitro-N′-nitrosoguanidine induced DNA sequence alteration: Non-random components in alkylation mutagenesis. *Mutat. Res.* **233**:95–103.

Gordon, A. J. E., Halliday, J. A., Horsfall, M. J., and Glickman, B. W. (1991). Spontaneous and 9-aminoacridine-induced frameshift mutagenesis—2nd-site frameshift mutation within the N-terminal region of the Laci gene of Escherichia-coli. *Mol. Gen. Genet.* **227**:160–164.

Gorelick, N. J., O'Kelly, J. A., Gu, M., and Glickman, B. W. (1993). Mutational spectra in the LacI transgene from 7,12-dimethylbenzanthracene (DMBA)-treated and control Big Blue mouse skin. *Environ. Mol. Mutagen.* **21(Suppl. 22)**:24

Gossen, J. A., De Leeuw, W. J. F., Tan, C. H. T., Zwarhoff, E. C., Berends, F., Lohman, P. H. M., Knook, D. L., and Vijg, J. (1989). Efficient rescue of integrated shuttle vectors from transgenic mice: A model for studying mutation in vivo. *Proc. Natl. Acad. Sci. USA* **86**:7971–7975.

Gossen, J. A., Molijn, A. C., Douglas, G. R., and Vijg, J. (1992). Application of galactose-sensitive E. coli strains as selective hosts for lacZ-plasmids. *Nucleic Acids Res.* **20**:3254.

Gunz, D., Shephard, S. E., and Lutz, W. K. (1993). Can nongenotoxic carcinogens be detected with the lacI transgenic mouse mutation assay? *Environ. Mol. Mutagen.* **21**:209–211.

Hansen, L. A., and Tennant, R. (1994). Focal transgene expression associated with papilloma development in v-Ha-ras-transgenic TG.AC mice. *Mol. Carcinogen.* **9**:143–154.

Holliday, R., and Grigg, G. W. (1993). DNA methylation and mutation. *Mutat. Res.* **285**:61–67.

Horsfall, M. J., Zeilmaker, M. J., Mohn, G. R., and Glickman, B. W. (1989). Mutational specificities of environmental carcinogens in the LacI gene of Escherichia coli: II A host-mediated approach to N-nitroso-N,N-dimethylamine and endogenous mutagenesis in vivo. *Mol. Carcinogen.* **2**:107–115.

Hozier, J., Sawyer, J., Moore, M., Howard, B., and Clive, D. (1981). Cytogenetic analysis of the L5178Y/TK+/− leads to TK−/− mouse lymphoma mutagenesis assay system. *Mutat. Res.* **84**:169–181.

Hu, Y. C., and Guttenplan, J. B. (1985). Evidence for a major premutagenic ethyldeoxythymidine–DNA adduct in an in vivo system: N-nitroso-N-ethylurea-treated *Salmonella typhimurium. Carcinogenesis* **6**:1513–1516.

International Agency for Research on Cancer. (1978). *IARC Monograph on the Evaluation of the Carcinogenic Risk of Chemicals to Humans*, IARC, Lyon, p. 125.

Jostes, R. F., Barnes, Y. C., Cross, F. T., Layton, A. D., Lutze, L. H., and Stillwel, L. (1995). Sequence analysis of lacI mutations obtained from lung cells of control and radon-exposed Big Blue transgenic mice. *Environ. Mol. Mutagen.* **25(Suppl. 25)**:25.

Kadlubar, F. F., Dooley, K. L., Teitel, C. H., Roberts, D. W., Benson, R. W., Butler, M. A., Bailey, J. R., Young, J. F., Skipper, P. W., and Tannenbaum, S. R. (1991). Frequency of urination and its effects on metabolism, pharmacokinetics, blood hemoglobin adduct formation, and liver and urinary bladder DNA adduct levels in beagle dogs given the carcinogen 4-aminobiphenyl. *Cancer Res.* **51**:4371–4377.

Kleina, L. G., and Miller, J. H. (1990). Genetic studies of the lac repressor. XIII Extensive amino acid replacements generated by the use of natural and synthetic nonsense suppressors. *J. Mol. Biol.* **212**:295–318.

Knoll, A., Jacobson, D. P., Kretz, P. L., Lundberg, K. S., Short, J. M., and Sommer, S. S. (1994). Spontaneous mutations in *lacI*-containing lambda lysogens derived from transgenic mice: The observed patterns differ in liver and spleen. *Mutat. Res.* **311**:57–67.

Knöll, A., Jacobson, D. P., Nishino, H., Kretz, P. L., Short, J. M., and Sommer, S. S. (1966). A selectable system for mutation detection in the Big Blue *lacI* transgenic mouse: what happens to the mutational spectrum over time. *Mutat. Res.*, in press.

Kohler, S. W., Provost, G. S., Fieck, A., Kretz, P. L., Bullock, W. O., Putman, D. L., Sorge, J. A., and Short, J. M. (1991a). Analysis of spontaneous and induced mutations in transgenic mice using a lambda ZAP/lacI shuttle vector. *Environ. Mol. Mutagen.* **18**:316–321.

Kohler, S. W., Provost, G. S., Fieck, A., Kretz, P. L., Bullock, W. O., Sorge, J. A., Putman, D. L., and Short, J. M. (1991b). Spectra of spontaneous and mutagen-induced mutations in the lacI gene in transgenic mice. *Proc. Natl. Acad. Sci. USA* **88**:7958–7962.

Kretz, P. L., Lundberg, K. S., Provost, G. S., and Short, J. M. (1992). The lambda/lacI transgenic mutagenesis systems: Comparisons between a selectable and a non-selectable system. *Environ. Mol. Mutagen.* **19(Suppl. 20)**:31.

Kretz, P. L., Lundberg, K. S., Wyborski, D. L., DuCoeur, L. C., and Short, J. M. (1993). Investigations of lacI selectable systems. *Environ. Mol. Mutagen.* **21(Suppl. 23)**:36.

Latt, S. A. (1974). Sister chromatid exchanges, indices of human chromosome damage and repair: Detection of fluorescence and induction by mitomycin. *C. Proc. Natl. Acad. Sci. USA* **71**:3162–3166.

Lavigueur, A., and Bernstein, A. (1993). p53 transgenic mice: Accelerated erythroleukemia induction by Friend virus. *Oncogene* **6**:2197.

Lavigueur, A., Maltby, V., Mock, D., Rossant, J., Pawson, T., and Bernstein, A. (1989). High incidence of lung, bone, and lymphoid tumors in transgenic mice overexpressing mutant alleles of the p53 oncogene. *Mol. Cell. Biol.* **9**:3982–3991.

LeClerc, J. E., Christensen, J. R., Tata, P. V., Christensen, R. B., and Lawrence, C. W. (1988). Ultraviolet light induces different spectra of lacI sequence changes in vegetative and conjugating cells of Escherichia coli. *J. Mol. Biol.* **203**:619–633.

Lee, A. T., DeSimone, C., Cerami, A., and Bucala, R. (1994). Comparative analysis of DNA mutations in lacI transgenic mice with age. *FASEB J.* **8**:545–550.

Lee, A. T., Plump, A., DeSimone, C., Cerami, A., and Bucala, R. (1995). A role for DNA mutations in diabetes-associated teratogenesis in transgenic embryos. *Diabetes* **44**:20–24.

Lee, J. M., Abrahamson, J. L., Kandel, R., Donehower, L. A., and Bernstein, A. (1994). Susceptibility to radiation-carcinogenesis and accumulation of chromosomal breakage in p53 deficient mice. *Oncogene* **9**:3731–3736.

Lefevre, P. A., Tinwell, H., Galloway, S. M., Hill, R., Mackay, J. M., Elcombe, C. R., Foster, J., Randall, V., Callander, R. D., and Ashby, J. (1994). Evaluation of the genetic toxicity of the peroxisome proliferator and carcinogen methyl clofenapate, including assays using Muta Mouse and Big Blue transgenic mice. *Hum. Exp. Toxicol.* **13**:764–775.

Lewis, S. E. (1994). A consideration of the advantages and potential difficulties of the use of transgenic mice for the study of germinal mutations. *Mutat. Res.* **307**:509–515.

Liegibel, U. M., Tinwell, H., Callander, R. D., Schmezer, P., and Ashby, J. (1992). Clastogenicity to the mouse bone marrow of the mouse germ cell genotoxin streptozotocin. *Mutagenesis* **7**:471–474.

Lundberg, K. S., Kretz, P. L., Provost, G. S., and Short, J. M. (1993). The use of selection in recovery of transgenic targets for mutation analysis. *Mutat. Res.* **301**:99–105.

Maron, D. M., and Ames, B. N. (1983). Revised methods for the Salmonella mutagenicity test. *Mutat. Res.* **113**:173–215.

Mazur, M., and Glickman, B. W. (1988). Sequence specificity of mutations induced by benzo[a]pyrene-7,8-diol-9,10-epoxide at endogenous aprt gene in CHO cells. *Som. Cell Mol. Genet.* **14**:393–400.

Medinsky, M. A., Sabourin, P. J., Lucier, G., Birnbaum, L. S., and Henderson, R. F. (1989). A physiological model for simulation of benzene metabolism by rats and mice. *Toxicol. Appl. Pharmacol.* **99**:193–206.

Meuth, M., Miles, C., Phear, G., and Sargent, G. (1990). Molecular patterns of Aprt gene rearrangements. *Mutat. Environ. Part A* **340**:305–314.

Michaels, M. L., Cruz, C., Grollman, A. P., and Miller, J. H. (1992). Evidence that MutY and MutM combine to prevent mutations by an oxidatively damaged form of guanine in DNA. *Proc. Natl. Acad. Sci. USA* **89**:7022–7025.

Miller, J. H. (1972). *Experiments in Molecular Genetics*. Cold Spring Harbor Laboratory Press, Cold Spring Harbor, NY.

Mirsalis, J. C. (1993). Dosing regimes for transgenic animal mutagenesis assays (letter). *Environ. Mol. Mutagen.* **21**:118–119.

Mirsalis, J. C., Hamer, J. D., O'Loughlin, K. G., Winegar, R. A., and Short, J. M. (1993a). Effects of nongenotoxic carcinogens on hepatic mutations in lacI transgenic mice. *Environ. Mol. Mutagen.* **21(suppl 22)**:48.

Mirsalis, J. C., Provost, G. S., Matthews, C. D., Hamner, R. T., Schindler, J. E., O'Loughlin, K. G., MacGregor, J. T., and Short, J. M. (1993b). Induction of hepatic mutations in lacI transgenic mice. *Mutagenesis* **8**:265–271.

Monroe, T. J., and Mitchell, M. A. (1993). In vivo mutagenesis induced by CC-1065 and adozelesin DNA alkylation in a transgenic mouse model. *Cancer Res* **53**:5690–5696.

Mullin, A. H., Rando, R., Esmundo, F., and Mullin, D. A. (1995). Inhalation of benzene leads to an increase in the mutant frequencies of a *lacI* transgene in lung and spleen tissues of mice. *Mutat. Res.* **327**:121–129.

Nakajima, T., Wang, R.-S., and Elovaara, E. (1993). Cytochrome P450-related differences between rats and mice in the metabolism of benzene, toluene and trichloroethylene in liver microsomes. *Biochem. Pharmacol.* **45**:1079.

Nelson, M. A., Futscher, B. W., Kinsella, T., Wymer, J., and Bowden, G. T. (1992). Detection of mutant Ha-ras genes in chemically initiated mouse skin epidermis before the development of benign tumors. *Proc. Natl. Acad. Sci. USA* **89**:6398–6402.

Nicklas, J. A., O'Neill, J. P., and Albertini, R. J. (1986). Use of T-cell receptor gene probes to quantify the in vivo hprt mutations in human T-lymphocytes. *Mutat. Res.* **173**:67–72.

Nicklas, J., Hunter, T., O'Neill, J., and Albertini, R. (1989). Molecular analyses of in vitro hprt mutations in human T-lymphocytes III. Longitudinal study of hprt gene structural alterations and T-cell clonal origins. *Mutat. Res.* **215**:147–160.

NTP Technical Report. (1978). Bioassay of o-anisidine hydrochloride for possible carcinogenicity. *NTP Technical Report* **89**.

Oberly, T. J., Bewsey, B. J., and Probst, G. S. (1984). An evaluation of the L5178Y TK+/− mouse lymphoma forward mutation assay using 42 chemicals. *Mutat. Res.* **125**:291–306.

Perera, F. P., Tang, D. L., O'Neill, J. P., Bigbee, W. L., Albertini, R. J., and Santella, R. (1993). HPRT and glycophorin A mutations in foundry workers in relationship to PAH exposure and to PAH-DNA adducts. *Carcinogenesis* **14**:969–973.

Piegorsch, W. W., Lockhart, A. M., Margolin, B. H., Tindall, K. R., Gorelick, N. J., Short, J. M., Carr, G. J., Thompson, E. D., and Shelby, M. D. (1994). Sources of variability in data from a lacI transgenic mouse mutation assay. *Environ. Mol. Mutagen.* **23**:17–31.

Piegorsch, W. W., Margolin, B. H., Shelby, M. D., Johnson, A., French, J. E., Tennant, R. W., and Tindall, K. R. (1995). Study design and sample sizes for a lacI transgenic mouse mutation assay. *Environ. Mol. Mutagen.* **25**:231–245.

Plummer, S. M. (1995). The mutation spectrum of fecapentaene-12 in Big Blue rat fibroblasts. *Carcinogenesis* **36**:162.

Provost, G. S., and Short, J. M. (1994). Characterization of mutations induced by ethylnitrosourea in seminiferous tubule germ cells of transgenic B6C3F1 mice. *Proc. Natl. Acad. Sci. USA* **91**:6564–6568.

Provost, G. S., Kohler, S. W., Putman, D. L., and Short, J. M. (1992). ENU induced germ cell mutations in lambda/lacI C57BL/6 and B6C3F1 transgenic mice. *Environ. Mol. Mutagen.* **19(Suppl. 20)**:51 (Abstract).

Provost, G. S., Kretz, P. L., Hamner, R. T., Matthews, C. D., Rogers, B. J., Lundberg, K. S., Dycaico, M. J., and Short, J. M. (1993). Transgenic systems for in vivo mutation analysis. *Mutat. Res.* **288**:133–149.

Provost, G. S., Rogers, B. J., Dycaico, M. J., Mirsalis, J. C., and Short, J. M. (1994). Evaluation of mutagenic and nonmutagenic compounds using lacI transgenic rodents. *Environ. Mol. Mutagen.* **23(Suppl. 23)**:55.

Quintanilla, M., Brown, K., Ramsden, M., and Balmain, A. (1986). Carcinogen-specific mutation and amplification of Ha-ras during mouse skin carcinogenesis. *Nature* **322**:78–80.

Recio, L., Bond, J. A., Pluta, L. J., and Sisk, S. C. (1993). Use of transgenic mice for assessing the mutagenicity of 1,3-butadiene in vivo. *IARC Sci. Publ.* 235–243.

Rice, J. M., Diwan, B. A., Donovan, P. J., and Perantoni, A. O. (1987). *Banbury Report* **26**:137–153.

Rogers, B. J., Provost, G. S., Young, R. R., Putman, D. L., and Short, J. M. (1995). Intralaboratory optimization and standardization of mutant screening conditions used for a lambda/lacI transgenic mouse mutagenesis assay (I). *Mutat. Res.* **327**:57–66.

Russell, L. B., and Russell, W. L. (1992). Frequency and nature of specific-locus mutations induced in female mice by radiation and chemicals: A review. *Mutat. Res.* **296**:107–127.

Russell, L. B., Selby, P. B., Von Halle, E., Sheridan, W., and Valcovic, L. (1981). The mouse specific-locus test with agents other than radiation: Interpretation of data and recommendations for future work. *Mutat. Res.* **86**:329–354.

Sakumi, K., Furuichi, M., Tsuzuki, T., Kakuma, T., Kawabata, S., Maki, H., and Sekiguchi, M. (1993). Cloning and expression of cDNA for a human enzyme that hydrolyzes 8-oxo-dGTP, a mutagenic substrate for DNA synthesis. *J. Biol. Chem.* **268**:23524–23530.

Sargentini, N. J., and Smith, K. C. (1986). Quantitation of the involvement of the recA, recB, recC, recF, recJ, recN, lexA, radA, radB, uvrD, and umuC genes in the repair of X-ray-induced DNA double-strand breaks in Escherichia coli. *Radiat. Res.* **107**:58–72.

Schaaper, R. M., and Dunn, R. L. (1987). Spectra of spontaneous mutations in Escherichia coli strains defective in mismatch correction: The nature of in vivo DNA replication errors. *Proc. Natl. Acad. Sci. USA* **84**:6220–6224.

Schaaper, R. M., and Dunn, R. L. (1991). Spontaneous mutation in the Escherichia coli lacI gene. *Genetics* **129**: 317–326.

Schaaper, R. M., Danforth, B. N., and Glickman, B. W. (1986). Mechanisms of spontaneous mutagenesis: An analysis of the spectrum of spontaneous mutation in the Escherichia coli lacI gene. *J. Mol. Biol.* **189**:273–284.

Schmezer, P., Eckert, C., and Liegibel, U. M. (1994). Tissue-specific induction of mutations by streptozotocin in vivo. *Mutat. Res.* **307**:495–499.

Shane, B. S., Winston, G. W., Reilly, P. A., Schaeffer, P. A., Battista, J. R., Swenson, D. H., Chang, S. H., and Lee, W. R. (1993). Mutation frequency of benzo[a]pyrene in rapidly dividing cells of the liver of C57Bl/6 transgenic mice. *Environ. Mol. Mutagen.* **21(Suppl. 22)**:64.

Shephard, S. E., Lutz, W. K., and Schlatter, C. (1994). The lacI transgenic mouse mutagenicity assay: Quantitative evaluation in comparison to tests for carcinogenicity and cytogenetic damage in vivo. *Mutat. Res.* **306**:119–128.

Shephard, S. E., Gunz, D., and Schlatter, C. (1995). Genotoxicity of agaritine in the lacI transgenic mouse mutation assay: Evaluation of the health risk of mushroom consumption. *Food Chem. Toxicol.* **33**:257–264.

Singer, B., Bodell, W. J., Cleaver, J. E., Thomas, G. H., Rajewsky, M. F., and Thon, W. (1978). Oxygens in DNA are the main targets for ethylnitrosourea in normal and uxeroderma pigmentosum fibroblasts and fetal rat brain cells. *Nature* **276**:5–88.

Sisk, S. C., Pluta, L. J., Bond, J. A., and Recio, L. (1994a). Molecular analysis of lacI mutants from bone marrow of B6C3F1 transgenic mice following inhalation exposure to 1,3-butadiene. *Carcinogenesis* **15**:471–477.

Sisk, S. C., Preston, R. J., and Recio, L. (1994b). Determination of circulating micronuclei and mutant frequency in lung, spleen, and germ cells of male B6C3F1 lacI transgenic mice after inhalation exposure to ethylene oxide. *Environ. Mol. Mutagen.* **23(Suppl. 23)**:62.

Skopek, T. R., Kort, K. L., and Marino, D. R. (1995). Dose-response of ENU mutagenesis at the endogenous hprt gene and lacI transgene in generic and Big Blue B6C3F1 mice. *Environ. Mol. Mutagen.* **25(Suppl. 25)**:49.

Spalding, J. W., Momma, J., Elwell, M. R., and Tennant, R. W. (1993). Chemically induced skin carcinogenesis in a transgenic mouse line (TG.AC) carrying a v-Ha-ras gene. *Carcinogenesis* **14**:1335–1341.

Tao, K. S., Urlando, C., and Heddle, J. A. (1993). Mutagenicity of methyl methanesulfonate (MMS) in vivo at the Dlb-1 native locus and a lacI transgene. *Environ. Mol. Mutagen.* **22**:293–296.

Tennant, R. W., French, J. E., and Spalding, J. W. (1995). Identifying of chemical carcinogens and assessing potential risk in short term bioassays using transgenic mouse models. *Environ. Health Perspect.* **103:**942–950.

Thompson, E. D., Gorelick, N. J., Binder, R. L., Myhr, B. C., and Putman, D. L. (1992). Interlaboratory comparison of dimethylbenzanthracene-induced mutations in skin of Muta-Mouse and the Big Blue mouse. *Environ. Mol. Mutagen.* **19(Suppl. 20)**:64.

Tinwell, H., Lefevre, P. A., and Ashby, J. (1994). Mutation studies with dimethyl nitrosamine in young and old lac I transgenic mice. *Mutat. Res.* **307**:501–508.

Ushijima, T., Hosoya, Y., Ochiai, M., Kushida, H., Wakabayashi, K., Suzuki, T., Hayashi, M., Sofuni, T., Sugimura, T., and Nagao, M. (1994). Tissue-specific mutational spectra of 2-amino-3,4-dimethylimidazo[4,5-f]-quinoline in the liver and bone marrow of lacI transgenic mice. *Carcinogenesis* **15**:2805–2809.

Vericat, J. A., Cheng, S. C., and Dipple, A. (1991). Absolute configuration of 7,12-dimethylbenz[a]anthracene–DNA adducts in mouse epidermis. *Cancer Lett.* **57**:237–242.

Vrieling, H., Venema, J., Van Rooyen, M. L., Van Hoffen, A., and Menichini, P. (1991). Strand specificity for UV induced DNA repair and mutations in the Chinese hamster HPRT gene. *Nucleic Acids Res.* **19**:2411–2416.

Wild, D. (1978). Cytogenetic effects in the mouse of 17 chemical mutagens and carcinogens evaluated by the micronucleus test. *Mutat. Res.* **56**:319–327.

Winegar, R. A., Lutze, L. H., Hamer, J. D., O'Loughlin, K. G., and Mirsalis, J. C. (1994). Radiation-induced point mutations, deletions and micronuclei in lacI transgenic mice. *Mutat. Res.* **307**:479–487.

Wyborski, D. L., Malkhosyan, S., and Short, J. M. (1995). Development of a rat cell line containing stably integrated copies of a lambda/lacI shuttle vector. *Mutat. Res.* **334**:161–166.

Zhang, L., Robertson, M. L., Kolachana, P., Davison, A. J., and Smith, M. T. (1993). Benzene metabolite, 1,2,4-benzenetriol, induces micronuclei and oxidative DNA damage in human lymphocytes and HL60 cells. *Environ. Mol. Mutagen.* **21**:339–348.

Zielenska, M., Ahmed, A., Pienkowska, M., Anderson, M. W., and Glickman, B. W. (1993). Mutational specificities of environmental carcinogens in the LacI gene of Escherichia coli. VI: Analysis of methylene chloride-induced mutational distribution in Uvr+ and UvrB− strains. *Carcinogenesis* **14**:789–794.

Chapter 31

Genotypic Mutation Assay (RFLP/PCR)

Fernando Aguilar and Peter Cerutti

1. INTRODUCTION

Mutational spectrum analysis of DNA has gained considerable interest among molecular toxicologists as an attempt to correlate DNA alterations caused by endogenous and exogenous mutagens with epidemiological studies of cancer risk factors (Shields and Harris, 1991). Mutational abnormalities in cancer-related genes may suggest mechanisms of carcinogenesis in specific organs or tissues. Furthermore, mutations in proto-oncogenes and tumor suppressor genes in premalignant tissues may become a useful parameter for risk assessment (Greenblatt *et al.*, 1994; Harris, 1993).

The multistage model of molecular carcinogenesis implies that a mutated gene must be present at early stages in a minute minority of cells in essentially normal tissues. In most cases somatic mutations do not give rise to a functional change in the mutated cell which would allow its identification or expansion *in vitro*. Therefore, methods are required which do not select mutated cells on the basis of an altered phenotype but rather detect few altered DNA sequences from 10^6 to 10^7 copies of the corresponding wild-type sequence in the presence of cellular DNA. Such genotypic mutation analysis requires large numbers of cells at the outset to compensate for the very low frequencies of mutation expected on exposure to a mutagen at recognition sequences of 4–6 base pairs (Rossiter and Caskey, 1990; Zijlstra *et al.*, 1990). On the other hand, methods for genotypic mutation analysis have the advantage that there are no silent mutations which escape detection.

The principal steps of the RFLP/PCR protocol are shown in Fig. 31.1 and comprise: (1) elimination of wild-type codons 249 of the p53 gene by extensive restriction with *Hae*III enzyme and enrichment of mutated codon 249 sequences by gel electrophoresis; (2) high-fidelity amplification of p53 exon VII fragments containing codon 249 mutants; (3) construction of synthetic single-base-pair mutants of p53 exon VII at *Hae*III site 14,072–14,075; and (4) determination of the mutant composition by oligonucleotide hybridization of RFLP/PCR products cloned into λgt10 bacteriophages.

The RFLP/PCR protocol can be subject to experimental variation due to the multiple steps involved and therefore it gives relative, rather than absolute, mutation frequencies. Absolute

Fernando Aguilar and Peter Cerutti • Nestec SA, 1000 Lausanne 26, Switzerland.

Technologies for Detection of DNA Damage and Mutations, edited by Gerd P. Pfeifer, Plenum Press, New York, 1996.

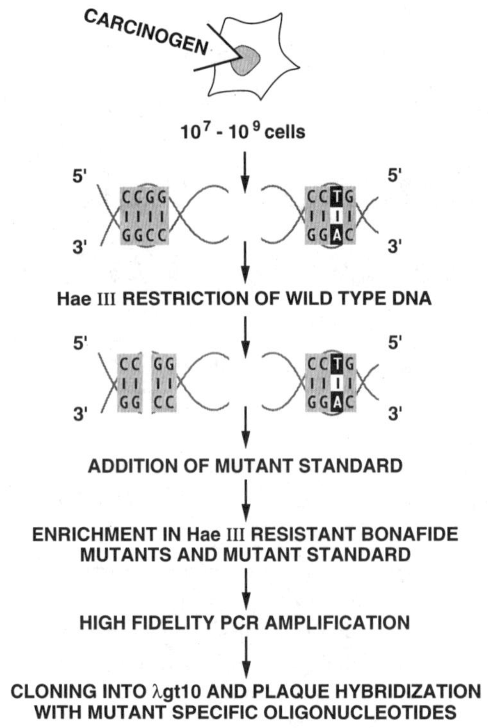

Figure 31.1. Genotypic mutation analysis by *Hae*III RFLP/PCR.

mutation frequencies are obtained by calibration with an internal mutant standard (MS). This MS is homologous to the amplified gene fragment except for three base pair changes located, one in the target restriction sequence and two more changes outside of this sequence. This approach has been applied *in vivo* and *in vitro* to study mutagenesis of p53 tumor suppressor gene and c-H-ras1 proto-oncogene in human cells by many potential human carcinogens (Aguilar *et al.*, 1993, 1994; Amstad *et al.*, 1994; Cerutti *et al.*, 1994; Chiocca *et al.*, 1992; Hussain *et al.*, 1994a,b; Pourzand and Cerutti, 1993).

2. METHODS

It is particularly important, due to the high sensitivity of the RFLP/PCR method, that several measures be taken to minimize the risk of carryover of amplified DNA. A laminar flow hood with a set of PCR positive-displacement pipettors, tubes, racks, etc. should be dedicated to the preparation of stock solutions, to the aliquoting of solutions for PCR and the preparation and distribution of pre-PCR reaction mixes. No unamplified DNA nor amplified DNA should ever be brought into this hood. All solutions to be used in PCR reactions like water, buffer, primers, deoxynucleotides (dNTPs), enzymes, etc. should be aliquoted and stored at −20°C. The unused portion of the aliquot should be discarded. Discardable aerosol resistant tips should be used to manipulate unamplified DNA and this manipulation should be done in a laboratory area where postamplified DNA is never brought. Whenever one leaves any defined working area, gloves should be changed. Finally, it is advisable that the thermocycling machine be placed in a different laboratory where unamplified DNA is manipulated.

2.1. *Hae*III Restriction of Codon 249 DNA Fragments

When the RFLP/PCR analysis was done in human cultured cells, high-molecular-weight DNA from $5–10 \times 10^7$ cells was extracted using the guanidium isothiocyanate method described by Sambrook *et al.* (1989). When human tissue samples were analyzed, DNA from approximately 1 g of tissue was isolated using the standard proteinase K–phenol–chloroform procedure (Ausbel *et al.*, 1994). High-molecular-weight DNA was dissolved in double-distilled H_2O (ddH_2O) and stored at 4°C.

A 400-μl solution containing 200 μg DNA, 5 U *Hae*III enzyme/μg DNA (40 U/μl, Boehringer-Mannheim, Rotkreuz, Switzerland), and 40 μl 10× enzyme buffer M was incubated overnight in an incubation oven at 37°C. After incubation the DNA was precipitated with 2.5 M NH_4acetate and 2.5 vol of ethanol. Subsequently, the DNA pellet was resuspended in ddH_2O generating a final volume of 400 μl including 2.5 U *Hae*III enzyme/μg DNA and 40 μl of 10× buffer M. The tubes were incubated overnight at 37°C. After incubation the samples were precipitated as before and resuspended in 50 μl of ddH_2O. The DNA concentration was determined spectrophotometrically at 260 nm using 1 A_{260} unit ~50 μg/ml double-stranded DNA.

2.2. Enrichment of Codon 249 Mutants

To approximately 2×10^7 copies of the p53 gene (~114 μg DNA) a small number of copies of the MS were added. The ideal copy number of MS which possess a length of 117 bp (Fig. 31.2) depends on the expected mutation frequencies, but typically is around 25 copies. The samples were adjusted to 1× gel loading buffer (0.04% bromophenol blue, 5% glycerol) and were incubated at 65°C for 10 min. The samples were loaded onto 0.1 × 2.5-cm wells in a 14-cm 2% agarose gel (Sigma, St. Louis, MO) and the gel was run overnight at 21 V in 1× TBE buffer (40 mM Tris-borate, 1 mM Na_2EDTA, pH 8.2). Five micrograms of pBR322 DNA digested with 1 U *Hae*III enzyme/μg DNA was used as molecular weight marker. Subsequently, a gel slide of a population containing 100- to 200-bp fragments was isolated using the Qiaex extraction kit (Qiagen, Studio City, CA) as indicated by the supplier. This fragment population contains the mutated 159-bp p53 segments which extend from flanking 5′ *Hae*III site (13,981) to the flanking 3′ *Hae*III site (14,139) as well as the 117-bp MS fragment. The cleavage of wild-type p53 sequences at *Hae*III site (14,072–14,075) gives rise to two short fragments of 66 and 93 bp which are eliminated. Samples were eluted from the Qiaex reagent with 4 × 50 μl of TE buffer (10 mM Tris-Cl, 1 mM Na_2EDTA, pH 7.5). Eluted samples were adjusted with ddH2O to a final volume of 250 μl including 20 U *Hae*III enzyme and 25 μl of 1× buffer M. Samples were incubated overnight in an incubation oven at 37°C. After incubation the samples were extracted three times with 1 vol of phenol:chloroform (1:1) and chloroform and the DNA was precipitated with 2.5 M NH_4acetate and 2.5 vol of ethanol. The organic extraction step is important to eliminate the remaining Qiaex reagent from the DNA solution which interferes with the subsequent PCR reactions. Care should be taken not to disturb the organic/inorganic interphase since Qiaex accumulates at the interphase.

2.3. High-Fidelity Amplification of Codon 249 Mutants

The DNA pellet from above was resuspended in 42 μl of ddH_2O by incubating the solution at 65°C for 5 min and vortexing thoroughly. Resuspended DNA was mixed with 20 pmole of each internal oligonucleotide primer, 200 μM of each dNTP (Pharmacia, Uppsala, Sweden), and 2.5 U

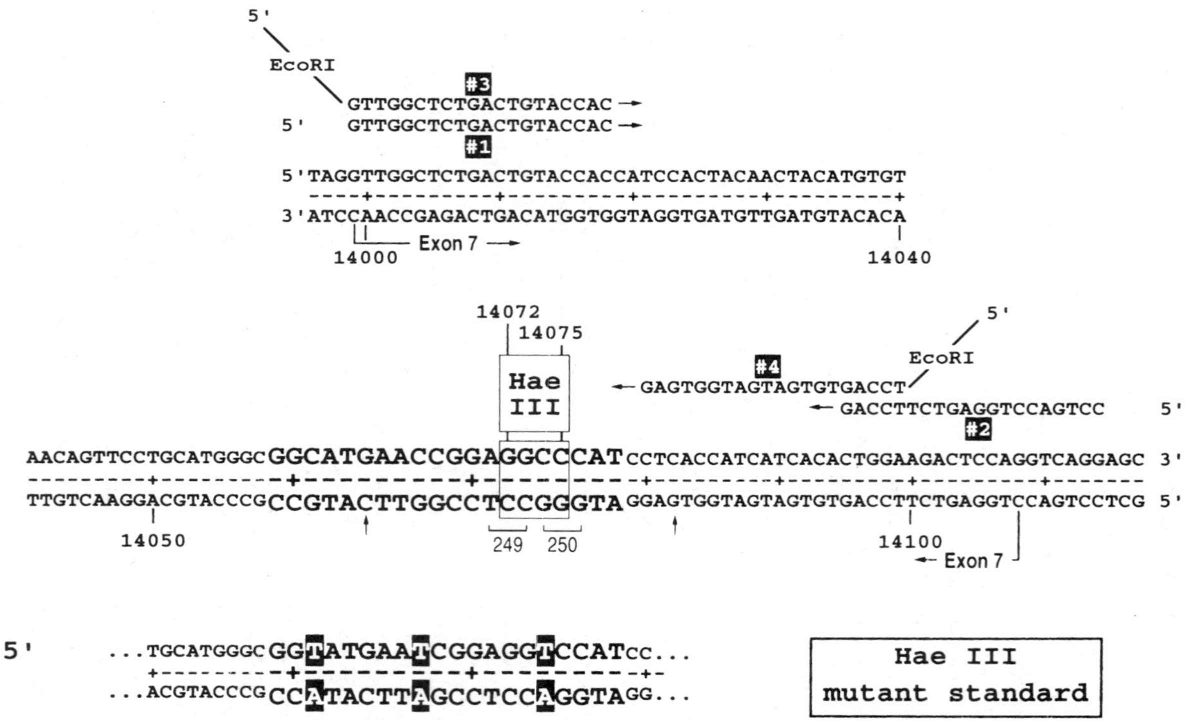

Figure 31.2. DNA sequence of exon VII of the p53 tumor suppressor gene. Information is given for the *Hae*III site (14,072–14,075), the mutant standard, and the four primers used in the RFLP/PCR analysis of codons 249 and 250. (Modified from Aguilar and Cerutti, 1994.)

of *Pyrococcus furiosus* (*Pfu*) DNA polymerase (Stratagene, Zurich, Switzerland) to a final volume of 50 μl in the buffer supplied with the enzyme, without bovine serum albumin (BSA) [20 mM Tris-Cl pH 8.2, 10 mM KCl, 6 mM $(NH)_4SO_4$, 2 mM $MgCl_2$, 0.1% Triton X-100]. The custom-synthesized oligonucleotides consisted of: sense primer No. 1, 5′ GTTGGCTCTGACTGTACCAC 3′ (residues 13,999–14,018), and antisense primer No. 2, 5′ CCTGACCTGGAGTCTTCCAG 3′ (residues 14,114–14,095) (Fig. 31.2).

The amplification mixtures were heated for 5 min at 96°C and 40 cycles of PCR was performed with the Perkin–Elmer Cetus Thermal Cycler 480 (Hoffman–LaRoche, Basel, Switzerland) as follows: denaturation, 30 sec at 94°C; annealing, 1 min at 45°C. After the last cycle of PCR a final 10-min elongation step at 72°C was performed.

A 25-μl aliquot from the amplified material was extracted once with phenol:chloroform (1:1) and the DNA was precipitated with 2.5 M NH_4acetate and 2.5 vol of ethanol as indicated before. The rest of the sample was stored at −20°C. The DNA pellet was resuspended in ddH_2O in a final volume of 25 μl containing 1× enzyme buffer M and 10 U *Hae*III enzyme. Samples were incubated overnight in an incubation oven at 37°C.

A 2-μl aliquot from the above restriction reaction served as template in a second PCR using oligonucleotide sense primer No. 3, 5′ GTTGGCTCTGACTGTACCAC 3′ (residues 13,999–14,018), and antisense primer No. 4; 5′ TCCAGTGTGATGATGGTGAG 3′ (residues 14,099–14,080) (Fig. 31.2). The amplification reaction contained 20 pmole of each oligonucleotide, 1 mM of each dNTP, and 1 U of *Taq* polymerase (Hoffman–La Roche) using the high-fidelity amplification buffer defined by Eckert and Kunkel (1990) (20 mM Tris-Cl pH 7.3 at 70°C, 50 mM KCl, and 10 mM $MgCl_2$). This second set of primers included a 5′ *Eco*RI restriction site in order to facilitate further cloning (Fig. 31.2). The PCR was done in the Perkin–Elmer DNA Thermal Cycler 480 and consisted of ten cycles of 1-min denaturation at 94°C, 2-min annealing at 45°C, and a last final 10-min elongation step at 72°C.

2.4. Preparation of Authentic Single-Base-Pair Mutants and Mutant Standard at *Hae*III Site 14,072–14,075 of Human p53 Gene

Authentic mutants and MS are used to determine the selective washing temperatures for the hybridization procedure and they routinely served as positive controls to ascertain the specificity of the hybridization. The synthesis of the *Hae*III authentic mutants by PCR should be done with extra precaution to minimize cross-contaminations. It is strongly recommended that all manipulations of pre- and postamplified mutants should be carried out in a laboratory different from the one where RFLP/PCR analysis is routinely done. Importantly, a different set of pipettors should be used to manipulate authentic mutants and MS. The reader is referred to recent books concerning basic PCR technologies (Innis *et al.*, 1990; Mullis *et al.*, 1994).

Twelve different 19-mers (residues 14,064–14,082) containing all possible base changes in the *Hae*III site were used as antisense primers for a first PCR reaction with the sense primer No.3 (Fig. 31.2). This primary PCR was carried out for 35 cycles with *Taq* polymerase and wild-type p53 as template under the high-fidelity amplification conditions outlined before. The primary PCR products consisting of 92-bp fragments were gel purified on 2% agarose by electroelution onto NA-45 DEAE membrane (Schleicher & Schuell, Dassel, Switzerland) according to standard procedures (Sambrook *et al.*, 1989) and then used as sense primers in a secondary PCR. The second amplification was done with oligonucleotide No. 4 as antisense primer and wild-type p53 as template for ten cycles of PCR using *Pfu* polymerase under the conditions described before, except that a final concentration of 6% dimethyl sulfoxide (Merck, Germany) was added to the amplification reaction.

The resulting 125-bp fragments were restricted with 5 U *Eco*RI enzyme (Boehringer-

Mannheim) and the restricted DNAs were gel purified on 2% agarose as before. After purification, authentic mutants were cloned into λgt10 *Eco*RI arms as suggested by the supplier (Promega, Zurich, Switzerland). *Escherichia coli* C600 *hfl* were used as host cells and the recombinant λgt10 phages were lifted onto Colony/Plaque Screen NEF-978 membranes as indicated by the supplier (New England Nuclear–DuPont, Boston, MA). The membranes containing recombinant phages were hybridized with authentic mutant-specific ^{32}P-end-labeled oligonucleotides and one positive plaque from each authentic mutant was plaque purified on *E.coli C600 hfl* cells and stored at 4°C.

The MS was prepared by an analogous procedure, using as antisense primer a 20-mer oligonucleotide (residues 14,058–14,077) containing the three base pair changes shown in Fig. 31.2. One of the changes is located at position 14,074 (C→T) which renders the MS resistant to *Hae*III cleavage. The two additional C→T changes (positions 14,061 and 14,067) were introduced to allow easy distinction of MS from *Hae*III bona fide mutants.

2.5. Specific Oligonucleotide Plaque Hybridization of RFLP/PCR Mutants

Selective washing temperatures were determined with the help of authentic mutants immobilized as slot blots on NEN Gene Screen membranes (Dupont, Boston, MA) and hybridized individually with every ^{32}P-end-labeled mutant-specific oligonucleotide. The autoradiograms were evaluated at different washing temperatures to ascertain that only the corresponding mutant construct is recognized by the specific oligonucleotide probe. We found that the selective washing temperatures for single-base-pair mutations in *Hae*III site 14,072–14,075 were 60°C for the MS, 59°C for wild type, and 62 and 64°C for the authentic mutants.

The RFLP/PCR products from the amplified DNA were precipitated with 2.5 M NH_4acetate and 1 vol of isopropanol. The DNA pellets were resuspended with ddH_2O and restricted with 5 U *Hae*III enzyme. The restricted RFLP/PCR products were purified either on Qiagen-tip 5 columns or through QIA quick spin columns as outlined by the supplier (Qiagen, Germany). The purified samples were restricted with 5 U *Eco*RI enzyme to produce clonable ends and the expected 117-bp fragment was further gel purified on 2% agarose by electroelution onto NA-45 membranes as indicated before. Approximately 2 ng of purified RFLP/PCR DNA was ligated to 500 ng of λgt10 *Eco*RI arms. Aliquots of the ligation reaction were used to transform host strain *E. coli* C600 *hfl* and the recombinant phages were lifted onto Colony/Plaque Screen NEF-978 membranes. Oligonucleotide plaque hybridizations were done with each of the 12 mutant ^{32}P-end-labeled 19-mer probes, with a wild-type 19-mer probe, and with an MS-specific 20-mer probe.

In order to obtain statistically significant data we analyzed for each RFLP/PCR product 1–2 $\times 10^3$ λ plaques on 5 to 10 petri dishes (8.5 cm). Care was taken that none of the plaques was counted more than once. The content of a particular mutation is expressed as percentage of the total number of identified plaques. Usually, 5–10% of the plaques remained unidentified. Sequencing of the λ inserts of unidentified plaques indicated the presence of small deletions and insertions which affect specific restriction as well as primer multimers.

2.6. Calibration of Absolute Mutation Frequencies Using the Mutant Standard

The bona fide mutant content of the RFLP/PCR product is determined by comparison to the percentage of plaques produced by the known number of MS copies added at the outset of the experiment. The MS sequence is identical to the genomic DNA with the exception of a few base pair changes and it has been checked that it is amplified with the same efficiency. Nevertheless, since there is no guarantee that a precise proportionality between bona fide mutants and MS is maintained throughout the RFLP/PCR protocol, absolute mutation frequencies obtained in this way have to be considered estimates rather than precise data.

3. DISCUSSION

Among the limitations that determined the sensitivity of the RFLP/PCR protocol are the completeness of the removal of wild-type sequences and the inherent error rate of the DNA polymerase at a particular base pair. A large fraction of uncut wild-type sequences after DNA restriction increases the frequency of DNA polymerase-induced background errors and the final content of wild-type sequences in the RFLP/PCR product.

The content of wild-type sequences varies for different restriction sites and depends on the activity of the particular endonuclease enzyme. The wild-type content of human p53 gene *Hae*III site (14,072–14,075) RFLP/PCR products varied from 2 to 10% of the total identified plaques (Aguilar *et al.*, 1994). We found that for human p53 gene *Msp*I site 14,067–14,070, which covers hot spot codon 248, the wild-type content was 22–32% (Hussain *et al.*, 1994b; Aguilar *et al.*, 1993). For the human H-*ras* 1 gene *Pvu*II (1727–1732) and *Msp*I (1695–1698) restriction sites, the residual wild-type content was 14–23 and 35–40%, respectively (Felley-Bosco *et al.*, 1991).

The presence of residual wild-type sequences in the RFLP/PCR products may be due to the formation of heteroduplexes between wild-type and bona fide mutant strands which are refractory to restriction by endonucleases. The residual wild-type sequences in the final PCR product can be markedly reduced by adding a thermostable endonuclease during the PCR. This has been demonstrated for the human H-*ras* 1 gene *Taq*I site 2508–2511 where wild-type sequences accounted for less than 1% of the total identified plaques when *Taq*I endonuclease was included in the PCR (Chiocca *et al.*, 1992; Sandy *et al.*, 1992). Therefore, background mutation arising from residual wild-type sequences can be highly reduced if thermostable endonucleases specific for the restriction site are incorporated during the PCR amplification. Unfortunately, many thermostable endonucleases which recognize common restriction sites are not commercially available. We have substantially increased the sensitivity of the assay by using the high-fidelity *Pfu* DNA polymerase isolated from the hyperthermophilic archaebacterium *Pyrococcus furiosus*. Unlike *Taq* polymerase, *Pfu* polymerase possesses a proofreading 3′–5′ exonuclease activity, which reduces the error rate per nucleotide by a factor > 10 compared to *Taq* polymerase (Lundberg *et al.*, 1991).

The RFLP/PCR protocol is highly sensitive and specific for genotypic mutations without phenotypic selection. On the other hand, the RFLP/PCR protocol can only be applied to sequences which harbor restriction recognition sequences of 4–6 bp and therefore the distribution of mutations in an entire gene cannot be determined even if many restriction sites are being analyzed.

The most time-consuming part of the RFLP/PCR protocol is the determination of mutations by specific oligonucleotide hybridization. Major mutations in the RFLP/PCR product can also be measured by quantitative sequence analysis or by slot-blot hybridization (Hussain *et al.*, 1994b; Amstad *et al.*, 1994). Quantitative sequence analysis yields reliable data for major mutations which represent at least 10% of the total bona fide mutants and when applicable the method is rapid, economical, and amenable to automation. The slot-blot hybridization avoids cloning of the RFLP/PCR products into λgt10 bacteriophages but it also avoids the *in vivo* amplification of mutants limiting the analysis to mutations which represent more than 10% of bona fide mutations.

ACKNOWLEDGMENTS. This work was supported by the Swiss National Science Foundation, the Swiss Association of Cigarette Manufacturers, and the Association for International Cancer Research.

REFERENCES

Aguilar, F., and Cerutti, P. (1994). Genotypic mutation analysis by RFLP/PCR. *Methods Toxicol.* **1B**:237.

Aguilar, F., Hussain, S. P., and Cerutti, P. (1993). Aflatoxin B_1 induces the transversion of G→T in codon 249 of the p53 tumor suppressor gene in human hepatocytes. *Proc. Natl. Acad. Sci. USA* **90**:8586.

Aguilar, F., Harris, C. C., Sun, T., Hollstein, M., and Cerutti, P. (1994). Geographic variation of p53 mutational profile in nonmalignant human liver. *Science* **264**:1317.

Amstad, P., Hussain, S. P., and Cerutti, P. (1994). Ultraviolet B light-induced mutagenesis of p53 hotspot codons 248 and 249 in human skin fibroblasts. *Mol. Carcinogen.* **10**:181.

Ausbel, F. M., Brent, R., Kingston, R. E., Moore, D. D., Seidman, J. G., Smith, J. A., and Struhl, K., eds. (1994). *Current Protocols in Molecular Biology*, Volume 1, Greene Publishing Associates and Wiley, New York.

Cerutti, P., Hussain, P., Pourzand, C., and Aguilar, F. (1994). Mutagenesis of the H-*ras* protooncogene and the p53 tumor suppressor gene. *Cancer Res.* (Suppl). **54**:1934s.

Chiocca, S. M., Sandy, M. S., and Cerutti, P. A. (1992). Genotypic analysis of N-ethyl-N-nitrosourea-induced mutations by Taq I restriction fragment length polymorphism/polymerase chain reaction in the c-H-*ras*1 gene. *Proc. Natl. Acad. Sci. USA* **89**:5331.

Eckert, K. A., and Kunkel, T. A. (1990). High fidelity DNA synthesis by the Thermus aquaticus DNA polymerase. *Nucleic Acids Res.* **18**:3739.

Felley-Bosco, E., Pourzand, C., Zijlstra, J., Amstad, P., and Cerutti, P. (1991). A genotypic mutation system measuring mutations in restriction recognition sequences. *Nucleic Acids Res.* **19**:2913.

Greenblatt, M. S., Bennett, W. P., Hollstein, M., and Harris, C. C. (1994). Mutations in the p53 tumor suppressor gene: Clues to cancer etiology and molecular pathogenesis. *Cancer Res.* **54**:4855.

Harris, C. C. (1993). The p53 tumor suppressor gene: At the crossroads of molecular carcinogenesis, molecular epidemiology and cancer risk assessment. *Science* **262**:1980.

Hussain, S. P., Aguilar, F., Amstad, P., and Cerutti, P. (1994a). Oxy-radical induced mutagenesis of hotspot codons 248 and 249 of the human p53 gene. *Oncogene* **9**:2277.

Hussain, S. P., Aguilar, F., and Cerutti, P. (1994b). Mutagenesis of codon 248 of the human p53 tumor suppressor gene by N-ethyl-N-nitrosourea. *Oncogene* **9**:13.

Innis, M. A., Gelfand, D. H., Sninsky, J. J., and White, T. J., eds. (1990). *PCR Protocols. A Guide to Methods and Applications*, Academic Press, London.

Lundberg, K. S., Shoemaker, D. D., Adams, M. W. W., Short, J. M., Sorge, J. A., and Mathur, E. J. (1991). High-fidelity amplification using a thermostable DNA polymerase isolated from Pyrococcus furiosus. *Gene* **108**:1.

Mullis, K. B., Ferré, F., and Gibbs, R. A., eds. (1994). *The Polymerase Chain Reaction*. Birkhäuser, Boston.

Pourzand, C., and Cerutti, P. (1993). Mutagenesis of H-*ras* codons 11 and 12 in human fibroblasts by N-ethyl-N-nitrosourea. *Carcinogenesis* **14**:2193.

Rossiter, B., and Caskey, C. T. (1990). Molecular scanning methods of mutation detection. *J. Biol. Chem.* **265**:12753.

Sambrook, J., Fritsch, E. F., and Maniatis, T. (1989). *Molecular Cloning. A Laboratory Manual*, 2nd ed., Cold Spring Harbor Laboratory Press, Cold Spring Harbor, NY.

Sandy, M. S., Chiocca, S. M., and Cerutti, P. A. (1992). Genotypic analysis of mutations in Taq I restriction recognition sites by restriction fragment length polymorphism/polymerase chain reaction. *Proc. Natl. Acad. Sci. USA* **89**:890.

Shields, P. G., and Harris, C. C. (1991). Molecular epidemiology and the genetics of enviromental cancer. *J. Am. Med. Assoc.* **266**:681.

Zijlstra, J., Felley-Bosco, E., Amstad, P., and Cerutti, P. (1990). A mammalian mutation system avoiding phenotypic selection: The RFLP/PCR approach, in: *Mutagens and Carcinogens in the Diet* (M. Pariza, ed.), Wiley–Liss, New York, p. 187.

Index